Biological Magnetic Resonance
Volume 13

EMR of Paramagnetic Molecules

A Continuation Order Plan is available for this series. A continuation order will bring delivery of each new volume immediately upon publication. Volumes are billed only upon actual shipment. For further information please contact the publisher.

Biological Magnetic Resonance
Volume 13

EMR of Paramagnetic Molecules

Edited by

Lawrence J. Berliner

Ohio State University
Columbus, Ohio

and

Jacques Reuben

Hercules Incorporated
Research Center
Wilmington, Delaware

Springer Science+Business Media, LLC

The Library of Congress has cataloged the first volume of this series as follows:

Library of Congress Cataloging in Publication Data

Main entry under title:

Biological magnetic resonance:

Includes bibliographies and indexes.
1. Magnetic resonance. 2. Biology—Technique. I. Berliner, Lawrence, J. II. Reuben, Jacques.
QH324.9.M28B56 574.19′285 78-16035
AACR1

ISBN 978-1-4613-6253-1 ISBN 978-1-4615-2892-0 (eBook)
DOI 10.1007/978-1-4615-2892-0

Personal computer software discussed in Chapter 1, Section 2.8 of this volume will be found on a diskette mounted inside the back cover.

If your diskette is defective in manufacture or has been damaged in transit, it will be replaced at no charge if returned within 30 days of receipt to Managing Editor, Springer Science+Business Media, LLC.

The publisher makes no warranty of any kind, expressed or implied, with regard to the software reproduced on the diskette or the accompanying documentation. The publisher shall not be liable in any event for incidental or consequential damages or loss in connection with, or arising out of, the furnishing, performance, or use of the software.

Additional material to this book can be downloaded from http://extras.springer.com.

Originally published by Plenum Press, New York in 1993

Contributors

William E. Antholine • National Biomedical ESR Center, Biophysics Research Institute, Medical College of Wisconsin, Milwaukee, Wisconsin, 53226

Riccardo Basosi • Department of Chemistry, University of Siena, 53100 Siena, Italy

Christopher L. Berger • Department of Biochemistry, University of Minnesota Medical School, Minneapolis, Minnesota 55455

Marcelino Bernardo • Exxon Corporate Research Laboratory, Annandale, New Jersey 08801

Peter G. Debrunner • Department of Physics, Loomis Laboratory of Physics, University of Illinois at Urbana-Champaign, Urbana, Illinois 61801

Victoria J. DeRose • Department of Chemistry, Northwestern University, Evanston, Illinois 60208-3113

Peter E. Doan • Department of Chemistry, Northwestern University, Evanston, Illinois 60208-3113

Piotr G. Fajer • Department of Biochemistry, University of Minnesota Medical School, Minneapolis, Minnesota 55455

Betty J. Gaffney • Department of Chemistry, The Johns Hopkins University, Baltimore, Maryland 21218-2685

Ryszard J. Gurbiel • Department of Chemistry, Northwestern University, Evanston, Illinois 60208-3113

Brian M. Hoffman • Department of Chemistry, Northwestern University, Evanston, Illinois 60208-3113

Andrew P. Houseman • Department of Chemistry, Northwestern University, Evanston, Illinois 60208-3113

Jürgen Hüttermann • Fachrichtung Biophysik and Physikalische Grundlagen der Medizin, Universität des Saarlandes, 6650 Homburg/Saar, Germany

James S. Hyde • National Biomedical ESR Center, Biophysics Research Institute, Medical College of Wisconsin, Milwaukee, Wisconsin 53226

Anna Iannone • Istituto di Patologia generale, Università degli Studi di Modena, 41100 Modena, Italy

Scott M. Lewis • Department of Biochemistry, University of Minnesota Medical School, Minneapolis, Minnesota 55455

James E. Mahaney • Department of Biochemistry, University of Minnesota Medical School, Minneapolis, Minnesota 55455

Klaus Möbius • Department of Physics, Free University of Berlin, D-1000 Berlin 33, Germany

E. Michael Ostap • Department of Biochemistry, University of Minnesota Medical School, Minneapolis, Minnesota 55455

Harris J. Silverstone • Department of Chemistry, The Johns Hopkins University, Baltimore, Maryland 21218-2685

Joshua Telser • Department of Chemistry, Northwestern University, Evanston, Illinois 60208-3113

Hans Thomann • Exxon Corporate Research Laboratory, Annandale, New Jersey 08801

David D. Thomas • Department of Biochemistry, University of Minnesota Medical School, Minneapolis, Minnesota 55455

Aldo Tomasi • Istituto di Patologia generale, Università degli Studi di Modena, 41100 Modena, Italy

Preface

This is the second part of our two-part series devoted to the magnetic resonance of paramagnetic molecules. The first part, presented in Volume 12, deals with NMR, whereas this volume is devoted to EPR (ESR), ENDOR, and Mössbauer spectroscopy as applied to paramagnetic molecules of biological interest.

In the opening chapter, Betty Gaffney and Harris Silverstone present a thorough treatment on the simulation of iron EPR spectra. Their computer program and user instructions are included on a Macintosh-compatible diskette that accompanies this volume. The second chapter, by Peter Debrunner, is dedicated to Mössbauer spectroscopy of iron proteins. Several applications of ESR of copper at multiple frequencies are presented by Basosi, Antholine, and Hyde. A number of chapters deal with ENDOR and pulsed EPR applications. Hoffman and co-workers discuss the active site structure of metalloenzymes as studied by ENDOR and pulsed ENDOR techniques. The ENDOR spectroscopy of randomly oriented metalloproteins is specifically addressed by Jürgen Hüttermann. Möbius demonstrates the application of EPR as well as ENDOR to studies of bioorganic systems. Hans Thomann and Marcelino Bernardo discuss state-of-the-art applications of pulsed ENDOR and ESE. Examples of spectra of nitroxide spin labels in the slow tumbling regime occurring in muscle proteins are presented by Thomas and co-workers. In the concluding chapter, Tomasi and Iannone critically discuss spin-trapping of biological free radicals in model systems.

We are very proud of this volume and of our authors, who are at the forefront of this subject. Again, we are looking forward to the reader's comments, criticisms, and suggestions.

Lawrence J. Berliner
Jacques Reuben

Contents

Chapter 1

Chapter 2

Mössbauer Spectroscopy of Iron Proteins

Peter G. Debrunner

Chapter 3

Multifrequency ESR of Copper: Biophysical Applications

Riccardo Basosi, William E. Antholine, and James S. Hyde

Chapter 4

Metalloenzyme Active-Site Structure and Function through Multifrequency CW and Pulsed ENDOR

Brian M. Hoffman, Victoria J. DeRose, Peter E. Doan, Ryszard J. Gurbiel, Andrew L. P. Houseman, and Joshua Telser

Chapter 7

Pulsed Electron Nuclear Double and Multiple Resonance Spectroscopy of Metals in Proteins and Enzymes

Hans Thomann and Marcelino Bernardo

Chapter 8

Transient EPR of Spin-Labeled Proteins

David D. Thomas, E. Michael Ostap, Christopher L. Berger, Scott M. Lewis, Piotr G. Fajer, and James E. Mahaney

Chapter 9

ESR Spin-Trapping Artifacts in Biological Model Systems
Aldo Tomasi and Anna Iannone

1

Simulation of the EMR Spectra of High-Spin Iron in Proteins

Betty J. Gaffney and Harris J. Silverstone

1. INTRODUCTION

Very detailed information about the energy levels and orientations of *d* orbitals in heme proteins has been obtained by combining EMR studies of paramagnetic samples with other structural information from X-ray crystallography and optical studies. As a result, the chemistry of heme enzymes can be discussed in detail. While the aim of this chapter is to review progress in bringing the chemistry of mononuclear iron centers in nonheme proteins to a similar level of knowledge, our understanding of line shape analysis for high-spin iron is dominated by the vast literature on heme samples. We begin this introduction with some of the history of EMR spectroscopy of methemoglobin and metmyoglobin.

Orientations of the heme planes in crystals of metmyoglobin and methemoglobin relative to crystal axes were determined by EMR measurements in the early stages of solving the crystal structures of these proteins (Bennett and Ingram, 1956; Ingram *et al.*, 1956; Kendrew and Parrish, 1956; Bennett *et al.*, 1957, 1961; Helcké *et al.*, 1968). Orientation assignments were based on the angular dependence of EMR resonances as the crystals were rotated in the magnetic field. Both resonant fields and linewidths varied

---

Betty J. Gaffney and Harris J. Silverstone • Department of Chemistry, The Johns Hopkins University, Baltimore, Maryland 21218-2685.

Biological Magnetic Resonance, Volume 13: EMR of Paramagnetic Molecules, edited by Lawrence J. Berliner and Jacques Reuben. Plenum Press, New York, 1993.

with orientation. The origin of the linewidth variation was attributed to variations in the parameters characteristic of the crystal field and the crystal structure (E. Feher, 1964; Eisenberger and Pershan, 1966; Helcké *et al.*, 1968). Refinements of these original EMR studies were reviewed by G. Feher (1970), and further analyses that include distributions in spin Hamiltonian parameters as sources of linewidth variation are continuing today (Fiamingo *et al.*, 1989). Figures 1 and 2 provide the geometric relationships important for demonstrating the sensitivity in structural analysis that can be achieved by EMR spectroscopy of high-spin ferric iron. The EMR spectra of many metmyoglobin and methemoglobin derivatives reflect axial symmetry at the high-spin ferric iron site. Figure 1 shows how the position of the resonance, expressed as a g' value [Eq. (2) below], varies with the orientation of molecular coordinates in the magnetic field. The figure is appropriate for heme samples where the heme normal is the molecular z axis. Figure 2 shows how stereograms are developed to show

1. the orientation of heme normals within the crystal, and
2. the g' values observed at different orientations of the crystal in the magnetic field.

The angular resolution of g' values from EMR of metmyoglobin crystals is high: the measured standard deviation of normals was 1.6° (Helcké *et al.*, 1968). This value has been refined to 1.4° (Hampton and Brill, 1979). In sections below, we will show some examples of simulations made using parameters taken from the heme literature. Some additional contributions to the breadth of lines in heme samples will be discussed in the section on line shapes.

There is a convenient way to refer to the location of certain distinctive features of EMR spectra, already used in the preceding paragraph. The resonance condition for an isolated spin $\frac{1}{2}$ is a strict proportionality between

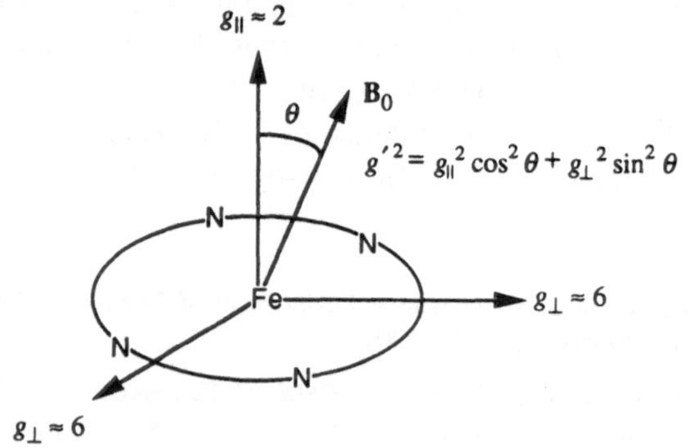

Figure 1. Principal magnetic axes of heme and the g' factor for axial symmetry.

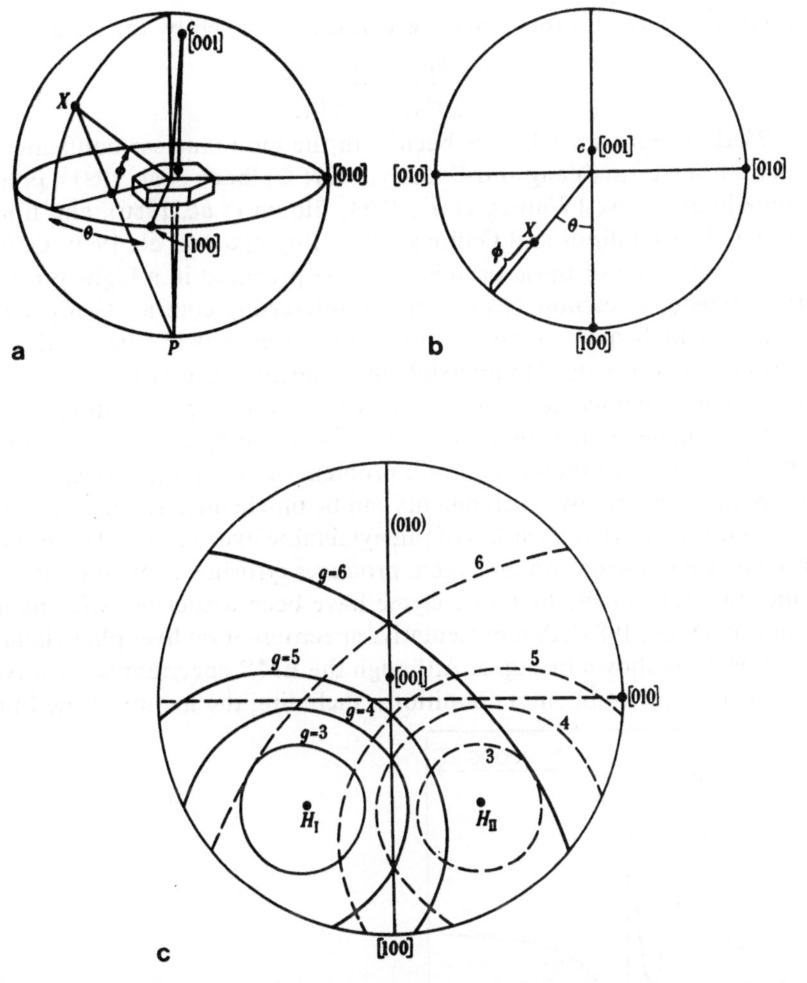

Figure 2. Development of stereograms showing how g' varies with heme orientation in $\mathbf{B}_0$. (a) Method of projection of orientations onto a plane. (b) The stereogram resulting from (a). (c) Stereogram representing g' value variation. Contours of constant g' are shown in solid and dashed lines for two separate heme orientations. Reproduced from Bennett *et al.* (1957).

the microwave frequency ν_0 and the resonance field B_{res},

$$h\nu_0 = \Delta E(B_{res}) = g_e \beta B_{res} \tag{1}$$

For $S > \frac{1}{2}$, the addition of spin-orbit coupling and spin-spin interaction terms to the Hamiltonian leads to a large spread in the range of the resonance fields. Because experimental spectra versus magnetic field are recorded at different frequencies, it is useful to have terminology for discussing resonance that is at least partially frequency-independent. Accordingly, one combines the spectrometer frequency ν_0 and the field B_{res} at which the resonance occurs into the "effective g value" g' given by the ratio

$$g' = \frac{h\nu_0}{\beta B_{\text{res}}} = \frac{\Delta E(B_{\text{res}})}{\beta B_{\text{res}}} \tag{2}$$

Work in our own labs has been with the mononuclear nonheme iron proteins transferrin (Yang and Gaffney, 1987; Dulbach *et al.*, 1991), phenylalanine hydroxylase (Wallick *et al.*, 1984; Bloom *et al.*, 1986), and lipoxygenase (Mavrophilipos and Gaffney, 1985; Boyington *et al.*, 1990; Gaffney *et al.*, 1993). Each of these proteins may be prepared in a high-spin ferric form. Transferrin carbonate provides an interesting contrast to myoglobin because the high-spin iron in this form of transferrin is in a nearly rhombic environment, as opposed to an axial environment in the case of myoglobin. The common characteristic of the EMR spectra of frozen solutions of all of these proteins is that there are several overlapping spectra. Assignment of the EMR spectra then becomes a problem in quantitative simulation of line shapes. Quantitative assignments can be biochemically very important. An example comes from studies of phenylalanine hydroxylase. This enzyme, with a pterin cofactor and dioxygen, produces tyrosine. A number of single point mutations in the human enzyme have been associated with phenylketonuria (Woo, 1989). A representative spectrum of rat liver phenylalanine hydroxylase is shown in Fig. 3. Although the EMR spectrum is sensitive to a number of environmental conditions, including the nature of the buffer,

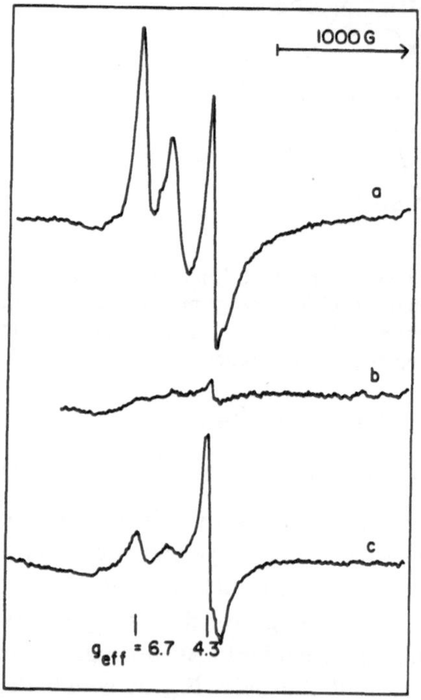

Figure 3. (a) X-band EMR of phenylalanine hydroxylase (PAH) in Tris buffer. (b) PAH after reduction of iron, showing cavity background. (c) The sample in (b) after exposure to air. Reproduced in part from Bloom *et al.* (1986).

important features of the iron center could be addressed for samples in Tris buffer (Wallick *et al.*, 1984; Bloom *et al.*, 1986).

The (effective g) g' values corresponding to the prominent features of the spectrum are at 6.7, 5.4, 4.3, and 2.0. Theoretically, it is unlikely that these g' values arise from a single species; g' values of 6.7, 5.4, and 2.0 suggest iron of one symmetry and the signal at $g' = 4.3$ suggests a different symmetry. After many samples had been studied, it became clear that the $g' = 4.3$ signal was bigger in samples with lower specific activity. Spectral simulations were used to show that the fraction of iron contributing the $g' = 4.3$ signal correlated with the fraction of inactive enzyme, and that all of the iron in the sample contributed to the observed resonance. This result is shown in Fig. 4. It is not known if any of the mutants of human phenylalanine hydroxylase has a defective iron center, but the studies on the enzyme from rat suggest that it may be possible to address this question by EMR.

To date, there is very little single-crystal EMR data from nonheme iron proteins to complement the studies of frozen solutions, although some early work has been done on crystalline transferrin (Blumberg, 1967). However, the structures of a number of nonheme iron proteins have been or are being solved, so it is reasonable to expect rapid progess in relating magnetic coordinates to structural ones in the near future. Table 1 summarizes progress on X-ray analysis of the structures of some mononuclear nonheme iron sites in proteins. Even in the absence of single crystals, structural information can be obtained by the orientation-selective double resonance techniques ESEEM (Singel, 1989; see also Chapter 7 of this volume) and ENDOR (Hoffman, 1991; Snetsinger *et al.*, 1990). In these latter approaches, theoretical analysis of spectra is an important element of planning the

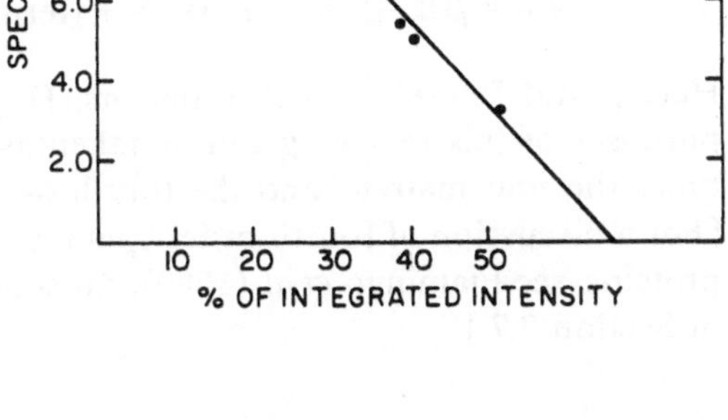

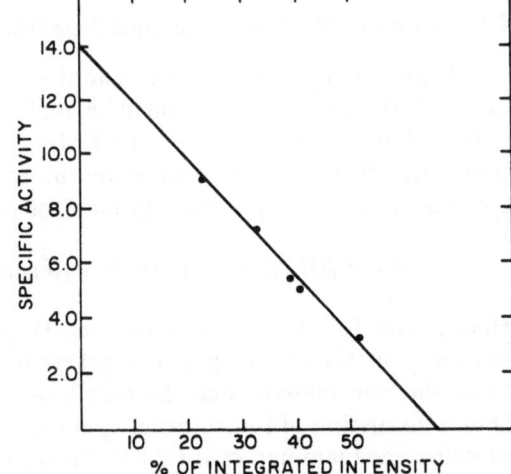

Figure 4. The specific activity of PAH from various preparations plotted as a function of the estimated concentration of spectral component giving rise to the broad $g' = 4.3$ signal. Reproduced in part from Wallick *et al.* (1984).

TABLE 1
Summary of Progress on X-ray Structures of Mononuclear, High-Spin, Nonheme Iron in Proteins[a]

Protein	Resolution Å	Symmetry	Metal ligands[b]	Reference
Lactoferrin	2.2	oct	tyr, tyr, his, asp anion2	Baker *et al.* (1990)
Transferrin	3.3	oct	tyr, tyr, his, asp, anion2	Bailey *et al.* (1988)
Iron superoxide dismutase	2.1	tbp	his, his, his, asp, H_2O	Carlioz *et al.* (1988); Stoddard *et al.* (1990)
Protocatechuate dioxygenase	2.8	tbp	tyr, tyr, his, his, (H_2O)	Ohlendorf *et al.* (1988)
Ferritin[c]	2.3	thd	glu, glu, his, H_2O	Yewdall *et al.* (1990)
Lipoxygenase	2.6	(note[d])	his, his, his, c-term carboxyl	Boyington *et al.* (1990; 1993)
Reaction centers	2.8	oct	his, his, his, his, glu2	Allen *et al.* (1988); Michel *et al.* (1986)

[a] Both ferric and ferrous proteins are included.
[b] The number "2" after a ligand means this ligand contributes bidentate coordination.
[c] A mononuclear binding site for iron in ferritin has been idcntified and may be the ferroxidase site.
[d] Distorted octahedron with two empty sites.

experiment, especially as applications are extended to $S > \frac{1}{2}$. Progress on biochemistry and spectroscopy of some mononuclear nonheme iron proteins was reviewed recently by Howard and Rees (1991).

2. THEORETICAL BACKGROUND FOR SIMULATIONS

2.1. Formal Structure of the Spin Hamiltonian

High-spin Fe^{3+} has the configuration $3d^5(t_{2g}^3 e_g^2)$. An extensive account of the spin physics for $S = \frac{5}{2}$ iron in hemoglobin has been given by Weissbluth (1967). A later monograph on the EMR spectroscopy of transition metals (Pilbrow, 1990) describes more recent work on iron in this and other configurations. The spin Hamiltonian for this configuration is given by

$$\mathcal{H}_S = \beta \mathbf{B} \cdot \mathbf{g} \cdot \mathbf{S} + \mathbf{S} \cdot \mathbf{D} \cdot \mathbf{S} + \text{(terms fourth-order in spin)} \qquad (3)$$

Here **g** and **D** are symmetric tensors. **D** gives the fine structure. For the purposes of this review, **g** can be taken as isotropic—that is, a constant g times the unit matrix—and the fourth-order spin terms can be neglected. [For a discussion of fourth-order spin terms in EMR spectroscopy of heme proteins, see Fiamingo *et al.* (1989). Anisotropy of **g** will be discussed briefly in Section 2.7.]

It is customary to put $\mathcal{H}_S$ into a simpler standard form. First choose the xyz axes to be the principal axes for the **D** tensor:

$$\mathcal{H}_S = g\beta\mathbf{B}\cdot\mathbf{S} + D_x S_x^2 + D_y S_y^2 + D_z S_z^2 \tag{4}$$

It is a minor algebraic maneuver to rewrite $\mathcal{H}_S$ in the form

$$\mathcal{H}_S = g\beta\mathbf{B}\cdot\mathbf{S} + D(S_z^2 - \tfrac{1}{3}\mathbf{S}^2) + E(S_x^2 - S_y^2) + \tfrac{1}{3}\mathbf{S}^2(D_x + D_y + D_z) \tag{5}$$

where

$$D = D_z - \tfrac{1}{2}D_x - \tfrac{1}{2}D_y, \qquad E = \tfrac{1}{2}D_x - \tfrac{1}{2}D_y \tag{6}$$

The last term in Eq. (5) is the constant $(35/12)(D_x + D_y + D_z)$ for high-spin Fe^{3+}, which has $S = \frac{5}{2}$. Consequently it is not observable by EMR and is customarily dropped from Eq. (5). This is equivalent to taking the matrix **D** to be traceless, $D_x + D_y + D_z = 0$. The EMR-observable fine structure terms in the spin Hamiltonian involve only two nonzero parameters, the tetragonal splitting constant D and the rhombic splitting constant E:

$$\mathcal{H}_S = g\beta\mathbf{B}\cdot\mathbf{S} + D(S_z^2 - \tfrac{1}{3}\mathbf{S}^2) + E(S_x^2 - S_y^2) \tag{7}$$

When **D** is traceless, Eq. (6) for D and E can be slightly simplified:

$$D = \tfrac{3}{2}D_z, \qquad E = D_x + \tfrac{1}{2}D_z = -D_y - \tfrac{1}{2}D_z, \qquad (\text{trace } \mathbf{D} = 0) \tag{8}$$

2.2. Choice of *xyz* Axes and the Values of *D* and *E*

The axis system is determined by **D**, but which axes to call x, y, and z is in a sense arbitrary. There can be up to six pairs of values of D and E, all of which describe the same physical spin Hamiltonian. These pairs, which result from labeling the coordinate axes (x, y, z), (y, x, z), (z, x, y), (z, y, x), (y, z, x), and (x, z, y), are (D, E), $(D, -E)$, $(-\frac{1}{2}D + \frac{3}{2}E, -\frac{1}{2}(D + E))$, $(-\frac{1}{2}D + \frac{3}{2}E, +\frac{1}{2}(D + E))$, $(-\frac{1}{2}D - \frac{3}{2}E, +\frac{1}{2}(D - E))$, and $(-\frac{1}{2}D - \frac{3}{2}E, -\frac{1}{2}(D - E))$. Blumberg (1967) proposed to resolve the ambiguity by choosing z so as to make D largest in magnitude and to distinguish x and y in such a way that E has the same sign as D. That is, given $D_x + D_y + D_z = 0$, choose the axes such that $|D_z| \geqslant |D_y| \geqslant |D_x|$. With this ordering, D_x and D_y must have the same sign, which is opposite to that of D_z. Further, $|D_x| \leqslant \frac{1}{2}|D_z|$, while $|D_y| \geqslant \frac{1}{2}|D_z|$, so that from Eq. (8) one sees that $0 \leqslant E/D \leqslant \frac{1}{3}$. Blumberg called this choice of axes the "proper coordinate system," but to us the name "Blumberg coordinate system" seems more appropriate.

Poole *et al.* (1974) proposed a similar resolution, except that they interchange x and y from Blumberg so that the sign of E is opposite to that of D.

We mention a third possibility, which might be called "rhombic" and is characterized by making D smallest in magnitude. That is, choose the axes such that $|D_x| \geq |D_y| \geq |D_z|$, interchanging x and z from Blumberg. Then it follows from Eq. (8) that $|E| \geq |D|$, that D and E have opposite signs, and that $-1 \leq D/E \leq 0$. The relation between the Blumberg and rhombic choices is symmetrical:

$$D_{\text{rhombic}} = -\tfrac{1}{2}D_{\text{Blumberg}} + \tfrac{3}{2}E_{\text{Blumberg}} \tag{9}$$

$$E_{\text{rhombic}} = +\tfrac{1}{2}D_{\text{Blumberg}} + \tfrac{1}{2}E_{\text{Blumberg}} \tag{10}$$

$$D_{\text{Blumberg}} = -\tfrac{1}{2}D_{\text{rhombic}} + \tfrac{3}{2}E_{\text{rhombic}} \tag{11}$$

$$E_{\text{Blumberg}} = +\tfrac{1}{2}D_{\text{rhombic}} + \tfrac{1}{2}E_{\text{rhombic}} \tag{12}$$

The Blumberg coordinate system is "natural" for axial symmetry ($D_x = D_y$), for which $D_{\text{Blumberg}} \neq 0$, $E_{\text{Blumberg}} = 0$ and

$$\mathcal{H}_S = g\beta\mathbf{B}\cdot\mathbf{S} + D(S_z^2 - \tfrac{1}{3}\mathbf{S}^2)$$

$$\text{(Blumberg coordinate system, axial symmetry)} \tag{13}$$

but for which $D_{\text{rhombic}} = -E_{\text{rhombic}} \neq 0$. The rhombic system is more natural for discussing rhombic symmetry ($D_z = 0$, $D_x = -D_y$), for which $D_{\text{rhombic}} = 0$, $E_{\text{rhombic}} \neq 0$, but for which $D_{\text{Blumberg}} = 3E_{\text{Blumberg}} \neq 0$:

$$\mathcal{H}_S = g\beta\mathbf{B}\cdot\mathbf{S} + E(S_x^2 - S_y^2)$$

$$\text{(rhombic coordinate system, rhombic symmetry)} \tag{14}$$

$$\mathcal{H}_S = g\beta\mathbf{B}\cdot\mathbf{S} + D[(S_z^2 - \tfrac{1}{3}\mathbf{S}^2) + \tfrac{1}{3}(S_x^2 - S_y^2)]$$

$$\text{(Blumberg coordinate system, rhombic symmetry)} \tag{15}$$

2.3. Origin of the Fine Structure: Second-Order Spin–Orbit Interaction

The fine-structure terms $\mathbf{S}\cdot\mathbf{D}\cdot\mathbf{S}$ in a spin Hamiltonian originate primarily from the spin-spin interaction in first order and spin-orbit interaction in second order of perturbation theory. Perhaps the simplest way to understand the case of high-spin Fe^{3+} is through the crystal-field model

(Griffith, 1964; Kotani, 1968): the Fe^{3+} is treated quantum mechanically as if it were a free ion perturbed by an electric field arising from the ligands surrounding it *in situ.* The lowest term of the ground configuration ($3d^5$) of Fe^{3+} is ^6S. Since $L = 0$, the first-order spin–spin interaction and spin–orbit interaction vanish, and the main energy contribution is from the spin–orbit interaction in second order. Of the four quartets of the $3d^5$ configuration—^4G, ^4P, ^4D, and ^4F—spin–orbit coupling can couple ^6S directly only to ^4P. If the spin–orbit coupling term for one electron in $3d^5$ is denoted by $\zeta_{3d}\mathbf{l}_i \cdot \mathbf{s}_i$, then the second-order contribution to the energy of the ^6S states that comes from coupling with ^4P is the same as the first-order contribution of the effective (nontraceless) Hamiltonian

$$\mathscr{H}_{\text{s-o}} = \frac{1}{2}\zeta_{3d}^2\left(\frac{S_x^2 - \frac{25}{4}}{E_{^4P_x} - E_{^6S}} + \frac{S_y^2 - \frac{25}{4}}{E_{^4P_y} - E_{^6S}} + \frac{S_z^2 - \frac{25}{4}}{E_{^4P_z} - E_{^6S}}\right) \tag{16}$$

From Eq. (6) one then gets for D and E

$$D = \frac{1}{4}\zeta_{3d}^2\left(\frac{2}{E_{^4P_z} - E_{^6S}} - \frac{1}{E_{^4P_x} - E_{^6S}} - \frac{1}{E_{^4P_y} - E_{^6S}}\right) \tag{17}$$

$$E = \frac{1}{4}\zeta_{3d}^2\left(\frac{1}{E_{^4P_x} - E_{^6S}} - \frac{1}{E_{^4P_y} - E_{^6S}}\right) \tag{18}$$

In the isolated ion, the x, y, and z components of the ^4P are degenerate. Then both D and E vanish, and the second-order spin–orbit interaction shifts all the ^6S states by the same constant,

$$\frac{1}{2}\zeta_{3d}^2\left(\frac{\frac{35}{4} - \frac{75}{4}}{E_{^4P} - E_{^6S}}\right) = -10 \times \frac{1}{2}\zeta_{3d}^2\left(\frac{1}{E_{^4P} - E_{^6S}}\right) \tag{19}$$

In this discussion, the labels x, y, z should be regarded as tentative, since we do not know the magnitudes of the zero-field parameters in advance: reassignment to conform to the Blumberg choice can be made later. The molecular environment—regarded as a "crystal field"—lifts the degeneracy of the $3d$ orbitals of Fe^{3+} and ultimately of the ^4P levels. Suppose first that the iron in hemoglobin is at a site of octahedral symmetry. In group theory nomenclature, the fivefold degenerate $3d$ level splits into the triply degenerate t_{2g} and the doubly degenerate e_g representations of the octahedral group. The ^6S atomic term belongs to the $^6A_1$ representation, and the ^4P terms belong to $^4T_1$. There is a competition between the interelectronic repulsion and the crystal field in determining the $^4T_1$ wave function. If the crystal field is weak relative to the interelectronic Coulomb repulsion, then the lowest $^4T_1$ is primarily the atomic ^4P, which is a mixture of three crystal-field configurations. Symbolically,

$$3d^5\,{}^4P = \tfrac{2}{5}t_{2g}^3 e_g + \tfrac{2}{5}t_{2g}^3 + \tfrac{1}{5}t_{2g}^2 e_g^3 \tag{20}$$

On the other hand, if the crystal field is strong (strong field approximation), as is to be the case for iron in hemoglobin, then the lowest $^4T_1$ is primarily the single $t_{2g}^4 e_g$ crystal field configuration, which is in turn described as a mixture of the atomic ^4P, ^4F, and ^4G configurations. Recall that only the ^4P atomic configuration has nonzero spin–orbit coupling with ^6S, so that its effect in the strong crystal field case is diluted by the ^4F and ^4G. The consequence is that the right-hand sides of Eqs. (16)–(19) need to be multiplied by 2/5, the fraction of the strong-field $^4T_1$ that is ^4P. An octahedral crystal field does not split $^4T_1$. If the symmetry is not lower than octahedral, then there would be a constant second-order spin–orbit shift of all the $^6A_1$ states by the same amount, but no splitting; D and E would vanish.

If there is tetragonal distortion (axial symmetry) in the z direction, then the spatially threefold degenerate $^4T_1$ level is split into a spatially nondegenerate level $^4A_2$ and a spatially twofold degenerate level ^4E. From Eqs. (17) and (18), one sees that $D \neq 0$, and $E = 0$, so that D is appropriately called the tetragonal splitting constant. One can see further from Eq. (17) that $D > 0$ if the $^4A_2$ level is lower in energy than the ^4E level. Combining Eqs. (17) and (18) with the 2/5 factor and changing from atomic to crystal-field notation, one finds that

$$\mathcal{H}_{s\text{-}o} = \frac{2}{5} \times \frac{1}{2} \zeta_{3d}^2 \left(\frac{S_x^2 - \frac{25}{4}}{E_{^4E_x} - E_{^6A_1}} + \frac{S_y^2 - \frac{25}{4}}{E_{^4E_y} - E_{^6A_1}} + \frac{S_z^2 - \frac{25}{4}}{E_{^4A_2} - E_{^6A_1}} \right) \tag{21}$$

$$D = \frac{2}{5} \times \frac{1}{4} \zeta_{3d}^2 \left(\frac{2}{E_{^4A_2} - E_{^6A_1}} - \frac{1}{-E_{^6A_1}} - \frac{1}{E_{^4E_y} - E_{^6A_1}} \right) \tag{22}$$

and

$$E = \frac{2}{5} \times \frac{1}{4} \zeta_{3d}^2 \left(\frac{1}{E_{^4E_x} - E_{^6A_1}} - \frac{1}{E_{^4E_y} - E_{^6A_1}} \right) \tag{23}$$

If there is rhombic distortion in the xy plane, then the symmetry is lower than tetragonal, and the spatially degenerate ^4E level is split into two nondegenerate levels. We denote them ^4E_x and ^4E_y for simplicity. From Eq. (23) one can see that $E \neq 0$, and thus E is appropriately associated with rhombic distortion. If we suppose that the energy of ^4E_x is lowered by the same small amount that ^4E_y is raised, and that the energy of $^4A_2$ is unchanged, then from Eq. (22), one sees that $D \sim 0$. For this reason, $D \equiv 0$,

$E \neq 0$ is called the "purely rhombic" case, and for the same reason we have called this coordinate system the "rhombic coordinate system." We shall use the rhombic system for the discussion of the EMR spectra for rhombic symmetry in Section 2.6.3. Otherwise we will use the Blumberg system. Recall that in Blumberg's coordinate system, the *xyz* axes would be relabeled *zyx* to make D nonzero, in this case resulting in $D_{\text{Blumberg}} = \frac{3}{2}E_{\text{rhombic}}$, and $E_{\text{Blumberg}} = \frac{1}{2}E_{\text{rhombic}}$. The condition for purely rhombic symmetry in Blumberg's coordinate system is then $E_{\text{Blumberg}}/D_{\text{Blumberg}} = \frac{1}{3}$, as contrasted to $D = 0$ in the "rhombic system." Note that rhombic symmetry of the Hamiltonian leads to the approximate equality $D \sim 0$ $[E_{\text{Blumberg}}/D_{\text{Blumberg}} \sim \frac{1}{3}]$, but not to exact equality. On the other hand, tetragonal symmetry of the Hamiltonian *requires* that E vanish exactly.

Fiamingo *et al.* (1989) have used values for the quantities in Eqs. (21)-(23) for hemoglobin that are typically $\zeta_{3d} = 400\ \text{cm}^{-1}$, $E(^4\text{A}_2) - E(^6\text{A}_1) \sim 2000\ \text{cm}^{-1}$, and $E(^4\text{E}_x) - E(^6\text{A}_1) = E(^4\text{E}_y) - E(^6\text{A}_1) \sim 6000\ \text{cm}^{-1}$, and which lead to $E_{\text{Blumberg}} \sim 0$, $D_{\text{Blumberg}} \sim 10.67\ \text{cm}^{-1}$.

The crystal-field model described is sometimes called the four-level model, as illustrated in Fig. 5. This model has obvious oversimplifications: only one atomic configuration is considered, and the effect of the molecular environment is treated only by the lowest relevant orders of perturbation theory. Its attractiveness is that it gives the essence of the explanation of the origin of D and E with a minimum of work. Solomon and co-workers (Gebbhard *et al.*, 1990; Zhang *et al.*, 1991) have pointed out cases requiring refinement of the model; in particular, delocalization of the electronic wave

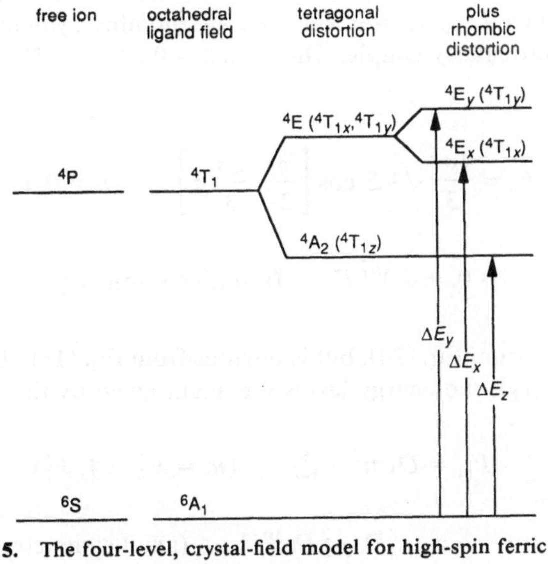

Figure 5. The four-level, crystal-field model for high-spin ferric iron.

function from ion to ligands (covalency) dilutes the x, y, and z contributions in Eqs. (22) and (23) differently. For instance, as the numerators 2 and -1 change individually, the formula (22) for D, with energy denominators that would yield a positive value, instead can yield a negative value.

2.4. Energy Levels of the Spin Hamiltonian in Zero Field

Although there are six $S = \frac{5}{2}$ states when the magnetic field strength B is zero, there are only three energy levels, as follows from the Kramers theorem. The three energy levels of the zero-field spin Hamiltonian [Eq. (7)] are given by the formula

$$E_n = \frac{4\sqrt{7}}{3} D^* \cos\left[\frac{1}{3}\arccos\left(\frac{10}{7^{3/2}}\cos 3\delta\right) + \frac{2\pi}{3}n\right] \qquad (n = 0, 1, 2) \qquad (24)$$

where

$$\cos\delta = D/D^*, \qquad \sin\delta = \sqrt{3}\,E/D^*, \qquad D^* = \sqrt{D^2 + 3E^2} \qquad (25)$$

Note that D^* is independent of the labeling of the xyz axes (Section 2.2). Although the separation of the zero-field energy levels is often associated with the size of D, it is perhaps better thought of as a function of D^*, since the relative values of D and E depend on the choice of axes, while D^* does not. When $D_{\text{rhombic}} = 0$, the case of rhombic symmetry, the energy formula is particularly simple. Then $\cos\delta = 0$ and $\cos 3\delta = 0$. The energy levels are just

$$E_n = \frac{4\sqrt{7}}{3}\sqrt{3}\,E\cos\left[\frac{\pi}{2} + \frac{2\pi}{3}n\right] \qquad (n = 0, 1, 2) \qquad (26)$$

$$= 0, \pm 2\sqrt{7}\,E \qquad \text{(rhombic symmetry)} \qquad (27)$$

It also follows from Eq. (24), but is obvious from Eq. (13), that when $E = 0$ (axial symmetry), the energy levels are again given by the simple formula

$$E_m = D\left(m^2 - \tfrac{35}{12}\right) \qquad \left(m = \pm\tfrac{1}{2}, \pm\tfrac{3}{2}, \pm\tfrac{5}{2}\right) \qquad (28)$$

$$= -\tfrac{8}{3}D, -\tfrac{2}{3}D, \tfrac{10}{3}D \qquad \text{(axial symmetry)} \qquad (29)$$

2.5. Energy Levels of the Spin Hamiltonian, Transition Dipoles, and EMR Spectra

The dependence of the energy levels of the spin Hamiltonian [Eq. (7)] on the values of *D*, *E*, and the magnetic-field strength *B*, as well as on the direction of **B** (angles θ and ϕ), can be complicated. To illustrate this point, we plot in Fig. 6 the energy levels for the polar angles $\theta = 0°$, 30°, 60°, and 90° for an axially symmetric case ($D > 0$, $E = 0$). (In axial symmetry, the levels are independent of the angle ϕ, which simplifies the plotting.) For concreteness, we have set $D = 10.7\ \text{cm}^{-1}$, typical of heme proteins. The same graphs apply to $D = 1.07\ \text{cm}^{-1}$ or $0.107\ \text{cm}^{-1}$: it is only necessary to

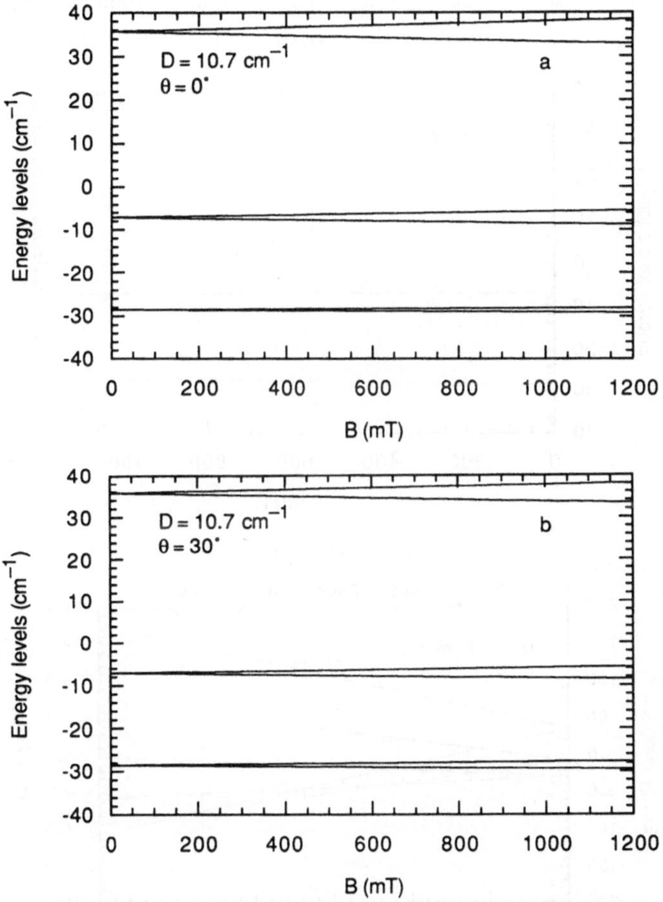

Figure 6. Energy versus field for axial symmetry. $D = 10.7\ \text{cm}^{-1}$ (typical for heme proteins), but energies and field scale proportionally to use graphs for other values of *D*. (a)–(d) Energy levels for small *B* at $\theta = 0°$, 30°, 60°, and 90°. (e)–(h) Energy levels for intermediate and large *B* at the same angles. For (e)–(h), Energy/*D* and *B*/*D* are given at the right and top axes.

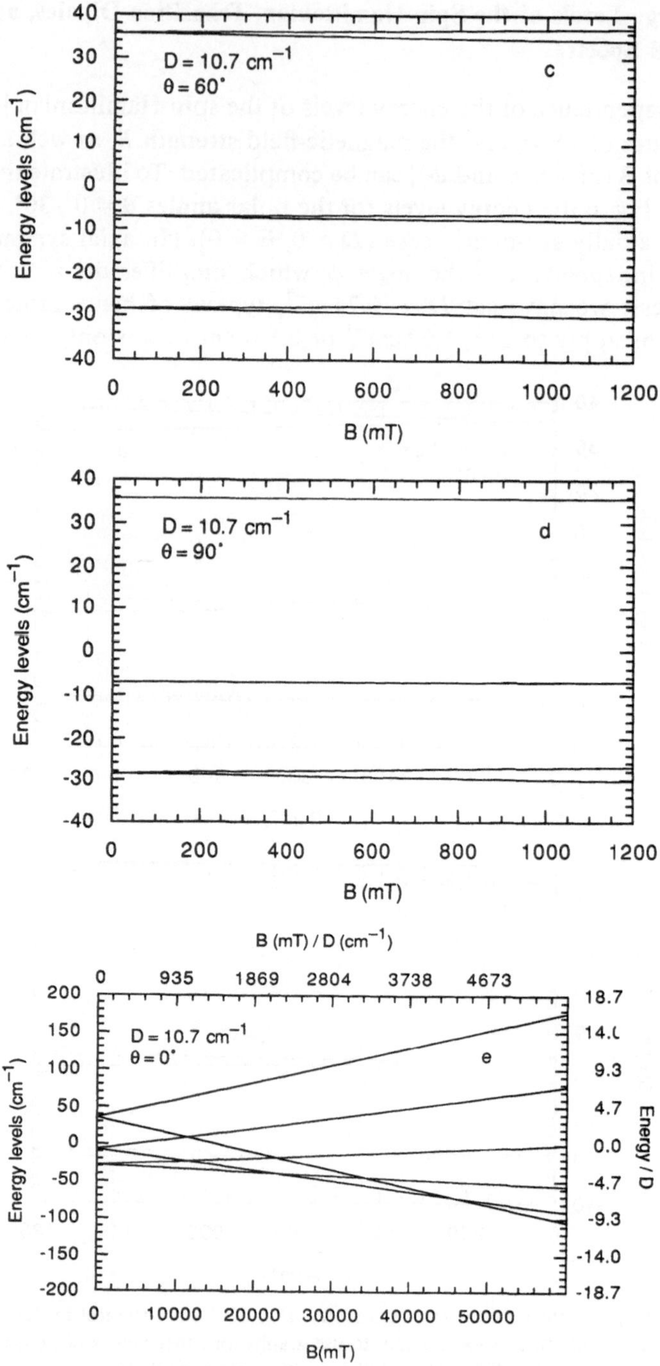

Figure 6. (*Continued*)

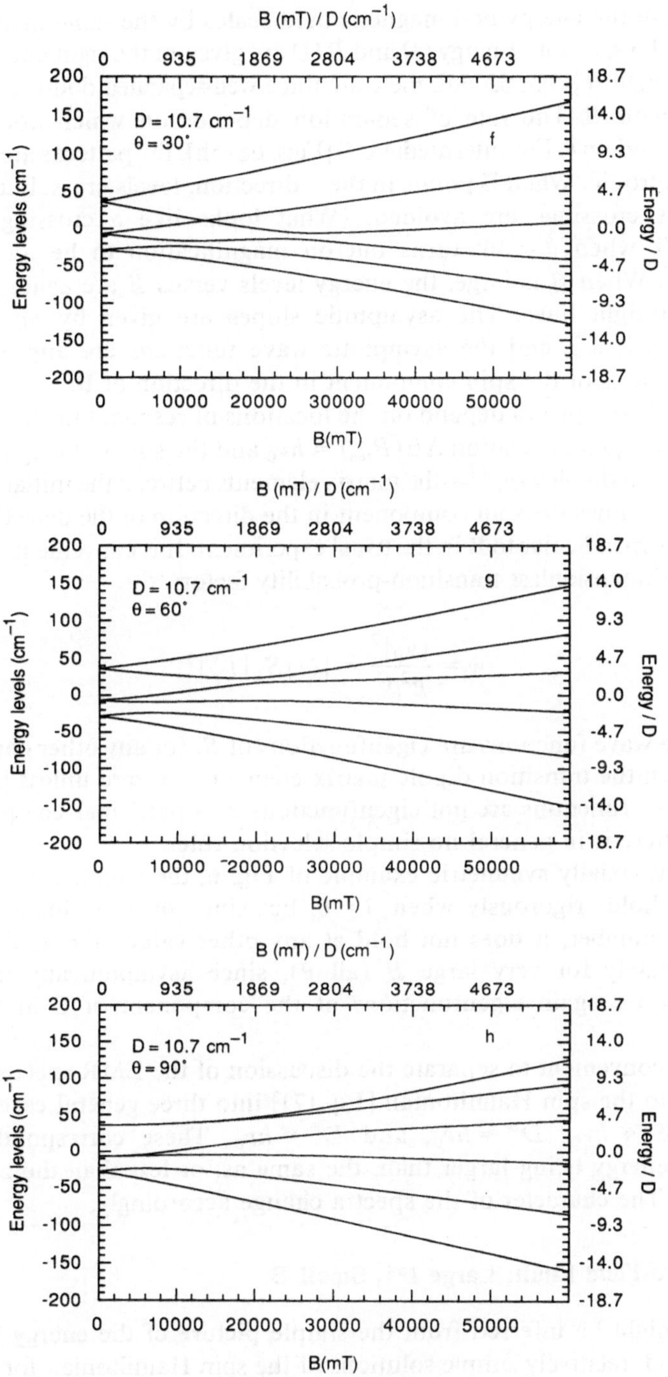

Figure 6. *(Continued)*

divide both the energy and magnetic field scales by the same factor (10 or 100). For Figs. 6e–6h, Energy$/D$ and B/D are given at the right and top axes.

For small B [Figs. 6a–6d], there are three well-separated doublets that split linearly with B. The rate of separation depends on which doublet one considers and on θ. For intermediate B [Figs. 6e–6h], the patterns are intricate and anisotropic. When $\mathbf{B}$ points in the z direction, levels cross, but in other directions crossings are avoided. (What looks like a crossing around 13,100 mT when $\theta = 30°$ turns out on magnification to be an avoided crossing.) When B is large, the energy levels versus B are again approximately straight lines. The asymptotic slopes are given by $mg\beta$, where $m = \pm\frac{1}{2}, \pm\frac{3}{2}, \pm\frac{5}{2}$; and the asymptotic wave functions are approximately eigenfunctions of the spin component in the direction of $\mathbf{B}$.

The EMR spectra depend on the locations of resonant fields, given via the Bohr frequency relation $\Delta E(B_{\text{res}}) = h\nu_0$ and the squared magnitudes of the transition dipoles $|\mu_{if}|^2$—the matrix elements between the initial and final states of $g\beta$ times the spin component in the direction of the detection field, which is perpendicular to $\mathbf{B}$ in the usual experiment. It is convenient to denote by w the dimensionless transition-probability factor:

$$w = \frac{|\mu_{if}|^2}{g^2\beta^2} = |\langle i|S_\perp|f\rangle|^2 \tag{30}$$

When the wave functions are eigenfunctions of S_z (or any other component of $\mathbf{S}$), then the transition dipole matrix elements are zero unless $|\Delta m| = 1$. If the wave functions are not eigenfunctions of a particular component of $\mathbf{S}$, then there is in general no simple selection rule.

In the axially symmetric example of Fig. 6, the simple selection rule $|\Delta m| = 1$ holds rigorously when $\theta = 0$, but since m is no longer a good quantum number, it does not hold at any other value of θ. It does hold approximately for very large B (all θ), since asymptotically the wave functions are again eigenfunctions of the component of $\mathbf{S}$ in the field direction.

It is convenient to separate the discussion of the EMR spectra corresponding to the spin Hamiltonian [Eq. (7)] into three general cases: $D^* = \sqrt{D^2 + 3E^2} \ll h\nu_0$, $D^* \approx h\nu_0$, and $D^* \gg h\nu_0$. These correspond to the Zeeman energy being larger than, the same as, or less than the zero-field energies. The character of the spectra change accordingly.

2.6. Zero-Field Limit: Large D^*, Small B

As might be inferred from the simple picture of the energy levels in Figs. 6a–6d, relatively simple solutions of the spin Hamiltonian for $S = 5/2$ result when $D^* \gg h\nu_0$, and when B does not increase much beyond $h\nu_0/2\beta$. The role of D and E is primarily to segregate the six levels into three

Kramers doublets. As B increases from 0, the degeneracy of each of the three Kramers doublets is lifted, but the shift of the energy levels is small compared with D^*. Low-order degenerate Rayleigh-Schrödinger perturbation theory (RSPT) applies, and the doublets do not mix significantly. This is easy to visualize in the example of Fig. 6, ($D = 10.7\ \text{cm}^{-1}$, $E = 0$), at X-band frequencies ($9\ \text{GHz} = 0.3\ \text{cm}^{-1} \ll 10.7\ \text{cm}^{-1}$). The relevant range for B is 0 to 400 mT. There are three possible EMR transitions, one within each doublet. Depending on relative values of D and E and on the direction of $\mathbf{B}$, the transition probability for one or more of the three transitions may vanish.

2.6.1. Axial Symmetry—Oriented Crystal

Consider first the limiting case of axial symmetry when B is small. In the axial symmetry example of Fig. 6, the zeroth-order subspaces are characterized by $|m|$ [Eqs. (28) and (29)]. The degeneracies of the $m = \pm\frac{5}{2}$ states and of the $m = \pm\frac{3}{2}$ states are lifted by $\mathbf{B}$, but transitions that have $\Delta m = 5$ and 3 have vanishing transition probability. Only the $|m| = \frac{1}{2}$ doublet gives an observable transition.

The RSPT first-order spin Hamiltonian in the two-dimensional ($|m| = \frac{1}{2}$) subspace is

$$\mathscr{H}_S + \tfrac{8}{3}D = 2B\beta(S_z \cos\theta + S_x \sin\theta) \tag{31}$$

$$= B\beta\begin{pmatrix} \cos\theta & 3\sin\theta \\ 3\sin\theta & -\cos\theta \end{pmatrix} \tag{32}$$

One can take $\phi = 0$, since for axial symmetry the energy eigenvalues do not depend on ϕ. The simplicity of Eq. (32) permits one to obtain closed-form solutions. The two energy levels, apart from the additive constant $-8D/3$, are given by

$$\pm B\beta\sqrt{9 - 8\cos^2\theta} \tag{33}$$

which leads to the equation for the resonance fields,

$$B_{\text{res}} = \frac{h\nu_0}{2\beta\sqrt{9 - 8\cos^2\theta}} \tag{34}$$

The g' factor, $2\sqrt{9 - 8\cos^2\theta}$, varies from 2 at $\theta = 0$ (z direction) to 6 at $\theta = 90°$ (xy plane). The dimensionless squared transition dipole, averaged over polarization, can also be calculated in closed form:

$$|\langle\text{lower}|S_\perp|\text{upper}\rangle|^2 = \frac{9}{4} - \frac{9\sin^2\theta}{1 + 8\sin^2\theta}. \tag{35}$$

Thus one expects to see EMR signals at the fields given by Eq. (34) and with the intensities given by Eq. (35). The "shape" of a signal depends on the experimental details. Typical absorption and "derivative of absorption" spectra (measured directly in the usual EMR experiment) are shown in Fig. 7a for $\theta = 0°$, 30°, 60°, and 90°. These low-order RSPT results for the resonance fields and intensities do not involve the parameter *D*, whose effect is only to isolate the three Kramers doublets from one another. Higher-order terms involve *D*, but as a consequence it is not easy to obtain *D* from EMR spectra when $D \gg h\nu_0$.

2.6.2. Axial Symmetry—Powder

With the exception of high-spin ferric hemoglobin and myoglobin, most studies of high-spin iron have been carried out on powder or amorphous samples. Powder spectra are the average over angle of oriented-crystal spectra. Because the resonance fields and transition probabilities may vary substantially and nonuniformly with angle, powder spectra can appear quite unlike crystal spectra. The powder spectrum for our axially symmetric

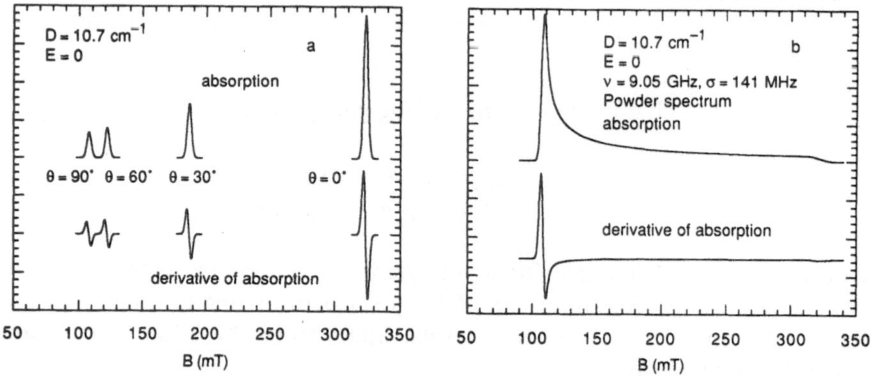

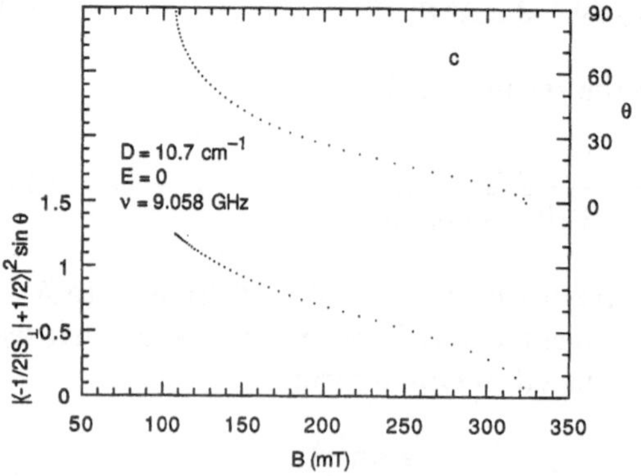

Figure 7. Crystal and powder simulated EMR spectra for a case of axial symmetry. (a) Oriented crystal at selected angles. (b) Powder pattern resulting from summation of spectra at all points, including the transition-probability weighting factor shown in (c). (c) Orientation (θ) and transition-probability weighting factor for "microcrystals" versus resonance field *B*. Successive dots are separated by 2° in θ.

example of a ferric ion with $D = 10.7\ \text{cm}^{-1}$, $E = 0$, $S = \frac{5}{2}$ is shown in Fig. 7b, along with the crystal spectra of Fig. 7a. The transition probability increases as θ increases from 0° to 90°, and the density of resonant fields versus θ increases. The angular average, $(\frac{1}{4\pi}) \int_0^{2\pi} d\phi \int_0^{\pi} d\theta \sin\theta$, has a strong feature at $g' = 6$ (from 90°) and a weaker feature at $g' = 2$ (from 0°). This powder spectrum is the signature of axial symmetry and reveals little about the size of D except that it is large compared with $2\beta B$.

One way to visualize the weighted averaging process that produces the powder spectrum in Fig. 7b is to focus on the various contributions at each value of the magnetic field. We calculate the resonance field for the $m = \pm\frac{1}{2}$ doublet at equally spaced values of the angle θ. Then we plot versus the resonance magnetic field both the angle θ and the transition probability factor [Eq. (30) multiplied by $\sin\theta$, which comes from the volume element $\sin\theta\, d\theta$ when the average over angle is computed], as shown in Fig. 7c. The strong feature at $g' = 6$ ($B = 110$ mT) is the consequence of the pile-up of resonance fields at $\theta = 90°$. That is, the resonance field is stationary at $\theta = 90°$ ($dB_{res}/d\theta = 0$) so that near $g' = 6$ a larger fraction of the Fe^{3+} ions will be in resonance than at higher fields. As the magnetic field increases, the density of Fe^{3+} ions in resonance decreases slowly at first, then more rapidly. [The points in Fig. 7c are spaced every 2° in θ.] Also, the transition probability, weighted by the geometric factor $\sin\theta$, decreases slowly. At the $g' = 2$ end, the spectrum stops abruptly, with a minor pile-up compared with the $\theta = 90°$ end. The downward dip in the derivative is characteristic of the end of an absorption band; the derivative is negative as the absorption falls to zero.

The rate of change of B_{res} with θ, which is important for simulating and interpreting powder spectra as just noted, can be found in closed form for the axially symmetric case in first-order RSPT:

$$\frac{dB_{res}}{d\theta} = -\frac{h\nu_0}{2\beta}(9 - 8\cos^2\theta)^{-3/2}\, 8\sin\theta\cos\theta, \tag{36}$$

The derivative clearly vanishes at both $\theta = 0°$ and $\theta = 90°$, as observed in Fig. 7c.

2.6.3. Rhombic Symmetry

Even though the zero-field energy levels can be found in closed form [Eq. (24)], the associated wave functions are trivially simple only for axial symmetry. For rhombic symmetry, however, the zero-field eigenfunctions are still simple enough to yield useful information in closed form relatively easily.

For this calculation, we find the rhombic coordinate system, with the Hamiltonian given by Eq. (14), more convenient than Blumberg's system. We denote by $|m\rangle$ an eigenfunction of S_z with eigenvalue m, and by $\psi_{E_n(\pm)}$ the two (±) zero-field eigenfucntions with energy E_n (in units of the rhombic splitting constant E, for notational clarity) that are also the "correct zeroth-order wave functions" when **B** points in the z direction. The six $\psi_{E_n(\pm)}$ are

$$\left.\begin{aligned} \psi_{0(+)} &= \frac{3}{\sqrt{14}}\left|\frac{5}{2}\right\rangle - \sqrt{\frac{5}{14}}\left|-\frac{3}{2}\right\rangle \\ \psi_{0(-)} &= -\frac{3}{\sqrt{14}}\left|-\frac{5}{2}\right\rangle + \sqrt{\frac{5}{14}}\left|\frac{3}{2}\right\rangle \end{aligned}\right\} \tag{37}$$

$$\left.\begin{aligned} \psi_{2\sqrt{7}(+)} &= \sqrt{\frac{5}{28}}\left|\frac{5}{2}\right\rangle + \sqrt{\frac{1}{2}}\left|\frac{1}{2}\right\rangle + \sqrt{\frac{9}{28}}\left|-\frac{3}{2}\right\rangle \\ \psi_{2\sqrt{7}(-)} &= \sqrt{\frac{5}{28}}\left|-\frac{5}{2}\right\rangle + \sqrt{\frac{1}{2}}\left|-\frac{1}{2}\right\rangle + \sqrt{\frac{9}{28}}\left|\frac{3}{2}\right\rangle \end{aligned}\right\} \tag{38}$$

$$\left.\begin{aligned} \psi_{-2\sqrt{7}(+)} &= \sqrt{\frac{5}{28}}\left|\frac{5}{2}\right\rangle - \sqrt{\frac{1}{2}}\left|\frac{1}{2}\right\rangle + \sqrt{\frac{9}{28}}\left|-\frac{3}{2}\right\rangle \\ \psi_{-2\sqrt{7}(-)} &= \sqrt{\frac{5}{28}}\left|-\frac{5}{2}\right\rangle - \sqrt{\frac{1}{2}}\left|-\frac{1}{2}\right\rangle + \sqrt{\frac{9}{28}}\left|\frac{3}{2}\right\rangle \end{aligned}\right\} \tag{39}$$

The top and bottom Kramers doublets with $E_n = \pm 2\sqrt{7} \times E$ are anisotropic in both θ and ϕ, the angles that give the direction of **B**. In contrast with the axial symmetry case, transitions within all three Kramers doublets have significant transition probability.

The middle Kramers doublet with zero-field $E_n = 0$ is special in that it is isotropic (for small B). To understand the isotropy, recall that for small B one may treat the $g\beta\mathbf{B}\cdot\mathbf{S}$ term in Eq. (14) by RSPT within the two-dimensional space of the two functions $\psi_{0(\pm)}$. The corresponding 2×2 matrices of the spin operators turn out to be

$$S_z = \frac{15}{7}\begin{pmatrix} \frac{1}{2} & 0 \\ 0 & -\frac{1}{2} \end{pmatrix}, \qquad S_x = \frac{15}{7}\begin{pmatrix} 0 & \frac{1}{2} \\ \frac{1}{2} & 0 \end{pmatrix}, \qquad S_y = \frac{15}{7}\begin{pmatrix} 0 & -i/2 \\ i/2 & 0 \end{pmatrix} \tag{40}$$

That is, the matrices of the spin components are exactly $\frac{15}{7}$ times the matrices for a single, isotropic spin-$\frac{1}{2}$ electron. Consequently the Zeeman splitting of the middle Kramers doublet will be exactly $\frac{15}{7}$ times the single-electron value, with a single isotropic resonance line at $g' = 2 \times \frac{15}{7} = \frac{30}{7} = 4.2857 \cdots \sim 4.3$.

The spin matrices in the top and bottom doublets are also instructive:

$$S_z = \frac{3}{7}\begin{pmatrix} \frac{1}{2} & 0 \\ 0 & -\frac{1}{2} \end{pmatrix}, \qquad S_x = \frac{18 \pm 6\sqrt{7}}{7}\begin{pmatrix} 0 & \frac{1}{2} \\ \frac{1}{2} & 0 \end{pmatrix}$$

$$S_y = \frac{18 \mp 6\sqrt{7}}{7}\begin{pmatrix} 0 & -i/2 \\ i/2 & 0 \end{pmatrix} \tag{41}$$

Here the upper sign is for the $(E_n = +2\sqrt{7} \times E)$ zero-field doublet, and the lower sign is for the $(E_n = -2\sqrt{7} \times E)$ zero-field doublet. The top doublet is therefore equivalent to a spin-$\frac{1}{2}$ with anisotropic g factor $(g_x, g_y, g_z)_{+2\sqrt{7}} = [(36 + 12\sqrt{7})/7, (36 - 12\sqrt{7})/7, 6/7] \sim (9.678, 0.607, 0.857)$. The same holds for the bottom doublet, except that the values of g_x and g_y are interchanged: $(g_x, g_y, g_z)_{-2\sqrt{7}} \sim (0.607, 9.678, 0.857)$.

The powder spectra for the top and bottom doublets are identical (to first-order RSPT), since their principal equivalent g values are the same, except for interchange of axes. They have a characteristic pattern, as illustrated in Fig. 8a, that at X band stretches from roughly 65 to 1100 mT. In this case of rhombic high-spin iron, the initial feature, which occurs at $g' = 9.678$, originates from molecules whose y axes lie near the magnetic field direction for the bottom doublet (x axes for the top doublet), and the final feature, at $g' = 0.607$, comes from molecules whose $x(y)$ axes lie near the field direction. The strong absorption maximum at $g' = 0.857$ does not originate just from the region around the molecular z axis, however, but from a plane including the z axis and characterized by $\phi \sim 86.4°$. The g' value is constant in this plane. (A slightly finer angular mesh, or a slightly broader linewidth, would have removed the ripples apparent in the derivative of the absorption.)

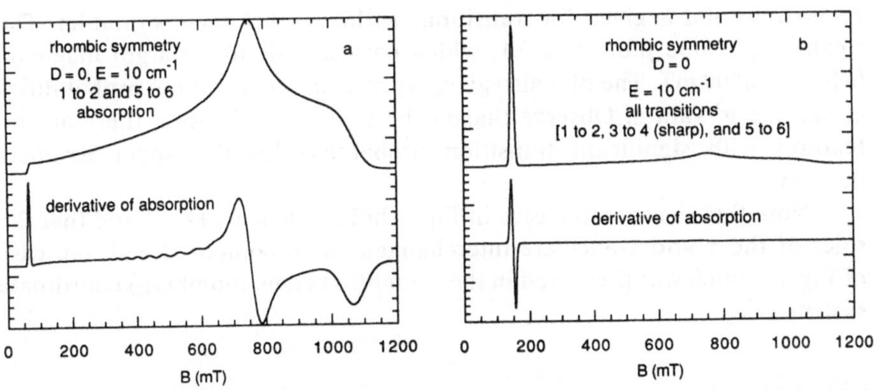

Figure 8. Powder spectra for the purely rhombic case. (a) The 1 → 2 and 5 → 6 transitions are shown as the absorption and its derivative. (b) The sum of the 3 → 4 transition and the two shown in (a). The ripples are computational artifact. Values given for D and E are D_{rhombic} and E_{rhombic} (see Section 2.2).

Since the middle doublet is isotropic, its intensity is not spread out over 1000 mT, and experimentally it always dominates the powder spectrum when there is rhombic symmetry, as can be seen in Fig. 8b.

The use of the rhombic coordinate system in this section not only facilitates discussion of the spin matrices but also simplifies specifying the invariant plane ($\phi \sim 86.4°$). In the Blumberg coordinate system the roles of the x and z axes are interchanged. The invariant plane could be described in words as containing the x axis and making an angle of ~86.4° with the z axis, but the mathematical description in the rhombic system, $\phi \sim 86.4°$, could hardly be simpler.

2.6.4. No Symmetry

The small B, large D^* case, when there is neither axial nor purely rhombic symmetry, is characterized in general by the superposition of three nonequivalent anisotropic doublets, each of which is analogous to the top or bottom doublet of the rhombic case. For the powder spectra, the positions of the beginning, maximum, and end, as well as the relative magnitudes, depend on D and E and differ from one another.

Since powder spectra are dominated by the directions for which the resonance fields (as a function of angle) are stationary, it is useful to know the directions and their associated g' values. At low fields, the stationary directions all lie along the principal axes, and the middle g' value holds approximately in a plane as well as along a principal axis. (At larger fields, there are always fifteen possible transitions, and there can be stationary directions within the xy, yz, and zx planes, as well as along the coordinate axes.) For illustration, we show in Fig. 9 a summary of the effective g values, g'_x, g'_y, and g'_z, at X band, for a range of values of E/D and $D = 2.0\ \text{cm}^{-1}$. A separate plot is given for transitions within each Kramers doublet. The smallest g' value plotted is 0.1, which corresponds to a (large) magnetic field of ~6500 mT. The plot also gives representative transition probabilities at various g' values. Observe that in the range $g' = 10$ to 2, there are no features with significant transition probability for the upper Kramers doublet.

Note that the rhombic case in Fig. 9 holds when $E/D = \frac{1}{3}$ and that the roles of the x and z axes are interchanged when compared with the case of Fig. 8, which was presented in the rhombic (versus Blumberg) coordinate system.

2.7. Moderate-to-Large B: B and D^* Terms Comparable

For $D^* \ll h\nu$, the Bohr frequency relation can be satisfied only for large B, so that the Zeeman term dominates the spin Hamiltonian. The

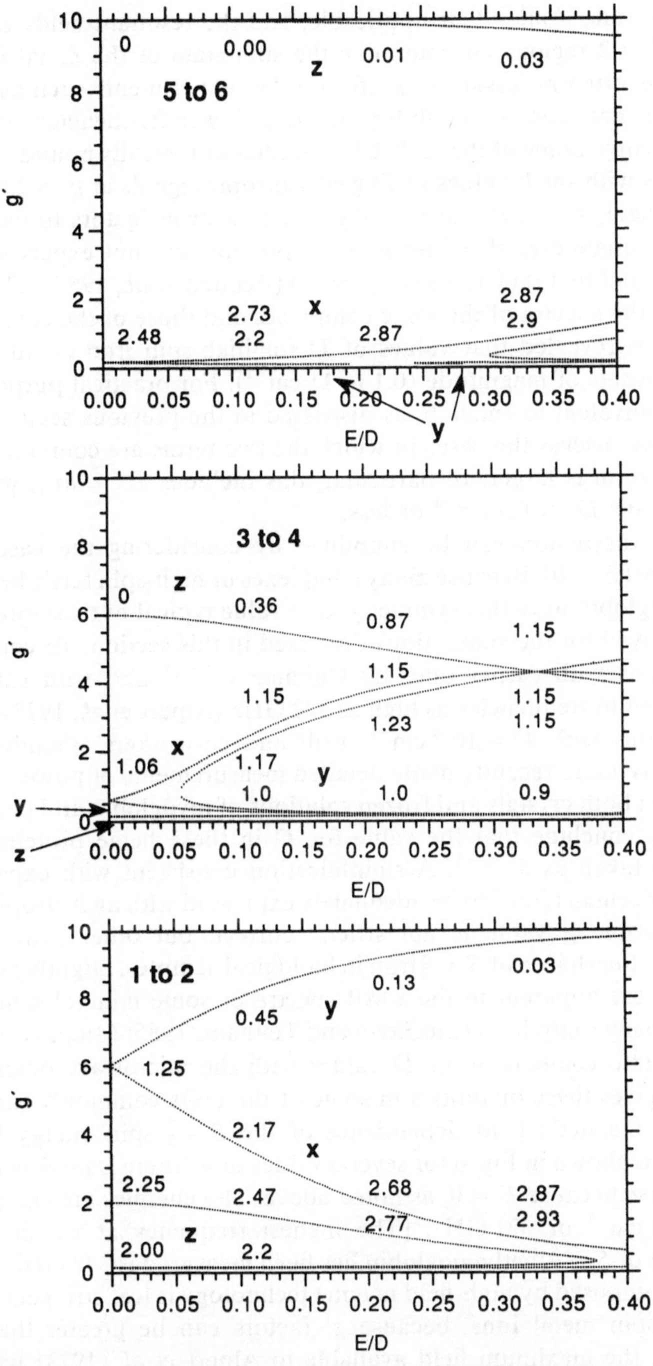

Figure. 9. Calculated *X*-band effective values of g_x, g_y, and g_z for the full range of E/D in the Blumberg coordinates (0 to $\frac{1}{3}$). The value of D used in the calculations is $2.0\ \text{cm}^{-1}$. Representative squared-transition-moments are indicated by the numbers.

selection rule, $|\Delta m| = 1$, is applicable, and the resonant fields correspond to the $g' = 2$ region. To determine the spin state of the metal ion in this case, it is often necessary to perform other experiments such as measurements of magnetic susceptibility, EMR at lower frequencies, and optical spectroscopy. Some of the early EMR studies of ionically bound $S = \frac{5}{2}$ ferric materials with small values of D gave isotropic signals at $g' = 2.0$ (Bleaney and Trenam, 1954). Because of this, the first investigators to examine the EMR of single crystals of ferric heme proteins did not expect to find the strong signal that they found at $g' = 6.0$ (Bennett *et al.*, 1957). The contrast between the spectra of the ionic complexes and those of the covalent heme complexes revealed that values of D for high-spin iron could vary over several orders of magnitude (0.1 to 11 cm^{-1}). For practical purposes, large D^* is equivalent to small B, as discussed in the previous sections. In this section we discuss the cases in which the two terms are comparable or the Zeeman term is larger. In particular, this includes $D^* \cong h\nu_0$, which at X band means $D^* \cong 0.3\ cm^{-1}$ or less.

The discussion can be simplified by considering the case of axial symmetry ($E = 0$). Because many complexes of high-spin ferric hemoglobin and myoglobin have this symmetry, a D value typical of these proteins will be employed for the simulations discussed in this section. To compare our simulations with earlier ones by Coffman (1975) and with experiments carried out to frequencies as high as 372 GHz (Alpert *et al.*, 1973), we show calculations with $D = 10.7\ cm^{-1}$. Brill and co-workers (Fiamingo *et al.*, 1989) have more recently made detailed measurements of power saturation and T_1 on both crystals and frozen solutions of myoglobin and hemoglobin, and they conclude that the value for D in these heme proteins is more properly taken as 8 cm^{-1}. A simplification consistent with experiment is that the Zeeman term can be adequately expressed with an isotropic g value. (An isotropic g value is not strictly correct, but other factors usually dominate lineshapes of $S = \frac{5}{2}$ iron in biological samples. Slightly anisotropic g values are apparent in the EMR spectra of some mineral samples with exceptionally sharp lines (McGavin and Tennant, 1985; Minge *et al.*, 1990).) To facilitate comparison of D values with the microwave quantum $h\nu_0$, Table 2 gives these quantities in some of the units commonly employed.

The magnetic field dependence of the $S = \frac{5}{2}$ spin energy levels for $D \cong h\nu_0$ is shown in Fig. 6 for several values of θ. There is no ϕ dependence in this case because $E = 0$, as noted above. The energies are calculated for $D = 10.7\ cm^{-1}$ or 321 GHz. [The highest frequency at which an EMR spectrum of $S = \frac{5}{2}$ methemoglobin has been measured is 372 GHz. The limit in NMR imposed by high-field magnet technology is less stringent for EMR of high-spin metal ions, because g' factors can be greater than 2. For example, the maximum field available to Alpert *et al.* (1973) was 5.75 T, but that was adequate to detect 372-GHz absorption for the transition at

TABLE 2
Values of *D* and *hν* in Various Units

Units	*D* (characteristic of heme)	*hν* (*X* band)	*hν* (*Q* band)	*hν* (*W* band)
GHz	320.8	9.2	35	90
cm^{-1}	10.7[a]	0.31	1.17	3.00
K	15.4	0.44	1.68	4.32

[a] Alpert *et al.* (1973).

lowest field. The frequency of 372 GHz was also used in simulations by Coffman (1975).] These values may appear irrelevantly large for most laboratories. Note, however, that D, E, and B all enter the spin Hamiltonian linearly, so that any plot versus B for a given D and E is identical with that for these three parameters scaled by the same factor. Indeed, it is often the practice to scale such plots by dividing all quantities by D (Aasa, 1970). This observation is intended to facilitate comparison with experiments on samples with different values of D and at different frequencies.

Because of the multiple crossings of energy levels, transitions among a number of levels are possible. As a result of avoided crossings, the resonance condition may be met more than once between two levels for a particular angle. For the purpose of illustration, we plot in Fig. 10 the transition energies, i.e., the separations of the energy levels shown in Fig. 6, for the angle $\theta = 30°$. The separation is given in GHz. Thus, to find the resonance fields for a given spectrometer frequency, one draws a horizontal line at that frequency, and the resonance fields occur where the separation between two levels crosses the horizontal spectrometer-frequency line. At 9 GHz, five transitions are encountered. If the states are labeled in increasing order of energy in the 30° plot in Fig. 6, then the transitions, in the order encountered, are $5 \rightarrow 6$, $3 \rightarrow 4$, $1 \rightarrow 2$ (the three Kramers doublets) and then $2 \rightarrow 3$ twice. These latter two transitions are on either side of the avoided crossing near 13,200 mT and would not be encountered if the microwave frequency were less than 6 GHz. The figure shows that at 300 GHz the additional lines, $4 \rightarrow 5$ twice and $2 \rightarrow 4$ twice, have just come into range. (As remarked in Section 2.6.1, at low fields the $3 \rightarrow 4$ and $5 \rightarrow 6$ transitions have vanishing small transition probabilities.)

A main point of Fig. 10 is to illustrate the complexity that is possible even in a simple case (axial symmetry). As remarked above, the identical figure would apply if D were much smaller, say 0.107 cm^{-1}, as long as the vertical scale and horizontal scale were also divided by 100. Thus, the magnetic field would range from 0 to 600 mT, and the GHz scales would range from 0 to 100 GHz and 0 to 4 GHz. The transition frequencies that for most of Fig. 10a are not accessible when $D = 10.7$ cm^{-1} are all accessible when $D = 0.1$ cm^{-1}.

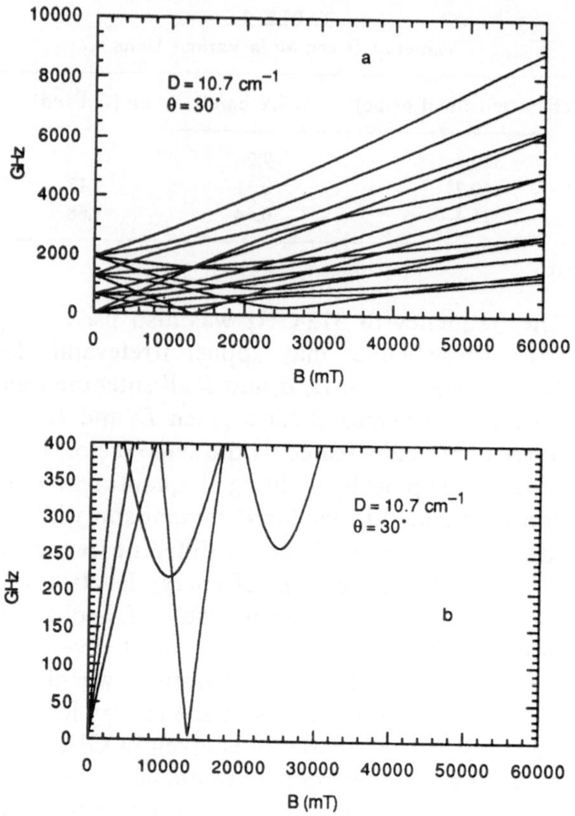

Figure 10. Variation of transition frequencies with magnetic field for an axially symmetric case. (a) All transitions. (b) Magnification of the 0–400 GHz range of (a).

2.8. Computational Methods for Spectral Simulation

There are four basic steps to the computer simulation of an EMR spectrum:

1. Calculate the energy levels of the spin Hamiltonian [Eq. (7)].
2. Determine the resonance magnetic fields from the Bohr frequency relation

$$E_{\text{upper}}(B_{\text{res}}) - E_{\text{lower}}(B_{\text{res}}) = h\nu_0 \tag{42}$$

3. Calculate the transition probability factor for each resonance [Eq. (30)].
4. Add up the contributions to the spectrum for each resonance.

For powder spectra, there is a fifth step: average over orientation. To account for statistical heterogeneity would require an additional step.

2.8.1. Calculation of Energy Levels and Resonance Fields

The 6 × 6 matrix of the spin Hamiltonian for spin-$\frac{5}{2}$ Fe^{3+} is straightforward to calculate, and formulas for the matrix elements can be found, for instance, in Yang and Gaffney (1987) and in Pilbrow (1984). The energy levels can be found by matrix diagonalization, although RSPT may be more efficient for special cases. Several excellent computer programs for diagonalization are widely available; we used the EISPACK routines in particular (Smith *et al.*, 1976), because the FORTRAN source code is published and can be implemented on any computer without special license.

If EMR experiments were carried out at fixed magnetic field, then the energy levels would immediately give the transition frequencies. But, practically, the microwave frequency at which transitions occur is kept constant, and the static magnetic field is swept. Theoretical determination of the resonance magnetic fields then requires solving the implicit Eq. (42) for B_{res}. This can be done in two steps. First, calculate the transition frequencies at a sequence of magnetic fields to locate two fields at which the calculated transition frequencies using Eq. (42) bracket the experimental ν_0. Then find the root that lies between these two values of B by an iterative procedure, such as iterative bisection, which converges linearly and always works, or Newton–Raphson (secant) iteration, which converges quadratically, but which sometimes does not converge properly. The first step, to bracket B_{res} approximately, may be trivial, as in the case of Fig. 6a, or it may require some care, as might be the case in Fig. 6b, to make sure that the initial survey fields are sufficiently close together to insure that no more than one resonance falls between a consecutive pair of them. To save computer time for the second step, we used the secant form of the Newton–Raphson method. On those (rare, easily trapped) occasions on which the Newton-Raphson method jumped to a root outside the two initial bracketing fields, we automatically switched to iterative bisection.

2.8.2. Calculation of Transition Probabilities

When the resonance field has been determined, it is then necessary to have the diagonalization routine give the eigenvectors of the spin Hamiltonian to calculate the transition moment [Eq. (30)]. If the Zeeman field $\mathbf{B}_0$ is in the direction $\mathbf{e}_0 = (\sin\theta\cos\phi, \sin\theta\sin\phi, \cos\theta)$, then the microwave field will lie in the plane perpendicular to $\mathbf{B}_0$ in the direction $\mathbf{e}_\chi = (\cos\chi\cos\theta\cos\phi - \sin\chi\sin\phi, \cos\chi\cos\theta\sin\phi + \sin\chi\cos\phi, -\cos\chi\sin\theta)$ that requires one additional angle variable χ to specify the direction in the plane. For the transition probability factor [Eq. (30)] one calculates the weighting factor w from the eigenvectors

$$w_{\text{lower-upper}} = |\langle\text{lower}|S_\perp|\text{upper}\rangle|^2 = |\langle\text{lower}|\mathbf{e}_\chi \cdot \mathbf{S}|\text{upper}\rangle|^2 \tag{43}$$

In the case of powder spectra, an average is taken over χ (as well as over θ and ϕ). The average of Eq. (43) over χ can be taken explicitly to give a weighting factor (with i for the lower and j for the upper state in the transition)

$$w = |\langle i|\mathbf{e}_\chi \cdot \mathbf{S}|j\rangle|^2_{\text{average over }\chi} = \tfrac{1}{2}(|\langle i|S_x|j\rangle|^2 + |\langle i|S_y|j\rangle|^2 + |\langle i|S_z|j\rangle|^2 - |\langle i|\mathbf{e}_0 \cdot \mathbf{S}|j\rangle|^2) \quad (44)$$

Equation (44) is particularly convenient to program because only the last term is explicitly angle-dependent, and the matrix of $\mathbf{e}_0 \cdot \mathbf{S}$ is already needed for the Zeeman terms in the Hamiltonian and need not be recalculated.

2.8.3. Summation of Contributions of Individual Resonances

Once the resonance fields and transition-probability factors Eq. (44) have been calculated, one can add up the contribution from each resonance to obtain the spectrum

$$S(B) = \sum_{\text{resonant fields } B_n} w_n \frac{\sigma_n}{(\sigma_n)_{\text{frequency}}} \frac{f(|B - B_n|/\sigma_n)}{\sigma_n} \quad (45)$$

Here w_n can denote either the nonaveraged transition-probability factor of Eq. (43) or the averaged factor of Eq. (44). The factor $\sigma_n/(\sigma_n)_{\text{frequency}}$ will be explained in the next paragraph. The *normalized* contribution of a single resonance, denoted by the line-shape function $f(|B - B_n|/\sigma_n)/\sigma_n$, is usually taken to be one of two standard forms, Lorentzian or Gaussian, and the parameter σ_n is a "width" parameter, sometimes called the "field-swept" or "field-domain" linewidth:

$$f_{\text{Lorentzian}}(|B - B_n|/\sigma_n)/\sigma_n = \frac{\sigma_n}{2\pi} \frac{1}{(B - B_n)^2 + \sigma_n^2/4} \quad (46)$$

$$f_{\text{Gaussian}}(|B - B_n|/\sigma_n)/\sigma_n = \frac{1}{\sigma_n \sqrt{2\pi}} e^{-(B-B_n)^2/2\sigma_n^2} \quad (47)$$

More general line-shape functions are possible; here we treat only the simplest and most widely used. We have chosen a notation that puts a "reduced variable" into the function $f(x)$, which itself satisfies $\int_{-\infty}^{\infty} f(x)\, dx = 1$. The $1/\sigma_n$ placed "outside" causes the product $(1/\sigma_n)f(y/\sigma_n)$ to be normalized to unity with respect to the nonreduced variable y.

The factor $\sigma_n/(\sigma_n)_{\text{frequency}}$ in Eq. (45) multiplies the field-normalized shape function [Eqs. (46) and (47)] to take into account how the experiments are usually carried out. Aasa and Vänngård (1975) pointed out that the formula one obtains for the absorption of radiation from time-dependent perturbation theory pertains to "frequency-swept" spectra and is normalized so that the integral of the line-shape function over *frequency*, rather than over field, is unity. Let $f_{\text{frequency}}(|\nu - \nu_{\text{res}}|/\sigma_{\text{frequency}})/\sigma_{\text{frequency}}$ denote the "frequency-swept" line-shape function, normalized to unity with respect to frequency:

$$\int_{-\infty}^{\infty} \frac{f_{\text{frequency}}(|\nu - \nu_{\text{res}}|/\sigma_{\text{frequency}})}{\sigma_{\text{frequency}}}\, d\nu = 1 \tag{48}$$

In the usual field-swept experiment, ν is held fixed at, say ν_0, and ν_{res} is varied by sweeping B. One can express $f_{\text{frequency}}(|\nu - \nu_{\text{res}}|/\sigma_{\text{frequency}})/\sigma_{\text{frequency}}$ in terms of B via the following:

$$\text{resonance field } B_0\text{:} \qquad E_{\text{upper}}(B_0) - E_{\text{lower}}(B_0) = h\nu_0 \tag{49}$$

$$\nu_{\text{res}} \equiv \frac{[E_{\text{upper}}(B) - E_{\text{lower}}(B)]}{h} \tag{50}$$

$$= \frac{[E_{\text{upper}}(B_0) - E_{\text{lower}}(B_0)]}{h} + (B - B_0)\frac{d[E_{\text{upper}}(B_0) - E_{\text{lower}}(B_0)]}{dB_0}\frac{1}{h} + \cdots \tag{51}$$

$$= \nu_0 + (B - B_0)\frac{d[E_{\text{upper}}(B_0) - E_{\text{lower}}(B_0)]}{dB_0}\frac{1}{h} + \cdots \tag{52}$$

Then the frequency-normalized line-shape function, as a function of field B, becomes

$$\frac{f_{\text{frequency}}(|\nu - \nu_{\text{res}}|/\sigma_{\text{frequency}})}{\sigma_{\text{frequency}}}$$

$$= \frac{1}{\sigma_{\text{frequency}}} f_{\text{frequency}}\left(\frac{|B - B_0|}{h\sigma_{\text{frequency}}}\frac{d[E_{\text{upper}}(B_0) - E_{\text{lower}}(B_0)]}{dB_0} + \cdots\right) \tag{53}$$

$$= \frac{1}{\sigma_{\text{frequency}}} f_{\text{frequency}}(|B - B_0|/\sigma_{\text{field}}) + \cdots \tag{54}$$

Equations (53) and (54) contain two results. The first is that the appropriate width σ_{field} to use for field-swept spectra is obtained from the frequency-swept width $\sigma_{\text{frequency}}$ via

$$\sigma_{\text{field}} = \sigma_{\text{frequency}} \frac{h}{d[E_{\text{upper}}(B_0) - E_{\text{lower}}(B_0)]/dB_0} \tag{55}$$

The second result, which follows from Eq. (54), is that the line-shape function derived for the field-swept spectrum (to leading order in $B - B_0$) has the same functional form as for the frequency-swept spectrum, but it has the "wrong" $1/\sigma$ multiplying it. That is, Eq. (54) will *not* integrate to unity with respect to B. In Eq. (54), the prefactor $1/\sigma_{\text{frequency}}$ would have to be $1/\sigma_{\text{field}}$ for the integral to equal unity. The integrated intensity [of Eq. (54)] with respect to B is therefore $\sigma_{\text{field}}/\sigma_{\text{frequency}}$, with

$$\frac{\sigma_{\text{field}}}{\sigma_{\text{frequency}}} = \frac{h}{d[E_{\text{upper}}(B_0) - E_{\text{lower}}(B_0)]/dB_0} \tag{56}$$

In the low-field case [cf. Figs. 6a–6d], the energy levels are linear functions of B, and

$$E_{\text{upper}}(B_0) - E_{\text{lower}}(B_0) = h\nu_0 \equiv g'\beta B_0 \tag{57}$$

and

$$\frac{1}{h}\frac{d[E_{\text{upper}}(B_0) - E_{\text{lower}}(B_0)]}{dB_0} \sim \frac{g'\beta}{h} = \frac{\nu_0}{B_0} \tag{58}$$

When Eq. (58) is valid, Eq. (55) simplifies to the Aasa–Vänngård $1/g'$ factor:

$$\frac{\sigma_{\text{field}}}{\sigma_{\text{frequency}}} \sim \frac{B_0}{\nu_0} = \frac{h}{g'\beta} \tag{59}$$

There are two important points for field-swept spectra:

1. If $\sigma_{\text{frequency}}$ is independent of B_0, the resonances get broader as B_0 increases.
2. The field-normalized intensity is multiplied by $\sigma_{\text{field}}/\sigma_{\text{frequency}} \sim h/g'\beta$.

Both these effects can easily be seen in Fig. 11. Notice especially how the heights of the absorption curves stay the same while the areas under them increase with resonance field. This conversion from frequency to field variables had been ignored in many earlier publications. The consequences of the frequency-to-field conversion have been discussed in detail by Pilbrow

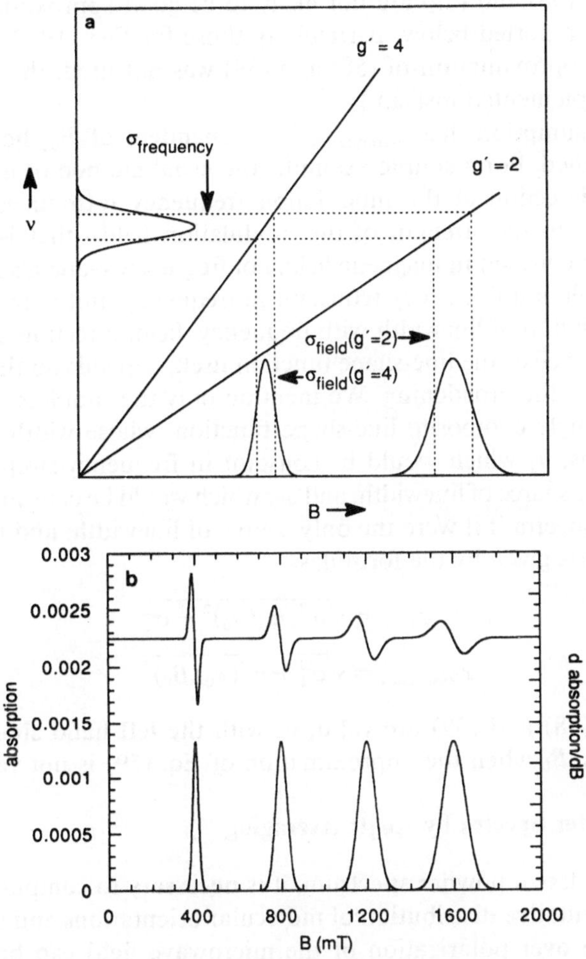

Figure 11. (a) A pictorial view in the style of Pilbrow (1990) of the relation of a fixed linewidth in frequency-swept spectroscopy to the corresponding variable linewidth in the field-swept case. (b) Four absorption (bottom) and derivative (top) lineshapes of equal transition probability and equal frequency-domain linewidths (300 MHz) as a function of magnetic field at 40, 80, 120, and 160 mT.

(1984, 1990), who also discusses other aspects of line shape, such as anisotropy. [Equation (58) is not an exact equality, because g' is not strictly independent of B_0. In fact, if B_0 lies to the right in Figs. 6a–6d, where the energy levels do not look like straight lines passing through the zero-field values, Eqs. (58) and (59) are not likely to be good approximations. In the calculations reported below, particularly those for Figs. 19–21, 28, and 30, the simple approximation of (58) and (59) was not used; the more general (55) was implemented instead.]

The assumption that $\sigma_{\text{frequency}}$ is independent of B_0, however, is not always justified. For a simple example, the usual method of modulating B_0 to permit detection at the modulation frequency introduces a linewidth proportional to the strength of the modulation field—that is, a linewidth contribution constant in magnetic field, not frequency—and also proportional to the modulation frequency (constant in frequency, not magnetic field). In general, variation of linewidth with frequency, field, direction, and other parameters, as well as the line-shape function itself, depends on the mechanism that produces the broadening. We mention only the simplest situation: that there is a single composite line-shape function, whose width contains two contributions; σ_1 which would be constant in frequency-swept spectra if it were the only source of linewidth, and σ_2, which would be constant in magnetic field-swept spectra if it were the only source of linewidth; and that the composite width is given by the formulas

$$\sigma_{\text{field}} = \sqrt{\sigma_1^2(B_0/\nu_0)^2 + \sigma_2^2} \tag{60}$$

$$\sigma_{\text{frequency}} = \sqrt{\sigma_1^2 + \sigma_2^2(\nu_0/B_0)^2} \tag{61}$$

when Eqs. (58) and (59) are valid, or with the left-hand side of Eq. (58) replacing ν_0/B_0 when the approximation of Eq. (59) is not valid.

2.8.4. Powder Spectra by Angle Averaging

To simulate a powder spectrum, it is necessary to compute the spectra for a representative distribution of molecular orientations and then average. The average over polarization of the microwave field can be carried out analytically [Eq. (44)], as discussed already in Section 2.8.2. It is usually convenient to use molecular-fixed axes and vary the orientation of the magnetic field $\mathbf{B}_0$. Conceptually, the angle average is simple, but since two angles specify an orientation, the number of orientations is proportional to the square of the fineness of the one-dimensional mesh, and computation times can become a problem. On the one hand, it is important that the directions that contribute strongly to the spectra be well sampled. On the other hand, directions that contribute weakly can be a numerical problem if individual lines are narrow, but their positions change rapidly with angle. (Cf.

the ripples in the derivative spectrum in Fig. 8a.) Care must be exercised to ensure convergence. This aspect of spectral simulation requires a good measure of data management.

When there are distributions in the parameters of the spin Hamiltonian, the approach that we have taken is to calculate the full spectrum for each of a set of spin-Hamiltonian parameters, for instance those that characterize a distribution in E/D, and then to add the calculated spectra with appropriate (e.g., Gaussian) weights with respect to the distributed variables.

2.8.5. Computer Program

Accompanying this volume is an 800 K $3\frac{1}{2}$-inch Macintosh diskette, which contains the FORTRAN programs we used to generate the simulated spectra in this article. These programs have been written in Language Systems FORTRAN, which runs within the MPW programming environment. Compiled, stand-alone applications for Macintosh computers with math coprocessors (Mac II and up) have also been included. The programs run in two steps. The first step is to calculate the resonance fields and transition-probability factors and is carried out by IronHS (high-spin iron). The second step is to put a Gaussian line-shape function at each resonance field and add up the contributions of all the resonances; speGIron (spectrum, Gaussian, for Iron) reads in the resonance output from IronHS and writes the simulated spectum to the file spectR.dat. This latter file is in a form that can be input into any standard plotting program, such as *KaleidaGraph.* The programs are self-documenting. The source code is extensively commented and can be read with any word processor. Also included are sample data files.

For earlier discussions on the simulation of EMR spectra, see Belford and Belford (1973) and van Veen (1978).

3. EXAMPLES OF SIMULATED LINESHAPES

Figure 12 shows the experimental, X-band EMR spectra of four representative high-spin iron proteins. Acid-methemoglobin (shown in Fig. 12a) gives a spectrum reflecting axial symmetry. The other proteins exhibit overlapping spectra from iron in more than one environment. For both transferrin (Fig. 12b) and phenylalanine hydroxylase (Fig. 12c), one component has symmetry that is near rhombic (at ~160 mT or $g' = 4.3$). The spectra of these proteins, as well as the spectrum of lipoxygenase (Fig. 12d), also have components reflecting symmetries intermediate between axial and rhombic. Simulations of each of these cases will be discussed below. For the samples with multiple components, the number of variables required for the simulation is large. An important tool in determining the simulation

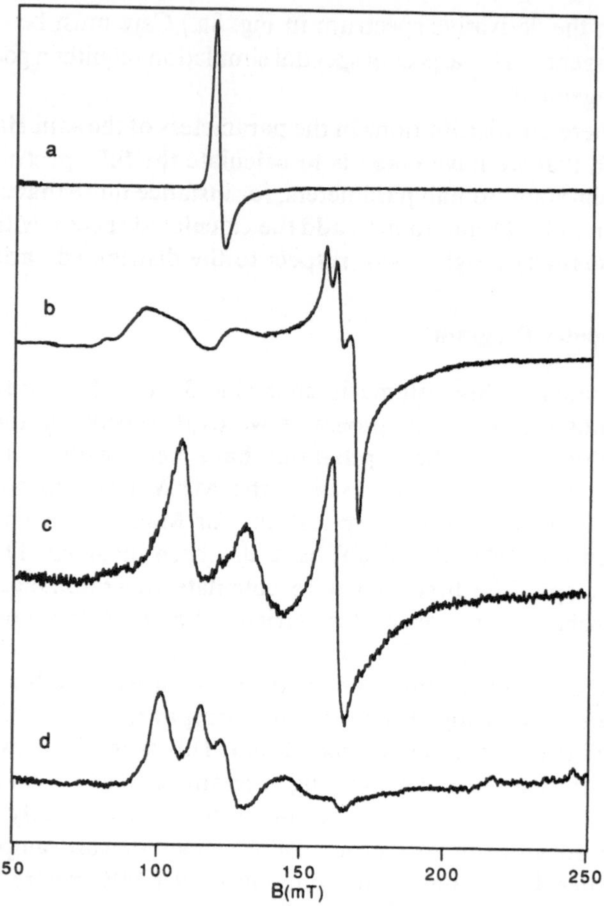

Figure 12. The low-field region of experimental EMR spectra representative of $S = \frac{5}{2}$ iron in environments of different symmetries. All spectra were recorded at 4 K. (a) Acid-methemoglobin (horse). (b) Hybrid oxalate–carbonate complex of diferric transferrin. The broad low-field region results from the oxalate complex; the feature at $g' = 4.3$ (~160 mT at X band) is from the carbonate complex. (c) PAH in Tris buffer. (d) Ferric soybean lipoxygenase-1 freshly prepared in 0.2-M phosphate buffer. The EMR spectra of each of these proteins is sensitive to the presence of small-molecule ligands. Although each spectrum has features at higher fields than the range shown in the figure, the higher-field features are all broad and hard to distinguish from the baseline.

parameters is combining biochemical changes in the sample with changes in the EMR spectra. The sections below will discuss how biochemical variations have been used in this way. The spectra of metmyoglobin and methemoglobin are understood in the most detail because both single-crystal data and results of frozen solution studies are available. As single-crystal EMR studies of nonheme iron samples become available, further refinements of simulations for these cases can be expected. It seems likely that the methods outlined below for simulating the EMR of several nonheme iron proteins would also be successful in simulating, for example, the pH-dependent EMR spectra of ferric superoxide dismutase (Fee *et al.*, 1981), a protein for which a crystal structure is available (Stoddard *et al.*, 1990).

3.1. Myoglobin and Hemoglobin at Multiple Frequencies: Symmetry near Axial

During the period 1966–1973, heme proteins were examined over a range of frequencies. The aim of those experiments was to compare the value of D from the frequency dependence of the g' values with the value obtained from other measurements (far IR, Mössbauer, and susceptibility) (Feher, 1970). The final set of experiments on human methemoglobin (Alpert *et al.*, 1973) was made at 1.5 K and over a range from 70 to 372 GHz. The resonant field of the low-field g-pependicular feature at the higher frequencies showed a small shift from linear dependence on frequency, and this shift was analyzed to give $D = 10.7\ \mathrm{cm}^{-1}$. Alpert *et al.* (1973) also show that the linewidth of the low-field feature depends, to a first approximation, on the square of the frequency. (See Section 3.3 for discussion of a different frequency dependence of linewidth for phenylalanine hydroxylase.)

Heme proteins provide an illustration of the complex contributions to line shapes that result from variations in the terms of the spin Hamiltonian and from heterogeneity in the crystalline environment. Calculations of X-band EMR powder spectra of high-spin iron in an axially symmetric environment are shown in Fig. 13b, c, and an experimental high-spin hemoglobin spectrum is given in Fig. 13a for comparison. Calculations employing two different "frequency-swept" linewidths (13c, d) and one with a "field-swept" linewidth (Fig. 13b) are shown. In fact, the calculation with the "field-swept" linewidth fits the methemoglobin experimental spectrum best. There are other examples, such as a ferric tetrathiolate model compound (Gebbhard *et al.*, 1990), for which a "field-swept" linewidth might be relevant.

A clue to the origin of the linewidths in heme samples was provided by single-crystal studies (Eisenberger and Pershan, 1967; Hampton and Brill, 1979). Peak-to-peak linewidths measured on hemoglobin single crystals at different orientations in the magnetic field vary from 2.4 to 40 mT (Hampton and Brill, 1979). Clearly, disorder in the crystal must be a major factor in the origins of the angle-dependent linewidths in the crystal. To evaluate contributions to linewidths arising from spatial misorientation in crystals, it is useful to compare spectral parameters deduced from oriented samples with those obtained from randomly oriented samples such as

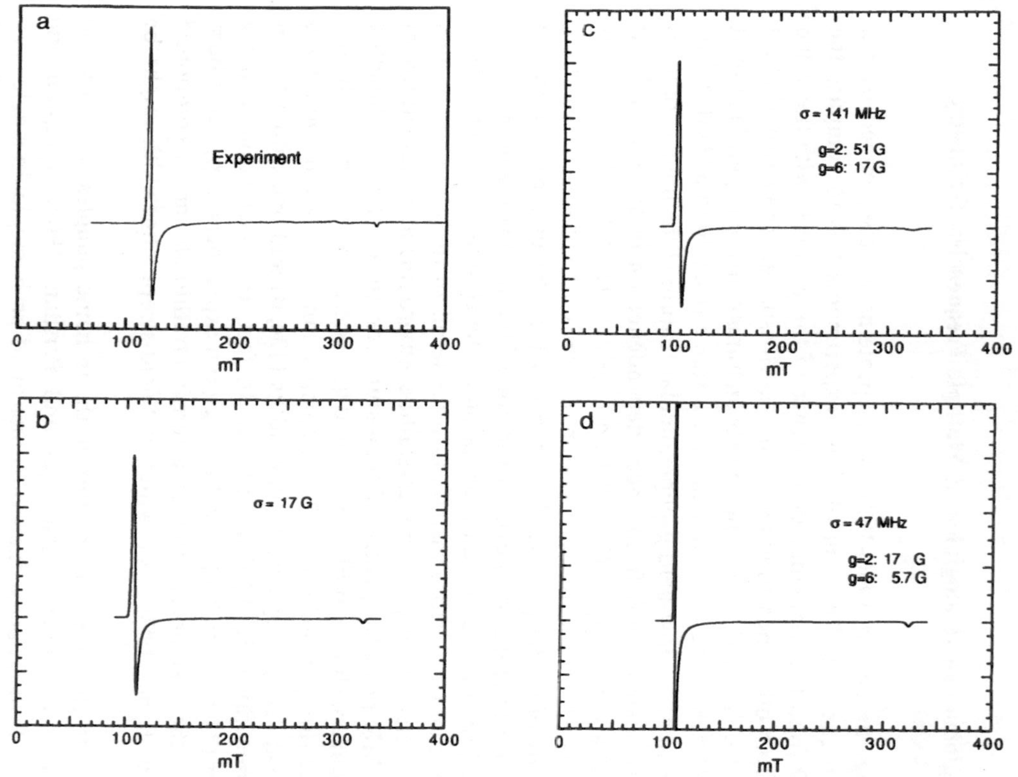

Figure 13. An experimental EMR spectrum of a frozen solution of $S = \frac{5}{2}$ methemoglobin (4 K) (a) is compared with three simulations of spectra for axial symmetry (b)–(d). Gaussian line-shape functions with different linewidths were used in the simulations. In (b), the field-swept linewidth was kept constant at 1.7 mT; in (c) and (d), the frequency domain linewidths were kept constant at 141 and 47 MHz, respectively. The insets in (c) and (d) give the apparent field-domain linewidths at $g = 2$ and $g = 6$.

polycrystalline ones or frozen solutions. If the properly weighted sum of the spectra from oriented crystals does not give the same spectrum as a frozen solution, then slightly different parameters of the spin Hamiltonian must be used for simulating spectra of the two cases. Brill and co-workers have studied in depth the parameters that characterize the EMR of $S = \frac{5}{2}$ methemoglobin and metmyoglobin in crystals and in frozen solution, as outlined in the next paragraph.

Brill and co-workers have identified a number of contributions to the linewidths of the EMR spectra of $S = \frac{5}{2}$ hemoglobin and myoglobin samples. As we showed in Fig. 13 and has been shown by more detailed line-shape analysis (Brill, 1978), normal, frequency-swept line shapes do not fit the experimental spectrum of a frozen solution sample of methemoglobin. Instead, distributions in the parameters E/D and the spin-orbit coupling parameter gave good quantitative fits to the low-field feature ($g' = 6$) of the experimental lineshapes (Brill, 1978). The method used was to calculate the g' value corresponding to each increment in θ and ϕ for a single choice of E/D and spin-orbit coupling parameter. The distribution was then introduced by calculating a linewidth ΔB from the spread in g' for each orientation specified by a θ and a ϕ. In further studies, the basic linewidth (component linewidth) was determined from T_1 measurements (Levin and Brill, 1988; Fiamingo *et al.*, 1989). Hyperfine contributions to the linewidth from the nitrogens in the heme plane and in the proximal histidine were based on ENDOR measurements (Scholes, 1970; Scholes *et al.*, 1982). Distributions in E/D and the spin-orbit coupling parameter are the dominant terms in the derived linewidth expressions (Fiamingo *et al.*, 1989). In ferric, aquo- and alcohol-hemoglobin samples, the linewidth contributions of the α and β chains were determined separately (Brill, 1986, p. 141). The spreads in E/D were attributed to distributions in the axial ligand orientations. The x axis in aquomyoglobin crystals was found to have an rms fluctuation of 18° from analysis of angle-dependent line shapes in the crystal while the deviation of the heme normal (z axis) was only 1.4°.

Aasa (1970), Coffman (1975), and Sweeney *et al.* (1973), using a value of D typical of heme, pointed out that the transitions between levels 2 and 3 and levels 4 and 5 need to be considered in simulations when the D value has a magnitude similar to that of the microwave quantum. The latter two papers calculate the 372 GHz hemoglobin spectra that would result if high enough fields were available to measure the complete spectra experimentally.

3.2. Lactoferrin and Transferrin Complexes

Lactoferrin and transferrin are among the most studied metalloproteins because they are abundant and because other metals can be substituted for iron in the metal binding sites. These proteins form complexes with iron that are reddish as a result of charge transfer between iron and two tyrosine ligands. There are two very similar lobes (N- and C-lobes) of the protein with an iron binding site in each. It has long been known that there are two distinct species contributing to the EMR spectrum of diferric transferrin. Although it was first thought that each lobe had a different spectrum, it was later shown with proteolytically cleaved half molecules that there are two conformational states for each lobe (Princiotto and Zapolski, 1975). The relative proportions of the two conformers are a function of salt concentration, the nature of the anions present and pH; changes in the EMR spectra accompany the conformational changes. Four of the ligands to iron are formed by side chains of amino acid residues (Table 1). The other two are provided by anions. Carbonate, the physiological anion, has recently been shown by X-ray diffraction to bind in a bidentate fashion to iron in lactoferrin (Baker *et al.*, 1990). Other bidentate anions can substitute for carbonate, and the structure of copper transferrin complexed with oxalate is known (Smith *et al.*, 1991). For instance, the affinity of oxalate for diferric transferrin is about one-thirtieth that of carbonate. Interaction of diferric transferrin with a large number of bidentate ligands has been examined by EMR and optical spectroscopy. For these samples, EMR is highly sensitive to the nature of the anion while optical spectroscopy in the visible region is not (Aisen *et al.*, 1972; Schlabach and Bates, 1975; Dubach *et al.*, 1991).

Approximate simulations of EMR line shapes have been calculated for diferric transferrin carbonate (Yang and Gaffney, 1987) and for the diferric complex with other bidentate anions (Dubach *et al.*, 1991). While the symmetry at the iron center changes with the bidentate anion bound, the broad distribution of parameters that characterize the EMR spectra of frozen solutions of transferrin is a feature of all anionic complexes of transferrin.

3.2.1. Rhombic Symmetry: Carbonate Complex

The EMR spectrum of transferrin carbonate has sharp features at $g' = 4.3$ and extremely broad tails extending over several dozen millitesla to high and low field from the sharp features. Figure 9 showed why the $g' = 4.3$ signal is, in general, so prominent for rhombic iron ($E/D = \frac{1}{3}$). There is very little anisotropy in the transition in the middle Kramers doublet, while the the anisotropy is large for transitions in the other two. The value of $D/h\nu$ used in preparing Fig. 9 is an order of magnitude larger than that of transferrin. Transferrin EMR at X band should have transitions, in addition to those shown in Fig. 9, between non-Kramers pairs (Aasa, 1970). Figure 14 shows the X-band EMR resonant frequencies and transition probabilities that would result from a high-spin iron center with the magnetic parameters of the rhombic component of transferrin carbonate.

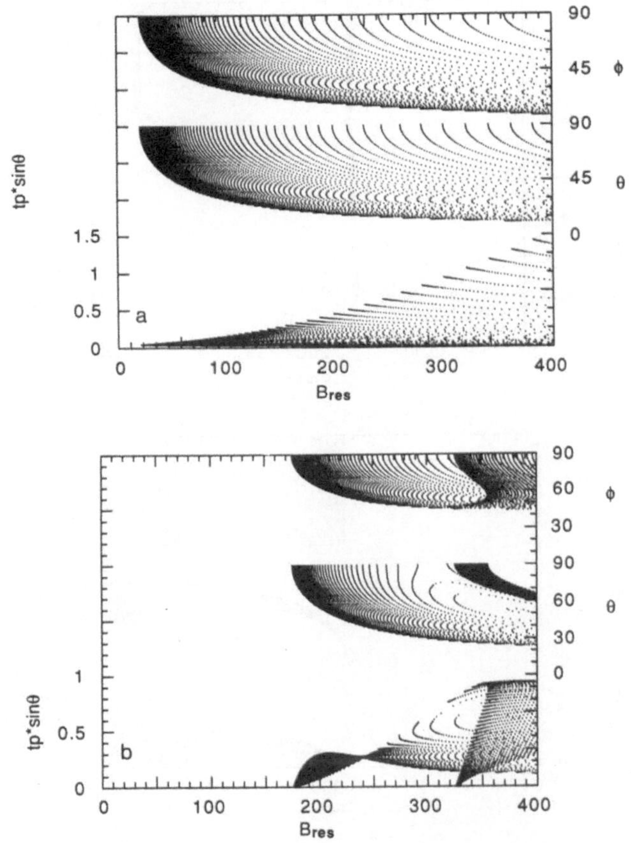

Figure 14. A scatter plot of the resonance fields as a function of angle and of the transition-probability factors for parameters characteristic of transferrin (near-rhombic symmetry): $D = 0.25\ \mathrm{cm}^{-1}$, $E/D = 0.326$, at X band. The significant transitions are in levels (a) $1 \rightarrow 2$, (b) $2 \rightarrow 3$, (c) $3 \rightarrow 4$, and (d) $4 \rightarrow 5$. The "tp" stands for the quantity w given in Eq. (44).

The calculated spectrum in Fig. 15 (calculated with $E/D = 0.326$ and $D = 0.25\ \mathrm{cm}^{-1}$) is compared with an experimental X-band spectrum of transferrin (Yang and Gaffney, 1987). Although the sharp peaks near $g' = 4.3$ are calculated, the wide tails of the experimental spectrum are not. A simulation of wider width begins to approximate the breadth of the tails, but no longer has sharp peaks. Another difficulty presented by the experimental spectrum is that the position of the maximum of the low-field peak, at $g' = 9.15$, is consistent with a $1 \rightarrow 2$ transition and $E/D \sim 0.20$, while the sharp peaks at $g' = 4.3$ should arise from a $3 \rightarrow 4$ transition with E/D near 0.333.

There are two approaches to solving the problem of simulating the broad tails of the $g' = 4.3$ portion of the signal of transferrin. One is to

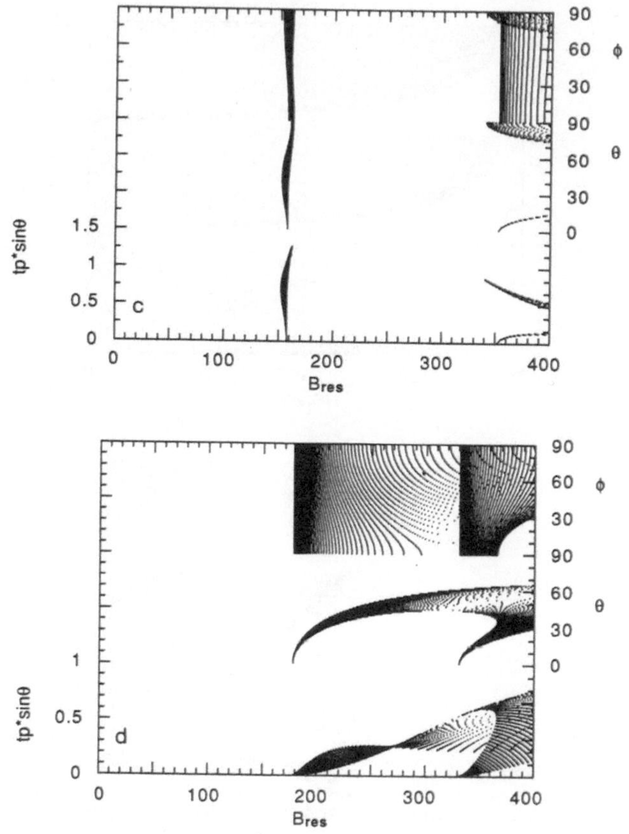

Figure 14. *(Continued)*

employ a distribution in values of E/D, and the other is to use a distribution of linewidths. Although either approach apparently could be used for the region on either side of $g' = 4.3$, the feature at low field ($g' \sim 9.15$) can be simulated only with the first approach.

To refine the simulation, the low-field region was considered in detail (Yang and Gaffney, 1987). Fortunately, transferrin EMR spectra are sensitive to salt concentration. As a result, simulations invoking multiple protein conformations can be tested experimentally. The low-field regions of the experimental spectra of diferric transferrin carbonate at three salt concentrations are compared in Fig. 16 (solid lines). The figure also shows the two simulated components (dashed lines) that are added to give good fits to the experiment. Using the same parameters used to obtain Fig. 16a, a good fit to the middle portion of the same spectrum was also obtained (Fig. 17). Besides the extended field range in Fig. 17, the only other difference in the

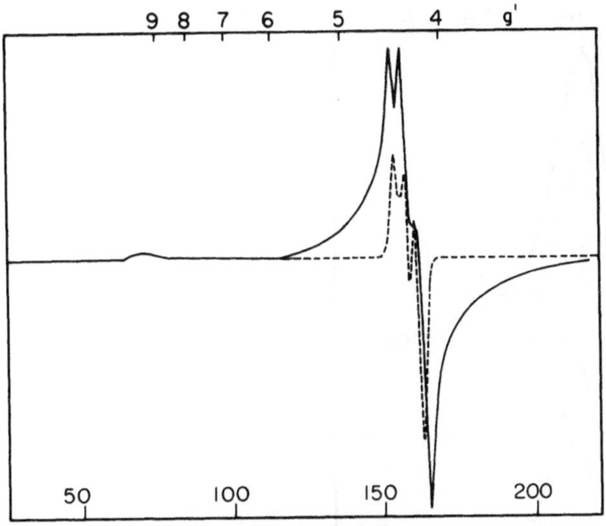

Figure 15. The experimental *X*-band spectrum of the carbonate complex of diferric transferrin (solid line) is shown compared with a simulated spectrum (dashed line) for the 3 → 4 transition. The simulated spectrum was calculated with the *D* and *E* values characteristic of this form of transferrin, but without line-broadening terms. A linewidth of 1.4 mT gave the required sharp peaks but not the broad tails. Simulation parameters: $E/D = 0.326$, $D = 0.25$, field-swept Gaussian linewidth of 1.4 mT. Reproduced from Yang and Gaffney (1987).

simulations between Figs. 16 and 17 is that component III, representing 6% of the iron, was added to account for the sharp $g' = 4.3$ peaks. Because of the high anisotropy, this component would give a negligible contribution to the lower doublet region shown in Fig. 16.

3.2.2. Intermediate Symmetry: Bidentate Ligands

Many bidentate ligands can substitute for carbonate in the anion binding sites of transferrin or lactoferrin (Schlabach and Bates, 1975). The bidentate ligands have one carboxyl group and a second, electron-donor group. The crystal structure of the carbonate complex of diferric lactoferrin has been solved, and Baker *et al.* (1990) have modeled the geometry of the N-terminal lobe for bidentate substitution of carbonate by oxalate (Fig. 18). However, subsequent crystallographic studies of copper(II)- and oxalate-substituted human lactoferrin have shown that oxalate binds preferentially to the C-lobe in the copper derivatives (Smith *et al.*, 1991). The mode of binding of oxalate to C-lobe copper is bidentate, and changes in the protein structure, compared to the other known structures, are small and local. Thus, the model in Fig. 18 is likely to convey the significant general features of oxalate coordination to either lobe.

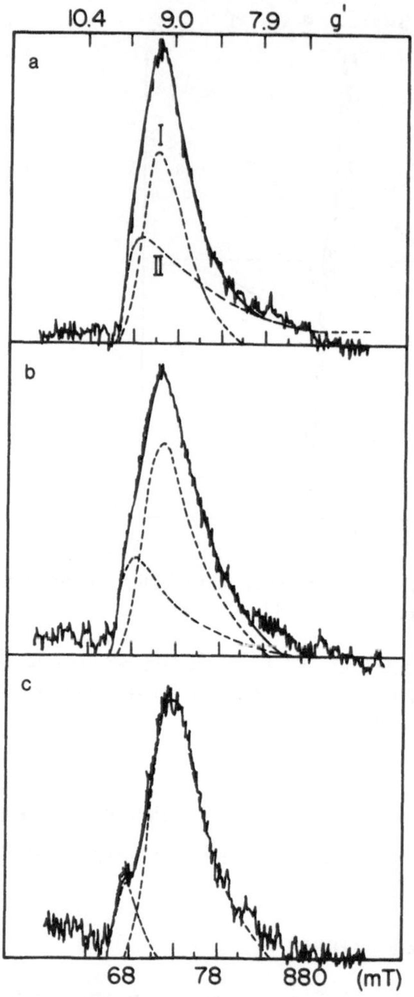

Figure 16. The low-field region of experimental and simulated *X*-band spectra of diferric transferrin at three salt concentrations is shown for (a) no salt, (b) 0.19-M sodium chloride, and (c) 0.5-M sodium chloride. The samples were 2.5, 2.3, and 2.1 mM in transferrin, respectively, in 50 mM Tris, 50 mM sodium bicarbonate at pH 7.0. The dashed lines show the two major components of a simulation, each of which was calculated using a Gaussian distribution of species with differing E/D. Parameters characterizing the distribution were the central value of E/D and the width of the distribution. Reproduced from Yang and Gaffney (1987).

The EMR spectra of transferrin oxalate and other noncarbonate complexes were shown to be shifted and even broader than the spectra of transferrin carbonate (Aisen, 1972). Recent simulations have shown that the broad spectra can be assigned as arising from features corresponding to transitions among several different energy levels (Dubach *et al.*, 1991). The symmetry of these complexes is intermediate between rhombic and axial. Figure 19 shows a calculation of all of the significant transitions at low field for cases of intermediate symmetry with $E/D = 0.1$ to 0.2. The D value known for transferrin carbonate was used in the calculation since this value has not been measured specifically for complexes with other anions. Figure 20 breaks down the calculation for $E/D = 0.1$ into its

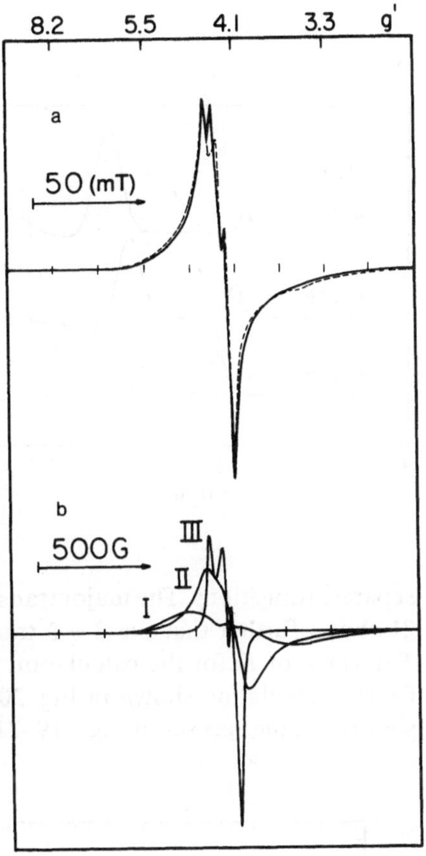

Figure 17. A portion of the experimental spectrum of differic transferrin (solid line) is compared in (a) with a simulation based on transitions between levels 3 and 4 (the middle Kramers doublet, dashed line). The simulation in (a) is the sum of the three calculations shown in (b). Conditions for the experimental spectrum are given for the spectrum in Fig. 16a. Components I and II in the simulated spectrum were calculated using a distribution of E/D values. The central E/D value was 0.22 for component I and 0.325 for component II. The width of the Gaussian distribution of E/D values was 0.04 for component I and 0.12 for component II. Frequency-swept linewidths were 524 and 313 MHz, respectively, for components I and II. Component III is the same spectrum as the simulated spectrum shown in Fig. 15. Reproduced from Yang and Gaffney (1987).

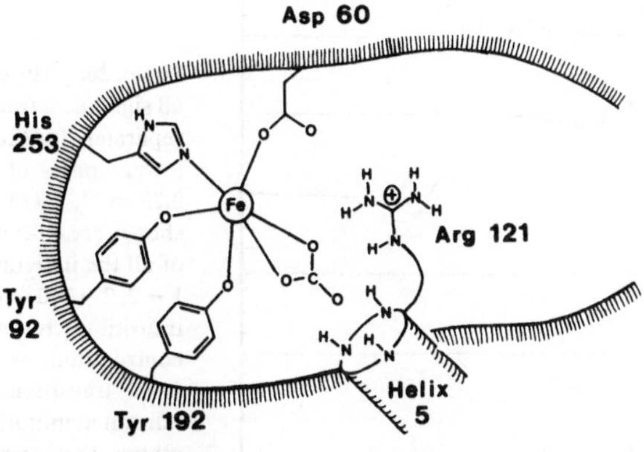

Figure 18. The general scheme of the metal and synergistic anion binding to lactoferrin. The numbering is for the N-lobe site. Reproduced from Baker *et al.* (1990).

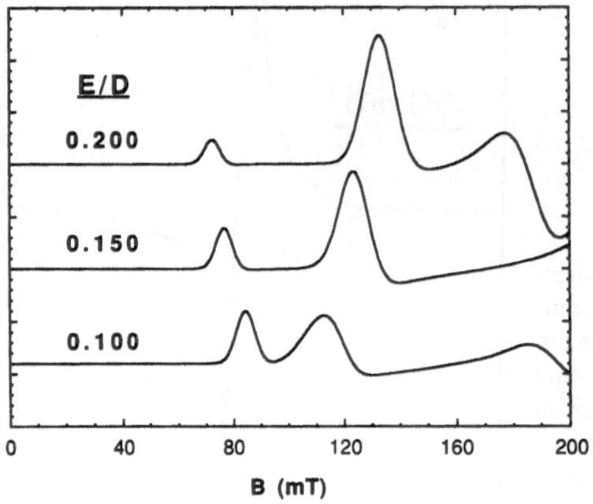

Figure 19. The low-field region of hypothetical simulated $S = \frac{5}{2}$ EMR spectra at X band is shown for $D = 0.25\ \text{cm}^{-1}$ and E/D ratios from 0.1 to 0.2. The maximum at low field is the g_y feature (Blumberg system) of the transition between levels 1 and 2. The feature on the right has two contributions, transitions between levels 2 and 3 and between 3 and 4. The frequency-swept linewidth for each calculation was 350 MHz. This figure is similar to Fig. 4 of Dubach *et al.* (1991) except that the contribution of the $2 \rightarrow 3$ transition is included here. A distribution of E/D was also used by Dubach *et al.* (1991), but no distribution was used in calculating the figure here.

separate transitions. The major transitions are $1 \rightarrow 2$, $2 \rightarrow 3$, and $3 \rightarrow 4$. Figure 21 shows further that the $2 \rightarrow 3$ transition is not significant when $D > h\nu$. The value of D for the calculation in Fig. 21 is four times larger than that for the calculation shown in Fig. 20. The dE/dB formula (55) was used in simulating the spectra in Figs. 19–21 (Doctor *et al.*, 1993).

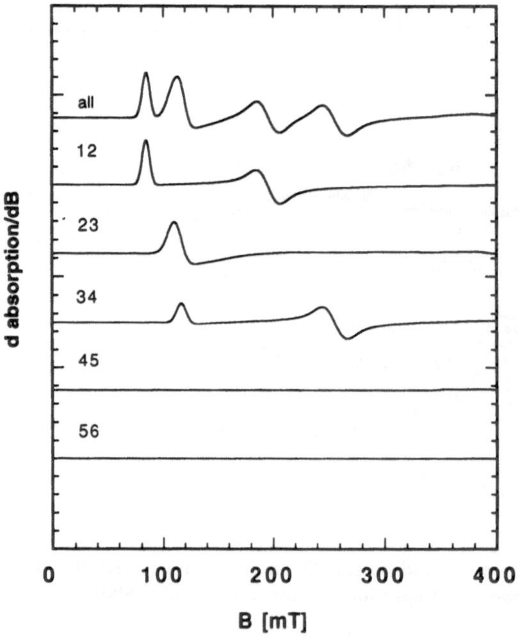

Figure 20. The contributions from all significant transitions are shown separately for the case given in the lower part of Fig. 18 ($D = 0.25\ \text{cm}^{-1}$, $E/D = 0.1$). The traces shown are, from the top, the sum of all the individual traces and the $1 \rightarrow 2$, $2 \rightarrow 3$, $3 \rightarrow 4$, $4 \rightarrow 5$, and $5 \rightarrow 6$ transitions, respectively. Although contributions from the $4 \rightarrow 5$ and $5 \rightarrow 6$ transitions are not of significant amplitude compared to the others, these transitions do have features in the region shown that are apparent upon hundredfold amplification.

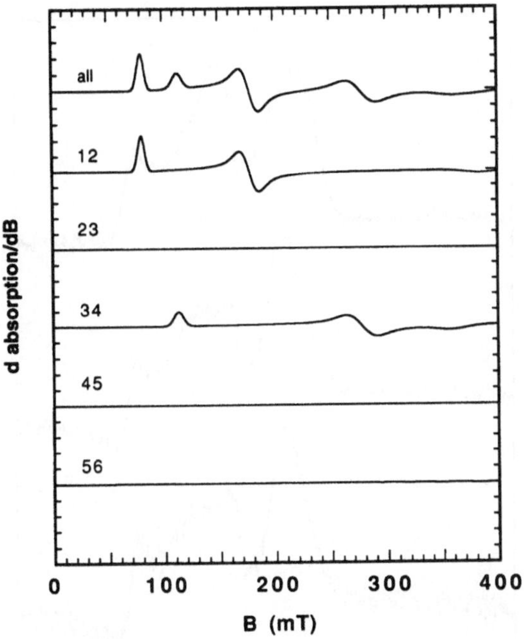

Figure 21. The contributions from all transitions are shown separately for a case that differs from that given in Fig. 20 by having a larger value of D ($D = 1.0\ \mathrm{cm}^{-1}$, $E/D = 0.1$). Note that, in comparison with the results shown in Fig. 20, there is now no significant $2 \rightarrow 3$ transition.

As with the carbonate complex, variations in salt concentration were used to show that the EMR spectra of the dioxalate complex of diferric transferrin were composed of two separate spectra (Dubach *et al.*, 1991). Each of these spectra, in turn, has multiple features arising from transitions between several pairs of levels. Figure 22 shows the low-field region of the experimental *X*-band EMR spectra of Tf-oxalate at three salt concentrations. Approximate simulations of the EMR spectra of transferrin oxalate included only the transitions between levels 1 and 2 and that between 3 and 4 (Dubach *et al.*, 1991). The transitions between other levels were not included in the simulations, to minimize computation time. As with the carbonate complex of transferrin, it was necessary to use a distribution of E/D values in the fit of computed to experimental spectra for the transferrin-oxalate complex. The paper by Dubach *et al.* (1991) also contains a simulation of the spectrum of the malonate complex of diferric transferrin. The possible contribution to these spectra of the $2 \rightarrow 3$ transition, for a calculation that includes a range of values of E/D, more recently has been considered (Doctor *et al.*, 1993).

3.3. Phenylalanine Hydroxylase

Phenylalanine hydroxylase (PAH) converts phenylalanine to tyrosine in the liver. It consists of two subunits of 50 kD each. Kaufman and co-workers (Fisher *et al.*, 1972) showed that the resting state of the enzyme gives an EMR spectrum consistent with high-spin iron and demonstrated

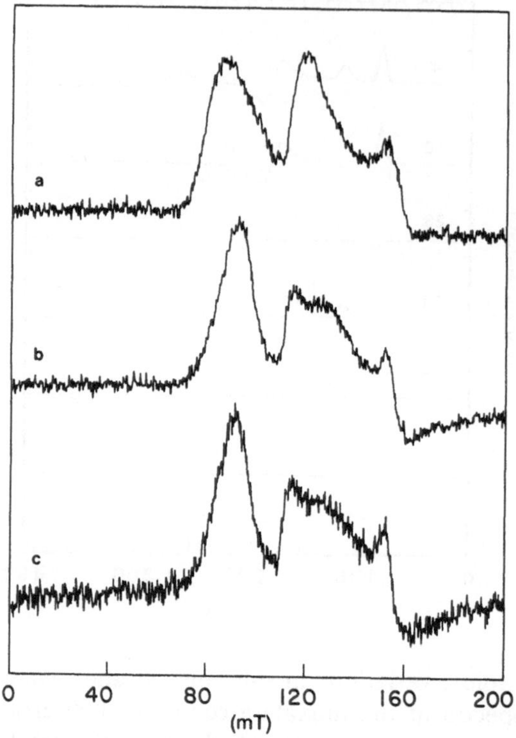

Figure 22. Experimental EMR spectra for FeTf-oxalate in buffers containing (a) 0, (b) 0.5, and (c) 1.0 M NaCl. The concentrations of FeTf-oxalate were (a) 1.0, (b) 0.75, and (c) 0.50 mM. Reproduced from Dubach *et al.* (1991).

that iron is required for activity. Gottschall *et al.* (1982) confirmed and extended this work and showed that there is one iron per subunit, that iron can be removed from the enzyme and restored with recovery of activity, and that enzymatic activity is proportional to iron content. We were thwarted in our first attempts to find EMR evidence for the involvement of iron in known catalytic steps of phenylalanine hydroxylase because, as we later found, glycerol used in the usual buffer has an effect on the EMR spectrum. In the absence of glycerol, a spectrum with features at $g' = 6.7$, 5.4 and ~1.9 was found (Wallick *et al.*, 1984). The more active the enzyme, the more intense this signal was. The EMR spectrum of phenylalanine hydroxylase was presented as Fig. 3, and the correlation with activity was shown in Fig. 4. It is also possible to remove iron selectively from the active site and replace it, demonstrating that the two types of site do not interconvert (Bloom *et al.*, 1986). The enzyme has both a substrate and an activator site for phenylalanine. Addition of phenylalanine to the enzyme, in the absence of the pterin cofactor necessary for turnover, causes a reversible conversion of the signal corresponding to the active fraction of the enzyme to a signal that overlaps strongly with the signal of inactive enzyme (Wallick *et al.*, 1984). The phenylalanine analog, 4-fluorophenylalanine, gives a similar result. The alternate substrate, tryptophan, is known to bind to the catalytic site but not to the activation site and it does not change the EMR signal of PAH. Thus, the spectral changes due to phenylalanine and 4-fluorophenylalanine seem to be the result of activation of the enzyme, not to direct interaction of the substrate with the iron (Bloom *et al.*, 1986).

Simulations of the EMR spectra of phenylalanine hydroxylase were worked out by combining the results of the biochemical variations mentioned above (correlations with activity, iron removal, and Phe activation) with additional EMR measurements at three microwave frequencies (Yang and Gaffney, 1987). Figures 23 and 24 show the results of simulations of the *X*-band EMR of native (Fig. 23) and activated (Fig. 24) PAH. More than 90% of the iron giving rise to the experimental spectra is found in two overlapping signals labeled I and II in the figures. The components labeled III and IV are small contributions from rhombic iron. As in the arguments for sharp $g' = 4.3$ signals in transferrin, it is not clear whether signals III

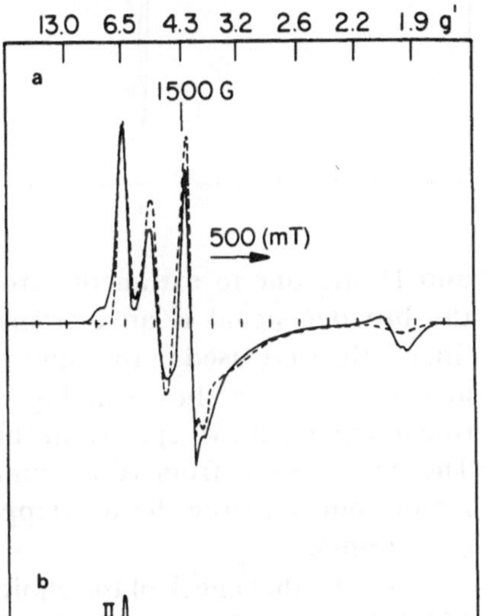

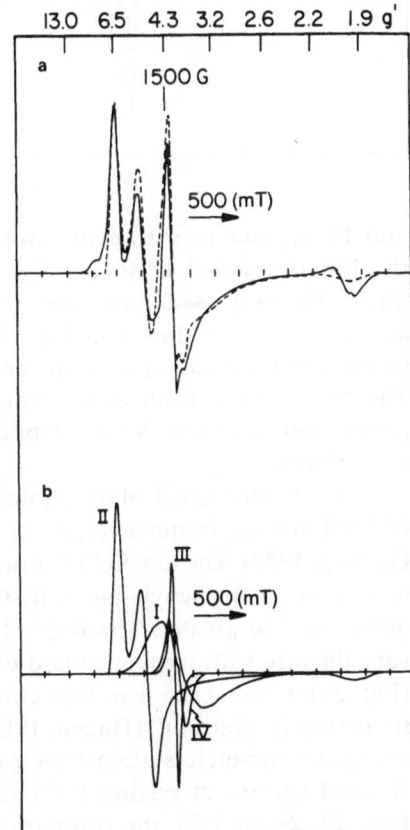

Figure 23. Experimental and simulated EMR spectra of PAH in the resting state. The experimental spectrum (solid line) was redrawn from Wallick *et al.* (1984). The mole fraction of components contributing to the simulated spectrum (dashed line) were 0.51 (component I), 0.41 (component II), 0.03 (component III), and 0.05 (component IV). Reproduced from Yang and Gaffney (1987).

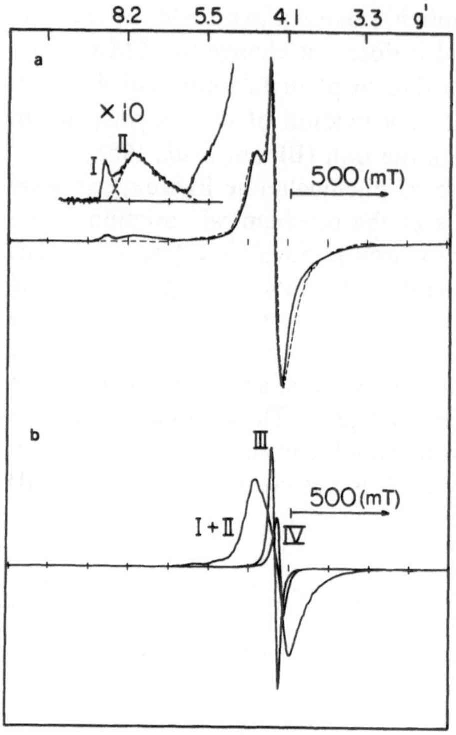

Figure 24. A simulation (dashed line) of the EMR spectrum of PAH in the presence of inhibitor (4-fluorophenylalanine) compared with experiment (solid line). The inset in (a) shows a tenfold amplification of the low-field region of the spectrum. This region was used to determine the relative amounts of the two major ferric species in this protein. The same proportions of major components were then used to simulate the absorption around $g' = 4.3$. Small amounts of sharp components were also added to the simulation. The individual components of the simulated spectrum are shown in (b). Reproduced from Yang and Gaffney (1987).

and IV are due to extraneous iron or whether they could be included in the broader signal if an extremely fine angular mesh and very narrow linewidths were used in the simulations. The simulation of spectra from the activated enzyme shown in Fig. 24 was facilitated by the separate $1 \rightarrow 2$ transitions from two species in the low-field region (see inset in Fig. 24a). The contribution from each component could be evaluated fully in this region, but not from the overlapping spectra of the $g' = 4.3$ region ($3 \rightarrow 4$ transitions).

As a further check of the choice of simulation parameters, EMR spectra of PAH at three frequencies, *L*-, *S*- and *X*-band, were recorded (Yang and Gaffney, 1987). The low-field feature occurred at $g' = 6.7$ in all three spectra, confirming that phenylalanine hydroxylase has a value of D on the order of 0.6 cm^{-1} or greater. The linewidths (low-field half-width at half-height) vary linearly with frequency and extrapolate to 66 MHz at zero frequency (Fig. 25). One mechanism that could explain dependence of linewidth on frequency is "g strain" (Hagen, 1989). The *S*-band spectra were fitted using computer parameters almost the same as those that gave a good fit to the *X*-band spectra of resting PAH (Fig. 26). For the simulations shown in Figs. 23, 24, and 26, the proportion of signal attributed to active enzyme

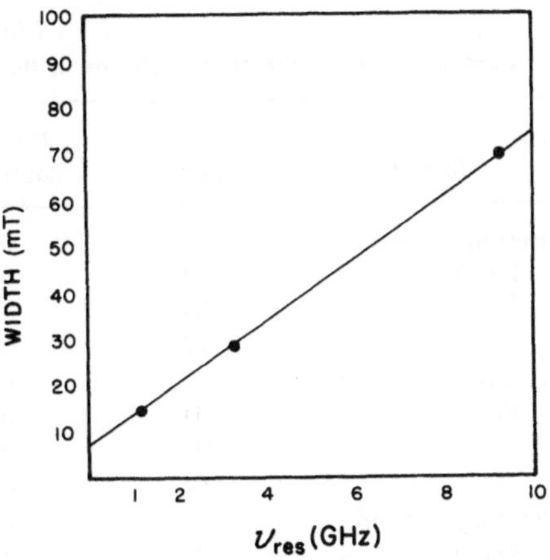

Figure 25. The linewidth (low-field width at half-height) of the lowest-field maximum in the EMR spectra of PAH is plotted versus frequency for spectra recorded at *L*, *S*, and *X* bands.

was 38–41%. The samples used for these figures had specific activities of ~5.0, which is 36% the maximal activity. Table 3 lists the simulation parameters for the calculations the results of which are shown in Figs. 23–26.

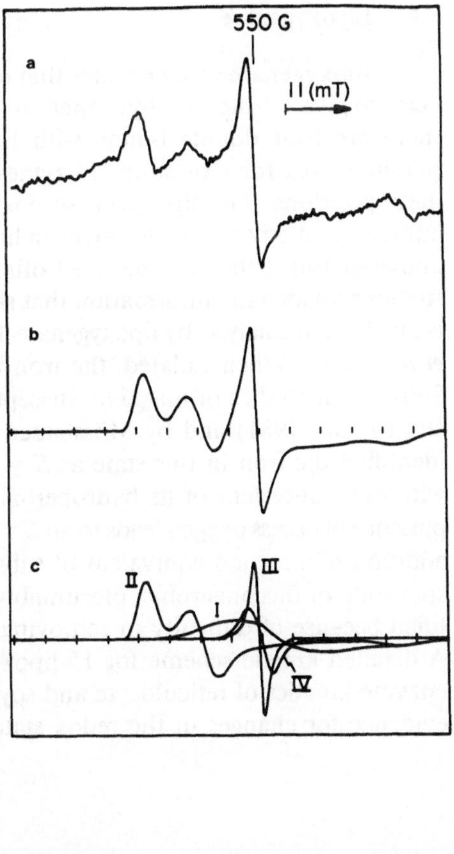

Figure 26. (a) The experimental *S*-band EMR spectrum of PAH in the resting state. (b) A simulated spectrum. (c) The individual components that contribute to (b). The simulation parameters were the same as those used for the spectra shown in Fig. 23 except for the microwave frequency, which was 3.27 GHz, and the linewidth, which was reduced by a factor of three. This spectrum was obtained with a different sample than that of Fig. 23. The ratios of components I-IV were, in this case, 49, 40, 5, and 6, respectively. Reproduced from Yang and Gaffney (1987).

TABLE 3
Parameters for Computer Simulation of the EMR Spectra of Phenylalanine Hydroxylase[a]

Sample	Component	Fractional contribution	E/D	D (cm^{-1})	$\Delta\nu$ (MHz)
Resting, X-band, Fig. 23	I	0.51	0.200[b]	0.60	524
	II	0.41	0.032	0.60	447
	III	0.03	0.333	0.37	208
	IV	0.05	0.302	0.28	313
Activated, X-band, Fig. 24	I	0.57	0.333[b]	0.60	524
	II	0.38	0.110[b]	0.60	524
	III	0.03	0.333	0.37	104
	IV	0.02	0.333	0.27	157
Resting, S-band, Fig. 26	I	0.49	0.200[b]	0.60	175
	II	0.40	0.032	0.60	149
	III	0.05	0.333	0.37	69
	IV	0.06	0.302	0.28	104

[a] Data from Yang and Gaffney (1987).
[b] A distribution of E/D values was used in the calculation. The distribution was Gaussian with the following half-widths: 0.05 for component I of resting PAH at both X and S band; 0.08 for component I of activated PAH; 0.05 for component II of activated PAH.

3.4. Lipoxygenase

Lipoxygenases are enzymes that oxidize 1,4-dienes in unsaturated fatty acids to give 1,3-dieno-5-hydroperoxo- products. In arachidonic acid, where there are four double bonds with intervening methylenes, there are six possible sites for oxidation. Enzymes are known that oxidize at each of these positions. The lipoxygenase enzymes are common in plants, corals, humans, and other animals. Soybean lipoxygenase-1 (an arachidonic acid 15-lipoxygenase) is the enzyme most often used in EMR studies. These EMR studies provided the information that showed that a single nonheme iron site is involved in catalysis by lipoxygenase-1 (L-1) (de Groot *et al.*, 1975; Pistorius *et al.*, 1976). When isolated, the iron in the enzyme gives no EMR signal. Subsequent studies of magnetic susceptibility (Slappendel *et al.*, 1982b; Axelrod *et al.*, 1981) and by Mössbauer spectroscopy (Dunham *et al.*, 1990) identified the iron in this state as $S = 2$. Interaction of the resting enzyme with one equivalent of its hydroperoxide product or of linoleic acid in the presence of excess oxygen leads to an $S = \frac{5}{2}$ state and an EMR signal. Anaerobic addition of a second equivalent of substrate quenches the EMR signal. The spin state of this anaerobic, presumably reduced enzyme has not been identified because of difficulty in removing fatty acid byproducts anaerobically. A detailed kinetic scheme for 15-lipoxygenases is based on constants from enzyme kinetics of reticulocyte and soybean lipoxygenases and on the EMR evidence for changes in the redox state of iron (Schewe *et al.*, 1986). This scheme has recently been refined for L-1 to emphasize the importance of oxygen in normal lipoxygenase kinetics (Schilstra *et al.*, 1992).

Variations in the EMR spectra of lipoxygenase were evident in the earliest reports (Egmond *et al.*, 1975; de Groot *et al.*, 1975; Pistorius *et al.*, 1976). Attempts to correlate the spectral variations with the biochemistry of the enzyme have been made (Slappendel *et al.*, 1982a; Mavrophilipos, 1986). The observed g' values for the low-field features of the two major components of ferric lipoxygenase in borate buffer are ~7.47 and 6.13 at X band (9.195 GHz) and 7.44 and 5.89 at Q band (34.66 GHz) (Slappendel *et al.*, 1980, 1981). The approximate value of D for these species was derived using Eq. (62) from the third-order perturbation solution of the secular equation (using the Blumberg coordinate system):

$$g'_{y,x} = 3g_{\text{electron}}\left[1 \pm 4\left(\frac{E}{D}\right) - 118\left(\frac{E}{D}\right)^2 - \frac{1}{2}\left(\frac{g_{\text{electron}}\mu_B B}{D}\right)^2\right] \tag{62}$$

The g'-shift method of calculating D was supplemented by calculation of D based on the temperature dependence of the signal intensity. Combined data gave a range of $4.4\ \text{K} > D > 1.8\ \text{K}$ for the more rhombic species and $3.0\ \text{K} > D > 1.5\ \text{K}$ for the axial species. Early simulations of the EMR spectra of lipoxygenase in borate buffer were based on these values of D (Slappendel *et al.*, 1981). Anisotropic linewidth terms were used in the simulation and no attempt was made to assign a basis for this anisotropy. The relative proportions of spins in the rhombic and axial regions for the sample studied by Slappendel *et al.* (1981) were approximately equal, although two near-axial components were included. D values on the order of 2 cm^{-1} for lipoxygenase samples with both symmetries have been confirmed in recent EPR and MCD studies (Zhang *et al.*, 1991; Whittaker and Solomon, 1988).

In our laboratory, we have sought to understand the role of oxygen in preparing ferric lipoxygenase samples and to establish biochemical conditions that favor a preponderance of the more rhombic species (Mavrophilipos, 1986; Gaffney, 1990; Gaffney *et al.*, 1993). This has been accomplished by carrying out the oxidation from Fe^{2+} to Fe^{3+} at low protein concentration to assure saturation with oxygen and by passing lipoxygenase, immediately after oxidation, through an HPLC gel-filtration column to remove fatty acid products. The samples purified in this manner exhibit EMR spectra that depend on time held at 4 °C and on buffer composition. Figure 27 shows a representative result for a sample freshly prepared in phosphate buffer. The upper curve is an in-phase second-harmonic (with respect to external field modulation at 100 kHz) spectrum, and the lower one is the

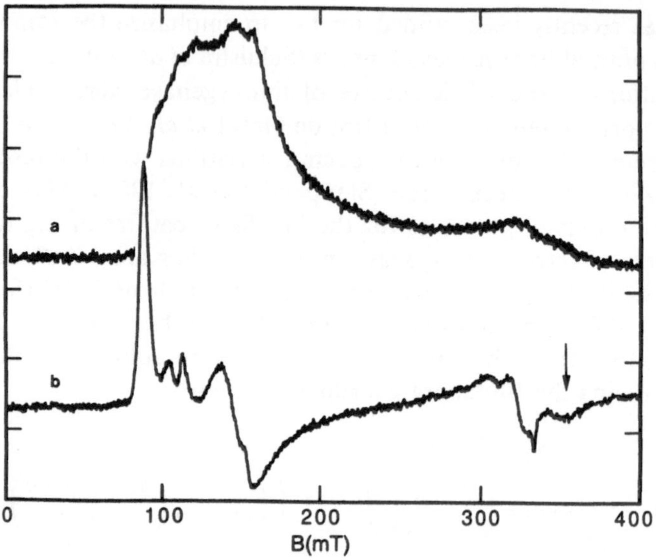

Figure 27. Experimental EMR spectra of the form of ferric lipoxygenase in which the signal at $g' = 7.3$ is dominant. The protein concentration is 27 μM in pH-6.8 sodium phosphate, 0.2 M; spectra were recorded at 4 K. The EMR spectrum was recorded immediately after oxidation and HPLC elution. Instrument settings for the lower (absorption, derivative) spectrum were 10G modulation amplitude, 8000 gain, and 1 mW power; and for the upper (in-phase second-harmonic) spectrum, 10G modulation amplitude, 10,000 gain, and 10 mW power. The arrow on the lower figure indicates the position of the broad g_z maximum from lipoxygenase. The other features near $g = 2$ are from cavity background.

normal absorption, derivative spectrum. Note that a baseline has not been subtracted from the spectra because of uncertainty about phase relationships for the second-harmonic spectrum of the enzyme and the baseline; the feature near $g' = 2$ is primarily due to cavity background. How second-harmonic signals resembling the absorption arise was detailed by Weger (1960): when the modulation frequency is similar in magnitude to the relaxation rate, the signal amplitude is greatest at each end of the modulation cycle, that is, twice per period.

Assignment of the features in the low-field region of the lower spectrum in Fig. 27 was made by spectral simulation using the methods outlined above. The results are given in Figs. 28 and 29. The experimental linewidth (outer half-width at half-height) for the $g' = 7.3$ peak is 3.2 mT. A g_y value at 7.3 corresponds to a ratio of zero-field splitting parameters E/D of about 0.065. The accompanying g_x and g_z features are expected at $g' \sim 4.6$ and 1.9, respectively, when the microwave quantum is larger than the D value. Attempts were made to simulate the major component in the EMR spectrum of Fig. 27 without recourse to distributed E/D values, but the relative

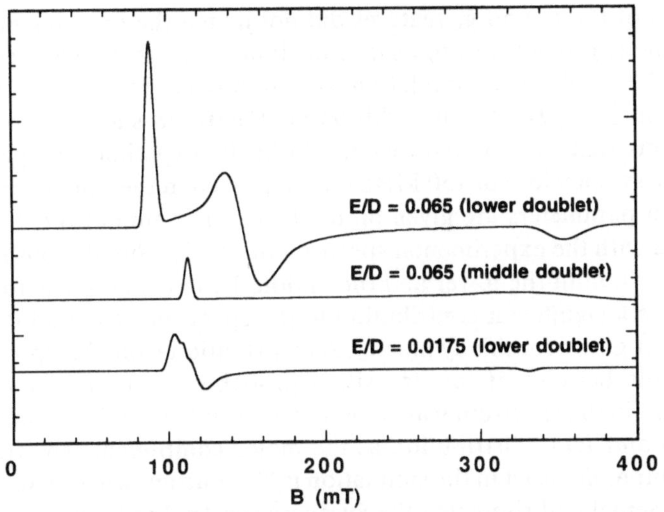

Figure 28. The three simulated spectra that make major contributions to the spectrum in Fig. 27b are shown separately here.

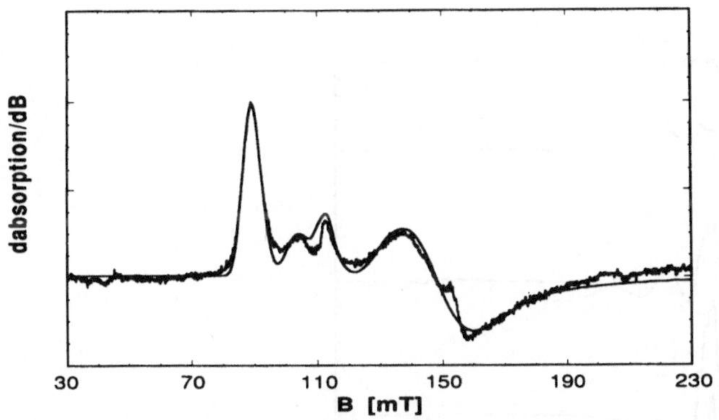

Figure 29. Simulated (smooth line) and experimental (noisy line) spectra are compared for the low-field region of the $g' = 7.3$ lipoxygenase-1 signal. The simulated spectrum is the sum of contributions from transitions in both lower ($1 \rightarrow 2$) and middle ($3 \rightarrow 4$) Kramers doublets for a species with $E/D = 0.065$ and from a lower doublet transition with $E/D = 0.0175$. The species with $E/D = 0.065$ accounts for 93% of the iron in the simulation. The other simulation parameters for each component of the spectrum are frequency-swept linewidth, 150 MHz; half-width of Gaussian distribution in E/D, 0.011; angular mesh, 1°. The simulations were normalized and added assuming equal population of the lower and middle doublets with $E/D = 0.065$. Only the lower doublet contributes significant intensity for the $E/D = 0.0175$ species. Instrumental settings are the same as those of Fig. 27, excepting scan width. Reproduced from Gaffney *et al.* (1993).

intensities of the g_y and g_x features did not match the experimental result; the feature at $g' = 4.6$ was too sharp in simulations that gave a good fit to the $g' = 7.3$ peak. The simulation was improved by using a Gaussian distribution in E/D of half-width 0.011. Figure 28 shows the individual components that were added together to fit the experimental spectrum. A frequency-swept width of 150 MHz was employed in the calculations. Other simulation parameters are given on the figure. The sum of these spectra is compared with the experimental spectrum in Fig. 29. Simulations show that transitions in both the lower and the middle Kramers doublets for $E/D =$ 0.065 make a significant contribution to the spectrum. The middle doublet feature is at $g' = 6$ for this symmetry. A contribution from the upper doublet is negligible because of low transition probability. The minor near-axial component in the spectrum was simulated with $E/D = 0.0175$ and the same linewidth and E/D distribution as the major component. The fraction of minor component used in the simulation is 7%. Further work on comparison of experimental and simulated lipoxygenase spectra has been done (Gaffney *et al.*, 1993). Figure 30a extends the simulations to other line shapes that result when the ratio of components with $E/D = 0.065$ and 0.0175 varies from 1/10 to 10/1. Figure 30b gives a shape closest to the mixture of components that was the subject of earlier simulation studies (Slappendel *et al.*, 1981).

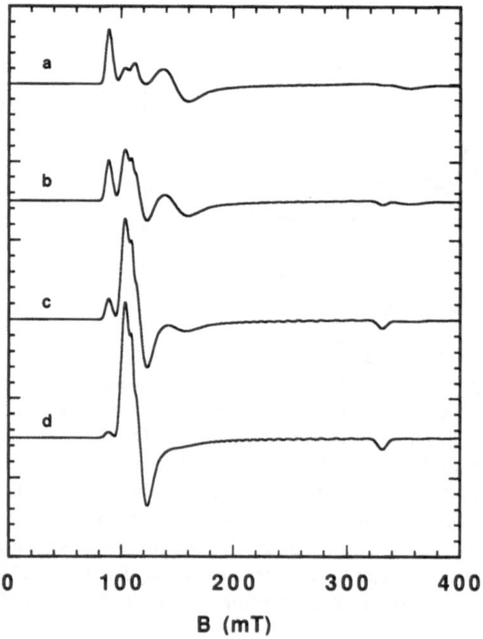

Figure 30. Simulated spectra for species with E/D equal to 0.065 and 0.0175 are added in different proportions. The ratio of species with $E/D =$ 0.065 to that with $E/D = 0.0175$ is (a) 10:1, (b) 2:1, (c) 1:2, and (d) 1:10.

ACKNOWLEDGMENTS. We would like to thank Kutbuddin S. Doctor for incorporating into the computer simulation program the derivative dependence of the field-swept linewidth, Eq. (56), rather than the Aasa-Vänngård approximation, Eq. (59), and for then generating Figs. 19–21, and 28–30. We gratefully acknowledge the support of The National Institutes of Health, grant number R01 GM36232.

REFERENCES

Aasa, R., 1970, *J. Chem. Phys.* **52**:3919–3930.

Aasa, R., and Vänngård, T., 1975, *J. Magn. Res.* **19**:308–315.

Aisen, P., Pinkowitz, R. A., and Leibman, A., 1972. *Ann. New York Acad. Sci.* **222**:337–346.

Allen, J. P., Feher, G., Yeates, T. O., Komiya, H., and Rees, D. C., 1988, *Proc. Natl. Acad. Sci. USA* **85**:8487–8491.

Alpert, Y., Couder, Y., Tuchendler, J., and Thomé, H., 1973, *Biochim. Biophys. Acta* **322**:34–37.

Axelrod, B., Cheesbrough, T. M., and Laakso, S., 1981, *Meth. Enzymol.* **71**:441–451.

Bailey, S., Evans, R. W., Garratt, R. C., Gorinsky, B., Hasnain, S., Horsburgh, C., Jhoti, H., Lindley, P. F., Mydin, A., Sarra, R., and Watson, J. L., 1988, *Biochem.* **27**:5804–5812.

Baker, E. N., Anderson, B. F., Baker, H. M., Haridas, M., Norris, G. E., Rumball, S. V., and Smith, C. A., 1990, *Pure Appl. Chem.* **62**:1067–1070.

Belford, R. L., and Belford, G. G., 1973, *J. Chem. Phys.* **59**:853–854.

Bennett, J. E., and Ingram, D. J. E., 1956, *Nature* **275**:275–276.

Bennett, J. E., Gibson, J. F., and Ingram, D. J. E., 1957, *Proc. Roy. Soc. (London) A* **240**:67–82.

Bennett, J. E., Gibson, J. F. and Ingram, D. J. E., Haughton, T. M., Kerkut, G. A., and Munday, K. A., 1961, *Proc. Roy. Soc. (London) A* **262**:395–408.

Bleaney, B., and Trenam, R. S., 1954, *Proc. Roy. Soc. (London) A* **223**:1.

Bloom, L. M., Benkovic, S. J., and Gaffney, B. J., 1986, *Biochem.* **25**:4204–4210.

Blumberg, W. E., 1967, The EPR of high spin Fe^{3+} in rhombic fields, in *Magnetic Resonance in Biological Systems* (A. Ehrenberg, B. G. Malmström, and T. Vänngård, eds.), Pergamon, Oxford, pp. 119–133.

Boyington, J. C., Gaffney, B. J., and Amzel, L. M., 1990, *J. Biol. Chem.* **265**:12771–12773.

Boyington, J. C., Gaffney, B. J., and Amzel, L. M., 1993 (submitted).

Brill, A. S., Fiamingo, F. G., and Hampton, D. A., 1978, Characterization of high-spin ferric states in heme proteins, in *Frontiers of Biological Energetics,* Vol. 2 (P. L. Dutton *et al.*, eds.), Academic Press, NY, pp. 1025–1033.

Brill, A. S., Fiamingo, F. G., and Hampton, D. A., 1986, *J. Inorg. Biochem.* **28**:137–143.

Carlioz, A., Ludwig, M. L., Stallings, W. C., Fee, J. A., Steinman, H. M., and Touati, D., 1988, *J. Biol. Chem.* **263**:1555–1562.

Coffman, R. E., 1975, *J. Phys. Chem.* **79**:1129–1136.

de Groot, J. J. M. C., Veldink, G. A., Vliegenthart, J. F. G., Boldingh, J., Wever, R. and van Gelder, B. F., 1975, *Biochim. Biophys. Acta* **377**:71–79.

Doctor, K. S., Gaffney, B. J., Alvarez, G., and Silverstone, H. J., 1993, *J. Phys. Chem.* **97**:3028–3033.

Dubach, J., Gaffney, B. J., More, K., Eaton, G. R., and Eaton, S. S., 1991, *Biophys. J.* **59**:1091–1100.

Dunham, W. R., Carroll, R. T., Thompson, J. F., Sands, R. H., and Funk, M. O., 1990, *Euro. J. Biochem.* **190**:611–617.

Egmond, M. R., Finazzi-Agrò, A., Fasella, P. M., Veldink, G. A., and Vliegenthart, J. F. G., 1975, *Biochim. Biophys. Acta* **397**:43–49.

Eisenberger, P., and Pershan, P. S. 1966, *J. Chem. Phys.* **45**:2832–2835.
Eisenberger, P., and Pershan, P. S., 1967, *J. Chem. Phys.* **47**:3327–3333.
Fee, J. A., McClune, G. J., Lees, A. C., Zidovetski, R. and Pecht, I., 1981, *Israel J. Chem.* **21**:54–58.
Feher, E., 1964, *Phys. Rev.* **136A**:145–157.
Feher, G., 1970, *Electron Paramagnetic Resonance with Applications to Selected Problems in Biology* (*Les Houches Lectures, 1969*), Gordon and Breach, New York.
Fiamingo, F. G., Brill, A. S., Hampton, D. A., and Thorkildsen, R., 1989, *Biophys. J.* **55**:67–77.
Fisher, D. B., Kirkwood, R., and Kaufman, S., 1972, *J. Biol. Chem.* **247**:5161–5167.
Gaffney, B. J., 1990, Contribution of iron to protein folding and catalysis in lipoxygenase, in *Abstracts of the XIV International Conference on Magnetic Resonance in Biological Systems* (Warwick, England), p. P9-6.
Gaffney, B. J., Mavrophilipos, D. V., and Doctor, K. S., 1993, *Biophys. J.* **64**:773–783.
Gebbhard, M. S., Deaton, D. C., Koch, S. A., Millar, M., and Solomon, E. I., 1990, *J. Am. Chem. Soc.* **112**:2217–2231.
Gottschall, D., Dietrich, R. F., Benkovic, S. J., and Shiman, R., 1982, *J. Biol. Chem.* **257**:845–849.
Griffith, J. S., 1964, *Biopolymers* **S1**:35–46.
Hampton, D. A., and Brill, A. S., 1979, *Biophys. J.* **25**:301–312.
Helcké, G. A., Ingram, D. J. E., and Slade, E. F., 1968, *Proc. Roy. Soc.* (*London*) **169**:275–288.
Hoffman, B. M., 1991, *Acct. Chem. Res.* **24**:164–170.
Howard, J. B., and Rees, D. C., 1991, *Adv. Protein Chem.* **42**:199–280.
Ingram, D. J. E., Gibson, J. F., and Perutz, M. F., 1956, *Nature* **178**:906–908.
Kendrew, J. C., and Parrish, R. G., 1956, *Proc. Roy. Soc.* (*London*) **238**:305–324.
Kotani, M., 1968, *Adv. Quantum Chem.* **4**:227–266.
Levin, P. D., and Brill, A. S., 1988, *J. Phys. Chem.* **92**:5103–5110.
Mavrophilipos, D. V., 1986, Characterization of the iron environment of lipoxygenase (Ph.D. thesis), Johns Hopkins University, Baltimore, Maryland.
Mavrophilipos, D. V., and Gaffney, B. J., 1985, *Biophys. J.* **47**:401a.
McGavin, D. G., and Tennant, W. C., 1985, *J. Magn. Res.* **61**:321–332.
Michel, H., Epp, O., and Diesenhofer, J., 1986, *EMBO J.* **5**:2445–2451.
Minge, J., Mombourquette, M. J., and Weil, J. A., 1990, *Phys. Rev. B* **42**:33–36.
Ohlendorf, D. H., Lipscomb, J. D., and Weber, P. C., 1988, *Nature* **336**:403–405.
Pilbrow, J. R., 1984, *Magn. Res.* **58**:186–203.
Pilbrow, J. R., 1990, *Transition Ion Electron Paramagnetic Resonance*, Clarendon Press, Oxford.
Pistorius, E. K., Axelrod, B. and Palmer, G., 1976, *J. Biol. Chem.* **251**:7144–7148.
Poole, C. P., Jr., Farach, H. A., and Jackson, W. K., 1974, *J. Chem. Phys.* **61**:2220–2221.
Princiotto, J. V., and Zapolski, E. J., 1975, *Nature* **255**:87–88.
Schilstra, M. J., Veldink, G. A., Verhagen, J., and Vliegenthart, J. F. G., 1992, *Biochemistry* **31**: 7692–7699.
Schlabach, M. R., and Bates, G. W., 1975, *J. Biol. Chem.* **250**:2182–2188.
Scholes, C. P., 1970, *J. Chem. Phys.* **52**:4890–4895.
Scholes, C. P., Lapidot, A., Mascarenhas, R., Inubushi, T., Isaacson, R. A., and Feher, G., 1982, *J. Am. Chem. Soc.* **104**:2724–2735.
Schewe, T., Wiesner, T., and Rappoport, S. M., 1986, *Adv. Enzymol. Related Areas Mol. Biol.* **58**:192–271.
Singel, D. J., 1989, Multifrequency ESEEM: Perspectives and applications in *Advanced EPR: Applications in Biology and Biochemistry* (A. J. Hoff, ed.), Elsevier, Amsterdam, pp. 119–134.
Slappendel, S., Veldink, G. A., Vliegenthart, J. F. G., Aasa, R., and Malmström, B. G., 1980, *Biochim. Biophys. Acta* **642**:30–39.
Slappendel, S., Veldink, G. A., Vliegenthart, J. F. G., Aasa, R., and Malmström, B. G., 1981, *Biochim. Biophys. Acta* **667**:77–86.

Slappendel, S., Aasa, R., Malmström, B. G., Verhagen, J., Veldink, G. A., and Vliegenthart, J. F. G., 1982a, *Biochim. Biophys. Acta* **708**:259–265.
Slappendel, S., Malmström, B. G., Petersson, L., Ehrenberg, A., Veldink, G. A., and Vliegenthart, J. F. G., 1982b, *Biochem. Biophys. Res. Commun.* **108**:673–677.
Smith, B. T., Boyle, J. M., Dongarra, J. J., Garbow, B. S., Ikebe, Y., Klema, V. C., and Moler, C. B., 1976, *Matrix Eigensystem Routines—EISPACK Guide*, 2nd ed., Springer-Verlag, New York.
Smith, C. A., Baker, H. M., and Baker, E. M., 1991, *J. Mol. Biol.* **219**:155–159.
Snetsinger, P. A., Chasteen, N. D., and van Willigen, H., 1990, *Am. Chem. Soc.* **112**:8155–8160.
Stoddard, B. L., Howell, P. L., Ringe, D., and Petsko, G. A., 1990, *Biochem.* **29**:8885–8893.
Sweeney, W. V., Coucouvanis, D., and Coffman, R. E., 1973, *J. Chem. Phys.* **59**:369–379.
Thomann, H., and Bernardo, M., 1993, Pulsed electron nuclear double and multiple resonance spectroscopy of metals in proteins and enzymes (Chapter 7, this volume).
van Veen, G., 1978, *J. Magn. Res.* **30**:91–109.
Wallick, D. E., Bloom, L. M., Gaffney, B. J., and Benkovic, S. J., 1984, *Biochem.* **23**:1295–1302.
Weger, M., 1960, *Bell System Tech. J.* **39**:1013–1112.
Weissbluth, M., 1967, *Structure and Bonding* **2**:1–125.
Whittaker, J. W., and Solomon, E. I., 1988, *J. Am. Chem. Soc.* **110**:5329–5339.
Woo, S. L. C., 1989, *Biochemistry* **28**:1–7.
Yang, A.-S., and Gaffney, B. J., 1987, *Biophys. J.* **51**:55–67.
Yewdall, S. J., Lawson, D. M., Artymiuk, P. J., Treffry, A., Harrison, P. M., Luzzago, A., Cesarni, G., Levi, S., and Arosio, P., 1990, *Biochem. Soc. Trans.* **18**:1028–1029.
Zhang, Y., Gebbhard, M. S., and Solomon, E. I., 1991, *J. Am. Chem. Soc.* **113**:5162–5175.

2

Mössbauer Spectroscopy of Iron Proteins

Peter G. Debrunner

1. INTRODUCTION

Iron is not only the biologically most important transition metal, but, by fortunate circumstances, one of its isotopes, ^{57}Fe, is also the most useful Mössbauer nucleus. Accordingly, nuclear gamma resonance, commonly known as the Mössbauer effect, has become a significant tool in the study of iron proteins. It complements earlier methods, in particular magnetic resonance techniques, and a comparison of the approaches is therefore in order. Distinctive features of the Mössbauer effect are

1. the sensitivity to all ^{57}Fe regardless of the oxidation and spin state of the iron as long as the atoms are bound in a solid, and
2. the fact that, for a given sample, the experimentalist has control over temperature and magnetic field only.

The theoretical description of paramagnetic states and spin transitions was stimulated by the advent of EPR, and most of the formalism developed for EPR can readily be adapted to Mössbauer spectroscopy.

This review attempts to give a selective but representative overview of the field. Examples are chosen to illustrate the main types of Mössbauer spectra as well as spectra of a variety of proteins. Emphasis is on the physics that arises in the analysis of the spectra rather than on numerical results.

Peter G. Debrunner • Department of Physics, Loomis Laboratory of Physics, University of Illinois at Urbana-Champaign, Urbana, Illinois 61801.

Biological Magnetic Resonance, Volume 13: EMR of Paramagnetic Molecules, edited by Lawrence J. Berliner and Jacques Reuben. Plenum Press, New York, 1993.

For discussions of the implications for electronic structure and coordination geometry the reader is referred to the original papers.

2. MÖSSBAUER EFFECT: SPECTRAL PARAMETERS

2.1. Basics

Mössbauer spectroscopy (MS) (Greenwood and Gibb, 1971) is based on the resonant absorption of gamma rays by a probe nucleus, here ^{57}Fe. The first excited state of ^{57}Fe at an energy of $E_\gamma = 14.4$ keV has a lifetime of $\tau = 140$ ns and hence a Lorentzian width of $\Gamma = \hbar/\tau = 4.7$ neV, which is small enough to resolve electric and magnetic hyperfine interactions of the ^{57}Fe nucleus. Most Mössbauer experiments are done in transmission, using ^{57}Co as a source of 14.4 keV γ rays. The energy is scanned by Doppler shifting the source relative to the absorber with a velocity v, thus providing an energy shift

$$\delta E = E_\gamma \frac{v}{c} \tag{1}$$

where c is the speed of light. It is customary in MS to express all energies in terms of velocity v, with the conversion factors

$$1 \text{ mm/s} \leftrightarrow 11.6 \text{ MHz} \leftrightarrow 3.88 \times 10^{-4} \text{ cm}^{-1} \leftrightarrow 5.58 \times 10^{-4} \text{ K} \leftrightarrow 48.1 \text{ neV} \tag{2}$$

In a transmission experiment the shape of the source line is convoluted with the shape of the absorber line, and in the thin absorber limit the result is a Lorentzian of theoretical width (FWHM) $\Gamma_{\text{theor}} = 0.196$ mm/s. In practice, the minimum observable linewidth is typically $\Gamma_{\text{min}} \approx 0.22$ mm/s.

2.2. Recoilless Fraction and Other Dynamical Aspects

As a result of momentum- and energy-conservation laws, Mössbauer transitions of intrinsic width Γ can be observed only for nuclei bound in a solid. Moreover, only a fraction $f < 1$ of the bound nuclei undergo transitions without recoil energy loss, where the "recoil-free" fraction for harmonically bound nuclei is

$$f = \exp(-k^2\langle x^2 \rangle) \tag{3}$$

Here, $k = 2\pi/\lambda = E_\gamma/(\hbar c)$ is the wave vector of the gamma ray, and $\langle x^2\rangle$ is the mean square displacement of the ^{57}Fe nucleus in the direction of **k**. Since $\langle x^2\rangle$ increases monotonically with temperature T the recoil-free fraction $f(T)$ is largest at low temperatures, where $\langle x^2\rangle$ is given by zero-point vibrations only and leads to $f(4.2K) \sim 0.8$ for most iron proteins. At higher temperatures $f(T)$ drops at a rate that depends on the vibrational properties of the individual protein.

Other dynamical aspects of the Mössbauer effect are the second-order Doppler shift, which shifts all energies uniformly with temperature; spin-lattice coupling, which affects the magnetic hyperfine structure; and other, still poorly understood effects on the quadrupole splitting that may arise from anharmonicity or vibronic coupling.

The second-order Doppler shift or thermal redshift is given by

$$\delta_{\mathrm{SOD}} = -\frac{\langle v^2\rangle}{2c^2} E_\gamma \tag{4}$$

where $\langle v^2\rangle$ is the mean square velocity of the ^{57}Fe. It shifts the whole spectrum to slightly lower energies and has a classical high temperature limit of $\delta_{\mathrm{SOD}} = -(3kT/2Mc^2)E_\gamma$, which is equivalent to -7.29×10^{-4} mm s^{-1} K^{-1} in terms of Doppler shift.

Spin–lattice coupling causes transitions among the eigenstates of a spin multiplet in analogy to spin relaxation in EPR. MS differs from EPR in two points, however:

1. the spin system is always in thermal equilibrium with the lattice, and
2. all levels of a multiplet contribute to a spectrum as given by their Boltzmann factors.

A dynamic line shape model is needed in the regime of intermediate fluctuations, i.e., when the transition rate is comparable to the nuclear Larmor frequencies in any state of the spin multiplet (Clauser and Blume, 1971; Schulz *et al.*, 1987). In order to avoid the complexities of a dynamic model most experimentalists try to satisfy the limiting conditions of either slow fluctuation rates, typically obtained at the lowest temperatures, or the limit of fast fluctuations at higher temperatures.

2.3. Hyperfine Interactions of ^{57}Fe

Most applications of MS center on the determination of the hyperfine interactions; i.e., the nuclear moments are used to measure the electric and magnetic fields at the iron. The ground state of ^{57}Fe has spin $I = \frac{1}{2}$ and nuclear g factor $g_N = 0.181$, while the 14.4 keV state has $I^* = \frac{3}{2}$, $g_N^* = -0.103$, and a positive quadrupole moment $Q \sim 8 \times 10^{-26}$ cm^2. In addition, the two states have different mean square charge radii $\langle R^2\rangle > \langle R^2\rangle^*$, and the transition energy therefore has a term $-(2\pi/3)\, Ze^2(\langle R^2\rangle^* - \langle R^2\rangle)|\psi(0)|^2$, where $|\psi(0)|^2$ is the electron density at the nucleus. If $|\psi(0)|^2$ differs between absorber (A) and source (S), the spectrum as a whole shifts by

$$\delta = \frac{2\pi}{3} Ze^2(\langle R^2\rangle^* - \langle R^2\rangle)(|\psi(0)|_A^2 - |\psi(0)|_S^2) \tag{5}$$

the so-called isomer or chemical shift. The observed center shift is the sum of Eqs. (4) and (5) and is usually quoted with respect to a standard, generally taken to be iron metal at 300 K and indicated by the notation δ_{Fe}.

While the isomer shift as an electric monopole interaction is unique to MS, the quadrupole interaction is familiar from other forms of spectroscopy. It can be written as (Abragam and Bleaney, 1970)

$$\mathcal{H}_Q = \hat{I}\cdot\mathbf{P}\cdot\hat{I} = \frac{eQV_{zz}}{12}\left[3\hat{I}_z^2 - \frac{15}{4} + \eta(\hat{I}_x^2 - \hat{I}_y^2)\right], \qquad \eta = \frac{V_{xx} - V_{yy}}{V_{zz}} \tag{6}$$

where the V_{ii}, $i = x, y, z$, are the negative of the principal axis values of the electric field gradient given by the volume integrals

$$V_{zz} = \int \rho(r) r^{-3}(3\cos^2\theta - 1)\, dv \tag{7}$$

etc. Under the influence of $\mathcal{H}_Q$, the excited state with $I^* = \frac{3}{2}$ splits into two levels, whereas the ground state with $I = \frac{1}{2}$ does not split. The resulting MS is then a quadrupole doublet with splitting

$$\Delta E_Q = \frac{1}{2} eQ|V_{zz}|\sqrt{1 + \frac{\eta^2}{3}} \tag{8}$$

It is convenient to assign ΔE_Q the sign of V_{zz}.

The major contributions to V_{ii} come from the anisotropic charge distribution of the $3d$ and $4p$ valence electrons, which in turn distort the inner shells as given by the Sternheimer factors. Crystal or ligand field contributions to V_{ii} may also be important.

The magnetic hyperfine splitting, finally, is given by the term $\hat{S}\cdot\mathbf{A}\cdot\hat{I}$, where the hyperfine tensor $\mathbf{A}$ in general has an orbital, a spin dipolar, and an isotropic Fermi contact term (Abragam and Bleaney, 1970),

$$\mathcal{H}_{S,I} = \hat{S}\cdot\mathbf{A}\cdot\hat{I} = 2\beta\beta_N g_N \int dv \sum_\nu r_\nu^{-3}\left\{\hat{l}_\nu + \left[\frac{3(\mathbf{r}_\nu\cdot\hat{s}_\nu)\mathbf{r}_\nu}{r_\nu^2} - \hat{s}_\nu\right] - \kappa\hat{s}_\nu\right\}\cdot\hat{I} \tag{9}$$

For iron compounds the contact term generally dominates. The high-spin ferric ion, in particular, ideally has an isotropic hyperfine interaction with $A \approx -30$ MHz or, in units of internal field per unit spin, $A/g_N\beta_N \approx -22$ T. The minus sign indicates that an external field is opposed by the internal field $B_{int} = -\langle \mathbf{S}\rangle \mathbf{A}/(g_N\beta_N)$.

In the absence of quadrupole interaction the ground and excited states of ^{57}Fe split linearly with an applied field, but since the Mössbauer transition is of magnetic dipole character, only the six transitions with $\Delta m_I = 0, \pm 1$ are allowed. For combined magnetic and electric interaction, up to eight transitions are possible.

2.4. Electronic States of the Iron: Spin Hamiltonian

A few remarks are in order on the characterization of the electronic states of the iron, the most frequent concern of MS. In contrast to EPR, which is sensitive mainly to half-integer spin systems, MS sees all ^{57}Fe in a given sample including diamagnetic and paramagnetic, integer and half-integer spin states. The first step in the interpretation of a spectrum is the assignment of the oxidation state—Fe(II), Fe(III), or Fe(IV)—and the spin state $S = 0 \ldots S = \frac{5}{2}$ (Greenwood and Gibb, 1971). High-spin ferrous iron is always recognizable on the basis of its large isomer shift and quadrupole splitting. Kramers systems display spontaneous magnetic hyperfine splitting at sufficiently low temperatures, while at high temperatures the magnetic interaction averages out to leave a (typically broadened) quadrupole doublet. If B_{int}, ΔE_Q, and δ_{Fe} are not sufficient to identify the spin state of a Kramers system, EPR can solve the problem. Integer spin systems, on the other hand, show no spontaneous magnetic splitting, but a measurable internal field can always be induced by an externally applied field. For diamagnetic compounds, finally, the effective field at the nucleus equals the applied field.

Given a basic characterization of the iron in terms of oxidation and spin state, the next step is the parametrization of the Mössbauer spectra in terms of a suitable model. Typically, a spin Hamiltonian $\mathcal{H}_S$ is used for this purpose (Abragam and Bleaney, 1970),

$$\mathcal{H}_S = \hat{S}\cdot\mathbf{D}\cdot\hat{S} + \beta\hat{S}\cdot\mathbf{g}\cdot B \tag{10}$$

where the zero-field splitting is usually expressed in its principal axis system as

$$\hat{S}\cdot\mathbf{D}\cdot\hat{S} = D\left(\hat{S}_z^2 - \frac{S(S+1)}{3}\right) + E(\hat{S}_x^2 - \hat{S}_y^2) \tag{10'}$$

with $E/D \leq \frac{1}{3}$. Using a second-order perturbation treatment, the deviations of the **g** tensor from the spin-only value, $g_s = 2.0023$, can be related to D, E, and the spin-orbit coupling λ. Equation (10) provides an adequate parametrization for many iron proteins, but apart from additional terms that can usually be ignored (Abragam and Bleaney, 1970), the spin Hamiltonian will fail if more than one orbital state is thermally accessible (Kent *et al.*, 1979; see also Section 5.2). Under the latter conditions a crystal-field model may be used to parametrize the data. In this model a $(3d)^n$ multielectron wave function is calculated in an electrostatic potential given by a multipole expansion, and the coefficients of the expansion are treated as adjustable parameters. Although the model is unrealistic because it ignores delocalization, it may reproduce the magnetic properties of iron complexes of sufficiently high symmetry. For iron proteins of known structure it is possible, in principle, to calculate hyperfine tensors as well as other observables and to compare them with experimental values.

2.5. Spin Coupling

A metalloprotein often contains more than one paramagnetic center, and if the coupling between the centers is not very weak, the magnetic properties of the system may differ entirely from those of the parts. This section discusses some relations between the isolated and the coupled centers. For large separation the coupling between two spins S_1 and S_2 is of the dipole-dipole, through-space type and is relatively weak, while for centers connected by covalent bonds the coupling is usually modeled as an isotropic exchange

$$\mathcal{H}_{\mathrm{xch}} = J\hat{S}_1 \cdot \hat{S}_2 \tag{11}$$

where J may range from small negative to large positive values, e.g., $-5\ \mathrm{cm}^{-1} \lesssim J \lesssim 250\ \mathrm{cm}^{-1}$. Although the principles of exchange interactions are understood, quantitative predictions given the structure of a cluster are still impractically difficult.

The spin-coupled species of interest in MS can be a radical and an iron site as in photosynthetic reaction centers (Section 5.2), the primary compounds of peroxidases (Section 4.2), or the intermediate in the formation of ribonucleotide reductase (Section 6.3), or they can be two or more iron sites bridged by inorganic sulfides as in the iron–sulfur clusters (Sections 3.3 and 3.4), or an oxygen-bridged pair of iron sites as in hemerythrin and ribonucleotide reductase (Section 6.3). The iron–sulfur proteins represent a special case as clusters of 2, 3, 4, and 6 iron sites as well as complexes with molybdenum are known. The coupling scheme for three and more spins can be quite complex, and other special features arise from double exchange (Münck *et al.*, 1988).

The isotropic exchange between two spins S_1 and S_2, Eq. (11), has eigenstates of spin S, $\hat{S} = \hat{S}_1 + \hat{S}_2$, $|S_1 - S_2| \leq S \leq S_1 + S_2$, with eigenvalues $E(S) = (J/2)[S(S+1) - S_1(S_1+1) - S_2(S_2+1)]$; hence energy splittings are $E(S+1) - E(S) = J(S+1)$ and for $J > 0$ the state of smallest S, $S = |S_1 - S_2|$, is the ground state. If $\mathcal{H}_i = \hat{S}\cdot\mathbf{D}_i\cdot\hat{S}_i + \beta\hat{S}_i\cdot\mathbf{g}_i\cdot\hat{B}$, $i = 1, 2$ represents the spin Hamiltonian of the isolated spins S_1 and S_2, then the total Hamiltonian of the coupled system is

$$\mathcal{H} = \mathcal{H}_1 + \mathcal{H}_2 + \mathcal{H}_{\mathrm{xch}} \tag{12}$$

For $|J/D_i| \gg 1$, $i = 1, 2$, the net spin S is still a good quantum number (strong coupling limit), and for each multiplet S an effective Hamiltonian can be written as (Scaringe *et al.*, 1978)

$$\mathcal{H}_S^{\mathrm{eff}} = \hat{S}\cdot\mathbf{D}_S\cdot\hat{S} + \beta\hat{S}\cdot\mathbf{g}_S\cdot\mathbf{B} \tag{13a}$$

with

$$\mathbf{D}_S = \sum_{i=1,2} \mathbf{D}_i a_i; \qquad \mathbf{g}_S = \sum_{i=1,2} \mathbf{g}_i ci \tag{13b, c}$$

where the a_i are given by Scaringe *et al.* (1978) and the c_i are

$$c_1 = \frac{S(S+1) + S_1(S_1+1) - S_2(S_2+1)}{2S(S+1)},$$

$$c_2 = \frac{S(S+1) - S_1(S_1+1) + S_2(S_2+1)}{2S(S+1)} \tag{13d}$$

Similarly the magnetic hyperfine interaction can be written as

$$\mathcal{H}_{S,I} = \hat{S}\cdot \sum_{i=1,2} \mathbf{A}_{S,i}\cdot\hat{I}_i, \qquad \mathbf{A}_{S,i} = c_i\mathbf{A}_i \tag{13e}$$

For the reduced 2Fe-2S proteins the strong coupling limit is a good approximation, and Eqs. (13a–e) with $S_1 = \frac{5}{2}$, $S_2 = 2$, $S = \frac{1}{2}$ are valid. For oxygen-bridged Fe(III)–Fe(II) clusters, on the other hand, as well as for some other systems, the exchange interaction is comparable to the zero-field splitting $|J/D_i| \sim 1$. If the ground state is a Kramers doublet and no other states are populated, the spectra can still be parametrized by an effective spin $S = \frac{1}{2}$ Hamiltonian, but since Eqs. (13c–e) no longer apply, it may be necessary to use the $(2S_1 + 1) \times (2S_2 + 1)$-dimensional representation of Eq. (12) to deduce the intrinsic parameters of the isolated iron sites.

Another unexpected feature in exchange-coupled systems is that integer spin multiplets may show magnetic broadening in weak or even zero field. In the low-symmetry environments characteristic of iron proteins, integer spin systems are expected to have nondegenerate sublevels and hence vanishing spin expectation values $\langle S \rangle_i$ in zero field. It turns out, however, that higher spin values may have quasidegenerate sublevels that split in very small fields and may even split in zero field as a result of magnetic hyperfine coupling.

2.6. Calculation of Mössbauer Spectra

In order to extract a maximum of information from Mössbauer spectra, an algorithm for fitting or simulating the spectra is needed. The easiest case is that of diamagnetic iron as in low-spin ferrous or spin-coupled diferric compounds with $S = 0$ ground state. The Mössbauer spectra are readily calculated from the eigenvectors and eigenvalues of the nuclear Hamiltonian

$$\mathcal{H}_I = \hat{I} \cdot \mathbf{P} \cdot \hat{I} - g_N \beta_N \hat{I} \cdot \mathbf{B} \tag{14}$$

and the known transition probabilities for $M1$ radiation. Since iron proteins are normally measured as frozen solutions, an average is needed over all orientations of **P** relative to **B** and the γ-beam, and for thin absorbers each line is given a Lorentzian shape.

For paramagnetic compounds the calculation is usually based on the total Hamiltonian

$$\mathcal{H} = \mathcal{H}_S + \mathcal{H}_{S,I} + \mathcal{H}_I \tag{15}$$

and the diagonalization of $\mathcal{H}$ in the $\hat{S} \times \hat{I}$ and $\hat{S} \times \hat{I}^*$ bases is more demanding. The work can be simplified if $\mathcal{H}_S$ is decoupled from the other terms by a small field with $B \gtrsim 10$ mT. $\mathcal{H}_S$ is then much larger than $\mathcal{H}_{S,I}$, and is diagonalized first. Next, its $(2S + 1)$ eigenvectors are used to calculate the spin expectation values $\langle \mathbf{S} \rangle_i$, which are substituted in $\mathcal{H}_{S,I}$. The nuclear eigenstates and eigenvalues are finally calculated from $\langle \mathbf{S} \rangle_i \cdot \mathbf{A} \cdot \hat{I} + \mathcal{H}_I$, $i = 1, \ldots, 2S + 1$, and the Mössbauer spectrum is obtained as a sum of $(2S + 1)$ subspectra with appropriate Boltzmann factors. This treatment assumes stationary eigenstates of $\mathcal{H}$, and generally applies at low temperatures. In the opposite limit of fast fluctuation rates a thermal average $\langle S \rangle_T$ of all $\langle S \rangle_i$ values is substituted in $\mathcal{H}_{S,I}$, and a single spectrum is calculated for every molecular orientation.

For fluctuation rates comparable to nuclear Larmor frequencies, a dynamic line shape model is needed, as mentioned in Section 2.2, and for diagonal $\mathcal{H}_{S,I}$ a $(2S + 1) \times (2I + 1) \times (2I^* + 1)$-dimensional Hamiltonian has to be solved for each molecular orientation. Calculations of this type are limited to cases where the basic parameters of Eq. (15) are reasonably well known so that only the transition rate(s) between the eigenstates of $\mathscr{H}_S$ need to be adjusted.

In view of the large number of adjustable parameters in Eq. (15), it is highly desirable to have additional information—e.g., from EPR or susceptibility experiments—to further constrain the parameters.

3. IRON–SULFUR PROTEINS

3.1. Overview

Iron-sulfur proteins are ubiquitous in all forms of life, and their functions range from electron transfer to catalysis (Spiro, 1982). The basic structural unit of their active centers is a high-spin iron nucleus with almost tetrahedral sulfur coordination. In rubredoxins, which have a single iron site, all sulfurs are from cysteines suitably located in the polypeptide, whereas clusters of 2, 3, 4, 6, and more iron atoms have bridging sulfides in addition to the cysteines that anchor the cluster to the polypeptide. Some exceptions to these rules add to the variety of iron-sulfur proteins; other iron ligands or additional metals such as molybdenum may be involved. The exchange coupling mediated by the bridging sulfides leads to interesting magnetic properties in the Fe-S clusters. For the 2Fe-2S proteins the coupling $J\hat{S}_1 \cdot \hat{S}_2$ is strongly antiferromagnetic; thus, the ground states have spin $S = 0$ for the oxidized $Fe(III)_2$ and spin $S = \frac{1}{2}$ for the reduced Fe(III)–Fe(II) complex, the only two stable forms (Gibson *et al.*, 1966). The 4Fe–4S proteins have a cubane structure, and in the 3Fe–4S proteins one corner of the cube is unoccupied. For the clusters with more than two iron sites, the spin-coupling scheme is complicated by the competition between pairwise exchange interactions of comparable strengths and by the phenomenon of double exchange (Münck *et al.*, 1988), which manifests itself in delocalized valence states of some of the pairs. Most of the ground states then have spin $S = 0$ or $S = \frac{1}{2}$, but larger cluster spins have been documented. Protein structures of atomic resolution are known for all cluster sizes up to the 4Fe–4S type (Watenpaugh *et al.*, 1980; Tsukihara *et al.*, 1981; Stout *et al.*, 1988) and, at lower resolution, even for an enzyme as large as nitrogenase, which contains multiple nonstandard Fe–S centers (Kim and Rees, 1992). Cluster extrusion and reconstitution experiments have been an important source of information, as have been studies of synthetic models (Averill, 1988). Mössbauer spectroscopy has played a crucial role in the elucidation of the iron-sulfur active sites, largely based on its ability to observe all iron species quantitatively and to sort them out according to oxidation and spin state. The discussion below illustrates some of these applications for the case of rubredoxin, a 1Fe–sulfur protein; putidaredoxin a 2Fe–2S protein, and aconitase, a 3/4Fe–4S protein.

3.2. Rubredoxin, a $Fe(SR)_4$ Protein

Rubredoxin (Rd) from *Clostridium pasteurianum* is an electron-transfer protein with a single iron site coordinated to four cysteines. The structure of Rd is known to 1.2 Å (Watenpaugh *et al.*, 1980). The Mössbauer spectra (Schulz and Debrunner, 1976) of reduced Rd in zero field consist of a quadrupole doublet with a splitting of $\Delta E_Q = 3.25$ mm/s at 4.2 K that decreases to 3.21 mm/s at 200 K. The isomer shift $\delta_{Fe} = 0.70$ mm/s is typical of high-spin ferrous iron with S_4 coordination, but it is ~0.5 mm/s smaller than the values found for six-coordinate ionic compounds. The weak temperature dependence of ΔE_Q implies minimal admixture of higher orbital states and suggests that the spin Hamiltonian, Eq. (15), should provide a good description of the ground quintet. Figure 1 shows Mössbauer spectra recorded at 4.2 K in external fields of different strengths. The two lines of the quadrupole doublet in the zero-field data are seen to broaden and split with increasing field, and for $B \geq 1.5$ T the spectra consist of relatively well-resolved lines with intensities that depend little on the direction of the applied field. This behavior indicates

1. that the iron complex has an easy axis of magnetization, and
2. that the magnetization has reached half-saturation in a field on the order of 1 T.

Both points can be modeled by a spin Hamiltonian with high rhombicity, $E/D \lesssim \frac{1}{3}$. For finite E/D and in the absence of magnetic couplings, the spin expectation values are zero for any integer spin system. An external field B mixes the $(2S+1)$ sublevels $|i\rangle$ and produces spin expectation values $\langle \mathbf{S} \rangle_i$ that are proportional to B for small fields and then saturate. For $E/D = \frac{1}{3}$, in particular, the spin Hamiltonian reduces to $\mathcal{H}_S = (2D/3) \times (\hat{S}_z^2 - \hat{S}_y^2) + \beta\hat{S}\cdot\mathbf{g}\cdot\mathbf{B}$, and for $D > 0$ the lowest two levels have a splitting of $\sim 0.31D$ that is much smaller than the splitting of $\sim D$ to the next higher level. At 4.2 K only these low-lying levels are populated, and since they are highly polarizable along y, their properties explain the basic features of the Rd Mössbauer data.

As the solid lines in Fig. 1 indicate, computer simulations based on a Hamiltonian with high rhombicity match the data quite well. Surprisingly, the spin-state fluctuation rate turns out to be slow on the Mössbauer time scale, and in contrast to other high-spin ferrous compounds the fast fluctuation limit is not even reached at 150 K (Schulz *et al.*, 1987).

Oxidized Rd has spin $S = \frac{5}{2}$ and a zero-field splitting of high rhombicity $E/D \approx 0.25$, as evidenced by the strong EPR signal near $g = 4.3$ originating from the middle Kramers doublet. The low-temperature Mössbauer spectra

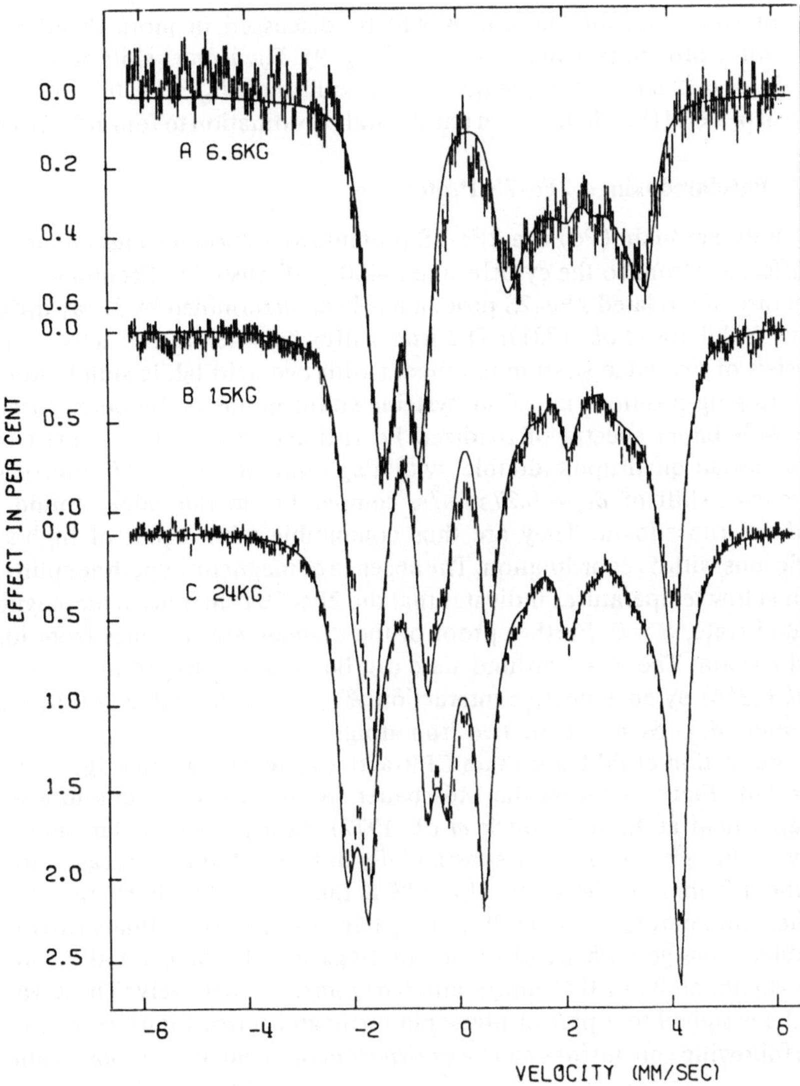

Figure 1. Mössbauer spectra of reduced rubredoxin at 4.2 K in different fields parallel to the γ beam. The solid lines are simulations in the slow fluctuation limit based on the parameters $\Delta E_Q = (-)3.25$ mm/s, $\eta = 0.65$, $\delta_{Fe} = 0.70$ mm/s, $D/k = 10.9$ K, $E/D = 0.28$, $\mathbf{g} = (2.11, 2.19, 2.00)$, $\mathbf{A}/(g_N\beta_N) = -(20.1, 8.3, 23)$ T. Reproduced from Schulz and Debrunner (1976).

(Schulz and Debrunner, 1976; not shown) display well-resolved magnetic hyperfine splittings that can be resolved into three component spectra arising from the three Kramers doublets of the spin sextet with temperature-dependent intensities. The same type of spectrum can be observed for any mononuclear high-spin ferric site of high rhombicity and sufficiently slow rate of spin state fluctuation, as will be discussed in more detail for a non-sulfur protein (see Section 5.3 and Fig. 9). The isomer shift of oxidized Rd, $\delta_{Fe} = 0.32$ mm/s, and the quadrupole splitting $\Delta E_Q = (-)0.5$ mm/s, are typical of Fe(III) with distorted tetrahedral coordination to four RS^- groups.

3.3. Putidaredoxin, a 2Fe–2S Protein

Putidaredoxin (Pd) is a 2Fe–2S protein from *Pseudomonas putida* that transfers electrons to the cytochrome P-450$_{cam}$ discussed in Section 4.3. The structure of a related 2Fe–2S protein has been determined by X-ray diffraction (Tsukihara *et al.*, 1981). The iron–sulfur complex of the active center consists of two edge-sharing tetrahedra with two acid-labile sulfide ions in the bridging positions and four cysteine sulfur atoms at the other corners. The Mössbauer spectra of oxidized Pd (Münck *et al.*, 1972) consist of a single broad quadrupole doublet with a splitting of $\Delta E_Q = 0.60$ mm/s and an isomer shift of $\delta_{Fe} = 0.27$ mm/s, comparable to the values found for oxidized rubredoxin. They are thus compatible with a pair of high-spin ferric ions with S_4 coordination. The absence of magnetic hyperfine splittings even at low temperatures indicates that the 2Fe–2S center has a diamagnetic ground state, $S = 0$. Further proof of the diamagnetism comes from high-field spectra. The $S = 0$ ground state can be modeled according to Gibson *et al.* (1966) by an exchange interaction $\mathcal{H}_{xch} = J\hat{S}_1 \cdot \hat{S}_2$ with $J > 0$ between the spins $S_1 = S_2 = \frac{5}{2}$ of the two iron atoms.

Reduction of Pd leads to an EPR-active state with g values $g_{x,y} = 1.94$, $g_z = 2.01$. Figure 2 shows the Mössbauer spectrum of this state in a weak parallel field at 4.2 K (Münck *et al.*, 1972). In a perpendicular field (not shown) the spectral shape is substantially different, but the overall splitting of about 7 mm/s is the same. Up to 80 K the spectra barely change, but at higher temperatures the overall splitting decreases, and eventually two broad doublets emerge with quadrupole splittings of ~0.6 mm/s and 2.7 mm/s and isomer shifts of 0.35 mm/s and 0.65 mm/s, respectively. These values can be assigned to a pair of high-spin ferric and ferrous ions, respectively. The following conclusions can be drawn from these qualitative observations:

1. It is plausible to assume that the spectra are a superposition of two component spectra arising from high-spin Fe(III) and Fe(II) sites at low as well as at high temperatures.
2. Both nuclei show magnetic hyperfine coupling to the resultant spin $S = \frac{1}{2}$ observed by EPR.
3. The spin fluctuation rate is slow for $T < 80$ K.
4. Given the 7-mm/s overall splitting of the low-temperature spectra, the maximum internal field $\mathbf{B}_{int} = -\langle S \rangle \mathbf{A}/g_N\beta_N$ must be ~20 T, which in turn implies a largest hyperfine coupling of $|A/g_N\beta_N| \sim$ 40 T.

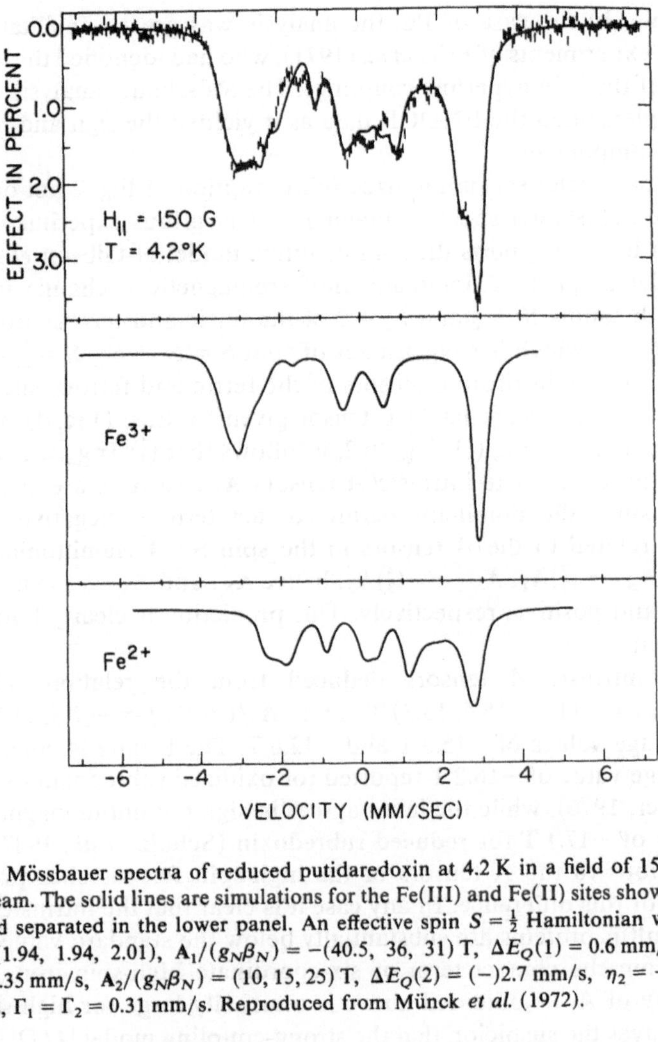

Figure 2. Mössbauer spectra of reduced putidaredoxin at 4.2 K in a field of 15 mT parallel to the γ beam. The solid lines are simulations for the Fe(III) and Fe(II) sites shown combined on top and separated in the lower panel. An effective spin $S = \frac{1}{2}$ Hamiltonian was assumed with $\mathbf{g} = (1.94, 1.94, 2.01)$, $\mathbf{A}_1/(g_N\beta_N) = -(40.5, 36, 31)$ T, $\Delta E_Q(1) = 0.6$ mm/s, $\eta_1 = 0.5$, $\delta_{Fe}(1) = 0.35$ mm/s, $\mathbf{A}_2/(g_N\beta_N) = (10, 15, 25)$ T, $\Delta E_Q(2) = (-)2.7$ mm/s, $\eta_2 = -3$, $\delta_{Fe}(2) = 0.65$ mm/s, $\Gamma_1 = \Gamma_2 = 0.31$ mm/s. Reproduced from Münck *et al.* (1972).

Based on these arguments, it should be possible to model the low-temperature spectra by the Hamiltonian $\mathcal{H} = \beta\hat{S}\cdot\mathbf{g}\cdot\mathbf{B} + \hat{S}\cdot\sum A_{S,i}\cdot\hat{I}_i + \sum \mathcal{H}_{I,i}$, $i = 1, 2$, where $\mathcal{H}_{I,1}$ and $\mathcal{H}_{I,2}$ are the nuclear Hamiltonians, Eq. (14), for the ferric and ferrous sites, respectively. As the solid line in Fig. 2 demonstrates, a parametrization in terms of this Hamiltonian is indeed possible with the parameter values given in the caption. The lower traces show the resolution of the spectrum into two components arising from the ferric site (1) and the ferrous site (2), respectively. It is clear from the complete overlap of the components that a meaningful parametrization requires additional information. The high-temperature data, for instance, yield an estimate of ΔE_Q and δ for both sites, and measurements in other fields provide additional constraints. In the case of Pd, the analysis was greatly facilitated by the ENDOR experiments of Fritz *et al.* (1971), who had identified the magnitude of some of the ^{57}Fe hyperfine couplings. The Mössbauer analysis confirmed and complemented the ENDOR data as it yielded the sign and magnitude of all A components.

The parameter set summarized in the caption of Fig. 2 not only reproduces the Mössbauer spectra obtained under various experimental conditions, but it also supports the spin-coupling model of Gibson *et al.* (1966). The model assumes a dominant antiferromagnetic exchange interaction between the spins $S_1 = \frac{5}{2}$ and $S_2 = 2$ of the ferric and ferrous iron, $\mathcal{H}_{\text{xch}} = J\hat{S}_1 \cdot \hat{S}_2$, $J > 0$, which has eigenstates of spin $\hat{S} = \hat{S}_1 + \hat{S}_2$, $S = \frac{1}{2}, \frac{3}{2}, \ldots, \frac{9}{2}$. If $\mathbf{g}_1$ and $\mathbf{g}_2$ are the intrinsic g tensors of the ferric and ferrous site, then the spin $S = \frac{1}{2}$ ground state has a g tensor given by Eqs. (13c, d) of $\mathbf{g}_S = (\frac{7}{3})\mathbf{g}_1 - (\frac{4}{3})\mathbf{g}_2$. With $g_1 \simeq 2$, $(\frac{1}{3})\,\mathrm{Tr}\,\mathbf{g}_2 \gtrsim 2$, it follows that $(\frac{1}{3})\,\mathrm{Tr}\,\mathbf{g}_S \lesssim 2$, as experimentally observed. The intrinsic A tensors $\mathbf{A}_1$ and $\mathbf{A}_2$ are expected to be negative since the dominant Fermi contact term is negative. They are similarly related to the A tensors in the spin $\mathbf{S} = \frac{1}{2}$ Hamiltonian by Eqs. (13d, e), $\mathbf{A}_{S,1} = (\frac{7}{3})\mathbf{A}_1$, $\mathbf{A}_{S,2} = -(\frac{4}{3})\mathbf{A}_2$, hence $\mathbf{A}_{S,1}$ and $\mathbf{A}_{S,2}$ are expected to be negative and positive, respectively. This prediction is clearly borne out by experiment.

The intrinsic A tensors deduced from the relations above are $\mathbf{A}_1/(g_N\beta_N) = -(17.4, 15.4, 13.3)$ T and $\mathbf{A}_2/(g_N\beta_N) = -(7.5, 11.3, 18.8)$ T, with average values of -15.3 T and -12.6 T. The former is comparable to the average value of -16.2 T reported for oxidized rubredoxin (Schulz and Debrunner, 1976), while the latter agrees in sign but not in magnitude with the value of -17.1 T for reduced rubredoxin (Schulz *et al.*, 1987). Orbital contributions to the A tensors of the high-spin ferrous sites presumably account for this difference. In any case it is clear that the intrinsic A values of iron–sulfur proteins are substantially below the standard value of -22 T arising from the contact term of six-coordinate high-spin iron. The 13% anisotropy of $\mathbf{A}_1$ as obtained above is unusually large for high-spin ferric iron; it leaves the suspicion that the strong-coupling model $|J/D| \gg 1$ is not adequate for reduced Pd and that part of the anisotropy comes from admixtures of states due to zero-field splittings as in the more extreme case of purple acid phosphatase discussed in Section 6.2 (Sage *et al.*, 1989).

3.4. Aconitase, a 3/4Fe–4S Protein

The last example of the iron–sulfur proteins to be discussed here is aconitase, an essential enzyme in the mitochondrial citric acid cycle (Emptage, 1988). The elegant and incisive Mössbauer studies of Münck and co-workers clarified many aspects of this 3/4Fe–4S protein. As isolated, aconitase is inactive and found to contain a $[3Fe\text{-}4S]^+$ cluster with an EPR signal at $g = 2.01$ from a system spin of $S = \frac{1}{2}$. The Mössbauer spectra at 85 K and above consist of a single quadrupole doublet with $\Delta E_Q = 0.71$ mm/s and $\delta_{Fe} = 0.27$ mm/s, consistent with the presence of indistinguishable high-spin ferric irons. Distinct hyperfine splittings are resolved at low temperatures, however, in line with observations on other 3Fe-4S proteins. Kent *et al.* (1980) showed how equal intrinsic hyperfine constants of the three Fe(III) sites can lead to the observed positive and negative A values as a consequence of the spin coupling.

Reduction of the protein as isolated leads to a paramagnetic $[3Fe\text{-}4S]^0$ cluster with two distinct quadrupole doublets of 2:1 intensity ratio, the first with $\Delta E_Q = 1.34$ mm/s and $\delta_{Fe} = 0.45$ mm/s, the second with $\Delta E_Q = 0.49$ mm/s and $\delta_{Fe} = 0.30$ mm/s. The parameters of the second species are characteristic of high-spin Fe(III), but those of the first are halfway between the characteristic Fe(III) and Fe(II) values and must arise from a delocalized valence pair. Mixed-valence pairs are found quite generally in 3Fe-4S and 4Fe-4S clusters, and Münck *et al.* (1988) have explained their properties by double exchange. The paramagnetism of the integer spin $[3Fe\text{-}4S]^0$ cluster in reduced aconitase is evident from the broadening of the Mössbauer doublets in weak external fields. Analogous behavior is found in other $[3Fe\text{-}4S]^0$ clusters, and good simulations of all the data are obtained with a spin $S = 2$ Hamiltonian.

Aconitase can be activated by treatment with Fe(II), and Mössbauer measurements of Emptage *et al.* (1983) on samples activated with ^{57}Fe or ^{56}Fe prove that this process converts the $[3Fe\text{-}4S]^0$ clusters to $[4Fe\text{-}4S]^{2+}$ clusters. As trace A of Fig. 3 illustrates, the activating ^{57}Fe gives rise to a quadrupole doublet with $\Delta E_Q = 0.80$ mm/s and $\delta_{Fe} = 0.45$ mm/s, while the original iron atoms, which are unenriched in this sample and hence barely noticeable, change their parameters to $\Delta E_Q = 1.30$ mm/s and $\delta_{Fe} = 0.45$ mm/s. Both isomer shifts are characteristic of mixed valence iron in diamagnetic $[4Fe\text{-}4S]^{2+}$ clusters. It is clear, then, that the extra iron atom is incorporated into the $[3Fe\text{-}4S]^0$ cluster, converting it to a "normal" $[4Fe\text{-}4S]^{2+}$ cluster.

When a fivefold excess of citrate is added to ^{56}Fe-activated aconitase, trace C of Fig. 3 is observed. Only the original three iron atoms of natural abundance are seen and, apart from a broad background akin to the spectrum of $[3Fe\text{-}4S]^+$, trace C consists of a majority doublet labeled b with $\Delta E_Q = 1.15$ mm/s and $\delta_{Fe} = 0.47$ mm/s. These values deviate only slightly from the values $\Delta E_Q = 1.30$ mm/s, $\delta_{Fe} = 0.45$ mm/s, obtained without substrate.

Trace B in Fig. 3, in contrast, shows the spectrum of ^{57}Fe-activated aconitase in the presence of excess citrate. In addition to a 20% fraction of the doublet from trace A, the spectrum now displays two radically different

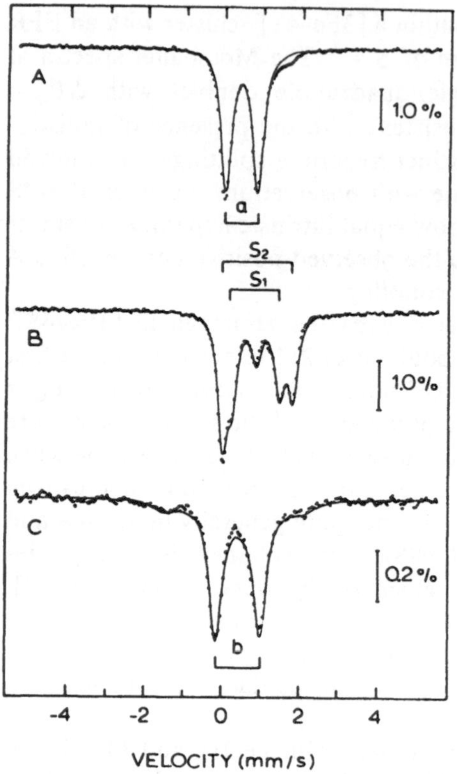

Figure 3. Mössbauer spectra of activated aconitase at 4.2 K in zero field (traces A, B), and in a 60-mT field parallel to the γ beam trace (C). Trace A: ^{57}Fe activated enzyme. The solid line traces the Fe_a doublet. Trace B: ^{57}Fe activated enzyme in the presence of excess citrate. The solid line is a superposition of three quadrupole doublets with intensities 20%, 40% (S_1) and 40% (S_2). For A and B the 7% contribution from natural-abundance ^{57}Fe in the Fe_b sites was subtracted. Trace C: ^{56}Fe activated enzyme in the presence of citrate. The solid line is the sum of a quadrupole doublet ($\Delta E_Q = 1.15$ mm/s, $\delta_{Fe} = 0.47$ mm/s) and the experimental spectrum of the oxidized 3Fe center. Reproduced from Emptage *et al.* (1983).

quadrupole doublets, labeled S_1 and S_2, each accounting for 40% of the total area. The new parameters are $\Delta E_Q = 1.26$ mm/s, $\delta_{Fe} = 0.84$ mm/s and $\Delta E_Q = 1.83$ mm/s, $\delta_{Fe} = 0.89$ mm/s for S_1 and S_2, respectively. The isomer shifts have thus moved much closer to the values observed for 5- or 6-coordinated high-spin Fe(II), and there is no doubt that the normal S_4 coordination of the activating iron site has changed to allow for substrate binding in two different conformations. This unexpected observation was the first indication that iron–sulfur clusters are capable of changing ligands and of forming substrate complexes.

Substrate binding is also suggested by EPR data on reduced, activated aconitase in the cluster state $[4Fe\text{–}4S]^+$, which has spin $S = \frac{1}{2}$. The g values of the enzyme decrease substantially on addition of substrates or inhibitors, (Emptage, 1988), but no interpretation at the atomic level is possible. ENDOR measurements (Telser *et al.*, 1986), on the other hand, have given detailed information on the coordination of substrate to the special iron sites of aconitase and have clearly confirmed the conclusions drawn above from Mössbauer spectroscopy.

4. HEME PROTEINS

4.1. Overview

The conspicuous color of porphyrins has long attracted attention, and optical techniques are still the primary tools in the study of heme proteins. EPR provided a first close look at the electronic state of the iron in ferric heme compounds, and the crystal-field model of Griffith (1957) as well as the "Truth Table" of Blumberg and Peisach (1971) still constitute the starting point for any analysis of low-spin ferric hemes. Mössbauer spectroscopy has added greatly to the present understanding of the heme iron as it applies not only to the Fe(III) states, but to Fe(II) and Fe(IV) as well. Examples of these physiologically relevant integer-spin states will be discussed in Sections 4.2 and 4.3.

Much of the pioneering work on heme compounds is due to G. Lang (1970), who adapted crystal-field models of the $3d$ electrons and spin Hamiltonians to the analysis of ^{57}Fe Mössbauer data. Quite generally, these models provide an adequate representation of the magnetic terms $\hat{S}\cdot\mathbf{g}\cdot\mathbf{B}$ and $\hat{S}\cdot\mathbf{A}\cdot\hat{I}$, which are ascribed to the spin and orbital angular momenta of the unpaired $3d$ electrons, but the crystal-field approximation fails to predict the experimentally observed quadrupole tensors $\mathbf{P}$ in Eq. (6). The failure of a crystal-field model for the heme iron is hardly surprising as the iron–pyrrole bonds and, to a large extent, also the axial bonds are highly covalent. At the very least, the $\langle r^{-3}\rangle$ values of the various $3d$ symmetry orbitals must be allowed to differ, and the resultant larger number of adjustable parameters can no longer be determined from experiment alone.

The porphyrin group brings about some unique properties of the central iron ion. In both ferric and ferrous porphyrins, intermediate spin states of $S=\frac{3}{2}$ and $S=1$ can be stabilized in addition to the usual high- and low-spin states. In ferric porphyrins, specifically, the spin-orbit coupling can mix the quartet and sextet to any degree (Maltempo and Moss, 1976). Quartet admixture leads to reduced g values, large zero-field splittings, and large, temperature-dependent quadrupole splittings, which may exceed 4 mm/s. The sign and symmetry of $\mathbf{P}$ implies a larger electron density in the heme plane than along the heme normal, as expected on the basis of the short $Fe\text{-}N_{pyrrole}$ bond lengths of ≈ 2 Å compared to the typically longer axial bonds.

As mentioned above, heme proteins can also stabilize the unusual ferryl state, Fe(IV)O, of spin $S=1$, which has been identified in reaction intermediates of peroxidases and oxidases and is widely considered to be an intermediate in cytochrome P-450. Mössbauer spectroscopy has played a crucial role in the elucidation of the electronic state of Fe(IV), as will be shown below.

4.2. Horseradish Peroxidase

Horseradish peroxidase (HRP) reacts with H_2O_2 to form a green primary compound (HRPI) that is two oxidizing equivalents above the resting, ferric state. A variety of electron donors can reduce HRPI in two one-electron steps first to a secondary compound (HRPII) and then to the native state. A combination of optical, EPR–Mössbauer, and ENDOR experiments has shown that HRPI is a ferryl porphyrin π-cation radical complex with weak exchange interaction between the spins $S = 1$ of the iron and $S' = \frac{1}{2}$ of the radical, $|J/D| \ll 1$. In HRPII the porphyrin is reduced to its normal, spin-paired state, while the iron has retained its ferryl $S = 1$ character. The heme iron of HRP has a histidine axial ligand as is found in cytochrome *c* peroxidase and the globins, whose crystal structures are known. In the following we first consider the native, high-spin ferric state and illustrate how spin-state fluctuations affect the Mössbauer spectra. Next, we will consider the case of HRPI.

The heme iron of HRP is easily perturbed by pH, ionic strength, and the presence of small molecules in solution (Smulevich *et al.*, 1991). A rather clean EPR signal with $g = (6.10, 5.46, 1.98)$ is observed in the benzohydroxamic acid complex, and the latter was therefore chosen for Mössbauer studies. Figure 4 shows the spectra of ferric HRP recorded in a moderate magnetic field perpendicular to the γ beam at various temperatures (Schulz *et al.*, 1987). At 4.2 K a well-resolved six-line spectrum is observed, which allows one to deduce the magnetic hyperfine and electric quadrupole interactions in the spin Hamiltonian formalism, Eq. (15). The quadrupole splitting, $\Delta E_Q = 1.65$ mm/s, is one of the largest reported for $S = \frac{5}{2}$ heme iron; it is indicative of some quartet admixture, in agreement with other features of HRP to be discussed shortly. At 10 K (not shown) the six lines broaden, and at 20 K the spectrum collapses into two broad unequal humps. At still higher temperatures the two lines sharpen and the lower energy line becomes deeper. The evolution of the spectra is reasonably well reproduced by a dynamic line shape model as shown by the solid lines in Fig. 4. The model is based on the spin Hamiltonian, Eq. (15), but allows transitions among the electronic eigenstates due to spin–lattice coupling. The transition rates are calculated by the same approximation used in the standard model of spin–lattice relaxation in EPR (Abragam and Bleaney, 1970), and a single parameter is introduced to adjust the overall rate. In contrast to EPR, Mössbauer spectroscopy observes the thermal equilibrium fluctuations of all sextet states rather than the relaxation of a single doublet,

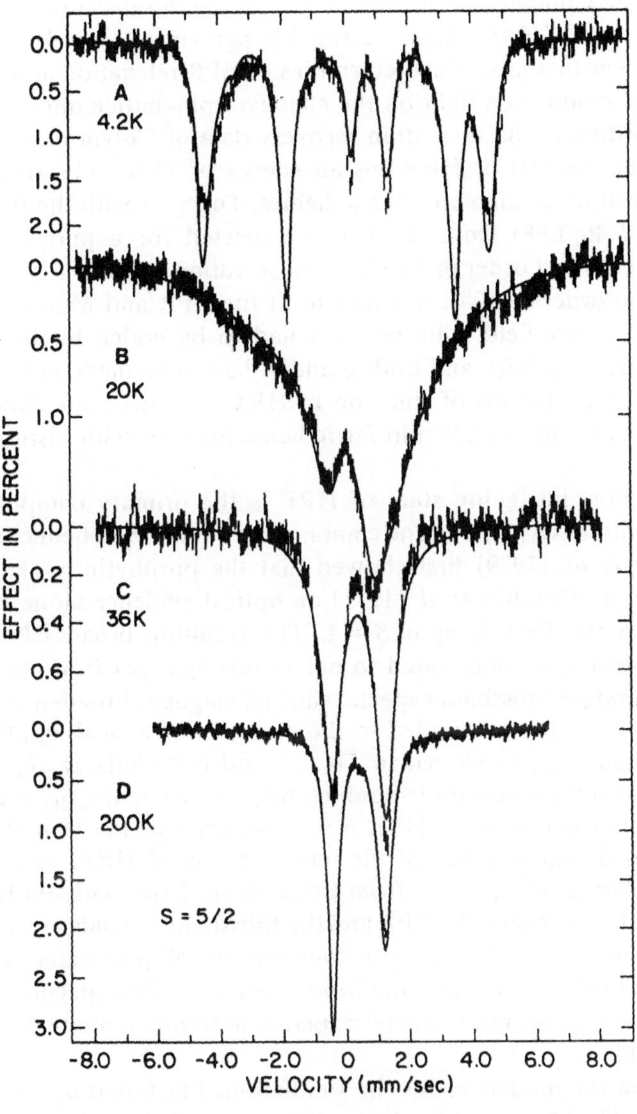

Figure 4. Mössbauer spectra of the benzohydroxamic acid complex of high-spin ferric horseradish peroxidase in a field of 180 mT perpendicular to the γ beam at the temperatures indicated. The solid lines are simulations based on a dynamic line shape model with the parameters $D/k = 30$ K, $E/D = 0.009$, $\mathbf{g} = (1.925, 1.925, 1.99)$, $\Delta E_Q = (+)1.656$ mm/s, $\eta = 0.25$, $\delta_{Fe} = 0.40$ mm/s, $\mathbf{A}/(g_N\beta_N) = -(19.5, 17.5, 20)$ T, $\Gamma = 0.30$ mm/s, $W_0 = 1.5 \times 10^3$ rad s^{-1} K^{-3}, and the term $F(7\hat{S}_z^4/36 - 95\hat{S}_z^2/72 + 63/64)$ with $F/k = 12.3$ K added to the spin Hamiltonian. Reproduced from Schulz *et al.* (1987.)

but the fluctuation model allows one to predict the relaxation rate T_1^{-1} for any doublet. Comparison of such predictions with saturation recovery data of Colvin *et al.* (1983) shows reasonable agreement (Schulz *et al.*, 1987), and it is clear that more detailed studies of EPR relaxation and Mössbauer fluctuation could shed light on the effective spin–lattice interaction.

According to the saturation recovery data of Colvin *et al.* (1983), the first excited doublet of HRP has an energy of 41 K, substantially higher than in most other high-spin ferric hemes. Together with the deviations of $\mathbf{g} = (6.10, 5.46, 1.98)$ from the values expected for a pure sextet, $g_{x,y} = 6 \pm 24E/D$ to first order in E/D, this observation suggests a quartet admixture on the order of 10%. In order to fit the EPR and Mössbauer data, a fourth-order zero-field splitting term had to be added to the $S = \frac{5}{2}$ spin Hamiltonian, Eq. (10), and both g and A had to be made anisotropic (see Fig. 4). The properties of the iron in HRP are thus quite different from those found in other high-spin ferric heme proteins with histidine coordination.

The most intriguing state of HRP is the primary compound in the reaction with H_2O_2, HRPI. The combined EPR and Mössbauer experiments of Schulz *et al.* (1979) first showed that the porphyrin π-cation radical postulated by Dolphin *et al.* (1971) on optical evidence alone was weakly coupled to the Fe(IV), spin $S = 1$. The resulting broad EPR spectrum centered near $g = 2$ integrated to nearly one spin per iron atom, and the low-temperature Mössbauer spectra showed magnetic broadening of a quadrupole doublet that responded to the direction of a weak applied field in the expected way, as shown in Fig. 5a and b (Schulz *et al.*, 1984). The parameters of the quadrupole doublet, $\Delta E_Q = 1.25$ mm/s, $\delta_{Fe} = 0.08$ mm/s, left no doubt about the Fe(IV), $S = 1$ character of the iron (Lang *et al.*, 1969), and definitive proof of the interpretation of HRPI as a ferryl porphyrin π-cation radical came from ENDOR (Roberts *et al.*, 1981a, b). Once the correlation between the EPR and the Mössbauer signals was established, the question arose of finding a common set of parameters that would reproduce both the EPR and the Mössbauer data. Two specific models will be discussed below, but a crucial experiment to differentiate between them is still lacking.

One of the models yields the simulations illustrated by the solid lines in Fig. 5, which shows Mössbauer spectra of HRPI recorded at various temperatures and fields. The model assumes an anisotropic exchange $\mathcal{H}_{xch} = \hat{S} \cdot \mathbf{J} \cdot \hat{S}'$ between the ferryl spin $S = 1$ and the radical spin $S' = \frac{1}{2}$, with $\mathbf{J}$ as given in the caption of Fig. 5. The zero-field splitting of the iron, $D/k = 37$ K, is known from EPR saturation recovery (Colvin *et al.*, 1983); it implies that at temperatures $T < 10$ K, where the EPR signal is observable, only the $m_S = 0$ level of the iron is populated. As a consequence, g_z of the coupled system must be essentially equal to the g value of the radical and corresponds

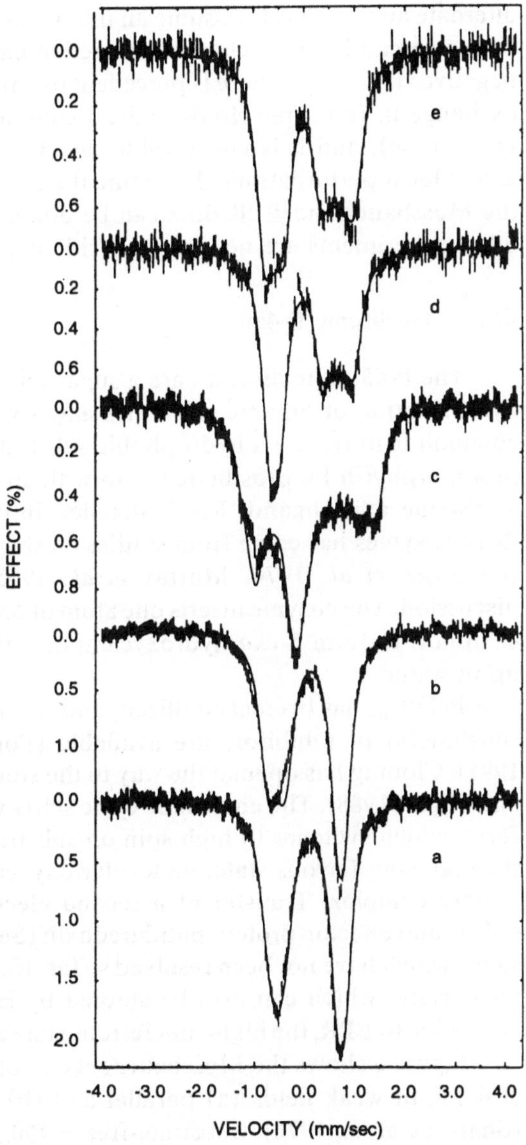

Figure 5. Mössbauer spectra of horseradish peroxidase compound I measured under the following conditions: (a) 1.5 K, 44-mT perpendicular field; (b) 1.5 K, 44-mT parallel field; (c) 4.2 K, 5.5-T parallel field; (d) 26 K, 3.0-T parallel field. The solid lines are simulations based on the spin Hamiltonian, Eq. (15), for $S = 1$, and the additional terms $Bg'\hat{S}' \cdot \mathbf{B} + \hat{S} \cdot \mathbf{J} \cdot \hat{S}'$, $S' = \frac{1}{2}$, with the parameters $D/k = 37$ K, $\mathbf{g} = (2.25, 2.25, 1.98)$, $g' = 2$, $\mathbf{J}/\mathbf{k} = (-4, -2, 6)$ K, $\mathbf{A}/(g_N\beta_N) = -(19.3, 19.3, 6)$ T, $\Delta E_Q = (+)1.25$ mm/s, $\delta_{\text{Fe}} = 0.08$ mm/s, $W_0 = 8.8 \times 10^4$ rad s^{-1} K^{-3}. Reproduced from Schulz *et al.* (1984).

to the peak of the EPR absorption near $g = 2$. The perpendicular g values of the coupled system, on the other hand, have first-order corrections proportional to $J_{x,y}/D$ and must correspond to the up- and down-field wings of the EPR absorption. Given the EPR data, it follows from this argument that $|J/D|$ must be of order 0.1. The simulations in Fig. 5 assume an anisotropic exchange with positive as well as negative components to account for effective g values $g < 2$ for $J > 0$ and $g > 2$ for $J < 0$. An alternate approach is to assume an inhomogeneous population of molecules with a spread in the exchange interaction encompassing positive as well as negative values of J. There is precedent for strong positive as well as negative exchange in ferryl porphyrin radical complexes (Boso *et al.*, 1983; Rutter *et al.*, 1984), and it is conceivable that HRP is a borderline case, $|J| \approx 0$, where local perturbations determine the sign of J. Adequate simulations of the Mössbauer and EPR data can be obtained by using either model, and other experiments are needed to decide which one is more realistic.

4.3. Cytochrome P-450

The P-450 cytochromes are a superfamily of enzymes that catalyze the hydroxylation of organic substrates and a variety of other reactions. Their common features are a hydrophobic substrate binding site next to a buried protoporphyrin IX prosthetic group with an iron atom covalently linked to a cysteine axial ligand. Much detailed information on the properties of these enzymes has come from studies on the bacterial cytochrome P-450$_{cam}$ (Gunsalus *et al.*, 1974; Murray *et al.*, 1985), the topic of the following discussion. The enzyme inserts one atom of oxygen from O_2 into its substrate, camphor, to form 5-*exo*-hydroxycamphor; the other atom of oxygen ends up in water.

P-450$_{cam}$ has been crystallized, and several structures with and without substrate(s) or inhibitors are available (Poulos *et al.*, 1987; Raag *et al.*, 1991). Cloning has opened the way to the study of P-450$_{cam}$ mutants (Atkins and Sligar, 1988). The enzymatic cycle starts with the resting, low-spin ferric form, which switches to high spin on substrate binding. After reduction to the high-spin ferrous state, molecular oxygen binds to form a diamagnetic ternary complex. Transfer of a second electron from the specific 2Fe–2S redox- and effector-protein putidaredoxin (Section 3.3) initiates the catalytic steps, which have not been resolved so far. Here, we will discuss the low-spin ferric state, which can also be studied by EPR, as well as two states not accessible to EPR, the high-spin ferrous state and the diamagnetic O_2 adduct.

Figure 6 shows the Mössbauer spectra of ferric, substrate-free P-450$_{cam}$ at 4.2 K in weak fields (a) parallel and (b) perpendicular to the γ beam (Sharrock *et al.*, 1976). Substrate-free P-450$_{cam}$ has a g tensor of $\mathbf{g} = (1.91, 2.26, 2.45)$, typical of low-spin hemes, but the spread in g values is smaller than for any other heme protein. The largest component, $g = 2.45$, is found approximately along the heme normal (Murray *et al.*, 1985). As is evident from Fig. 6, the Mössbauer spectra depend on the direction of the applied field, and they are broad, extending continuously from -5 mm/s to $+4$ mm/s, with few prominent features. The broad shape suggests that each molecular orientation has a distinctly different Hamiltonian, and given the modest anisotropy of $\mathbf{g}$, this means that the hyperfine interactions must be

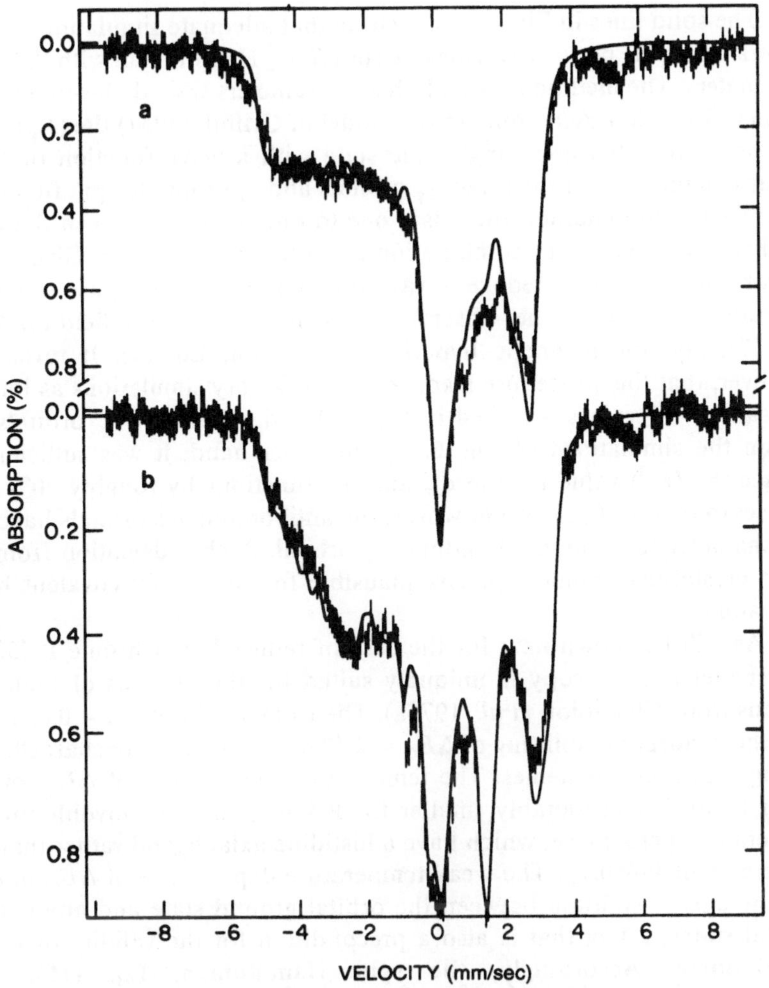

Figure 6. Mössbauer spectra of substrate-free, low-spin ferric cytochrome P-450_{cam} at 4.2 K in a field of 44 mT (a) parallel and (b) perpendicular to the γ beam. The solid lines are simulations with $\mathbf{g} = (1.91, 2.26, 2.45)$, $\mathbf{A}/(g_N\beta_N) = (-45, 10, 19)$ T, $\Delta E_Q = 2.85$ mm/s, $\eta = -1.80$, $\delta_{Fe} = 0.38$ mm/s, $\Gamma = 0.3$ mm/s. Reproduced from Sharrock *et al.* (1976).

quite anisotropic. Low-spin ferric heme iron in an environment of low symmetry is indeed expected to have an anisotropic Hamiltonian as **A** and **P** arise, in first approximation, from a single t_{2g} orbital missing in an otherwise filled $(t_{2g})^6$ subshell. Accordingly, this quadrupole splitting, which can be measured in the regime of fast spin-state fluctuations at $T > 130$ K, is relatively large, $\Delta E_Q = 2.85$ mm/s, and **A** has roughly the magnitude and symmetry expected from a single t_{2g} electron, specifically d_{yz}.

The solid lines in Fig. 6 demonstrate that adequate simulations can be generated on the basis of an effective spin $S = \frac{1}{2}$ Hamiltonian with the listed parameters. The method by which these parameters were deduced deserves further discussion. According to the model of Griffith (1957) the $(t_{2g})^5$ state can be represented as a single-hole state with a wave function that is a linear combination of the three t_{2g} orbitals and appropriate spin functions. In a given representation there is a one-to-one relation between the three coefficients of the linear combination and the three g values. Given the g tensor, one can hence deduce a wave function and, in turn, use this wave function to calculate other observables, such as the electric field gradient, Eq. (7), and the magnetic hyperfine interaction, Eq. (9). It turns out, however, that this procedure leads to unsatisfactory simulations as long as the same $\langle r^{-3}\rangle$ values are used in Eqs. (7) and (9) for all t_{2g} orbitals. To obtain the simulations of Fig. 6, on the other hand, it was sufficient to reduce the $\langle r^{-3}\rangle$ value for the d_{xz} and d_{yz} functions by roughly 20% with respect to that of d_{xy}; in other words, the antibonding d_π^* orbitals have less $3d$ character than the nonbonding d_{xy} orbital. Such a deviation from the strict crystal-field model appears plausible for the highly covalent heme iron atom.

As will be shown next for the case of reduced cytochrome P-450$_{\text{cam}}$, Mössbauer spectroscopy is uniquely suited for the analysis of high-spin ferrous iron (Champion *et al.*, 1975a). The isomer shift of $\delta_{\text{Fe}} = 0.83$ mm/s and the quadrupole splitting of $\Delta E_Q = 2.42$ mm/s at 4.2 K are characteristic of high-spin ferrous hemes. The temperature dependence of ΔE_Q, on the other hand, is considerably smaller for P-450$_{\text{cam}}$ than for myoblobin and horseradish peroxidase, which have a histidine axial ligand rather than the cysteinate of P-450$_{\text{cam}}$. The weak temperature dependence of ΔE_Q implies a large energy splitting between the orbital ground state and any excited orbital states, a fact that is also a precondition for the validity of a spin Hamiltonian. Accordingly, the spin Hamiltonian, Eq. (15), $\mathscr{H} = \hat{S}\cdot\mathbf{D}\cdot\hat{S} + \beta\hat{S}\cdot\mathbf{g}\cdot\mathbf{B} + \hat{S}\cdot\mathbf{A}\cdot\hat{I} + \hat{I}\cdot\mathbf{P}\cdot\hat{I} - \beta_N g N\hat{I}\cdot\mathbf{B}$, should be appropriate for the analysis of the Mössbauer data. Since each of the tensors **D**, **g**, **A**, and **P** requires the specification of the principal axis components as well as of the orientation angles, the number of adjustable parameters is very large, and Champion *et al.* (1975a) had to develop a strategy that allowed them to zero in on the proper parameters. The major determinants of the spectral shape are the quadrupole interaction $\hat{I}\cdot\mathbf{P}\cdot\hat{I}$ and the internal magnetic field $\mathbf{B}_{\text{int}} = -\langle\mathbf{S}\rangle\mathbf{A}/(g_N\beta_N)$. Because the magnitude of $\hat{I}\cdot\mathbf{P}\cdot\hat{I}$ is known from measurements in zero field, since $A/(g_N\beta_N)$ is likely to be of the order of -22T, a typical value for the contact term, and since the spin expectation value $\langle S\rangle$ can be varied continuously from $\langle S\rangle = 0$ in zero field to the saturation value $\langle S\rangle = 2$ in strong field, there is hope that **D**, **A**, and **P** can be determined completely from measurements as a function of field and of temperature. The Mössbauer spectra are not sensitive to **g**, on the other hand, as long as **g** is moderately isotropic. In the present case, second-order perturbation theory can be used to relate the orbital contribution of g to the zero-field splitting, Eq. (10′) (Abragam and Bleaney, 1970). The principal axes of **D** and **g** then coincide, and those of **A** coincide approximately with those of **D** as well.

For positive zero-field splitting, $D > 0$ in Eq. (10′), the lowest energy level is a singlet, and for small fields the spin expectation value is $\langle S_z \rangle \sim 0$, $\langle S_{x,y} \rangle \propto B_{x,y}/D$. Hence, Champion *et al.* (1975a) were able to extract the following information from the broadening of the two lines of the quadrupole doublet as a function of field and temperature:

1. The D value was estimated to be $D/k > 15$ K, in agreement with susceptibility data;
2. there is a rhombic contribution $E \neq 0$ to the zero-field splitting;
3. the largest component of the quadrupole tensor is positive and tilted away from the z-axis; and
4. the asymmetry parameter is large, $\eta \approx 0.8$.

Fortunately, the interpretation of the 4.2 K data depends little on whether fast or slow rates of spin state fluctuations are assumed since higher states have minimal population.

Figure 7 shows the Mössbauer spectrum in a field of 25 kG at 4.2 K and a simulation based on a $S = 2$ spin Hamiltonian with the listed parameters. At 2.5 T the spin expectation value is practically saturated; thus, the overall splitting has reached its maximum value. The spectrum shows considerable structure, which facilitates the analysis in terms of the spin Hamiltonian. The simulation is not perfect, but the parameter set used reproduces the main features of the spectrum and many others taken under different conditions. Moreover, similar parameter sets match the spectra of chloroperoxidase (Champion *et al.*, 1975b), another heme protein with an axial cysteine ligand, and of P-450 model compounds. It thus appears that the axial coordination of RS^- to the ferrous heme iron is essential for the particular features of ferrous P-450$_{cam}$;

1. the large and definitely rhombic zero-field splitting,
2. the weak temperature dependence of the quadrupole interaction, and
3. the large asymmetry of the quadrupole tensor and its rotation with respect to the principal axes of **A**.

In contrast to the case of cytochrome P-450$_{cam}$, myoglobin, reduced horseradish peroxidase and other ferrous heme complexes with an axial nitrogen base have strongly temperature-dependent quadrupole splittings, and a spin Hamiltonian was found to be inappropriate for Mössbauer simulations of their high-field spectra (Kent *et al.*, 1979).

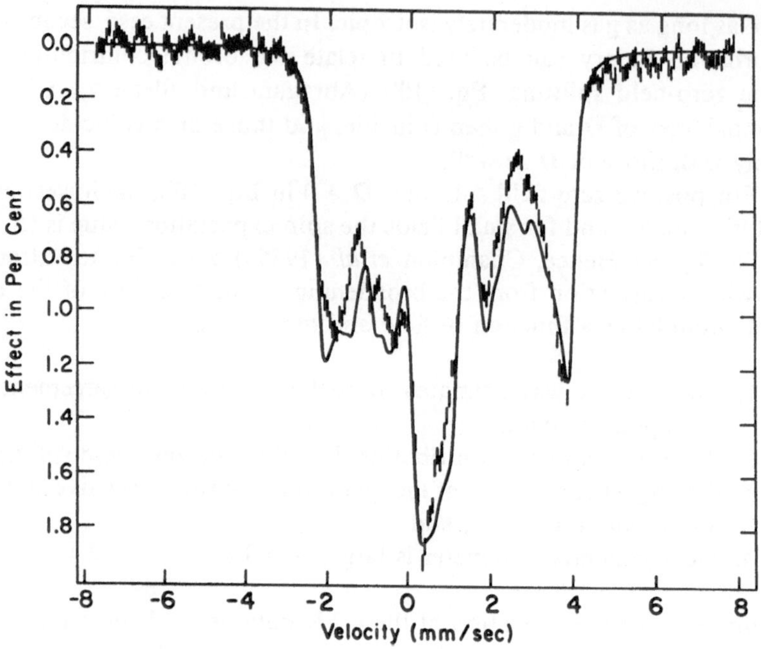

Figure 7. Mössbauer spectrum of reduced cytochrome P-450$_{cam}$ complexed with camphor at 4.2 K in a field of 2.5 T parallel to the γ beam. The solid line is a simulation based on the spin Hamiltonian, Eq. (15), with $S = 2$ and the parameters $D/k = 20$ K, $E/D = 0.15$, $\mathbf{g} = (2.24, 2.32, 2.00)$, $\Delta E_Q = (+)2.42$ mm/s, $\eta = 0.8$, $\mathbf{A}/(g_N\beta_N) = -(18, 12.5, 15)$, T. The quadrupole tensor was rotated with respect to **D** by the Euler angles $\alpha = 60°$, $\beta = 70°$, $\gamma = 0$. Reproduced from Champion *et al.* (1975a).

The last state of cytochrome P-450$_{cam}$ to be considered here is the ternary complex of the reduced enzyme with camphor and O_2. This diamagnetic complex has an isomer shift of $\delta_{Fe} = 0.31$ mm/s and a quadrupole splitting of $\Delta E_Q = (-)2.15$ mm/s at 4.2 K, a value that decreases to $(-)2.07$ mm/s at 200 K. While the isomer shift is typical for low-spin ferrous iron, the quadrupole splitting is unusually large, but matches that of oxymyoglobin in magnitude, sign, and temperature dependence. Evidently, the quadrupole interaction of these complexes is mainly determined by the covalent Fe(II)-O_2 bond, whereas the nature of the fifth axial ligand hardly affects the electronic structure of the iron. Again, Mössbauer spectroscopy provides unique information on the state of the oxyheme iron.

5. MONONUCLEAR IRON PROTEINS

5.1. Overview

Mononuclear iron proteins of the nonheme, nonsulfur type are found to have a variety of functions (Carlioz *et al.*, 1988; Whittaker and Solomon,

1988; Baldwin and Bradley, 1990; Butler *et al.*, 1980), but only a few of these proteins, in particular some dioxygenases (Lipscomb and Orville, 1992) have been studied systematically by MS. In most cases the structure of the protein was not known, and MS in conjunction with other methods has been useful in exploring the active-site ligands. The stability of the iron in these proteins suggests that most of the ligands are amino acids such as histidine, tyrosine, cysteine, and glutamate. The symmetry of the iron site is likely to be low, and the principal axes of the various tensors in Eqs. (9), (10), and (14) need not coincide. Of particular interest are the high-spin ferrous proteins, which are not normally accessible to EPR but can obviously be studied by MS. A special case is that of ferrous complexes that bind NO and thus become EPR active. Below we will discuss a few examples of this group.

5.2. Photosynthetic Reaction Center

The primary process in photosynthesis, the conversion of light energy to electrochemical potential, takes place in a membrane-bound protein complex known as the reaction center (RC). The simplest RC is found in photosynthetic bacteria; it has long been known to contain a single iron atom, which is the subject of the following discussion. The beautiful X-ray diffraction studies of Deisenhofer *et al.* (1985) have now revealed the coordination of the iron atom. It is located rather symmetrically at the interface of two subunits of the RC, deriving a bidentate glutamate and two histidine ligands from one subunit and two additional histidines from the other subunit. Moreover, the iron sits between the primary and the secondary electron acceptors, Q_A and Q_B, a tightly and a weakly bound quinone at distances of 11 Å and 8 Å, respectively (Allen *et al.*, 1987). In spite of the prominent location of the iron atom between Q_A and Q_B, its function is quite elusive. It affects some electron transfer rates, but the RC also functions without it (Debus *et al.*, 1986). Interestingly, an iron atom is also found in the analogous position of photosystem II of green plants (Petrouleas and Diner, 1986, 1990).

Early Mössbauer studies (Debrunner *et al.*, 1975; Boso *et al.*, 1981) on RCs from *Rhodobacter sphaeroides* grown in a ^{57}Fe-enriched medium showed that the iron is high-spin ferrous no matter whether the primary acceptor Q_S is in the native quinone or in the reduced semiquinone state. The latter displays a very unusual EPR signal that is very broad and observable at liquid helium temperatures only (Butler *et al.*, 1980). These unusual features arise from an exchange interaction, Eq. (11), between the spin $S = 2$ of the iron and the spin $S' = \frac{1}{2}$ of the semiquinone. The spin coupling is evident in the Mössbauer spectra as well. While the native RC exhibits a quadrupole doublet with relatively narrow lines at any temperature, the semiquinone complex shows two lines that broaden magnetically at 4.2 K even in the absence of an external field, a clear sign of a Kramers system. Butler *et al.* (1980) were able to explain the EPR spectra quantitatively using the Hamiltonian $\mathcal{H} = D(\hat{S}_z^2 - 2) + E(\hat{S}_x^2 - \hat{S}_y^2) + \beta(g\hat{S} + g'\hat{S}')\cdot\mathbf{B} + J\hat{S}\cdot\hat{S}'$, with $g' = 2.00$, $J = (0.2 \pm 0.1)\ \text{cm}^{-1}$, and the Fe(II) parameters $D = 5.5\ \text{cm}^{-1}$, $E/D = 0.266$, $g = 2.17$ derived from low-temperature susceptibility measurements on native RCs. The plausible assumption was made that the intrinsic parameters of the iron do not change upon electron transfer to the quinone. As will be shown below, however, a spin Hamiltonian is not adequate to simulate the high-field Mössbauer data over an extended temperature range.

Figure 8 shows Mössbauer spectra of native RCs in a field of 3.75 T at various temperatures. By and large, the spectra consist of two broad humps that derive from the two lines of the quadrupole doublet, but no sharp features comparable to those of Figs. 1 and 7 are present. The simulations shown as solid lines in Fig. 8 are based on the following approximation (Varret, 1976; Kent *et al.*, 1977). The fast fluctuation limit of the spin states is assumed to be valid, so that each iron has a certain internal field, $\mathbf{B}_{\text{int}} = -\mathbf{A}\langle\mathbf{S}\rangle_T/(\beta_N g_N)$, that depends, for a given molecular orientation, on the external field $\mathbf{B}$ and on temperature T only. It is further assumed that the relation between $\mathbf{B}_{\text{int}}$ and $\mathbf{B}$ can be expressed by the relation $\mathbf{B}_{\text{int}} = \boldsymbol{\omega}\mathbf{B}$, where $\boldsymbol{\omega}(T; B)$ is a symmetric second-rank tensor. The principal values $\omega_i(T)$, $i = 1, 2, 3$, of the tensor $\boldsymbol{\omega}$, deduced from the simulations based on this model are shown in the right panel of Fig. 8. At high temperatures, the empirical values of ω_i clearly follow a Curie law, $\omega_i \propto T^{-1}$, as indicated by the straight lines in the log–log plot, and they saturate below some transition temperatures that differ widely for the three principal components. The unusual sign of ω_3 implies that the corresponding component A_3 of the hyperfine tensor, Eq. (9), must be positive, i.e. the orbital and spin-dipolar contribution to A_3 must exceed the negative contact term. It should be noted that in the simulations of Fig. 8 the quadrupole splittings $\Delta E_Q(T)$ are taken from measurements in zero field, but the asymmetry η and the Euler angles α, β, γ that rotate from the principal axes of $\boldsymbol{\omega}$ to those of $\mathbf{P}$ are adjustable parameters. The best values of these parameters vary with temperature in the range of $\eta = 0.45 \pm 0.15$ and $\alpha = 50°$, $\beta = (35 \pm 10)°$; the third Euler angle γ was found to be negligible and was set equal to zero.

Given the results of the ω-tensor model, an attempt was made to reproduce the whole set of spectra by the Hamiltonian, Eq. (15). It turned out, however, that no single values of D and of E would fit the data for a fixed A tensor; or, *vice versa*, for fixed D and E, no single A tensor would fit all the data. Simulations as good as those in Fig. 8 could be obtained only if $\mathbf{A}$ were allowed to vary with temperature. These results imply that

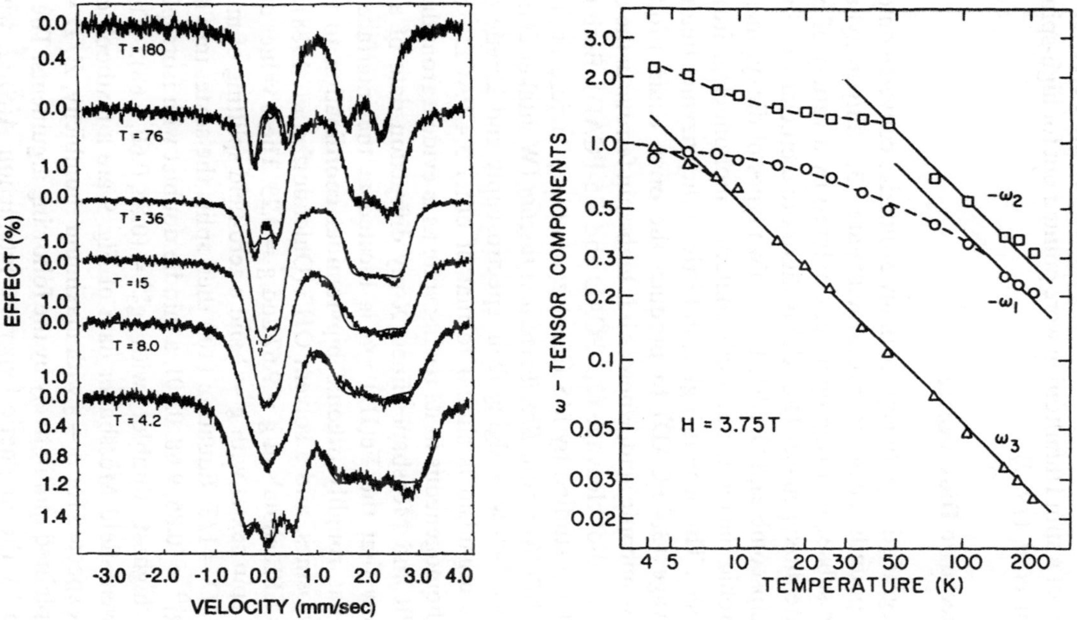

Figure 8. Mössbauer spectra of *R. sphaeroides* photosynthetic reaction centers in a field of 3.75 T parallel to the γ beam at the temperatures indicated. The solid lines are simulations based on the ω tensor model with the ω values shown in the panel on the right. Reproduced from Boso (1982).

the Hamiltonian, Eq. (15), provides an adequate parametrization over a limited temperature range only, presumably because the large orbital and dipolar contributions to **A** vary with *T*. More generally, a spin Hamiltonian is valid only if higher orbital states are far removed in energy from the ground state, a condition obviously not satisfied for the Fe(II) of the RC. Similar failures of a spin Hamiltonian were pointed out for high-spin ferrous hemes by Kent *et al.* (1979).

5.3. Protocatechuate Dioxygenase

The biodegradation pathways of many aromatics converge on one-ring aromatic compounds such as protocatechuate (3,4-dihydroxybenzoate, PCA), and the subsequent ring opening catalyzed by a variety of bacterial protocatechuate dioxygenases (PCAD) is therefore a crucial step in carbon metabolism (Lipscomb and Orville, 1992). Two types of dioxygenases have evolved in aerobic bacteria, and both contain a mononuclear iron center at each active site. This iron is high-spin ferric in the enzymes that catalyze intradiol cleavage (3,4-PCAD) to produce the open chain product, β-carboxy-*cis, cis*-muconic acid, whereas it is high-spin ferrous in the enzymes that catalyze extradiol cleavage (2,3-PCAD and 4,5-PCAD). Both types of PCADs have been studied by MS (Kent *et al.*, 1987; Arciero *et al.*, 1983), but only the 3,4-PCAD from *Brevibacterium fuscum* (Whittaker *et al.*, 1984) will be discussed below. Judging from spectroscopic and kinetic data the enzyme of this organism is similar to that of other 3,4-PACDs, but it is exceptionally homogeneous in its resting state as is evident from the sharpness of its EPR and Mössbauer lines. X-ray diffraction data of a related 3,4-PCAD show that the Fe(III) ion is located at the interface of two subunits and has roughly trigonal bipyramidal coordination to two histidines, two tyrosines, and a solvent OH^- (Ohlendorf *et al.*, 1988). Native 3,4-PCAD has EPR signals at $g = 9.67$ and $g = 4.23$. These values are those expected for a spin sextet with $g = 2$ and a zero-field splitting of maximum rhombicity, $E/D = 1/3$. Equation (10′) then splits the sextet into a lowest doublet with $\mathbf{g}_1^{\text{eff}} = (0.86, 9.68, 0.60)$, a middle doublet with isotropic $\mathbf{g}_2^{\text{eff}} = 4.28$, and a highest doublet with $\mathbf{g}_3^{\text{eff}} = (0.86, 0.60, 9.68)$. The low-temperature, weak-field Mössbauer data of Fig. 9 are a textbook example of the spectra expected for $S = \frac{5}{2}$ with maximum rhombicity, $E/D = \frac{1}{3}$, and they can be explained qualitatively by the following arguments. The spectra are dominated by the magnetic hyperfine interaction $A\langle\hat{S}\rangle_i \cdot \hat{I}$, where $i = 1, 2, 3$ labels the doublets and $A/g_N\beta_N = -21$ T is a scalar, since high-spin iron is an orbital singlet and has no orbital and spin dipolar contributions to **A**. Compared to $A\langle\hat{S}\rangle_i \cdot \hat{I}$, the quadrupole interaction with $\Delta E_Q = 0.40$ mm/s is a small perturbation. As Fig. 9 illustrates, the spectra are superpositions of three subspectra arising from the three doublets, each

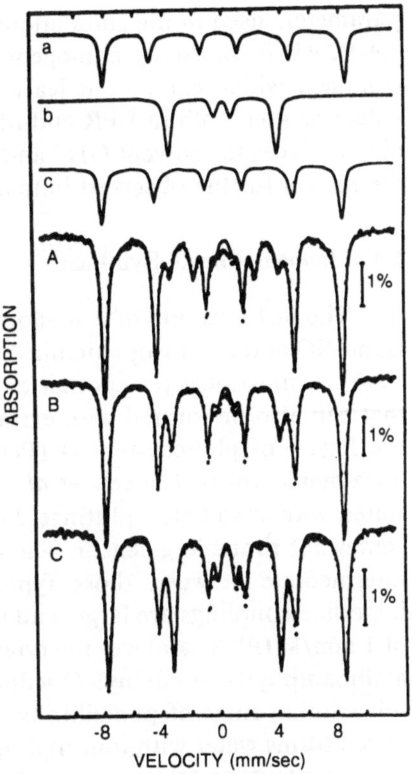

Figure 9. Mössbauer spectra of protocatechuate 3,4-dioxygenase at 4.2 K (traces A, B) and 13 K (trace C) in a field of 60 mT perpendicular (trace A) and parallel (traces B, C) to the γ beam. The solid lines a, b, c are simulated spectra of the lowest, middle, and upper Kramers doublet, respectively, in parallel field. The solid lines in the traces are appropriate superpositions based on the parameters $D/k = 2.0$ K, $E/D = 1/3$, $A/(g_N\beta_N) = -21$ T, $\Delta E_Q = 0.40$ mm/s, $\eta = -0.6$, $\delta = 0.44$ mm/s. Reproduced from Whittaker *et al.* (1984).

having an intensity given by its Boltzmann factor, $\exp(-E_i/kT)$. The three traces labeled a, b, and c at the top of Fig. 9 represent the three subspectra. The ground doublet has an easy axis of magnetization along y in a molecule-based coordinate system defined by Eq. (10′); hence the internal field is oriented along $\pm\hat{\mathbf{y}}$ for most molecules. Since these molecular axes are randomly oriented with respect to the γ beam, the intensity ratios of the Mössbauer lines are expected to be in the ratio of 3:2:1:1:2:3 as observed. The same arguments apply to the spectrum arising from the highest doublet, except that $\hat{\mathbf{z}}$ is the easy axis of magnetization rather than $\hat{\mathbf{y}}$. The quadrupole interaction, accordingly, is sampled along $\hat{\mathbf{y}}$ in the lowest doublet and along $\hat{\mathbf{z}}$ in the highest doublet; obviously, P_{yy} and P_{zz} have different signs and magnitudes, as is evident from the different splittings between lines (1, 2) and (5, 6), respectively. The middle doublet, in contrast, is magnetically isotropic, and the internal field is parallel or antiparallel to the external field. The intensity ratios are therefore 3:0:1:1:0:3 for an external field parallel to the γ beam and 3:4:1:1:4:3 for a perpendicular field. The maximum overall splittings are proportional to the maximum g values of the three doublets. The qualitative arguments given here are borne out by computer simulations of the spectra, shown as solid lines in Fig. 9. (The parameters used in the simulations are given with the figure.) While native 3,4-PCAD is unusually homogeneous, the opposite is true for its substrate complex, which exhibits at least five distinct species with different E/D values as seen both in EPR and Mössbauer spectra. The substrate presumably replaces the solvent OH^- and binds as a monodentate anion RO^-, but the reason for the observed inhomogeneity is not known.

5.4. Isopenicillin N Synthase

The last example of a mononuclear iron protein to be discussed here is the NO adduct of isopenicillin N synthase (IPNS) (Baldwin and Bradley, 1990). It illustrates yet another characteristic complex of high-spin Fe(II) that can also be formed with extradiol PCADs (Arciero *et al.*, 1983), with the Fe(II) of photosystem II (Petrouleas & Diner, 1990), and even with deoxyhemerythrin (Nocek *et al.*, 1988). These complexes are all quartet states with zero-field splittings $D \geq 10\ \text{cm}^{-1}$ and a ground doublet with somewhat rhombic g tensor. The isomer shifts are roughly 0.7 mm/s, i.e., intermediate between those typical of high- and low-spin Fe(II), the hyperfine couplings are large, and the quadrupole splittings are of the order of 1 mm/s. IPNS catalyzes the double ring closure of the tripeptide δ-(L-α-aminoadipoyl)-L-cysteinyl-D-valine (ACV) to form the β-lactam and thiazolidine rings of penicillin-type antibiotics. The reaction requires O_2, which forms water with four hydrogen atoms abstracted from the substrate. IPNS contains an essential iron atom and can be reconstituted with ^{57}Fe(II) from the apoprotein.

Chen *et al.* (1989) reported an isomer shift of $\delta_{Fe} = 1.30$ mm/s and quadrupole splitting of $\Delta E_Q = 2.70$ mm/s for the active enzyme, which changed to $\delta_{Fe} = 1.10$ mm/s and $\Delta E_Q = 3.40$ mm/s upon binding of ACV. Obviously, the coordination of the high-spin Fe(II) ion changed, and the authors suggested that the decrease in δ_{Fe} of 0.20 mm/s might be due to the substitution of a more covalent cysteine ligand in the substrate complex for one of the N/O ligands in the native enzyme. A similar decrease in the isomer shift is observed in heme proteins when an axial histidine, as found in deoxymyoglobin, $\delta_{Fe} = 0.96$ mm/s (Kent *et al.*, 1979), is replaced by cysteine as in ferrous cytochrome P-450$_{cam}$, $\delta = 0.84$ mm/s (Sharrock *et al.*, 1976). The lower isomer shifts of the iron–sulfur proteins compared to the corresponding spin states with N/O coordination have already been mentioned in Section 3. A question that now arises for IPNS is whether the cysteine of the substrate, ACV, coordinates to the iron or whether substrate binding moves one of the two intrinsic cysteines into the coordination sphere of the Fe(II) ion. In a recent paper Orville *et al.* (1992) prove that the first alternative is correct. Some of the Mössbauer data that lead to this conclusion are discussed below. Native IPNS happens to have a relatively low affinity for its substrate so that only about 65% complexation is practical. The NO adduct of IPNS, on the other hand, has higher substrate affinity, and it is therefore possible to prepare homogeneous samples of the ternary complex.

Figure 10 shows the low-temperature Mössbauer spectra of the NO adduct of IPNS with and without substrate (Orville *et al.*, 1992). All the data are taken in a field parallel to the γ beam; for traces A–C this field is 60 mT and serves to define the electronic spin state but does not produce any noticeable nuclear Zeeman splitting, while for trace D the field is 6 T and thus represents a noticeable fraction of the effective field at the iron nucleus. The NO adduct of IPNS has an EPR spectrum with g values of 4.09, 3.95, 2.0 characteristic of the $m_s = \pm\frac{1}{2}$ ground doublet of a quartet state, $S = \frac{3}{2}$, with small rhombicity, $E/D = 0.015$. The Mössbauer spectra of Fig. 10 are consistent with this interpretation. They basically show the six-line pattern expected for an internal field of ~25 T. Numbering the lines from left to right as 1 to 6, lines 2 and 5 are observed to be weak since the internal field is mainly in the xy-plane of easy magnetization and close to the direction of the γ beam. The hyperfine coupling is large; with $\mathbf{A}/g_N\beta_N = -(25, 24, 33)$ T it exceeds the Fermi contact term of -22 T and is anisotropic.

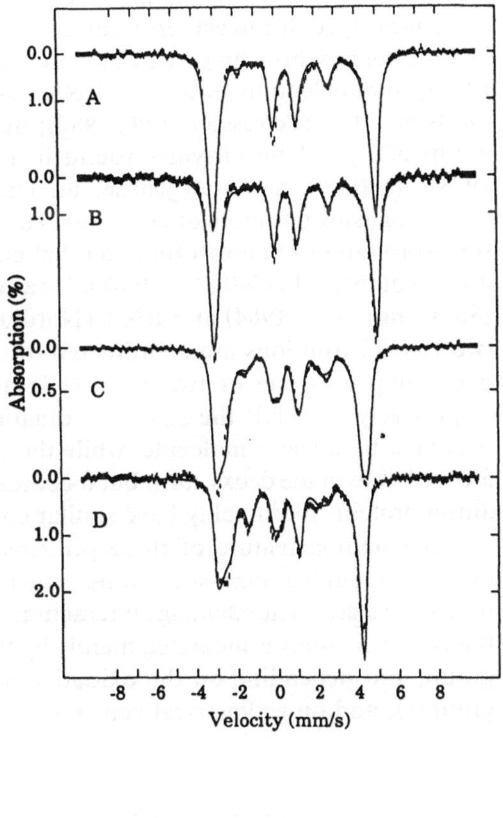

Figure 10. Mössbauer spectra of the NO complexes of isopenicillin N synthase with and without substrate at 4.2 K in fields parallel to the γ beam of 60 mT (traces A, B, C) and 6 T (trace D). Trace A: wild-type; trace B: double mutant Cys 106,255 → Ser; traces C, D: sample in trace B plus substrate. The solid lines are simulations based on a $S = \frac{3}{2}$ spin Hamiltonian with the parameters $D/k = 20$ K, $E/D = 0.015$, $\Delta E_Q = (-)1.0$ mm/s, $\eta = 0.1$, $\mathbf{A}/(g_N\beta_N) = -(25, 24, 33)$ T, $\delta_{Fe} = 0.75$ mm/s, $R(\alpha, \beta, \gamma) = (0, 8°, 0)$ for traces A and B and $D/k = 20$ K, $E/D = 0.035$, $\Delta E_Q = (-)1.2$ mm/s, $\eta = 1.0$, $\mathbf{A}/(g_N\beta_N) = -(22, 20, 25)$ T, $\delta_{Fe} = 0.65$ mm/s, $R(\alpha, \beta, \gamma) = (10°, 20°, 0)$ for traces C and D. Here, $R(\alpha, \beta, \gamma)$ is a set of Euler angles that rotates $\mathbf{A}$ and the quadrupole tensor relative to the zero-field splitting. Reproduced from Orville *et al.* (1992).

As pointed out earlier, the large A value and the unusual isomer shift are characteristic of these peculiar Fe(II)–NO quartet states.

The main conclusion of Orville *et al.* (1992) that the cysteine of the substrate ACV coordinates to the iron is well documented by the data in Fig. 10. The spectrum of wild type IPNS · NO in trace A is indistinguishable from that of a mutant IPNS · NO in trace B, which had both cysteines of the enzyme replaced by serine; hence the iron environment appears to be the same in both proteins. When ACV is added to the mutant lacking the cysteines in trace C, the isomer shift and the internal field both decrease as a result of the replacement of a Fe–O/N bond by a more covalent Fe–S bond. The putative sulfur ligand must therefore come from the cysteine of the tripeptide ACV. The high-field spectrum of the ternary NO/ACV complex of the mutant IPNS in trace D of Fig. 10, finally, helps to determine the parameters given in the caption.

6. OXYGEN-BRIDGED DINUCLEAR CLUSTERS

6.1. Overview

There has been growing interest over the last ten years in proteins with oxygen-bridged diiron centers (Que and True, 1990). Members of this group include hemerythrin, an O_2 carrier in marine invertebrates (Klotz and Kurtz, 1984); ribonucleotide reductase (RNR), an enzyme essential in the biosynthesis of DNA (Petersson *et al.*, 1980); purple acid phosphatases (PAP), a group of hydrolytic enzymes found in mammals (Antanaitis and Aisen, 1983); methane monooxygenase, an enzyme that converts methane to methanol using one atom of an O_2 molecule (Fox *et al.*, 1988); and rubrerythrin, a protein of unknown function that contains an FeS_4 as well as an oxo diiron complex (LeGall *et al.*, 1988). X-ray diffraction studies of hemerythrin (Stenkamp *et al.*, 1984) and RNR (Nordlund *et al.*, 1990) revealed that the two (ferric) iron ions are coordinated to six or seven amino acids with an oxo group and one or two carboxylic acids acting as bridging ligands, respectively. In RNR the sixth coordination position of each iron atom is occupied by a water molecule, while the O_2-binding site in hemerythrin is 5-coordinate in the deoxy form but 6-coordinate in the oxy adduct. The other diiron proteins presumably have similar coordination.

A common feature of these proteins is an active site with a pair of exchange-coupled high-spin irons either in the III/III, III/II or II/II oxidation states. The exchange interaction, $\mathcal{H}_{xch} = JS_1 \cdot \hat{S}_2$, Eq. (11), between the two iron atoms is mediated mainly by the bridging μ(hydr)oxo or μ-OR group, and depending on the oxidation state of the iron, on the bridging group(s), and on geometrical constraints, the exchange coupling may vary from large positive to negative values. As a result, the diiron proteins have unusual magnetic properties. As a rule, diferric complexes have diamagnetic ground states, but some diferrous complexes are paramagnetic. Of special interest in the present context are the mixed-valence Kramers systems with ground states of effective spin $S = \frac{1}{2}$ and g values that differ greatly from $g = 2$.

While the metal complexes of the oxo diiron proteins thus have similar structures, their functions are quite diverse, and of particular interest is the oxygen chemistry of hemerythrin, RNR, and methane monooxygenase. Control of the reactivity apparently lies in the amino acid environment of the metal complex, and it is a challenging problem to rationalize the relation between the structures and functions of these proteins. Since the number of potential cluster ligands is large, the chemistry of these systems is likely to be very rich.

Moreover, it was found the *bis* (μ-carboxylato) (μ-oxo) dinuclear complexes assemble spontaneously in aqueous solution in the presence of additional ligands, and various related compounds have been prepared and analyzed (Que and True, 1990). This work has opened up a new field of biomimetic chemistry that attempts to model the structure and function of dinuclear proteins. A fruitful interplay has emerged between studies of proteins and model complexes, and rapid further progress can be expected. The following discussion illustrates some contributions of Mössbauer spectroscopy to our emerging knowledge of oxo diiron proteins.

6.2. Purple Acid Phosphatase

Several purple acid phosphatases (PAP) with a diiron center are found in mammals, and similar enzymes with Fe/Zn or Mn complexes are found in plants (Antanaitis and Aisen, 1983; Que and True, 1990). In the mixed-valence (III/II) oxidation state these proteins show high phosphatase activity at low pH, while the oxidized (III/III) form is inactive, and no stable reduced (II/II) form exists. The enzyme from pig allantoic fluid, also known as uteroferrin, has been studied in most detail and will be discussed below. So far only a few of the iron ligands are known, and the bridging groups remain hypothetical. The known ligands include a tyrosine coordinated to Fe(III), recognizable by the ligand-to-metal charge-transfer band responsible for the purple color, and a histidine coordinated to each iron atom. Active enzyme can be reconstituted from apoprotein and Fe(II), and some activity is found in Zn/Fe mixed-metal clusters (Keough *et al.*, 1980; Beck *et al.*, 1984). PAP also forms complexes with phosphate and similar oxoanions.

Mössbauer spectra of oxidized PAP show the presence of two distinct high-spin ferric sites, which couple to a diamagnetic ground state

(Debrunner *et al.*, 1983). The quadrupole splittings are large, $\Delta E_Q =$ 1.69 mm/s and 2.17 mm/s, respectively. At temperatures above 80 K the quadrupole doublets show an unexpected broadening even in a field of less than 5 μT (Sage *et al.*, 1989). The broadening may arise from quasi-degenerate levels in the excited triplet at $J \approx 80\ \text{cm}^{-1}$ and quintet at $3J$, as alluded to in Section 2.5, where J is taken from the work of Lauffer *et al.* (1983).

Reduced PAP exhibits a low-temperature EPR signal with g values of 1.56, 1.72, and 1.93. At temperatures above 30 K the Mössbauer spectra consist of two quadrupole doublets with splittings and isomer shifts characteristic of high-spin ferric and high-spin ferrous iron. These features disappear completely when the temperature is lowered to 4.2 K. As shown in Fig. 11, the Mössbauer spectra then are broad and continuous, extending from −6 mm/s to +7 mm/s. Comparison of traces a and b in Fig. 11 indicates that the spectral shape depends on the direction of a weak applied field. Lowering of the temperature to 1.8 K does not change the spectra. These observations show that both iron atoms are coupled magnetically to the EPR-detectable effective spin of $S = \frac{1}{2}$, which must represent the ground doublet. Moreover, for the data shown the internal fields are stationary. The data can therefore be parametrized by an effective spin $S = \frac{1}{2}$ Hamiltonian

$$\mathcal{H} = \beta \hat{S}\cdot \mathbf{g}_S \cdot \mathbf{B} + \sum_{i=1,2} (\hat{S}\cdot \mathbf{A}_{S,i}\cdot \mathbf{I}_i + \hat{I}_i\cdot \mathbf{P}_i \cdot \hat{I}_i - \beta_N g_N \hat{I}_i \cdot \mathbf{B})$$

where the subscripts $i = 1, 2$ refer to the Fe(III) and Fe(II) site, respectively.

Simulations based on this $S = \frac{1}{2}$ model are shown by the solid lines in Fig. 11. They represent the data reasonably well, and the parameter set, which is given in the caption, deserves further discussion. It turns out that the outer wings of the spectra in Fig. 11 are entirely due to the ferric site, as the contribution of the ferrous site is confined to the range $-4\ \text{mm/s} < v <$ 4.5 mm/s. By focusing on the outer wings first, it is therefore possible to adjust the parameters $\mathbf{A}_{S,1}$ and $\mathbf{P}_1$ independently of $\mathbf{A}_{S,2}$ and $\mathbf{P}_2$. The only way to match these broad features is to choose a highly anisotropic hyperfine tensor $\mathbf{A}_{S,1}$ in spite of the expected isotropy of the intrinsic A value, $A_1/g_N\beta_N \approx -22$ T. If $\mathbf{A}_{S,1}$ is anisotropic, then the orientation of the quadrupole tensor $\mathbf{P}_1$ with respect to $\mathbf{A}_{S,1}$ becomes important, and the number of adjustable parameters increases further. Once the components of $\mathbf{A}_{S,1}$ and $\mathbf{P}_1$ are optimized, the parameters of the ferrous site are adjusted. For both sites the principal axes of the quadrupole tensors differ significantly from those of the magnetic hyperfine tensors.

Next, the question arises how to relate the empirical parameters of the effective spin $S = \frac{1}{2}$ Hamiltonian to the intrinsic parameters of the ferric

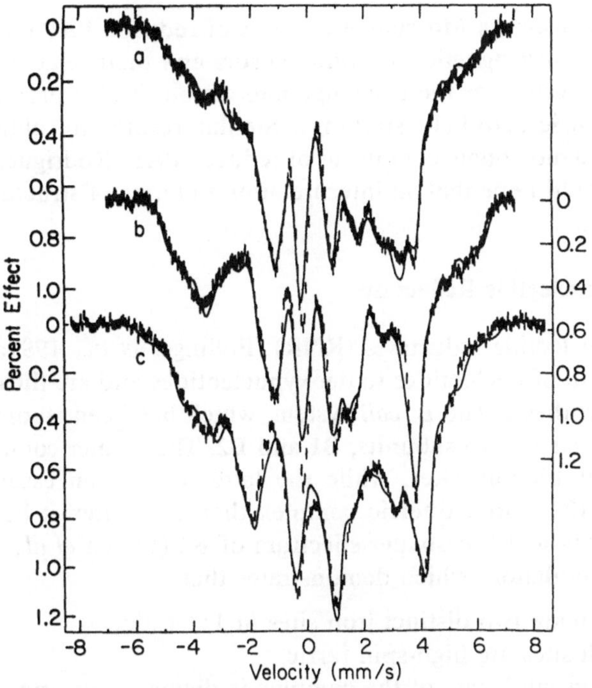

Figure 11. Mössbauer spectra of reduced purple acid phosphatase at 4.2 K in magnetic fields of 32 mT (a) perpendicular and (b) parallel to the gamma beam and (c) in a 3.7-T field parallel to the γ beam. The solid lines are simulations based on an effective spin $S = \frac{1}{2}$ Hamiltonian with the parameters $\mathbf{g} = (1.56, 1.72, 1.93)$ and $\mathbf{A}/(g_N\beta_N) = -(68, 45, 25)$ T, $\Delta E_Q = (-)1.85$ mm/s, $\eta = 0$, $\delta_{Fe} = 0.54$ mm/s, $R(\alpha, \beta, \gamma) = (60°, 60°, 0)$ for the ferric site, and $\mathbf{A}/(g_N\beta_N) = (26, 18, 10)$ T, $\Delta E_Q = (+)2.68$ mm/s, $\eta = 0.3$, $\delta_{Fe} = 1.24$ mm/s, $R(\alpha, \beta, \gamma) = (90°, 60°, 90°)$ for the ferrous site, where $R(\alpha, \beta, \gamma)$ is a set of Euler angles that rotate the principal axes of $\mathbf{A}$ to those of the quadrupole tensor. Reproduced from Sage *et al.* (1989).

and ferrous site. To rephrase the question, is it possible to find a reasonable parameter set for the $S_1 = \frac{5}{2}$ and $S_2 = 2$ Hamiltonian, $\mathcal{H}_1$ and $\mathcal{H}_2$ in Eq. (12), which will reproduce the spectra in the coupled $\frac{5}{2} \otimes 2$ representation when the known exchange interaction is included, $\mathcal{H}_{xch} = (20\ \text{cm}^{-1})\hat{S}_1 \cdot \hat{S}_2$ (Lauffer *et al.*, 1983; Day *et al.*, 1988). Of particular interest is the question of whether a ferric site with spin-only paramagnetism, i.e., with $g_1 = 2$ and isotropic hyperfine interaction $A_1/g_N\beta_N \approx -22$ T, can lead to an acceptable simulation. Sage *et al.* (1989) indeed found such a solution and demonstrated that the anisotropy of the empirical Fe(III) hyperfine tensor in the effective spin $S = \frac{1}{2}$ representation is the result of intermediate coupling, $|D/J| \sim 1$. Moreover, they derived a correction for Eqs. (13c–e) to first order in D/J. The g tensor of the ground doublet, $\mathbf{g} = (1.56, 1.72, 1.93)$, is explained by similar arguments.

In conclusion, a Mössbauer analysis of reduced PAP not only reveals the electric and magnetic hyperfine tensors and their relative orientations, but, given a value for the exchange interaction, it also provides estimates of the intrinsic zero-field splittings. Similar results are obtained for the arsenate and molybdate complexes of reduced PAP (Rodriguez *et al.*, 1992), and one would hope that an interpretation in terms of structure will eventually emerge.

6.3. Ribonucleotide Reductase

Ribonucleotide reductases (RNR) (Bollinger *et al.*, 1991a, b) catalyze the reduction of nucleotides to deoxynucleotides and are thus essential for DNA biosynthesis. The *E. coli* system, which has been studied in greatest detail, consists of two subunits, B1 and B2. The former contains substrate and effector binding sites, while the latter carries an essential tyrosine radical, ·Y122, and a diferric complex that is of interest here. Figure 12 shows a high-field Mössbauer spectrum of B2 (Lynch *et al.*, 1989) and an excellent simulation, which demonstrates that

1. there are two distinct iron sites in 1:1 ratio,
2. both sites are high-spin ferric,
3. the ground state of the complex is diamagnetic, and
4. the quadrupole interactions are unusually large for a nominal $^6A_1$ state of the iron and the asymmetry parameters are quite sizable.

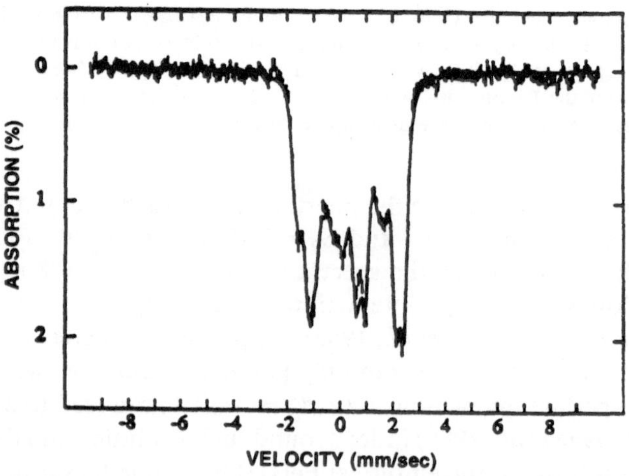

Figure 12. Mössbauer spectrum of native ribonucleotide reductase subunit B2 at 4.2 K in a 6.0-T field parallel to the γ beam. The solid line is a simulation of the diamagnetic, $S = 0$, ground state of the diferric cluster with $\Delta E_Q(1) = (-)1.62$ mm/s, $\eta(1) = 0.6$, $\delta_{Fe}(1) = 0.55$ mm/s; $\Delta E_Q(2) = -2.44$ mm/s, $\eta(2) = 0.2$, $\delta_{Fe}(2) = 0.46$ mm/s, $\Gamma = 0.27$ mm/s. Reproduced from Lynch *et al.* (1989.)

The large and asymmetric field gradient is in line with the known structure of the B2 subunit (Nordlund *et al.*, 1990), which shows that each iron atom has a highly distorted octahedral environment. The μ-oxo bridge is responsible for the strong exchange interaction, $J = 216\ cm^{-1}$ (Petersson *et al.*, 1980).

Active B2 can be reconstituted from apoprotein by incubation with Fe^{2+} in the presence of oxygen. The reaction mechanism that leads to the μ-oxo bridged diferric center and the ·Y122 radical is of great interest as it contrasts with the reactivity of other oxo-diiron proteins with O_2, viz., the reversible binding in hemerythrin and the hydroxylation of methane in methane monooxygenase. Recent optical, EPR, and Mössbauer measurements using stopped-flow and freeze-quench techniques revealed two novel intermediates in the reactivation of *apo* RNR B2 by Fe^{2+} and O_2 (Bollinger *et al.*, 1991a, b). Figure 13 shows low-field Mössbauer spectra of one of

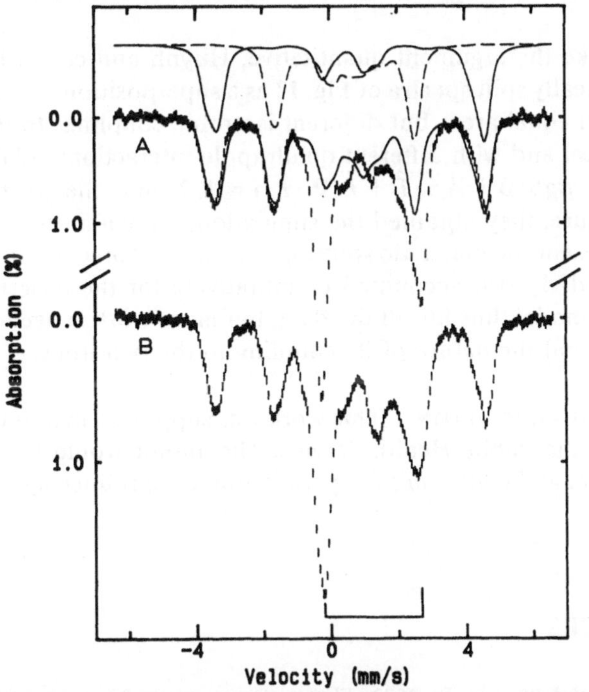

Figure 13. Mössbauer spectra of ribonucleotide reductase subunit B2 during reconstitution at 4.2 K in a field of 50 mT parallel to the γ beam. Intermediate X was stabilized by rapid quenching. Spectrum B: wild type; spectrum A: Tyr 122 Phe mutant. The solid line in spectrum A is a simulation assuming $S = \frac{1}{2}$ for sites (1) and (2) with the parameters $\Delta E_Q = (-)1.0$ mm/s, $\eta = 0.5$, $\delta_{Fe} = 0.55$ mm/s, $|A/g_N\beta_N| = 52$ T for site (1) and $\Delta E_Q = (-)1.1$ mm/s, $\eta = 0.7$, $\delta_{Fe} = 0.45$ mm/s, $|A/g_N\beta_N| = 24.5$ T for site (2). The two traces above A show the contributions of site (1) and (2). Reproduced from Bollinger *et al.* (1991a).

these intermediates, called X. Apo B2 was incubated with $^{57}Fe^{2+}$ in the presence of O_2 for 0.31 seconds and then quenched. Trace B is from wild type protein, while trace A is from a mutant, Y122F, that lacks tyrosine 122 and therefore cannot form the ·Y122 radical. The spectra are complex and differ somewhat between −1 mm/s and +3 mm/s, where the two quadrupole doublets of native diferric B2 are expected. The doublet of high-spin Fe^{2+}, $\Delta E_Q = 3.2$ mm/s, $\delta_{Fe} = 1.3$ mm/s, indicated by a bracket, falls into the same interval. Both spectra show prominent lines at −3.9, −1.8, 2.7, and 5 mm/s, however, which must arise from magnetic hyperfine splitting of high-spin ferric iron in an unusual Kramers system. EPR and optical data leave no doubt that the magnetic splitting is associated with the intermediate X. Freeze-quenched samples of the Y122F mutant reconstituted with ^{56}Fe display an isotropic EPR signal at $g = 2$ with a peak-to-trough width of 2.2 mT (not shown). Substitution of ^{57}Fe for the ^{56}Fe, on the other hand, produced hyperfine structure of the $g = 2$ signal with a peak-to-trough width of 4.4 mT and thus suggests that X is a radical species that couples to the iron atom.

To make the argument quantitative, Huynh and co-workers modeled the magnetically split spectra of Fig. 13 as a superposition of two component spectra with equal areas but different isotropic couplings to the spin $S = \frac{1}{2}$ of the radical and with different quadrupole interactions. With the Hamiltonian $\mathcal{H} = \beta g\hat{S}\cdot\mathbf{B} + A_i\hat{S}\cdot\hat{I}_i + \hat{I}_i\cdot\mathbf{P}_i\cdot\hat{I}$, $i = 1, 2$, and the parameters given with the figure, they obtained the simulations shown at the top of Fig. 13, which represent the magnetic splitting very well. Moreover, the same couplings A_1 and A_2 also accounted quantitatively for the observed EPR line shape. The model thus fits all the data, but more work is needed to identify the radical and the nature of its coupling to the two irons.

ACKNOWLEDGMENTS. This work was supported in part by Grant GM 16406 from the Public Health Service. The author would like to thank his collaborators and colleagues for permission to reproduce figures from their work.

REFERENCES

Abragam, A., and Bleaney, B., 1970, *Electron Paramagnetic Resonance of Transition Ions*, Oxford, Clarendon Press.

Allen, J. P., Feher, G., Yeates, T. O., Komiya, H., and Rees, D. C., 1987, *Proc. Natl. Acad. Sci. USA* **84**:5730.

Antanaitis, B. C., and Aisen, P., 1983, *Adv. Inorg. Biochem.* **5**:111.

Arciero, D. M., Lipscomb, J. D., Huynh, B. H., Kent, T. A., and Münck, E., 1983, *J. Biol. Chem.* **258**:14,981.

Atkins, W. M., and Sligar, S. G., 1988, *J. Biol. Chem.* **263**:18,842.

Averill, B. A., 1988, in "Metal clusters in proteins," ACS Symposium Series, vol. 372 (L. Que, Jr., ed.), ACS Washington, D.C., p. 258.
Baldwin, J. E., and Bradley, M., 1990, *Chem. Rev.* **90**:1079.
Beck, J. L., Keough, D. T., de Jersey, J., and Zerner, B., 1984, *Biochim. Biophys. Acta* **791**:357.
Blumberg, W. E., and Peisach, J., 1971, in *Probes of Structure and Function of Macromolecules and Membranes*, vol. 2 (B. Chance, T. Yonetani and A. S. Mildvan, eds.), Academic Press, New York, p. 215.
Bollinger, J. M., Stubbe, J., Huynh, B. H., and Edmondson, D. E., 1991a, *J. Am. Chem. Soc.* **113**:6259.
Bollinger, J. M., Jr., Edmondson, D. E., Huynh, B. H., Filley, J., Norton, J. R., and Stubbe, J., 1991b, *Science* **253**:292.
Boso, B., 1982, Ph.D. dissertation, U. of Illinois, unpublished.
Boso, B., Debrunner, P., Okamura, M. Y., and Feher, G., 1981, *Biochim. Biophys. Acta* **638**:173.
Boso, B., Lang, G., McMurray, T. J., and Grove, J. T., 1983, *J. Chem. Phys.* **79**:1122.
Butler, W. F., Johnston, D. C., Shore, H. B., Fredkin, D. R., Okamura, M. Y., and Feher, G., 1980, *Biophys. J.* **32**:967.
Carlioz, A., Ludwig, M. L., Stallings, W. C., Fee, J. A., Steinman, H. M., and Touati, D., 1988, *J. Biol. Chem.* **263**:1155.
Champion, P. M., Lipscomb, J. D., Münck, E., Debrunner, P., and Gunsalus, I. C., 1975a, *Biochemistry* **14**:4151.
Champion, P. M., Chiang, R., Münck, E., Debrunner, P. G., and Hager, L. P., 1975b, *Biochemistry* **14**:4159.
Chen, V. J., Orville, A. M., Harpel, M. R., Frolik, C. A., Surerus, K. K., Münck, E., and Lipscomb, J. D., 1989, *J. Biol. Chem.* **264**:21,677.
Clauser, M. J., and Blume, M., 1971, *Phys. Rev.* **B3**:583.
Colvin, J. T., Rutter, R., Stapleton, H. J., and Hager, L. P., 1983, *Biophys. J.* **41**:105.
Day, E. P., David, S. S., Peterson, J., Dunham, W. R., Bonvoisin, J. J., Sands, R. H., and Que, L., Jr., 1988, *J. Biol. Chem.* **263**:15,561.
Debrunner, P. G., Schulz, C. E., Feher, G., and Okamura, M. Y., 1975, *Biophys. J.* **15**:226a.
Debrunner, P. G., Hendrich, M. P., de Jersey, J., Keough, D. T., Sage, J. T., and Zerner, B., 1983, *Biochim. Biophys. Acta* **745**:103.
Debus, R. J., Feher, G., and Okamura, M. Y., 1986, *Biochemistry* **25**:2276.
Deisenhofer, J., Epp. O., Miki, K., Huber, R., and Michel, H., 1985, *Nature* **318**:618.
Dolphin, D., Forman, A., Borg, D. C., Fajer, J., and Felton, R. H., 1971, *Proc. Natl. Acad. Sci. USA* **68**:614.
Emptage, M. H., 1988, in *Metal Clusters in Proteins*, ACS Symposium Series, vol. 372 (L. Que, Jr., ed.), ACS Washington, D.C., p. 343.
Emptage, M. H., Kent, T. A., Kennedy, M. C., Beinert, H., and Münck, E., 1983, *Proc. Natl. Acad. Sci. USA* **80**:4674.
Fox, B. G., Surerus, K. K., Münck, E., and Lipscomb, J. D., 1988, *J. Biol. Chem.* **263**:10,553.
Fritz, J., Anderson, R., Fee, J., Petering, D., Palmer, G., Sands, R. H., Tsibris, J. C. M., Gunsalus, I. C., Orme-Johnson, W. H., and Beinert, H., 1971, *Biochim. Biophys. Acta* **253**:110.
Gibson, J. F., Hall, D. O., Thornley, J. H. M., and Whatley, F. R., 1966, *Proc. Natl. Acad. Sci. USA* **56**:987.
Greenwood, N. N., and Gibb, T. C., 1971, *Mössbauer Spectroscopy*, Chapman and Hall, London.
Griffith, J. S., 1957, *Nature* **180**:30.
Gunsalus, I. C., Meeks, R., Lipscomb, J. D., Debrunner, P. G., and Münck, E., 1974, in *Molecular Mechanisms of Oxygen Activation* (O. Hayaishi, ed.), Academic Press, New York, pp. 559–613.
Kent, T. A., Spartalian, K., Lang, G., and Yonetani, T., 1977, *Biochim. Biophys. Acta* **490**:331.
Kent, T. A., Spartalian, K., and Lang, G., 1979, *J. Chem. Phys.* **71**:4899.

Kent, T. A., Huynh, B. A., and Münck, E., 1980, *Proc. Natl. Acad. Sci. USA* **77**:6574.
Kent, T. A., Münck, E., Pyrz, J. W., Widom, J., and Que, L., Jr., 1987, *Inorg. Chem.* **26**, 1402.
Keough, D. T., Dionysius, D. A., de Jersey, J., and Zerner, B., 1980, *Biochem. Biophys. Res. Commun.* **94**:600.
Kim, J., and Rees, D. C., 1992, *Science* **257**:1677.
Klotz, J. M., and Kurtz, D. M., Jr., 1984, *Acct. Chem. Res.* **17**:16.
Lang, G., 1970, *Quart. Rev. Biophys.* **3**:1.
Lang, G., Asakura, T., and Yonetani, T., 1969, *J. Phys.* **C2**:2246.
Lauffer, R. B., Antanaitis, B. C., Aisen, P., and Que, L., Jr., 1983, *J. Biol. Chem.* **258**:14,212.
LeGall, J., Prickril, B. C., Moura, I., Xavier, A. V., Moura, J. J. G., and Huynh, B.-H., 1988, *Biochemistry* **27**:1636.
Lipscomb, J. D., and Orville, A. M., 1992, in *Metal Ions in Biological Systems*, vol 28 (H. & A. Sigel, eds.) Marcel Dekker, New York, p. 243.
Lynch, J. B., Juarez-Garcia, C., Münck, E., and Que, L. Jr., 1989, *J. Biol. Chem.* **264**:8091.
Maltempo, M. M., and Moss, T. H., 1976. *Quart. Rev. Biophys.* **9**:181.
Münck, E., Debrunner, P. G., Tsibris, J. C. M., and Gunsalus, I. C., 1972, *Biochemistry* **11**:855.
Münck, E., Papaefthymiou, V., Sureus, K. K., and Girerd, J.-J., 1988, in *Metal Clusters in Proteins*, ACS Symposium Series, vol. 372 (L. Que, ed), American Chemical Society, Washington, D.C., pp. 302–325.
Murray, R. I., Fisher, M. T., Debrunner, P. G., and Sligar, S. G., 1985, in *Metalloproteins, Part I* (P. M. Harrison, ed.), Macmillan, London, pp. 157–206.
Nocek, J. M., Kurtz, D. M., Jr., Sage, J. T., Xia, Y.-M., and Debrunner, P. G., 1988, *Biochemistry* **27**:1014.
Nordlund, P., Sjöberg, B.-M., and Eklund, H., 1990, *Nature* **345**:593.
Ohlendorf, D. H., Lipscomb, J. D., and Weber, P. C., 1988, *Nature* **336**:403.
Orville, A. M., Chen, V. J., Kriauciunas, A., Harpel, M. R., Fox, B. G., Münck, E., and Lipscomb, J. D., 1992, *Biochemistry* **31**:4602.
Petersson, L., Gräslund, A., Ehrenberg, A., Sjöberg, B.-M., and Reichard, P., 1980, *J. Biol. Chem.* **255**:6706.
Petrouleas, V., and Diner, B. A., 1986, *Biochim. Biophys. Acta* **849**:264.
Petrouleas, V., and Diner, B. A., 1990, *Biochim. Biophys. Acta* **1015**:131.
Poulos, T. L., and Howard, A. J., 1987, *Biochemistry* **26**:8165.
Que, L., Jr., and True, A. E., 1990, *Prog. Inorg. Chem. Bioinorg. Chem.* **38**:97.
Raag, R., Martinis, S. A., Sligar, S. G., and Poulos, T. L., 1991, *Biochemistry* **30**:11,420.
Roberts, J. E., Hoffman, B. M., Rutter, R., and Hager, L. P., 1981a, *J. Biol. Chem.* **256**:2118.
Roberts, J. E., Hoffman, B. M., Rutter, R., and Hager, L. P., 1981b, *J. Am. Chem. Soc.* **103**:7654.
Rodriguez, J. H., Xia, Y.-M., Debrunner, P. G., Holz, R. C., Que, L., Jr., Wang, P., and Day, E. P., 1992, *Bull. Am. Phys. Soc.* **37**:445.
Rutter, R., Hager, L. P., Dhonau, H., Hendrich, M., Valentino, M., and Debrunner, P. G., 1984, *Biochemistry* **23**:6809.
Sage, J. T., Xia, Y.-M., Debrunner, P. G., Keough, D. T., deJersey, J., and Zerner, B., 1989, *J. Am. Chem. Soc.* **111**:7239.
Scaringe, R. P., Hodgson, D. J., and Hatfield, W. E., 1978, *Mol. Phys.* **35**:701.
Schulz, C., and Debrunner, P. G., 1976, *J. Physique* **37(C6)**:153.
Schulz, C. E., Devaney, P. W., Winkler, H., Debrunner, P. G., Doan, D., Chiang, R., Rutter, R., and Hager, L. P., 1979, *FEBS Lett.* **103**:102.
Schulz, C. E., Rutter, R., Sage, J. T., Debrunner, P. G., and Hager, L. P., 1984, *Biochemistry* **23**:4743.
Schulz, C. E., Nyman, P., and Debrunner, P. G., 1987, *J. Chem. Phys.* **87**:5077.
Sharrock, M., Debrunner, P. G., Schulz, C. E., Lipscomb, J. D., Marshall, V., and Gunsalus, I. C., 1976, *Biochim. Biophys. Acta* **420**:8.
Smulevich, G., English, A. M., Mantini, A. R., and Marzocchi, M. P., 1991, *Biochemistry* **30**:772.

Spiro, T. G., Ed., 1982, *Metal Ions in Biology*: *Iron–Sulfur Proteins*, Wiley-Interscience, New York.

Stenkamp, R. E., Sieker, L. C., and Jensen, L. H., 1984, *J. Am. Chem. Soc.* **106**:618.

Stout, G. H., Turley, S., Sieker, L. C., and Jensen, L. H., 1988, *Proc. Natl. Acad. Sci. USA* **85**:1020.

Telser, J., Emptage, M. H., Merkle, H., Kennedy, M. C., Beinert, H., and Hoffman, B. M., 1986, *J. Biol. Chem.* **261**:4840.

Tsukihara, T., Fukuyama, K., Nakamura, M., Katsube, Y., Tanaka, N., Kakudo, M., Wada, K., Hase, T., and Matsubara, H., 1981, *J. Biochem.* **90**:1763.

Varret, F., 1976, *J. Phys. Chem. Solids* **37**:265.

Watenpaugh, K. D., Sieker, L. C., and Jensen, L. H., 1980, *J. Mol. Biol.* **138**:615.

Whittaker, J. W., and Solomon, E. I., 1988, *J. Am. Chem. Soc.* **110**:5329.

Whittaker, J. W., Lipscomb, J. D., Kent, T. A., and Münck, E., 1984, *J. Biol. Chem.* **259**:4466.

3

Multifrequency ESR of Copper
Biophysical Applications

Riccardo Basosi, William E. Antholine, and James S. Hyde

1. INTRODUCTION

1.1. Spin Hamiltonian Parameters of Copper

The treatise of Abragam and Bleaney (1970) covers most aspects of the theory of ESR of metal complexes and is the definitive reference. Practical knowledge of the ESR of copper complexes of biological relevance that has been most useful to the authors has been obtained from Goodman and Raynor (1970), Vänngård (1972), Boas *et al.* (1978), and Pilbrow (1988). In addition, the application of ESR spectroscopy to copper proteins has been outlined in several texts (i.e., Spiro, 1981; Low, 1960; Yen, 1969; Brill, 1977 and references therein). A comprehensive text by Pilbrow (1990) on ESR of metal complexes updates the work of Abragam and Bleaney and probably is the most useful for a graduate course. In this review, we limit our considerations to copper with the aim of demonstrating the value of using microwave frequency as a new experimental parameter. Most of the concepts outlined here can be extended to other transition-metal ions.

The spectral features of copper complexes without superhyperfine

---

Riccardo Basosi • Department of Chemistry, University of Siena, 53100 Siena, Italy. **William E. Antholine and James S. Hyde** • National Biomedical ESR Center, Biophysics Research Institute, Medical College of Wisconsin, Milwaukee, Wisconsin 53226.

Biological Magnetic Resonance, Volume 13: EMR of Paramagnetic Molecules, edited by Lawrence J. Berliner and Jacques Reuben. Plenum Press, New York, 1993.

structure are evident in Fig. 1. Consider first the liquid-phase spectrum of Fig. 1B. The relaxation times of copper complexes are long enough that ESR spectra are readily obtained at room temperature. Low molecular weight complexes (i.e., molecular weight less than 2000) tumble fast enough to average the magnetic parameters associated with the principle magnetic axes to give isotropic values. The unpaired electron in a d^9 square-planar configuration occupies the d_{x2-y2} orbital. The ESR signal is split into four lines because of the interaction of the unpaired electron with the copper $\frac{3}{2}$ nuclear spin.* To first order at 9 GHz, the four lines are equally spaced

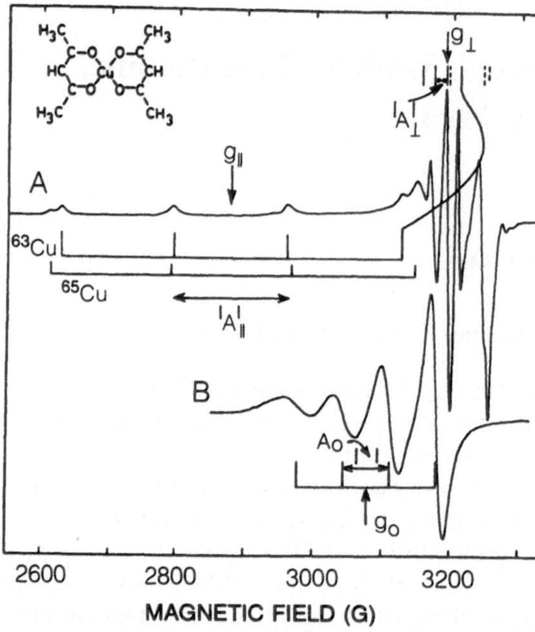

Figure 1. EPR spectra of a solution of a Cu^{2+} complex, (A) frozen at 77 K and (B) at room temperature. The g factors and hyperfine constants are measured as described in the text, and their values are $g_\parallel = 2.285$, $g_\perp = 2.060$, $g_0 = 2.135$, $A_\parallel = 167$ G, $A_\perp = 17$ G, and $A_0 = 67$ G. In (A) the dashed bars indicate the position of the overshoot lines, which occur from the particular angular dependence of the copper hyperfine lines as shown for one of them (the full curved line to the right). The small peaks at the highest field in (A) arise from a second complex existing in low concentration [*cf.* Fig. 3 in Gersmann and Swalen (1962)]. The sample is 5 mM in Cu^{2+} and 40 mM in acetylacetonate ($CH_3COCH_2COCH_3$) in a 1:1 water:dioxane solvent at alkaline pH. The same tube was used for both spectra with five times higher spectrometer gain in (B). Other settings were microwave power, 2 mW; frequency, 9.2 GHz; field modulation, 7 G; time constant, 0.3 s; and sweep time, 4 min. Reproduced from Vänngård (1972).

* There are two isotopes of copper, ^{63}Cu with natural abundance 69% and ^{65}Cu with natural abundance 31%. Each isotope has a nuclear spin of $\frac{3}{2}$. The ratio of the nuclear moments of ^{65}Cu and ^{63}Cu is 1.07. This doubles the number of lines in the spectrum. The linewidth for complexes with nitrogen donor atoms are often sufficiently large that the patterns for ^{65}Cu and ^{63}Cu are not resolved.

and the center of the four-line pattern is the isotropic g value, g_{iso}. Inclusion of second-order terms shifts the line positions substantially at low frequencies. These shifts in line position as well as changes of linewidth with frequency are multifrequency ESR parameters that can be used to determine the values of the rigid-limit ESR parameters. If the donor atoms from the ligands are equivalent nitrogens, each of the four lines is split into $2nI + 1$ superhyperfine lines, where $I = 1$ for ^{14}N and $\frac{1}{2}$ for ^{15}N and n is the number of nitrogen atoms.

If the motion of the copper complex is slowed by, for example, freezing the sample or increasing the molecular weight, values for the ESR parameters along the principle magnetic axes can often be obtained from the resulting powder-pattern ESR spectrum, Fig. 1A. The fast-tumbling and slow-tumbling or rigid-limit parameters are related by $g_{iso} = +\frac{1}{3}(g_{\parallel} + 2g_{\perp})$ and $A_{iso} = \frac{1}{3}(A_{\parallel} + 2A_{\perp})$ for axial symmetry and by $g_{iso} = \frac{1}{3}(g_1 + g_2 + g_3)$ and $A_{iso} = \frac{1}{3}(A_1 + A_2 + A_3)$ for rhombic symmetry. Superhyperfine structure from ligating atoms is usually resolved in the perpendicular region and often in the parallel region at low microwave frequency (2–4 GHz). For four equivalent nitrogens, 36 lines (9 superhyperfine lines on each of four copper hyperfine lines) plus "forbidden transitions" arising from quadrupole couplings plus additional features that are called overshoot lines (or hyperfine anomaly lines or extra absorption lines) comprise the $g_{\perp}$ region (Fig. 1); typically, however, only nine to eleven lines are resolved. These resolved lines are formed by the superposition of multiple lines. Figure 2 shows the transitions for a nucleus of spin $\frac{3}{2}$ including hyperfine, nuclear Zeeman, and nuclear quadrupole interactions. It illustrates the complexity of the spectrum. Partial resolution complicates the determination of the number of nitrogen donor atoms, especially if the symmetry is rhombic, which doubles the number of lines in this region. It has been our experience that the number of nitrogen donor atoms can be determined most easily for square-planar copper complexes (so-called Type 2) when the superhyperfine structure is resolved on one of the lines in the parallel region. The nitrogen superhyperfine structure in the $g_{\parallel}$ region is better resolved at low frequencies, and this provides a major rationale for the use of low frequency for cupric complexes.

Comparison of the ESR parameters from a cupric complex with the values in a Peisach–Blumberg plot of $g_{\parallel}$ versus $A_{\parallel}$ is the first step in identifying the donor atoms (Peisach and Blumberg, 1974). Typically, $g_{\parallel}$ increases as oxygen replaces a nitrogen or nitrogen replaces a sulfur donor atom. $A_{\parallel}$ usually increases as nitrogen replaces a sulfur donor atom, but decreases if oxygen replaces a nitrogen donor atom. If $A_{\parallel}$ is greater than about 120 G, the complex is likely to be a Type 2 square-planar complex or pyramidal square-planar complex with axial or almost axial symmetry.

If $A_{\parallel}$ is less than 120 G, with typical values ranging from about 60 to 90 G, the complex is classified as a "blue" or Type 1 complex. The extinction

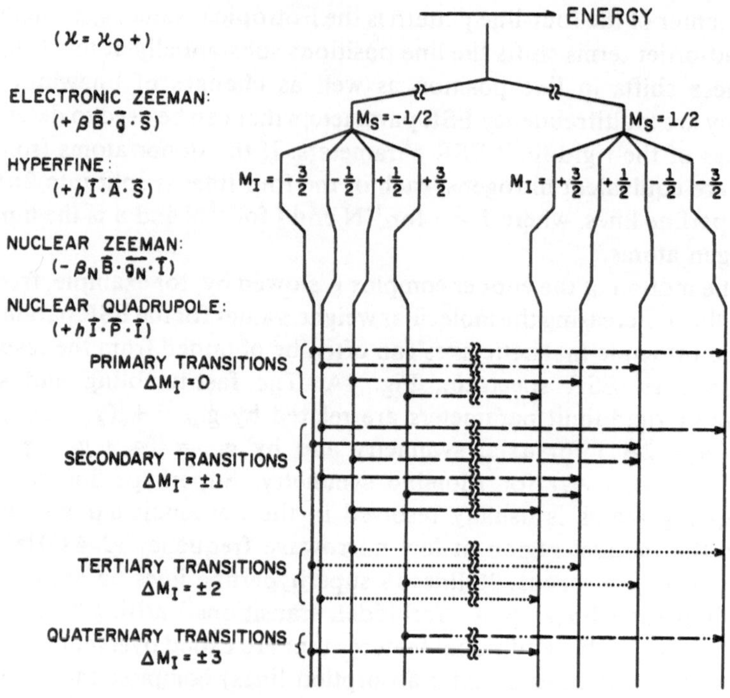

Figure 2. Schematic representation of energy levels and possible transitions for an $S = \frac{1}{2}$, $M_I = \frac{3}{2}$ system [such as $^{63}Cu(II)$] subject to an initial Hamiltonian $\mathscr{H}_0$ plus four spin Hamiltonian terms. The M_I quantum numbers are assigned in the conventional way so that the selection rule would be $\Delta M_I = 0$ if there were no nuclear Zeeman or quadrupole terms. The primary transitions are the four allowed transitions normally described in textbooks; the secondary and tertiary transitions are usually ignored or referred to as "forbidden," often an inappropriate name. The quaternary transitions normally are so weak as to merit the name "forbidden." The relative transition frequencies and intensities are highly dependent on magnetic field orientation. From Belford and Duan (1978).

coefficient at about 600 nm of a blue complex can be several orders of magnitude greater than for a nonblue or Type 2 complex. A typical configuration for the Type 1 complex consists of three donor atoms: two from the 3-nitrogen in the imidazole ring of two histidine residues, with a sulfur from a cysteine residue serving as the third, thereby defining a plane. The cupric ion lies in the plane at the center of the three donor atoms, or perhaps displaced slightly out of the plane (Baker, 1988; Guss *et al.*, 1986; Adman *et al.*, 1989; Adman, 1991). A fourth and possibly a fifth donor atom stabilize the configuration through long bonds perpendicular to this plane. Beta proton couplings (either one or two couplings) from the cysteine residue are of about the same magnitude as the nitrogen couplings. Distinguishing among superhyperfine patterns of three approximately equivalent nitrogen donor atoms and two equivalent nitrogen donor atoms plus two equivalent proton couplings is difficult without additional information. For Type 1 complexes, as for Type 2, an optimum frequency to resolve superhyperfine structure occurs at low microwave frequencies.

1.2. ESR Related to Frequency

Most conventional ESR spectroscopy is carried out at 9–10 GHz, in the *X*-band microwave region, and employs 100-kHz field modulation. Since the resonance condition can be fulfilled for a variety of frequency–field combinations, many other frequencies are possible. The frequencies used in the laboratories of the authors are shown in Table 1.

This chapter considers results obtained using the microwave bands of Table 1. Very promising results have been obtained by other authors (Budil *et al.*, 1989; Haindl *et al.*, 1985) at higher frequencies, where spectral dispersion provides resolution of small g-value differences and yields accurate magnetic tensors. At 250 GHz, it would also be possible to observe slow-motion spectra and to look for details of reorientational dynamics in low-viscosity fluids (Lebedev, 1987).

The main rationale for multifrequency analysis, both in the liquid and the solid states, is that fitting ESR spectra at several frequencies using the same spin Hamiltonian input parameters is a very stringent requirement for an unambiguous characterization of the system.

The evident advantages in the use of multifrequency ESR have been, until recently, eclipsed by the cost of equipment (microwave bridges, cavities, modulation coils, etc.) and the technical problems of sensitivity at lower frequencies. The introduction of the loop-gap resonator (Froncisz

TABLE 1
Microwave Bands for Multifrequency ESR[a]

Band	Frequency range (GHz); $h\nu = g\beta H$
L	1–2
S	2–4
C	4–8
X	8.5–10
K	18–25
Q	35

[a] Conventionally, the notations *L*, *S*, and *C* refer to ranges narrower than a factor of two and the notations *X* and *Q* to ranges greater than indicated.

and Hyde, 1982) has modified the ESR situation because of the enhanced product of ηQ_0 (where η is the filling function and Q_0 is the unloaded quality factor of the resonator), with as much as 200 times improvement at 3 GHz. [See Hyde and Froncisz (1989) for a review.] At the same time, octave-bandwidth microwave bridges can now be constructed readily from commercially available coaxial microwave components and tunable solid-state oscillators.

Microwave frequency is an important experimental parameter that can be used

1. to study the frequency dependence of relaxation times and linewidths,
2. to provide a critical test for theoretical simulations,
3. to provide a means of increasing spectral resolution by varying the interplay of Zeeman and hyperfine interaction (e.g., to separate the spectra of different species),
4. to study higher-order effects that depend on the magnitude of the applied field, and
5. to interpret forbidden transitions involving change in nuclear quantum number, permitting measurement of quadrupole couplings.

There exist several levels at which multifrequency ESR of copper complexes can yield useful structural information. At one level, answers are sought to questions such as the number of nitrogen ligands or perhaps an estimate of the degree of rhombicity. At another level, comparison with spectra of model compounds can permit statements of charge and ligation. Molecular weights can be estimated from liquid-phase spectra, and information on the anisotropy of rotational diffusion may be obtainable. At the highest level, one can hope for a proper molecular-orbital description of the complex. Many years ago, Maki and McGarvey (1958) made a first approach using data from optical and ESR spectroscopies to arrive at bonding parameters. The approach was extensively refined by Buluggiu and associates in the early 1970's (Buluggiu *et al.*, 1971, 1972; Buluggiu and Vera, 1973), who consider rhombic symmetry.

The overall strategy is apparent: obtain multifrequency spectra, employ a computer-driven fit of the spectra to a spin Hamiltonian in order to extract the magnetic parameters, and then arrive at the MO parameters using the methods of Buluggiu and his colleagues. This goal remains elusive, in part because computer-driven fits of theoretical models to multifrequency ESR data still have not been carried out. The authors believe success is within reach. As this methodology develops, it is likely to occur in the following stages:

1. testing using a powdered single crystal for which all paramagnetic parameters are known,

2. application to frozen solutions of known low molecular weight complexes,
3. application to frozen solutions of metalloproteins, and
4. application to metalloproteins at biologically relevant temperatures.

2. THE ROLE OF MICROWAVE FREQUENCY FOR DIFFERENT ROTATIONAL CORRELATION TIMES

2.1. Fast Motion

The relaxation mechanisms governing the ESR linewidth for Cu(II) in solution were first described by McConnell (1956), based on the early experimental work of McGarvey (1956) on Cu(II) acetylacetonate. The McConnell theory related the broadening of the hyperfine lines to the tumbling of a microcrystalline species (with an anisotropic *g* factor and anisotropic hyperfine coupling *A*). The linewidth could then be attributed to the incomplete averaging of the spin Hamiltonian parameters by the tumbling motion. Kivelson (1960; 1972; and references therein) refined the theory for the motionally narrowed limit.

After Kivelson's work, a number of papers were devoted to the study of the motional line-narrowing effect of copper complexes in organic or aqueous solution (Herring and Park, 1979; Kiejzers *et al.*, 1975a; Lewis *et al.*, 1966; Noack and Gordon, 1968). Difficulties have arisen in application of the theory to the experimental data. The correlation times have usually been found to be much shorter than those calculated by the Debye–Stokes formula. Furthermore, the theory failed in the elucidation of molecular dynamics in the slow-motional region. A significant refinement was the inclusion of the spin-rotational mechanism as an M_I-independent contribution to the linewidth (Atkins and Kivelson, 1966).

The general theories of magnetic resonance predict modification of the position of hyperfine lines in ESR spectra as a result of the effects of the line broadening and relaxation mechanisms (Bloch, 1956; Redfield, 1957; Abragam, 1970).

Fraenkel (1965, 1967) discussed the equation of motion of the density matrix in terms of the relaxation matrix, found the form of the spectral densities (for the case of a free radical), and calculated the dynamic shifts using the Redfield–Abragam approach. Bruno *et al.* (1977), using the stochastic–Liouville approach of Freed, analyzed experimentally and theoretically VO^{2+} complexes in the slow-tumbling limit. In the course of their analysis, they developed a complete expression for the dynamic second-order shifts in the fast-tumbling limit. Fraenkel (1967) and Bruno *et al.* (1977), having developed the shift analysis, dismissed it as a negligible effect. However, second-order shifts became quite significant at lower microwave frequency, and they can provide direct information on the magnetic parameters in the liquid phase at room temperature (see Section 4.2.1). These shifts were not discussed in the early literature on copper complexes in solution. Their study has been a significant aspect of our own work (Basosi *et al.*, 1984).

One might attempt to determine the line positions by calculation of the rigid limit interactions using Breit–Rabi analysis for second-order shifts and then averaging by rotational diffusion. Alternatively, one might first average motionally and then calculate the second-order shifts. Rigorously, the second-order shifts vary between these limits with an inflection in a plot of the second-order shifts as a function of rotational correlation time when $\omega_0 \tau_R = 1$, where ω_0 is the microwave frequency. Dynamic second-order shifts are expected to be pronounced at low microwave frequencies in systems with large hyperfine interactions, such as square-planar Cu^{2+} complexes. Furthermore, they give observable effects at *X*- and *Q* band (Basosi *et al.*, 1984).

The obvious effects on the ESR spectrum as the frequency varies are the changes in linewidths and line positions (Fig. 3). The range in the variation of the linewidth as the frequency is varied is evident from multifrequency ESR data and from three-dimensional plots of linewidth, frequency, and rotational correlation time, as illustrated in Fig. 4 (Hyde and Froncisz, 1982). An estimate of the correlation time can be obtained

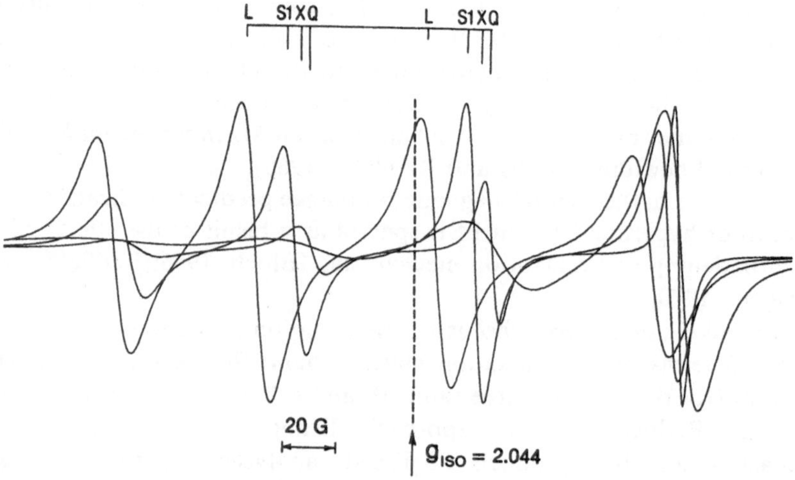

Figure 3. Simulated spectra at *L*, *S*, *X*, and *Q* bands (1.19, 3.44, 9.12, and 34.78 GHz) for $^{65}Cu(dtc)_2$ in *o*-toluidine at room temperature aligned at $g_{iso} = 2.044$. Note the magnitude of the changes in both line position and linewidth as the frequency is varied. From Basosi *et al.* (1984).

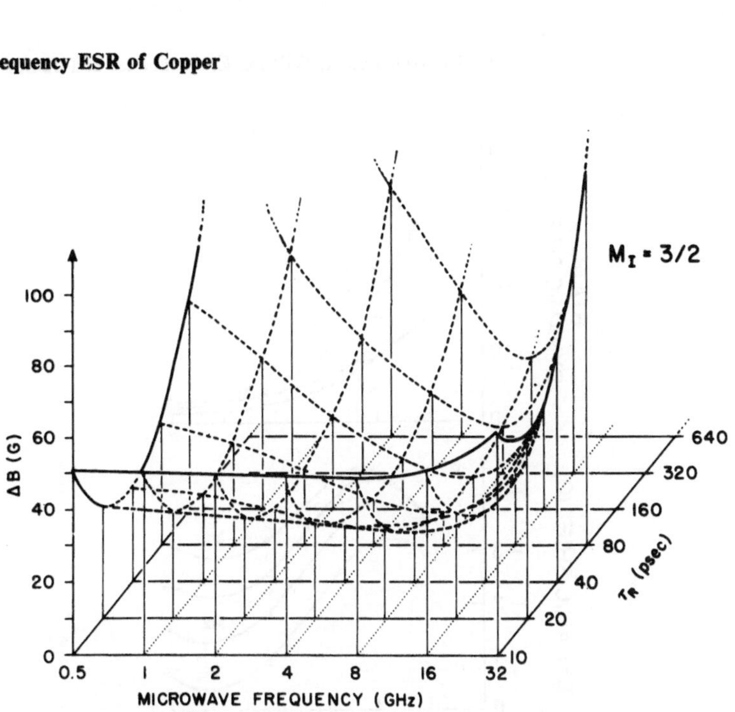

Figure 4. Peak-to-peak liquid-phase linewidths versus microwave frequency and correlation time for $M_I = +\frac{3}{2}$ of a typical copper complex, calculated according to the complete Kivelson theory. $A_\parallel = -594$ MHz, $A_\perp = -77.5$ MHz, $g_\parallel = 2.256$, and $g_\perp = 2.051$ [the values for Cu(catechol)$_2$]. From Hyde and Froncisz (1982).

from the frequency dependence of the linewidth. In addition to the linewidth, the line positions vary substantially as a function of microwave frequency, as is shown in the so-called "shift" equation:

$$\Delta B(M_1) = \frac{\hbar}{gB_e}\left\{\frac{[M_I^2 - I(I+1)]a^2}{2\omega_0} - \left[\frac{1}{15}B_0^2(\Delta\gamma)^2\frac{f}{\omega_0} + \frac{7}{90}(\Delta A)^2 I(I+1)\frac{f}{\omega_0} + \frac{2}{15}M_I B_0 \Delta\gamma \Delta A \frac{f}{\omega_0} - \frac{1}{90}M_I^2(\Delta A)^2\frac{f}{\omega_0}\right]\right\} \quad (1)$$

Here $\Delta A = (A_\parallel - A_\perp)$ rad s^{-1}, $\Delta g = g_\parallel - g_\perp$, $\Delta\gamma = (\Delta g \beta_e)\hbar$, $\omega_0 = 2\pi\gamma_0$, $a = \frac{1}{3}(2A_\perp + A_\parallel)$, $u = 1/(1 + \omega_0^2\tau_R^2)$, and $f = \omega_0^2\tau_R^2 u$. The static term is the first term in the shift equation. It can range from a few gauss to more than 20 G as the microwave frequency decreases (Fig. 5). The static shift is independent

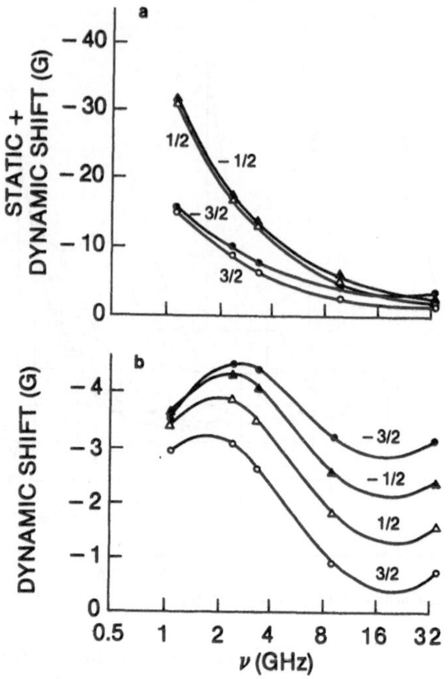

Figure 5. Static plus dynamic shift (a) and dynamic shift (b) versus frequency calculated for $Cu(dtc)_2$ in *o*-toluidine at room temperature. From Basosi *et al.* (1984).

of correlation time, whereas the dynamic shift at constant frequency is sigmoidal as the rotational correlation time varies. The magnitude of the dynamic shift at low frequencies is on the order of 5–10 G.

2.2. Slow Motion

ESR studies of rotational dynamics in liquids have been advanced primarily by the work of Freed and co-workers (Freed, 1987; Schneider and Freed, 1989a; Moro and Freed, 1981). In the slow-motion regime, ESR spectra are no longer simple Lorentzians. They contain information about the microscopic models of rotational dynamics. The utility of these methods for metal complexes became obvious from the temperature-dependent experimental and simulated spectra from the rapid motional to the rigid limit for $VO(acac_2(ph))$ in toluene (Campbell and Freed, 1980). However, there has been no application of the methods of Freed to copper systems.

The theoretical basis for simulation of ESR spectra in the slow-motional regime is the stochastic–Liouville equation (SLE). The SLE amounts to treating the spin degrees of freedom of the system in a quantum-mechanical fashion, while the orientation of the molecule is determined by a classical stochastic process. The quantum-mechanical spin degrees of freedom are coupled to the classical orientational degrees of freedom through the anisotropic part of the orientation-dependent spin Hamiltonian. The choice of stochastic process used to model the reorientation of the molecules affects the time evolution of the spin system and, therefore, the resulting ESR spectrum.

We applied a widely distributed personal computer version of Freed's slow-motion program (Freed, 1989; Schneider and Freed, 1989b) that was developed for spin labels to copper. Computed ESR spectra (based on parameters of Fig. 1) demonstrate the limits of this program as applied to simulation of multifrequency ESR spectra of a typical Type 2 copper complex (Fig. 6). The spectra for the top half of Fig. 6 have been generated by a program in which nonsecular contributions are not considered. These nonsecular terms become large at low frequencies and should be included. The bottom spectra in Fig. 6 have been obtained using the Kivelson theory for fast motion. The slow-motion program fails at $\tau_c = 10^{-12}$ s because spin-rotational terms are omitted, and the program based on Kivelson theory fails at long correlation times because of breakdowns in fast-tumbling assumptions. Comparisons of the slow-motion simulations with experimental spectra are favorable, but differences are especially evident at low frequencies where, according to theory, the nonsecular contributions cannot be neglected. In summary, existing programs can be used to understand the slow-motion behavior of copper complexes, but more comprehensive programs using more powerful computers are clearly desirable.

2.3. Rigid Limit

Froncisz and Hyde noted the change in linewidth in the $g_{\parallel}$ region as the microwave frequency decreased and accounted for the observed dependence of the linewidths by considering M_I-dependent terms and microwave frequency-dependent terms (Froncisz and Hyde, 1980; Hyde and Froncisz, 1982). This dependence can easily be inferred from Fig. 7, where widths of the $g_{\parallel}$ turning points are plotted against frequency for a Cu(catechol)$_2$ complex. The figure is calculated on the basis of the g- and A-strain theory developed by Hyde and Froncisz.

Maki and McGarvey (1958) showed that $g_{\parallel}$ and $A_{\parallel}$ are theoretically correlated for Type 2 complexes. Many workers have observed that for complexes with large g anisotropy, line-broadening mechanisms that arise from small variations in local-site geometry exist in the rigid limit. These mechanisms result in a distribution of magnetic parameters and a decrease in resolution as the microwave frequency increases. A key equation of g-

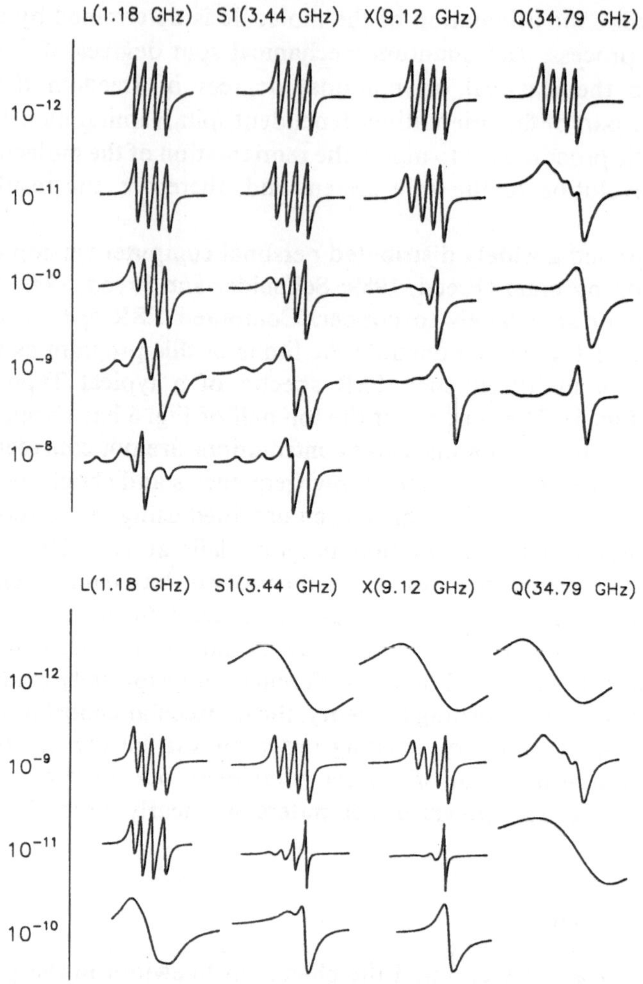

Figure 6. Stimulated ESR spectra (top) using the Freed program EPRLP, and parameters for copper acetylacetonate and (bottom) using CuKIVI which includes the fast-tumbling theory of Kivelson, Atkins and Kivelson's theory of spin-rotational relaxation (1966), and second-order dynamic shift equations of Bruno *et al.* (1977). Parameters are for copper acetylacetonate.

and A-strain is

$$\Delta B_{1/2} = \left[(\Delta B_{1/2}^{R})^2 + (M_I \sigma A_{\parallel})^2 + \left(\frac{h\nu}{g_{\parallel}^2 \beta_e} \sigma g_{\parallel} \right)^2 + \varepsilon \frac{2 M_I h\nu}{g_{\parallel}^2 \beta_e} \sigma g_{\parallel} \sigma A_{\parallel} \right]^{1/2} \quad (2)$$

Here $\Delta B_{1/2}$ is the observed width, $\Delta B_{1/2}^{R}$ the residual width, $\sigma A_{\parallel}$ and $\sigma g_{\parallel}$ are distributions of $g_{\parallel}$ and $A_{\parallel}$, and ε expresses the correlation between $A_{\parallel}$ and $g_{\parallel}$ distributions. The other terms have their usual meanings.

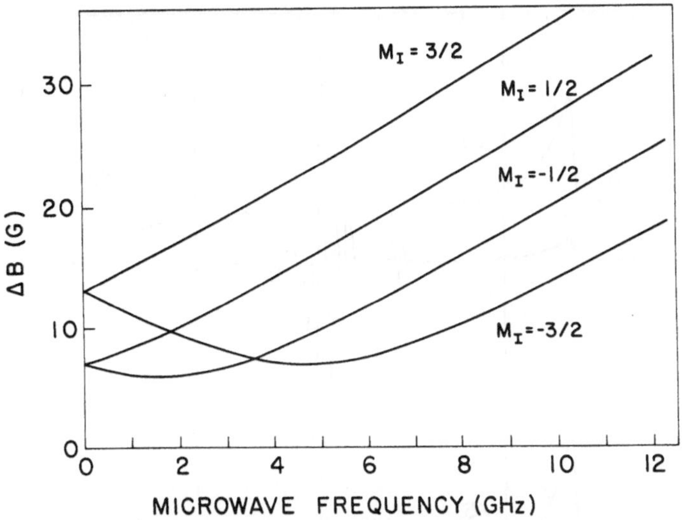

Figure 7. Widths of $g_\parallel$ turning points as calculated by the theory of Froncisz and Hyde, using the data of Cu(catechol)$_2$ to fix the adjustable parameters in the theory. From Hyde and Froncisz (1982).

Equation (2) predicts that there exists a frequency where the linewidth of the $M_I = -\frac{1}{2}$ transition in the parallel region is at a minimum value; similarly, another frequency exists where the $M_I = -\frac{3}{2}$ linewidth is minimum. If the correlation is perfect ($\varepsilon = 1$), these minimum linewidths equal the residual linewidth.

Narrower lines at low frequency in the $g_\parallel$ region account for the resolution of the nitrogen superhyperfine structure observed for many cupric complexes formed from biological ligands, especially the nitrogen donor atoms from the imidazole moiety of histidine (see Section 4.1).

The approach of Kivelson and Neiman (1961) was used by Froncisz and Hyde (1980) to arrive at an equation equivalent to Eq. (2) expressed in terms of distributions of bonding parameters and the correlations among these distributions. Multifrequency ESR permits the measurement of distributions of g and A values; such measurements can in turn lead to knowledge of distributions of bonding parameters. This new spectroscopic information can be used to characterize the system under investigation.

Belford and co-workers developed the use of multifrequency ESR to obtain quadrupole couplings based on an idea of Hyde and Froncisz (Rothenberger *et al.*, 1986). Low-frequency ESR (L band, 1–2 GHz) with parallel static and oscillating microwave fields was used to obtain quadrupole couplings through the measurement of forbidden ($\Delta M_S = \pm 1, \Delta M_I = \pm 1$) transitions (Fig. 8; Rothenberger *et al.*, 1986). If oxygen donor atoms form the cupric complex, there exists an excellent correlation with geometry

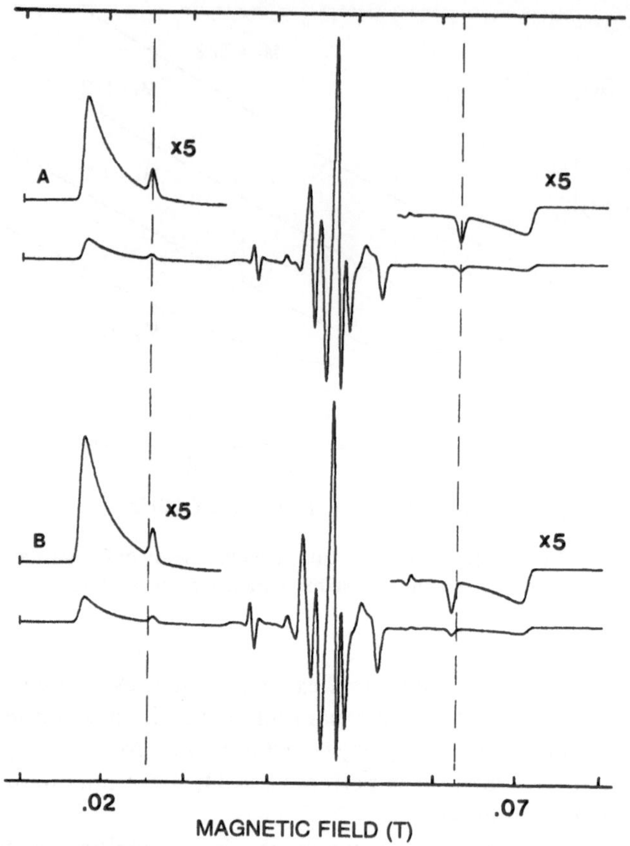

Figure 8. Computer simulations of $^{63}Cu(acac)_2$ using parameters from Table 1 of Rothenberger *et al.* (1986) except with variation of QD. (A) QD = 9 MHz, (B) QD = 0 MHz.

(e^2qQ is smaller for square-planar complexes and larger for octahedral complexes). Belford and Duan (1978) show that nuclear quadrupole constants contain substantial contributions from axial ligands (i.e., e^2qQ(solvate) = e^2qQ(complex) + ne^2qQ(axial ligands), where $n = 1$ or 2). This type of information is important in discerning the axial ligation of copper complexes in biological systems.

3. MULTIFREQUENCY APPARATUS AND METHODS

3.1. New Instruments for Multifrequency ESR

A synopsis of a workshop entitled, "The Future of ESR Instrumentation," held in Denver in August 1987, describes the state-of-the-art ESR instrument (G. R. Eaton and S. S. Eaton, 1988a, b). One conclusion of the symposium is that the spectrometer of the 1990s should be a single multi-frequency machine ranging from 3–15 GHz. An approximation to this instrument exists at the ESR Center in Milwaukee, where a single spectrometer station has seven bridges, four of which are in octave bandwidths (0.5–1, 1–2, 2–4, and 4–8 GHz). New resonators called loop-gap resonators have been designed for each of these frequency ranges (Froncisz and Hyde, 1982). The particulars of the loop-gap resonator are beyond the scope of this review, but Hyde and Froncisz (1986, 1989) have reviewed the subject. Of interest in the 1989 review is a section on multifrequency ESR. Of particular importance to the user is the design of a loop-gap resonator at low frequency that accommodates a standard 4-mm OD quartz sample tube. There is a tradeoff between low Q and high filling factor that maintains the signal-to-noise ratio within a factor of about three of an X-band spectrometer. Loop-gap technology is rapidly expanding and new applications are being found. Multifrequency saturation recovery (2–4, 9, and 19 GHz) and multifrequency ENDOR (2–8 and 9 GHz) became operational in 1991 at the ESR Center. The loop-gap resonator is the key technological advance that makes EPR in the range of 0.5 to 6 GHz practical. Cavity resonators are too large and amounts of sample too great in this frequency range using conventional resonators.

3.2. Sample Preparation

3.2.1. Isotopic Substitution (^{63}Cu and ^{65}Cu)

ESR spectra of a frozen solution of a Cu^{2+} complex with four oxygen donor atoms exhibit resolved lines for both the ^{63}Cu (69% abundance) and the ^{65}Cu (31% abundance) component (Fig. 1). Superposition of multiline superhyperfine patterns results in overlap of the patterns for the two components and loss of resolution of the superhyperfine structure. It has been our experience that the superhyperfine pattern for the $M_I = -\frac{1}{2}$ line is too complicated and too poorly resolved to determine the number of nitrogen donor atoms if both isotopes are present. (Cupric oxides of these isotopes have been purchased from Oak Ridge National Laboratory.) Removal of the metal from the protein can be a difficult problem. For example, apo fungal laccase was prepared by anaerobic dialysis against 0.2-M KCN at pH 10.3 for 48 h (Hanna *et al.*, 1988). Reconstitution is equally difficult. In that work, the native Cu binding sites were not restored upon incubation of apo fungal laccase with copper but a mixed-metal derivative of tree laccase for which Type 1 copper is replaced by Hg has been prepared (Hanna *et al.*, 1988, and references therein).

3.2.2. ^{15}N and ^{33}S

^{15}N compounds commercially available for substitution of ^{15}N for ^{14}N include nitrates, nitrites, nitrous gas, and several biologically relevant compounds, including most amino acids. Since the nuclear spin is $\frac{1}{2}$ for ^{15}N and 1 for ^{14}N, the ESR patterns change upon substitution. In addition, ^{15}N has a larger magnitude of the magnetogyric ratio (-0.27107×10^4 rad G^{-1}) than ^{14}N ($+0.19324 \times 10^4$ rad G^{-1}) and shifts in line positions occur. ^{33}S ($I = \frac{3}{2}$) would also provide information concerning ligation, especially for Type 1 copper complexes. However, high-purity ^{33}S (a maximum of 73%) is not readily available.

3.2.3. Optimum Samples

If all lines in the $g_{\parallel}$ region exhibit the same width at X band, there is essentially no further improvement on going to low frequency. If g- and A-strain exist, a minimum linewidth for the $M_I = -\frac{1}{2}$ line at $g_{\parallel}$ can be obtained by selection of the appropriate frequency. Sample preparation can help minimize the linewidth as follows:

1. A single isotope of copper, either ^{63}Cu or ^{65}Cu, should be used.
2. Deuterated solvents, especially if one or more of the equatorial sites is water, can improve resolution.
3. Aggregation or formation of dimers should be avoided.
4. Mixed ligation, i.e., multiple species, complicates analysis.
5. Good glasses are preferred.

Often mixed solvents are necessary to obtain adequate solubility and a good glass (for example, see the caption to Fig. 1). The concentration should be 1 mM for our S-band apparatus. Ligand concentration and pH should be varied through several orders of magnitude, if possible, to avoid dimers, effects of stoichiometry on ligation, and displacement of axial or equatorial ligands by —OH.

A conventional ESR spectroscopic approach is to freeze the sample and extract the magnetic parameters from analysis of the frozen solution powder pattern. There are many reasons why this is undesirable. Vänngård (1972) called attention to changes in coordination that can occur. Falk *et al.* (1970) observed a temperature dependence on equilibrium between two species. Aggregation can occur (for example, Saryan *et al.*, 1981) and often tedious empirical approaches to formulation of good glasses must be employed. In water, freezing can change the local pH (Falk *et al.*, 1970; Orii and Morita, 1977), and Wilson and Kivelson (1966) found that the isotropic g and A values of copper acetylacetonate are actually temperature dependent. All of these complications can be avoided through use of multifrequency ESR of copper complexes above 0 °C.

3.3. Processing of Data

3.3.1. Signal Averaging

It has been apparent for many years that signal averaging of repeated fast scans is preferable to a single slow scan. There are immediate practical advantages: one can stop scanning whenever the signal-to-noise ratio is deemed satisfactory, a particular scan can be deleted if some transient event occurs, or accumulation can be interrupted to tune the spectrometer or add cryogenic fluids.

With increasingly rapid analogue-to-digital (A–D) converters and increasingly convenient computer programs, there are some new rules of thumb for signal averaging:

1. Scan as rapidly as possible consistent with relaxation times of the spin system.
2. Use the shortest time constant available.
3. Average high-frequency noise after acquisition using a Gaussian smoothing filter (or, for good reason, some other filter).
4. Consider saving all separate scans and inspecting them, after acquisition, for transient events before signal averaging.

The first rule fundamentally reduces very-low-frequency noise. The second and third rules are based on the recognition that a postacquisition (noncausal) filter does not shift the line positions, as occurs if an excessively large time constant is used. It should be recognized that postacquisition filtering is always preferable to filtering during acquisition, assuming of course that a low-pass filter consistent with the Nyquist condition for the A–D converter is employed. Rule of thumb number four looks forward to the day when various objective computer-based criteria can be used to decide whether or not to save or delete particular scans.

3.3.2. Second Derivative

The second-derivative (or second harmonic) display emphasizes sharp features and discriminates against broad features. This display is particularly useful for analyzing superhyperfine patterns. Shoulders in the spectrum become peaks that are useful for estimation of coupling constants. The binding of pterin to phenylalanine hydroxylase from *Chromobacterium violaceum* illustrates the use of second-derivative spectra (Fig. 9). A cofactor analogue, 6,7-dimethyltetrahydropterin ($DMPH_4$), with either ^{14}N or ^{15}N substituted at the 5 position, was added to phenylalanine hydroxylase (Pember *et al.*, 1987). ESR spectra were taken at both *X* band and *S* band. Spectra were not particularly well resolved in the $g_{\parallel}$ region at low frequency

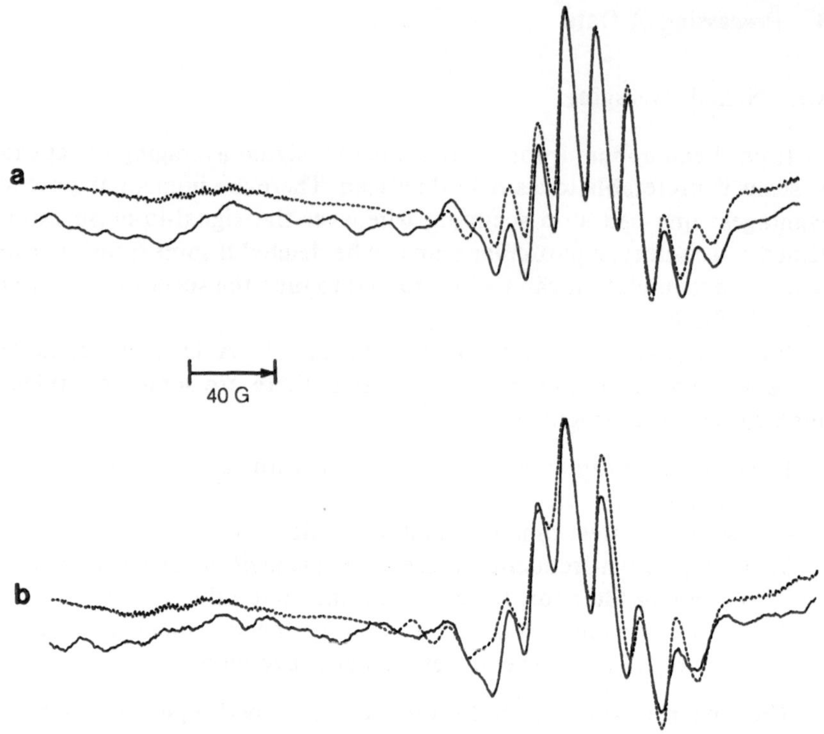

Figure 9. Experimental (solid lines) and simulated (dashed lines) *S*-band second-derivative ESR spectra for the cupric site in bacterial phenylalanine hydroxylase in the presence of the pterin cofactor in which the nitrogen in the 5 position is ^{14}N (a) or ^{15}N (b). ESR parameters for simulated spectra given in Fig. 6 of Pember *et al.* (1987).

because naturally abundant copper (69% ^{63}Cu and 31% ^{65}Cu) occupied the binding site. It was apparent that the $g_{\perp}$ region contained several resolved superhyperfine lines in the presence of either ^{14}N-5-$DMPH_4$ or ^{15}N-5-$DMPH_4$. Small differences, particularly in the intensities of the resolved lines, indicate that the superhyperfine pattern was changed upon substitution of ^{15}N for ^{14}N at the 5 position of $DMPH_4$. Second-derivative displays (Fig. 9) emphasize the superhyperfine features and confirm the differences attributed to substitution of ^{15}N for ^{14}N.

It is now well established that second-harmonic displays should never be obtained experimentally. Just as it is always better to filter spectra postacquisition, it is also better to *transform* spectra after acquisition. Taking the mathematical second derivative of an experimental spectrum enhances high-frequency noise, but the dual use of derivative plus Gaussian filter results in spectra that are as good as or better than—but never worse than—those acquired experimentally. The EPR experimentalist may well prefer to use pseudomodulation to obtain postacquisition second-harmonic spectra (Hyde *et al.*, 1990, 1992). This algorithm is a rigorous simulation of the experimental application of an additional field-modulation frequency following phase-sensitive detection at 100 kHz. The effects of overmodulation are the same, and the modulation amplitude can conveniently be optimized postacquisition. Pseudomodulation filters as well as transforms and has been shown to yield better spectra of copper complexes than are obtained experimentally using conventional techniques (Hyde *et al.*, 1992). It is concluded that experimental acquisition of second-harmonic spectra, which is a very time consuming process, has become obsolete.

The rational for second-harmonic displays is improved resolution. A variety of resolution-enhancement algorithms can now be studied conveniently by computer; see Hyde *et al.* (1992) and references therein. All of these algorithms are basically filters, and improved resolution implies reduction of low frequencies relative to high frequencies. There is a tradeoff between resolution and noise. The lower the noise in the initial spectrum, the better the resolution that is possible using postacquisition techniques. And, of course, the better the initial experimental resolution, the better the final result.

3.3.3. Simulations Using Minimization Programs

We have four computer simulation programs in our laboratories. A rigid limit program was obtained from Dr. John R. Pilbrow, Monash University, which accommodates copper ESR parameters as well as superhyperfine parameters from two nonequivalent nitrogen donor atoms and two nonequivalent atoms with spin $\frac{1}{2}$ (either ^{15}N or H). Three nonequivalent linewidths (three *g*- and *A*-strain parameters) are necessary to simulate both the position and the intensity of spectra for Type 1 copper sites. Pilbrow (1990) considers in detail computer simulations of powder spectra. A particularly good simulation of an EPR spectrum of a copper porphyrin bound to DNA is presented (Dougherty *et al.*, 1985). Pilbrow also discusses simulations at *X* and *S* band for which cobalt (II) dibarreleno porphyrazine spectrum is simulated quite satisfactorily at *X* band, but not at *S* band. The discrepancy was attributed to use of an oversimplified expression for the linewidths, which did not include the fourth-order spherical harmonic description of the orientational dependence.

The slow-motion program of Dr. Jack H. Freed (Cornell University) has been modified to include nonsecular contributions. Apparently good simulations without the nonsecular terms are obtained at *X* band for higher frequencies, but not at low frequencies. The reader is reminded that this program was developed for spin labels and not for copper.

A program written by Dr. Wojciech Froncisz, Jagiellonian University, Krakow, Poland, designated CuKivII, uses Kivelson's theory to simulate spectra of fast-tumbling low-molecular-weight copper complexes. It includes dynamic second-order shifts and the spin–rotation interaction.

The fourth program, QPOW, from Dr. R. Linn Belford, University of Illinois, is used to simulate *L*-band spectra. It includes the forbidden transitions from which quadrupole couplings are determined (see Section 5.2). Simulations with $H_1 \| H_0$ are possible. Unlike the Pilbrow program, the solution to the Hamiltonian is exact and not a perturbation treatment. Superhyperfine couplings, however, are treated by perturbation methods.

These programs are used to simulate rigid, slow, and fast motion, respectively. The simulations from the four programs should converge at the limits of their usefulness.

Several iterative procedures for fitting models to spectra, including a Monte Carlo and a damped-least square method, exist (Pasenkiewicz-Gierula *et al.*, 1987a, b). Typically, about 500 simulations are required to optimize the fit when the spectral features in the simulation are already close to the experimental data. Simulations at many frequencies are critical, and convergence to a single set of rigid limit parameters that fit all frequencies is often elusive.

Many computer-based transformation and filtering algorithms can be developed to improve the appearance of a spectrum as viewed by an expert ESR spectroscopist. If these same algorithms are also applied to simulated spectra, it would seem plausible that a computer-driven fit of a simulation to an experimental spectrum would also be enhanced. But more work is needed before this concept fully springs to life.

3.3.4. Sensitivity Analysis

Over a series of papers (Basosi *et al.*, 1984; Hyde *et al.*, 1985, 1989), the authors have developed the subject of sensitivity analysis and have applied it to spectra of square-planar copper complexes. Sensitivity analysis is applied only to simulated spectra of square-planar copper complexes, including simulated experimental spectra where noise has been added. It can thus be viewed as an approach to computer modeling of ESR spectroscopy.

If we let P_i be an input parameter in a spin Hamiltonian, then we can calculate the partial derivatives of a simulated spectrum $S(H)$ with respect to that parameter using typical values for all input parameters:

$$\frac{\partial S(H)}{\partial P_i} \tag{3}$$

This display may look somewhat like an ESR spectrum, but it could never be obtained experimentally. It shows at once which parts of a spectrum are sensitive to parameter P_i.

If we let P_k be another input parameter and calculate

$$\frac{\partial S(H)}{\partial P_k} \tag{4}$$

we can ask whether or not a spectral region that is sensitive to P_i is also *specifically* sensitive to that parameter. For example, it was discovered that overshoot lines are very sensitive to all input parameters, but not specific to any. These are often dominant features in copper spectra. This lack of specificity will tend to frustrate any computer-driven fit of a model to data. Sensitivity analysis teaches us to avoid overshoot lines.

A common criterion of excellence in judging the fit of a model to data is the integrated least squares. The parameter *sq* was defined in the context of sensitivity analysis as

$$sq = \int \left[\frac{\partial^2 S}{\partial H \partial P_i}\right]^2 dH \tag{5}$$

The old method of taking two spectra to a window, lining up the dominant features, and judging the quality of the fit by areas between the two curves is equivalent to estimating the integrated modulus. The parameter *m* is written

$$m = \int \left|\frac{\partial^2 S}{\partial H \partial P_i}\right| dH \tag{6}$$

Figure 10 shows plots of the parameters *m* and *sq* as functions of microwave frequency and rotational correlation time for a square-planar complex for $P_i = \Delta g$. These plots help us to find the best frequency for determining P_i if we decide to use *sq* or *m* as a criterion for the quality of the fit.

It is a familiar observation that when best fit is achieved, $\partial^2 S/\partial P_i^2$ is zero for all values of *i* at every value of *H*. An interesting use of sensitivity analysis is to assume a significant error in one parameter, say P_k, and ask what the resulting error would be in a best fit where just one of the other parameters was optimized. Such a multidimensional sensitivity analysis is illustrated in Fig. 11.

One can also estimate the error ΔP_i in determination of a parameter P_i in the presence of noise $n(H)$:

$$(\Delta P_i)^2 = \frac{\int [n(H)]^2\, dH}{\int \left(\frac{\partial S}{\partial P_i}\right)^2 dH} \tag{7}$$

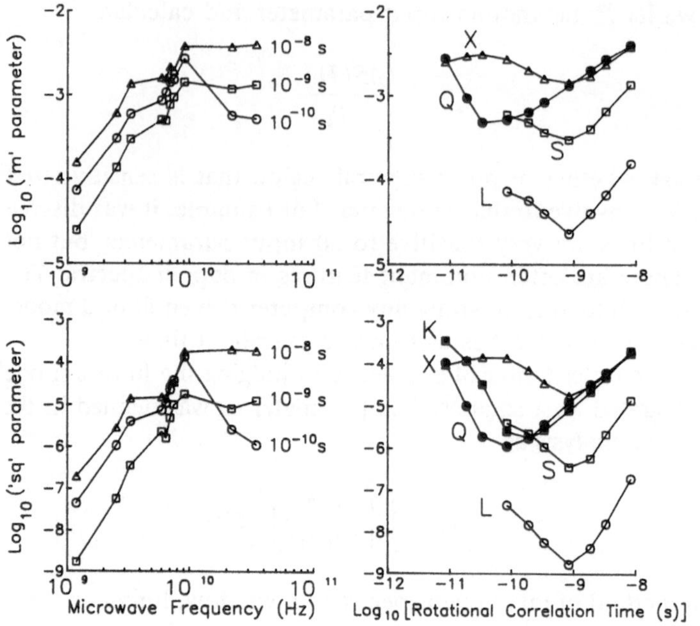

Figure 10. Reference data for magnetic parameters using the *sq* integrated χ^2 parameter (bottom) and the *m* (integrated modulus) parameter (top). From Hyde *et al.* (1989).

Sensitivity analysis is a natural adjunct to multifrequency ESR of copper complexes. It is time consuming to examine a sample at every frequency. Sensitivity analysis can be a significant aid in design of the experiment and selection of the optimum microwave frequency. It teaches that one ought in principle to weight the fitting of models to data much more cleverly than by using *sq* or *m*. For example, the $g_{\parallel}$ region is very specific and ought to be strongly weighted, whereas the $g_{\perp}$ region presents a very difficult problem. One might also weight various regions of spectra obtained at differing microwave frequencies in different ways. Sensitivity analysis remains under development.

4. BIOPHYSICAL APPLICATIONS OF MULTIFREQUENCY ESR

4.1. Improved Resolution of Superhyperfine Patterns

Improved resolution of superhyperfine lines in low-microwave-frequency spectra of immobilized cupric complexes in powders or frozen solutions is a consequence of the cancellation of a term dependent on frequency by a term that is M_I-dependent [see Fig. 7 and Eq. (2)]. The

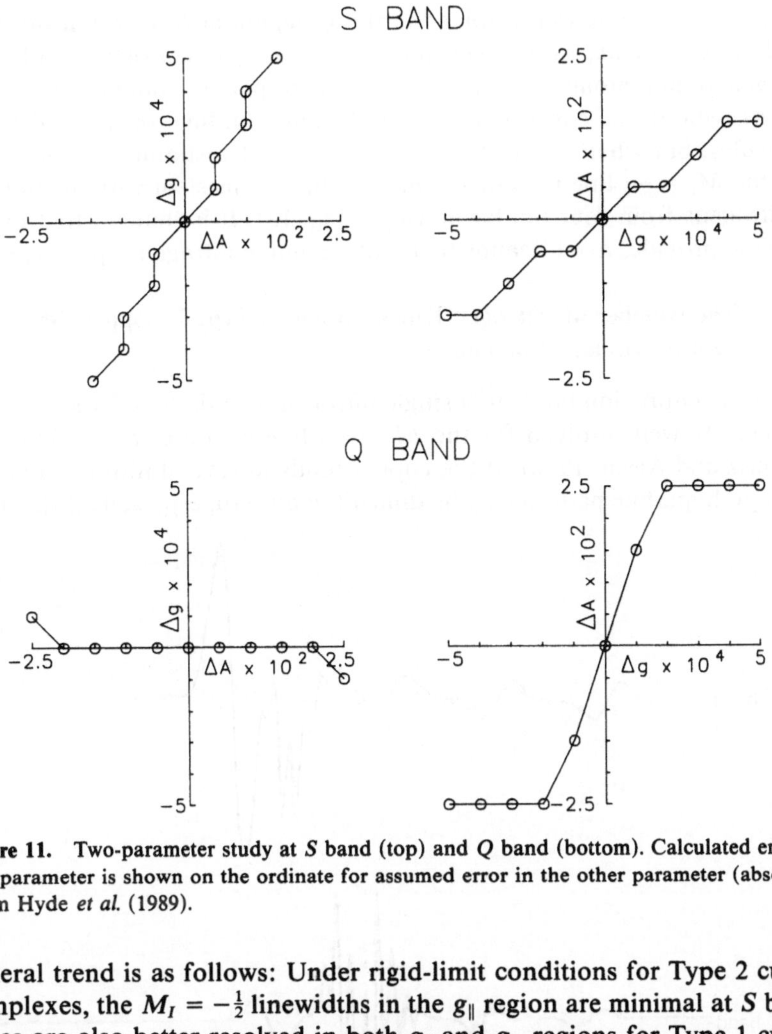

Figure 11. Two-parameter study at *S* band (top) and *Q* band (bottom). Calculated error in one parameter is shown on the ordinate for assumed error in the other parameter (abscissa). From Hyde *et al.* (1989).

general trend is as follows: Under rigid-limit conditions for Type 2 cupric complexes, the $M_I = -\frac{1}{2}$ linewidths in the $g_\parallel$ region are minimal at *S* band. Lines are also better resolved in both $g_\parallel$ and $g_\perp$ regions for Type 1 cupric complexes at low frequencies under rigid-limit conditions, although not enough experimental work has been completed to make a general statement. For fast-tumbling low-molecular-weight Type 2 cupric complexes, the high field ($M_I = +\frac{3}{2}$) linewidths are minimal at *X* band because of cancellation of positive linewidth terms proportional to Δg^2 and ΔA^2 by a negative cross term proportional to $M_I \Delta A \Delta g$.

4.1.1. Histidine Complexes

ESR spectra of the cupric ion in the presence of tenfold excess histidine illustrate the possible changes in structure that may occur upon freezing.

The $M_I = -\frac{1}{2}$ line is well resolved in the $g_{\parallel}$ region at *S* band but not at *X* band (Basosi *et al.*, 1986). The nine-line pattern is consistent with four nitrogen donor atoms comprising the square-planar configuration. This pattern would be consistent with bidentate binding of two histidine molecules, but when ^{14}N in the imidazole ring of histidine is replaced by ^{15}N, the $M_I = -\frac{1}{2}$ line is resolved into five lines consistent with the binding of nitrogens from four imidazole rings (Fig. 12). It is thought that excess histidine provides axial ligands that replace amine nitrogens upon freezing.

4.1.2. The Number of Nitrogen Donor Atoms in Type 2 Copper Sites of Transferrin and Albumin

If the cupric ion binds to a single nitrogen ligand, the ESR spectra are particularly well resolved for the $M_I = -\frac{1}{2}$ line in the $g_{\parallel}$ region (Fig. 13; Froncisz and Aisen, 1982). If the copper binds to several nitrogen ligands, the superhyperfine pattern may be difficult to discern, especially if the outer

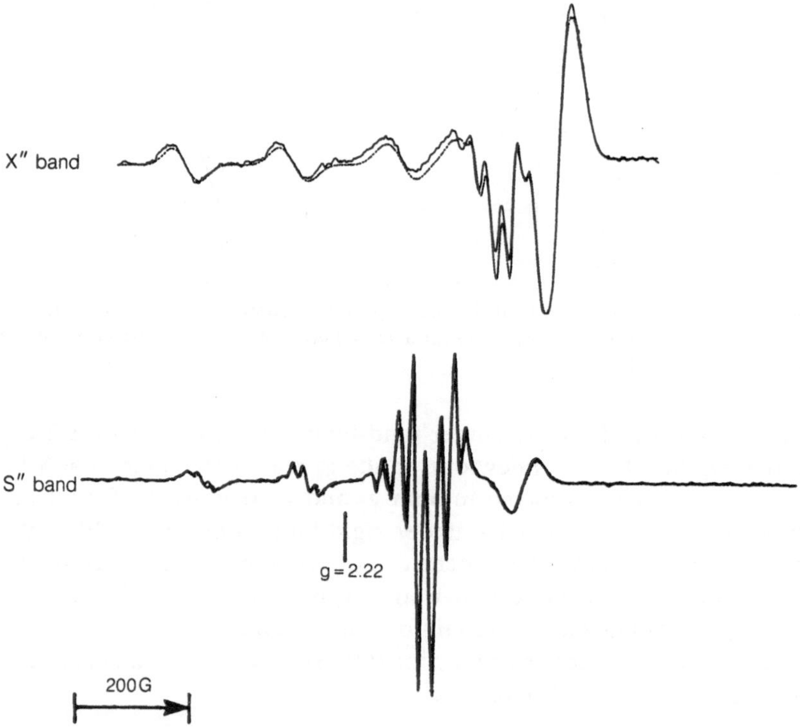

Figure 12. Experimental second-derivative (solid line) *X*-band (top) and *S*-band (bottom) spectra of $^{65}Cu(II)(L\text{-}His)_4$ in which ^{15}N is substituted for ^{14}N in the imidazole ring (histidine-1,3-^{15}N) in D_2O at pD 7.3 and 77 K compared with simulations (dotted lines). ESR conditions and parameters given in Figs. 1 and 2 of Basosi *et al.* (1986) except that the second modulation frequency is 270 Hz and modulation amplitude 10 G.

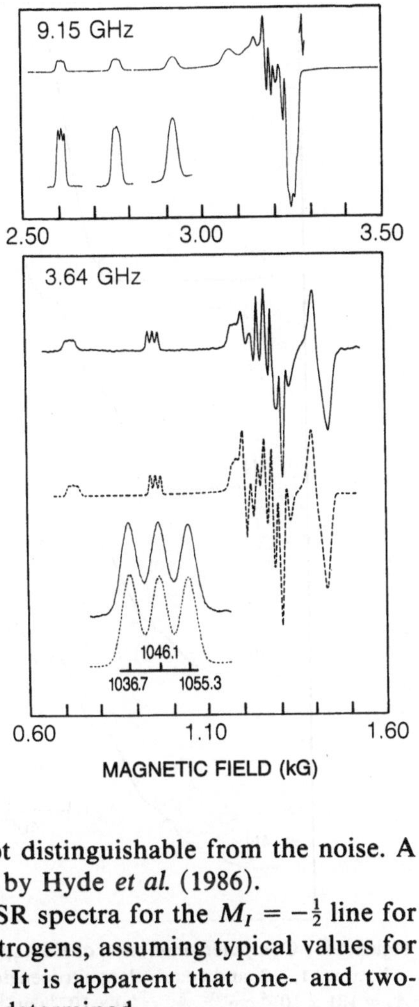

Figure 13. EPR spectra of 3.3-mM monocupric transferrin taken at 9.15 and 3.64 GHz. The temperature was −179 °C; microwave power, 3 mW; modulation amplitude, 2 G. Inset shows the $M_I = -\frac{1}{2}$ line taken at an eightfold increase in gain. Solid lines are the experimental spectra, and dashed lines are simulated. From Froncisz and Aisen (1982).

lines with the weakest intensity are not distinguishable from the noise. A solution to this problem was provided by Hyde *et al.* (1986).

Figure 14 shows simulations of ESR spectra for the $M_I = -\frac{1}{2}$ line for one, two, three, and four equivalent nitrogens, assuming typical values for the input parameters to the program. It is apparent that one- and two-nitrogen ligation is easily and reliably determined.

Distinguishing between three and four is a more difficult problem that has been and continues to be a particular goal as we refine our methodological approaches. The ratios of intensities of the superhyperfine lines in a stick diagram are

Line number	4	3	2	1	0	1	2	3	4
3N		1	3	6	7	6	3	1	
4N	1	4	10	16	19	16	10	4	1

The problem is that the ratio 6:7 is almost indistinguishable from 16:19, so the central three lines look almost the same. Of course, if the signal-to-noise ratio is good enough, the outside line for four N is definitive. Long

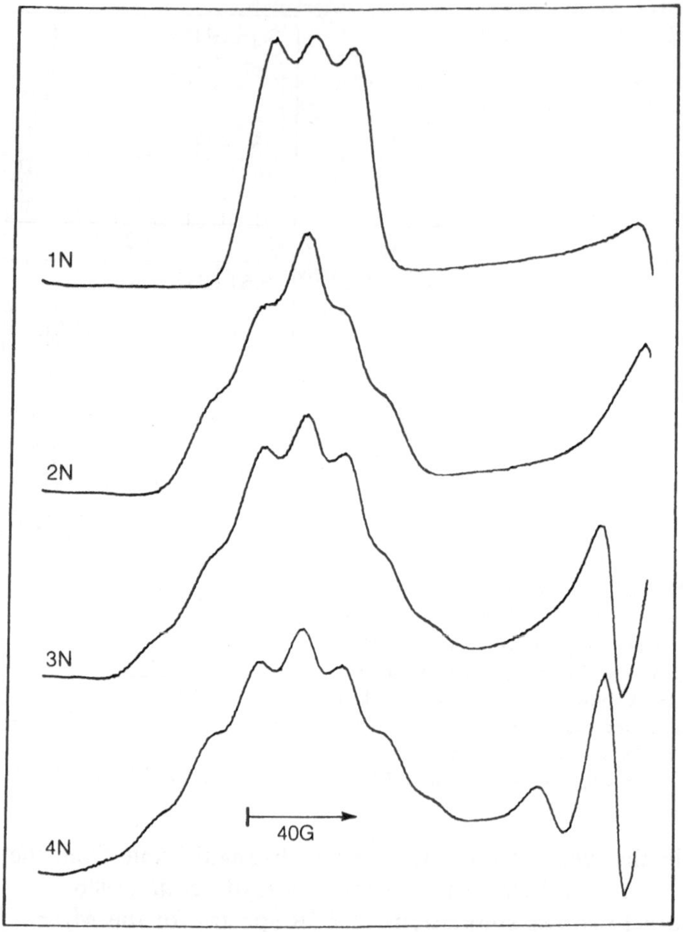

Figure 14. Computer simulation of ESR spectra for $M_I = -\frac{1}{2}$ line in $g_\parallel$ region assuming (top to bottom) 1, 2, 3, and 4 equivalent nitrogen donor atoms. ESR parameters: $g_\parallel = 2.22$; $g_\perp = 2.02$; $A_\parallel = 181 \times 10^{-4}\ \mathrm{cm}^{-1}$; $A_\perp = 20 \times 10^{-4}\ \mathrm{cm}^{-1}$; $A^N = 16 \times 10^{-4}\ \mathrm{cm}^{-1}$; linewidth, 6.5 G; microwave frequency, 3.7 GHz. From Hyde *et al.* (1986).

signal acquisition and signal averaging are highly desirable. The best ratios are, however, the intensities of the second and third lines compared to that of the center.

The cupric ion binding site in serum albumin is a notable example where it was determined that four nitrogen donor atoms are bound to cupric ion (Rakhit *et al.*, 1985). In this case, both the resolution of the nine-line pattern and a better fit to a simulation that includes four instead of three nitrogen donor atoms demonstrates the method (Fig. 15). Many examples of resolved hyperfine patterns for the $M_I = -\frac{1}{2}$ line in the $g_\parallel$ region at *S* band have been reported (Sealy *et al.*, 1985, and references therein).

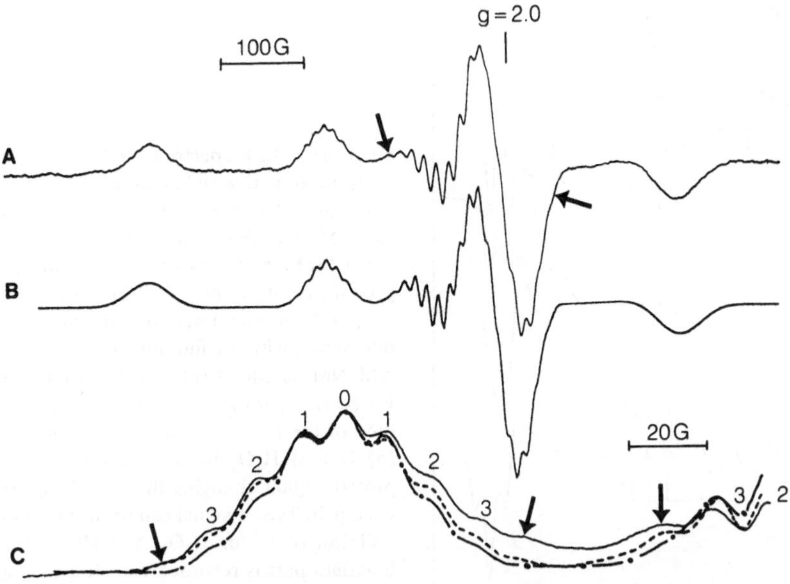

Figure 15. ESR spectra (*S* band) of frozen (−150 °C) solutions of 3.0-mM BSA and 2.0-mM $^{65}Cu^{2+}$ at pH 7.13. (A) Experimental: microwave frequency, 3.382 GHz; time constant, 1 s; microwave power, 20 dB. (B) Computer simulation assuming four equivalent nitrogens. EPR parameters are $g_x = g_y = 2.055$; $g_z = 2.1766$; $A_x^{Cu} = 16 \times 10^{-4}\ cm^{-1}$, $A_z^{Cu} = 217 \times 10^{-4}$ cm; $A_x^N = A_y^N = 13.0 \times 10^{-4}\ cm^{-1}$; $A_z^N = 12.2 \times 10^{-4}\ cm^{-1}$; H_{pp} for the $g_\perp$ region is 5.5 G. Note: g_z was taken from *X*-band data, but the center of the *g* spectrum at *S* band is about $g = 2.20$. *S*-band data not only is less sensitive to changes in the *g* value, but also may have second-order shifts of as much as 20 G. (C) Expanded region around $m_I = -\frac{1}{2}$ line in $g_\parallel$ region for BSA and $^{65}CuCl_2$ under conditions for A and B. (——) represents experimental data with a 10-s time constant, (– – –) represents a computer simulation assuming four equivalent nitrogens, and (– · –) a computer simulation assuming three equivalent nitrogens. Computer simulation parameters are as given, except $a = (4.8)^2\ G^2$, $b = 0$, $c = 0$, which gives $(\frac{1}{2}\Delta H_w) = 4.8$ G. From Rakhit *et al.* (1985).

4.1.3. A Strategy for Type 1 Blue Copper Sites in Laccase and Azurin

Blue Type 1 copper complexes are characterized by a small value for $A_\parallel$ compared with Type 2 copper complexes. Spectra of multiple cupric sites including both Type 1 and Type 2 sites could be separated at higher microwave frequencies, *Q* band, by taking advantage of the spread of the *g* values (Fig. 16). At *Q* band, the low-field M_I-dependent lines in the $g_\parallel$ region of the Type 2 site are isolated from the $g_\parallel$ region of the Type 1 site. Superhyperfine structure in the $g_\parallel$ region at *Q* band and *X* band (due to the nitrogen donor atoms or the protons from the methylene moiety of the cysteine residue) has not often been resolved using conventional ESR, but ENDOR has provided couplings in the axial direction for both nitrogens and protons (Roberts *et al.*, 1980; Werst *et al.*, 1991).

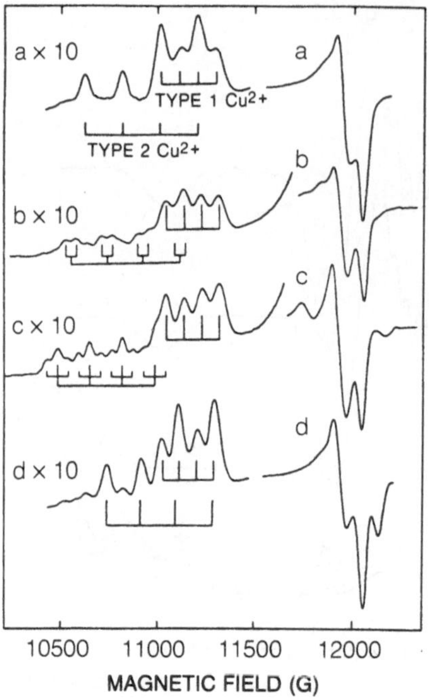

Figure 16. EPR spectra at 34.5 GHz of frozen solutions of native and modified fungal laccase. Spectrum (a) is obtained from the native protein (0.3 mM) and shows the two types of Cu^{2+}. In (b) 0.7-mM NaF has been added to 0.7-mM protein, resulting in a splitting of the Type 2 copper hyperfine lines into doublets due to interaction with the fluorine nucleus. When 14-mM NaF is added (c), equal coupling to two F occurs, yielding 1:2:1 triplets. The magnitude of the fluorine splitting is about 55 G. In (d) 10-mM H_2O_2 has been added to 0.3-mM protein, again changing the Type 2 signal. No change in Type 1 signal can be observed on the addition of F^- or H_2O_2. For all spectra the low-field part is recorded with 10 times higher spectrometer gain compared with the high-field part. From Brändén *et al.* (1971), Malkin *et al.* (1968), and Fig. 9-9 in Vänngård (1972).

Spectra obtained at lower microwave frequencies exhibit improved resolution of the superhyperfine structure although the number of examples is limited. This implies that strain is important for Type 1 as well as Type 2 complexes [see Eq. (2)]. The spectrum for the Type 1 site in Type 2-depleted laccase is well resolved at about 2.4 GHz (Fig. 17), in both the $g_\perp$ and $g_\parallel$ regions (Hanna *et al.*, 1988). However, the spectrum for the Type 1 site in azurin (data not shown) is better resolved only in the $g_\perp$ region at this same frequency. If the pattern for the $M_I = -\frac{1}{2}$ line in the $g_\parallel$ region is not well resolved, it still may be possible to simulate the entire spectrum when the axial parameters have been obtained from ENDOR. This strategy assumes that the nitrogen superhyperfine couplings are nearly isotropic. It is noted that the apparent ESR parameters may vary with temperature as reported for tree laccase (Morpurgo *et al.*, 1981). Fluid-phase spectra of copper complexes in high-molecular-weight proteins tend to resemble frozen-solution spectra, but evidence of motion may still be detectable. These motional effects should not be confused with temperature-dependent change in geometry.

4.1.4. An Unusual Copper Site in Cytochrome *c* Oxidase

The hyperfine lines in the ESR spectrum from an anomalous cupric site, typified by the ESR parameters for cytochrome *c* oxidase, are better

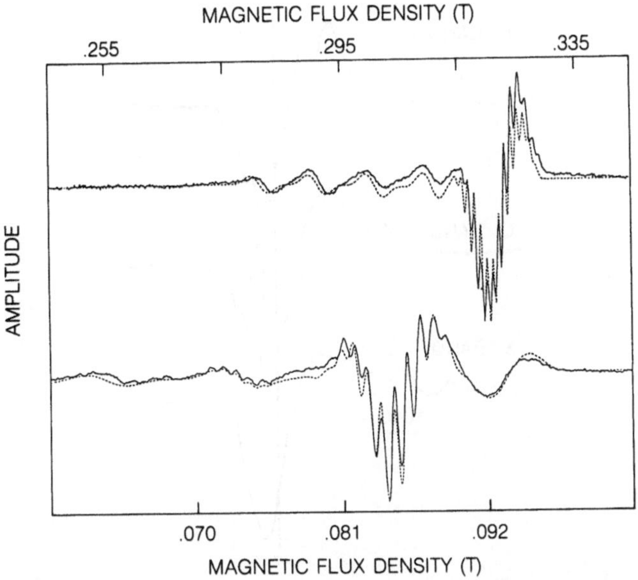

Figure 17. Multifrequency EPR of Type 2-depleted laccase in 0.1-M phsophate buffer, pH 6.0. (Top): Second-derivative spectrum (——) of Type 2-depleted laccase at 9.15 GHz from an average of four scans, and simulation (---) with hyperfine splitting from two N atoms and two H atoms. Experimental conditions: microwave power, 5 mW; modulation amplitude, 0.5 mT; modulation frequencies, 100 kHz and 1 kHz; temperature, −150 °C. (Bottom): Second-derivative spectrum (——) of Type 2-depleted laccase at 2.38 GHz from an average of four scans, and simulation (---) with hyperfine splitting from two N atoms and two H atoms. Experimental conditions: microwave power, 10 dB; modulation amplitude, 0.5 mT; modulation frequencies, 100 kHz and 1 kHz; temperature, −150 °C. From Hanna *et al.* (1988).

resolved at low frequency (Froncisz *e al.*, 1979; Dunham *et al.*, 1983). The *g* values (2.18, 2.00) are taken from *X*-band data (Aasa *et al.*, 1976). Resolution of the *X*-band $g_\perp$ value into 2.02 and 1.99 by these authors based on *Q*-band spectra is, in our opinion, doubtful in the absence of a full-scale simulation. The hyperfine coupling constants ($A_z^{Cu} = 90$ MHz, $A_y^{Cu} = 90$ MHz, and $A_x^{Cu} = 68$ MHz) (Hoffman *et al.*, 1980) are about one half the couplings in other Type 1 sites. Two proton couplings ($A_1^H \sim 12$ MHz and $A_2^H \sim 19$ MHz) and a nitrogen coupling ($A^N \sim 17$ MHz) are observed by ENDOR (Hoffman *et al.*, 1980; Stevens *et al.*, 1982).

Subsequently, Coyle *et al.* (1985) reported the ESR spectrum of a cupric site in nitrous oxide reductase; it also had a g_z value of 2.18. The seven-line pattern in the g_z region was compared to a seven-line pattern from the EPR signal in the violet Cu(I)Cu(II) acetate mixed-valence complex. Based on a comparison of *S*-band (2.8 GHz) and *C*-band (4.56 GHz) data for the cupric-detectable site in nitrous oxide reductase and cytochrome *c* oxidase (Fig. 18), it was argued that the Cu_A site in cytochrome *c* oxidase is a

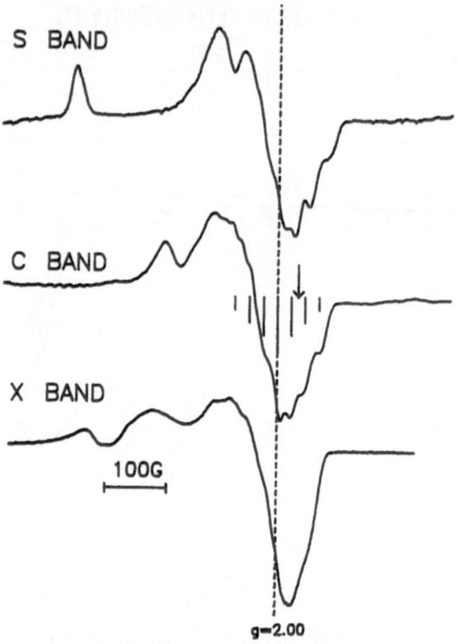

Figure 18. Multifrequency EPR data for nitrous oxide reductase (left) and cytochrome *c* oxidase (right). Note the superposition of the EPR signal from heme *a* in the spectra for cytochrome *c* oxidase. From top: *S* band (2.76 GHz), *C* band (4.53 GHz), and *X* band (9.13 GHz).

bimetallic center (Kroneck *et al.*, 1990, and references therein). The existence of a bimetallic center fits nicely with the couplings obtained by ENDOR, which are about one-half of the values expected for a Type 1 cupric site.

4.2. Motional Studies Using Multifrequency ESR

4.2.1. Dynamic Second-Order Shifts in a Cupric Complex of Dithiocarbamate

The expressions developed by Kivelson for the linewidths of copper complexes permit, by suitable detailed analysis, the determination of all spin Hamiltonian input parameters. However, the dominant terms are all proportional to the rotational correlation time τ_R, and in the real world of noise and experimental error, the less-dominant terms are difficult to measure. Thus one cannot usually obtain unique parameters for both the correlation time and the spin Hamiltonian input parameters.

In the early history of the ESR of copper, dynamic second-order shift contributions to the positions of the lines were ignored. With the development of the loop-gap resonator (Froncisz and Hyde, 1982) and new octave-bandwidth microwave bridges at low microwave frequencies, second-order effects became more pronounced. A suitable theoretical formulation is based on the work of Bruno *et al.* (1977). All dominant terms are proportional not to τ_R but to τ_R^2. If dynamic second-order shifts can be measured, then a full solution leading to unique determination of all spin Hamiltonian input parameters as well as the rotational correlation time is possible.

A computer simulation of dynamic shifts was developed as a function of microwave frequency and rotational correlation time (see Fig. 3). It was apparent that the shifts should, in fact, be readily determinable. Basosi *et al.* (1984) carried this regimen through in a multifrequency study. They extracted the linewidths and the line positions with respect to DPPH at five widely varying microwave frequencies (1.19, 2.51, 3.44, 9.12, and 34.7 GHz) using a low-molecular-weight complex, $Cu(dtc)_2$, in toluidine at room temperature. After algebraic manipulation using the linewidth and line-position data, values for the various parameters were determined together with error estimates. These values were then refined by computer simulation of the spectra (see Fig. 19). The data were not precisely consistent with theory, but the agreement was reasonably good.

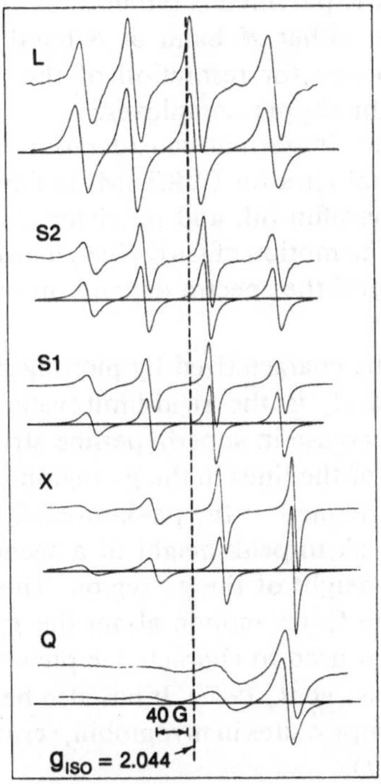

Figure 19. Experimental and simulated room-temperature spectra for $^{65}Cu(dtc)_2$ in *o*-toluidine at five frequencies. From Basosi *et al.* (1984).

4.2.2. Cupric Complexes of *bis*-Thiosemicarbazones as Spin Probes

The copper complex 3-ethoxy-2-oxobutyraldehyde *bis*(N^4,N^4-dimethylthiosemicarbazonato) copper (II), $CuKTSM_2$, a derivative of a potent antitumor agent, has proven to be a very effective ESR probe. The properties of $CuKTSM_2$ include a favorable partition coefficient into the lipid bilayer of membranes; a redox potential below that of glutathione (a prominent reducing agent in cells), which maintains intracellular $CuKTSM_2$ in the cupric state; and tetradentate coordination—a structure that is resistant to chemical exchange. It is a particularly inert compound. ESR spectra of $CuKTSM_2$ (Fig. 20; Antholine *et al.*, 1984a) at four different frequencies illustrate the following points:

1. Motional effects in the spectra of low-molecular-weight complexes are sensitive to frequency. Only the two high-field lines, particularly the $M_I = +\frac{3}{2}$, resemble spectra in the fast-motion domain at Q band. At L band, the four hyperfine lines are nearly equal in intensity, as is consistent with fast-motion conditions (Basosi, 1988).
2. The best resolution of the nitrogen superhyperfine structure is at X band for the ESR parameters typified by $CuKTSM_2$. In general, it is expected that either X band or S band (4–8 GHz) will be the optimum frequency for resolution of the $M_I = +\frac{3}{2}$ lines of low-molecular-weight copper complexes.
3. The solvent is an important consideration in defining the motion. Fast motion dominates for $CuKTSM_2$ in DMSO, slower motion for $CuKTSM_2$ in paraffin oil, and restricted motion for $CuKTSM_2$ in lipid bilayers. The motion of $CuKTSM_2$ in oriented artificial bilayers is anisotropic, and the spectra depend on orientation.

Slow motion can be characterized by plotting A_z^*/A_z, where A_z^* is the apparent $A_\parallel$ value and A_z is the rigid-limit value. Motion can also be characterized by the decrease in superhyperfine structure in the $g_\perp$ region. The relative resolution of the lines in the $g_\perp$ region can be measured by the Cu-motion parameter, defined as the peak-to-peak height of one superhyperfine line plus the peak-to-peak height of a second superhyperfine line divided by the overall height of the $g_\perp$ region. The Cu-motion parameter is believed to reflect the faster motion about the $g_\parallel$ axis in lipid bilayers. This parameter has been used to characterize phase transitions in artificial lipid bilayers (Subczynski *et al.*, 1987). It has also been found to be a useful indicator of motion of cupric sites in myoglobin, serum albumin, and laccase (Baffa and Hanna, 1987).

Translational motion of copper complexes can be deduced from collisions of these complexes with paramagnetic reporter species such as

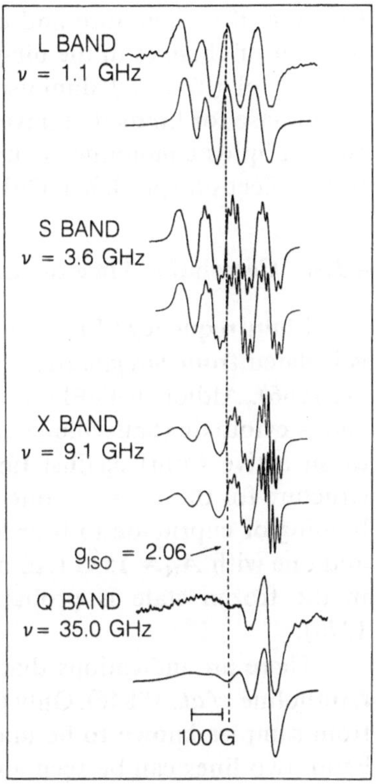

Figure 20. Multifrequency room-temperature (298 K) ESR spectra and simulated spectra for $^{65}CuKTSM_2$ in 50% Me_2SO–50% H_2O at four frequencies (from top) *L*-band, 1.093 GHz, *S* band, 3.579 GHz; *X* band, 9.094 GHz; *Q* band, 34.895 GHz. From Antholine *et al.* (1984a).

nitroxides. The method is based on Heisenberg exchange that shortens the effective spin-lattice relaxation time T_1 of spin labels (Hyde and Sarna, 1978). For example, consider acyl chains of a lipid bilayer that are spin-labeled at the 16-position near the bilayer center. Spin-label T_1s were measured in the presence and absence of $CuKTSM_2$ (Subczynski *et al.*, 1987, 1991). The difference in the inverse T_1, T_1^{-1} ($CuKTSM_2$) − T_1^{-1} (no $CuKTSM_2$), is proportional to the translational diffusion constant times the concentration of $CuKTSM_2$, i.e., the diffusion-concentration product of $CuKTSM_2$ at the center of the lipid bilayer:

$$\begin{aligned} W(\mathrm{CuKTSM_2}) &= T_1^{-1}(\mathrm{CuKTSM_2}) - T_1^{-1}(\mathrm{no\ CuKTSM_2}) \\ &= 4\pi pRD(\mathrm{CuKTSM_2})[\mathrm{CuKTSM_2}] \end{aligned} \tag{8}$$

Here *p* is a factor that allows for ineffective collisions and *R* is the interaction distance between the copper complex and the spin label during an encounter. The acyl chain of the bilayer can be labeled at different depths, and the $CuKTSM_2$ translational transport parameter *W* can be determined as a function of temperature and cholesterol content. $CuKTSM_2$ is a sensitive monitor of changes in the bilayer. More extensive studies have shown that cholesterol (27–30%) diminishes spin-label $CuKTSM_2$ interaction in the fluid phase for lipids that have saturated or *trans*-unsaturated alkyl chains. In host lipids containing *cis*-unsaturated alkyl chains, cholesterol has little or no effect on spin label–$CuKTSM_2$ interaction (Subczynski *et al.*, 1991).

4.2.3. Correlation Time for Cupric Bleomycin at Low Frequency

Multifrequency ESR has been useful to characterize CuBlm. Bleomycin is isolated from *Streptomyces verticillus* as a copper complex (Umezawa *et al.*, 1966). Although FeBlm is active in the DNA strand scission reactions and is cytotoxic (Sausville *et al.*, 1978a, b), CuBlm is active (or a precursor to an active form) against tumor cells (Rao *et al.*, 1980). Well-resolved structure for the $M_I = -\frac{1}{2}$ line in the $g_{\parallel}$ region at *S* band is consistent with binding of cupric ion to four nitrogen donor atoms, three with $A_N = 10$ G and one with $A_N = 15$ G (Antholine *et al.*, 1984b). The structure of CuBlm in the frozen state is probably pyramidal square-planar (Iitaka *et al.*, 1978).

There are indications that the structure changes at room temperature (Antholine *et al.*, 1984c). Only the high-field $M_I = +\frac{3}{2}$ line resembles spectra from samples known to be under fast-motion conditions at *X* band. At *S* band, two lines can be seen and a value for A_{iso} determined. The value at room temperature is 30 MHz less than the value of $\frac{1}{3}(A_{\parallel} + 2A_{\perp})$, where $A_{\parallel}$ and $A_{\perp}$ are obtained in frozen solution.

Another apparent change between room-temperature and frozen-solution data occurs in the presence of pyridine (Antholine *et al.*, 1984c). No difference between spectra in the presence and absence of pyridine could be detected for frozen samples, but at room temperature, changes are observed in linewidth and line position. It is suggested that the CuBlm is not spherical, but cigar shaped, and may have segmental flexibility or rotation about a hinge. The effect of pyridine on this motion would not have been detected in frozen solution.

4.3. Additional Applications

4.3.1. Multiple Spin Probes, Multifrequency Data, and Interspin Distance in Cu Nitroxyls

In fluid solution in the rapid-tumbling limit, the dipolar contribution to the ESR line shape is averaged to zero by the tumbling of the molecule, but the exchange information is retained (S. S. Eaton and G. R. Eaton,

1978, 1988; Eaton *et al.*, 1980). Values of the electron-electron coupling constant J have been obtained from the fluid solution spectra for a wide variety of spin-labeled transition-metal complexes. In the case of Cu(II), single-crystal ESR spectra permit a complete analysis of both the exchange and dipolar contributions to the spin–spin interaction. However, it is often difficult to obtain single crystals of the compounds of interest. It is, therefore, important to determine the extent to which information can be taken from frozen-solution ESR spectra. For cases where the g and A values must also be determined from the frozen-solution spectra, comparison of spectra taken at two or more microwave frequencies is necessary to separate the effects of spin–spin interaction from the determination of the Hamiltonian parameters (Francesconi *et al.*, 1981; Coffman and Beuttner, 1979; Eaton *et al.*, 1983). Furthermore, when J is large, the distance between the two unpaired electrons can be determined from the dipolar splitting.

An accurate method of analysis has been carried out by Pezeshk and Coffman (1986) in order to determine the values of the conformational parameters that characterize the interactions. The *L*-, *S*-, *X*-, and *Q*-band ESR spectra of a *bis*(diisopropylsalicylato)copper(II) pyridyl nitroxide complex have been analyzed by computer simulation. The low-frequency ESR spectra of this type were found to be of better quality for analysis than *X*- and *Q*-band spectra. Simulations of the spectra in the $g = 2$ region at *S* and *L* band were carried out. Excellent agreement of the interspin distance, to within ±0.005 Å, was demonstrated using two independent methods: line-shape simulation and direct determination from the relative intensities of the half-field line using Eaton's method. The values of the spin Hamiltonian parameters found from the least-squares analysis of the half-field lines were used to predict the broad doublet spectra at several microwave frequencies (see Fig. 21). The results shown in Fig. 21 demonstrate agreement with experiment at all three frequencies and allow precise measurement of a short distance, $R = 3.36$ Å, that can be explained on the basis of a dimeric model.

4.3.2. Characterization of the Local Environment of Copper Ions in Polymeric Systems

An analysis of ESR spectra of the cupric probe in cross-linked polyacrylamide gels at various microwave frequencies has been used to determine the distribution parameters as a function of the gel pore size and to quantify the gradual change in the water properties due to the porosity of the network (Rex and Schlick, 1987). In studies on perfluorination ionomers swollen by various solvents, multifrequency ESR was used to determine the contribution of g and hyperfine anisotropies to the linewidth and to isolate the broadening associated with paramagnetic neighbors; thus the interaction

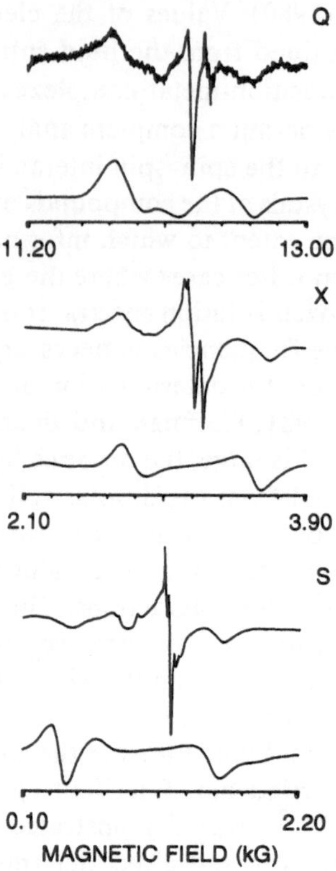

Figure 21. Comparison of experimental and computed broad-band, $S = 1$ doublet ($S1$) spectra corresponding to the Cu(II)-radical interaction with an interspin distance of 3.365 Å; top, 34.847 GHz; middle, 9.113 GHz; bottom, 3.4379 GHz. The sharp lines in the center belong to the $S2$ spectrum. From Pezeshk and Coffman (1986).

distance was determined (Schlick and Alonso-Amigo, 1987, 1989; Alonso-Amigo and Schlick, 1989). In these systems, the ligands were oxygen atoms.

The gradual replacement of oxygen ligands around Cu^{2+} with nitrogen ligands in Nafion soaked with acetonitrile has been studied using ESR spectra at *L*, *S*, *C*, and *X* bands combined with computer simulations. After two cycles of drying and soaking with the solvents, all four oxygen ligands in the equatorial plane of Cu^{2+} are replaced by nitrogen ligands. Multi-frequency ESR spectra allow determination of all ESR parameters (Fig. 22; Bednarek and Schlick, 1991). In particular, spectra at *C* band are critical for the determination of the ratio $\Delta g/\Delta A$ of the distribution parameters due to strain. Furthermore, the greater resolution of hyperfine structure in the $M_I = -\frac{3}{2}$ and $M_I = -\frac{1}{2}$ components at *C* band relative to *X* band and *S* band is crucial for simulation of experimental results (see Fig. 7). The higher resolution at the high-field edge of the $g_\parallel$ signal ($M_I = \frac{3}{2}$) of the *C*

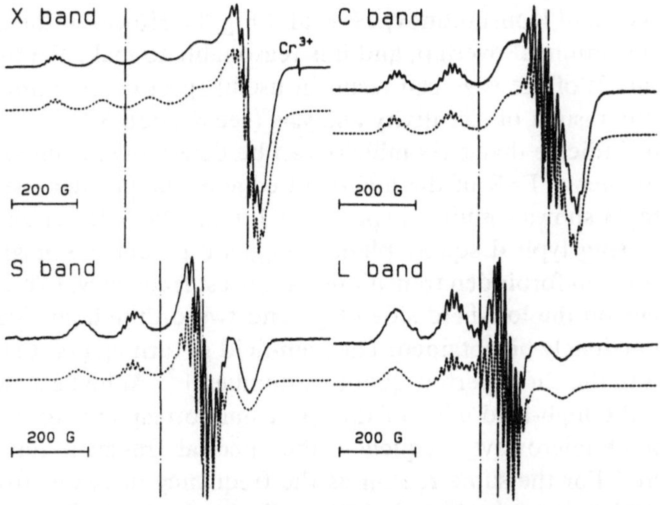

Figure 22. Experimental (——) and simulated (---) ESR spectra of Nafion/$^{63}Cu^{2+}$ (Cu^{2+} concentration 4%) soaked *twice* by CH_3CN at *X* band (9.36 GHz, top left) and 110 K and at *C* band (4.7 GHz, top right), *S* band (2.8 GHz, bottom left), and *L* band (1.2 GHz, bottom right) at 123 K. Vertical lines (- · -) indicate the position of $g_{\parallel}$ (low field) and $g_{\perp}$ (high field). From Bednarek and Schlick (1991).

band spectrum is an additional advantage. The example emphasizes the importance of multifrequency ESR analysis for disordered systems containing copper in the presence of strain.

5. PERSPECTIVES

5.1. Separation of g_x and g_y for Copper Complexes with Limited Rhombicity Using *K*-Band ESR

There are several technical difficulties in obtaining magnetic parameters for the ⊥ region of Type 2 copper complexes because of spectral overlap. Little is known of *g*- and *A*-strain phenomena in this region, leading to uncertainty in the linewidth. Belford and Duan (1978) have pointed out that the so-called secondary and tertiary transitions can be as intense as the allowed transitions at *X* band. These features depend on the copper quadrupole interaction, which varies greatly depending upon ligation. In addition, both *A* and *g* tensors can be rhombic. A common ligand is nitrogen, and the ^{14}N-hyperfine couplings are themselves somewhat anisotropic in the ⊥-plane (Rist and Hyde, 1969) and give rise to angularly dependent inhomogeneous linewidths. Finally, the ⊥-region is dominated in a wide range of conditions by so-called extra-absorption or overshoot lines

(Ovchinnikov and Konstantinov, 1978, and Fig. 1). However, the high-field edge exhibits minimal overlap, and it is reasonable to make the hypothesis that an analysis of the edge can result in useful spectroscopic information. Based on the insight of sensitivity analysis (see Section 3.3.5), preliminary simulations indicate that g-rhombicity can be determined from analysis of the multifrequency ESR of the high-field edge of the powder spectrum.

Figure 23 shows simulated spectra at 1.0 and 19 GHz for axial g and A tensors using typical square-planar magnetic parameters neglecting g- and A-strain and forbidden transitions. At lowest frequency, two $g_{\parallel}$ turning points occur on the low-field side of $g_{\perp}$ and two on the high side: $g_{\parallel}$ and $A_{\parallel}$ can immediately be obtained. The high-field $g_{\parallel}$ turning point at 1.0 GHz is assigned to the Cu hyperfine quantum number $+\frac{3}{2}$. At highest microwave frequency, the high-field edge of the spectrum corresponds to $g_{\perp}$, $+\frac{3}{2}$. As a function of microwave frequency, the spectral fragment for $+\frac{3}{2}$ turns "inside out." For the same reason as the frequency increases from a low value, the $+\frac{3}{2}$ $g_{\parallel}$ intensity in a derivative display begins to increase.

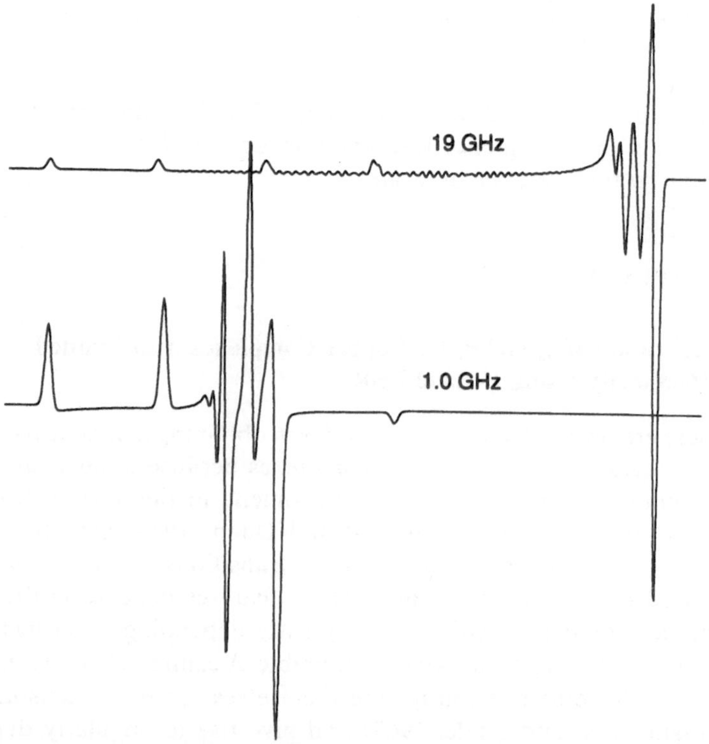

Figure 23. Simulated spectra at 1.0 (bottom) and 19 GHz (top) for axial g and A tensors using typical square-planar magnetic parameters, neglecting strain and forbidden transitions.

In particular, it has been found by computer simulation that the intensity of the high-field parallel feature is sensitive to g-tensor rhombicity. The ratio of the high-field $g_\perp$ peak to the low-field $g_\parallel$ peak in the case of square-planar symmetry with slight distortion is dependent on g-tensor rhombicity. Plots of this ratio as a function of microwave frequency for both axial and rhombic g tensors are shown in Fig. 24. Pogni *et al.* (1990) suggested that one can obtain $\perp$-information by comparison of high- and low-field peak intensities in low-frequency spectra, both of which are assigned to the $\parallel$-region turning points. It is particularly attractive in that the information comes from spectral extrema where overlap is minimal.

5.2. Determination of Quadrupole Coupling

Of particular note is the work of several groups, but predominantly Belford's (Rothenberger *et al.*, 1986), in extracting copper quadrupole interactions by computer simulations of the $g_\perp$ region of spectra obtained at two frequencies, usually X and Q band (see also Conesa and Soria, 1979). This information can aid in distinguishing sulphur and oxygen ligands and in obtaining information on axial ligation. The quadrupole-sensitive features in the spectrum arise from so-called secondary, tertiary, and quaternary forbidden transitions that are, in fact, not forbidden at all because of strong mixing of the nuclear states.

Belford and Duan (1978) observe that when "the neighboring atoms are oxygen, one can pretty well read off the general amount of site symmetry from the principal quadrupole coupling constant." For a given in-plane ligation, they observe that QD values increase with expected coordinating power of the solvents. This is one of the few methods that relates to axial ligation of copper complexes.

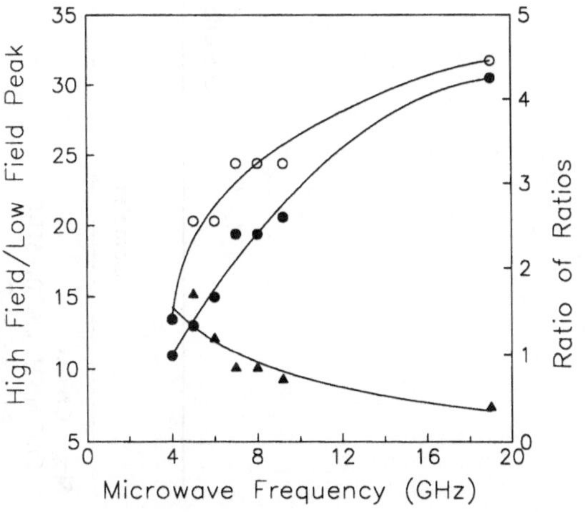

Figure 24. Plots of the ratio of the high-field $g_\perp$ peak to the low-field $g_\parallel$ peak as a function of microwave frequency for both axial and rhombic g tensors. ○ = square-planar; ▲ = rhombic; ● = ratio of ratios.

TABLE 2
Magnetic Parameters for Type 2 Copper Complexes Where QD Is Known

Complex	*QD* (MHz)	*QE* (MHz)	g_x	g_y	g_z	A_x (MHz)	A_y (MHz)	A_z (MHz)	Reference[a]
Single Crystal									
$Cu/Zn(NH_4)_3(SO_4)_3 \cdot 6H_2O$	33	—	—	—	—	—	—	—	1
$Cu/Pd(acac)_2$ 4-O	21	—	2.055	2.051	2.266	57	57	479	2
$^{63}Cu/Pd(accac)_2$ 4-O	10	—	2.055	2.051	2.266	57	57	479	3
$^{63}Cu/Pd(bzac)_2$ 4-O	7.8	—	—	—	—	—	—	—	3
$^{63}Cu/Cd(CH_3COO)_2 \cdot 2H_2O$ 7-O	29	3.6	2.0382	2.1644	2.4139	150	75.8	304.5	4
Cu(phthalocyanine)/metal-free phthalocyanine 4–N	18	—	2.050	2.050	2.179	57	57	605.5	5
Cu^{2+}/simple crystals of cytosine mono hydrate(3-O + 1-N)	15 ± 6	—	2.053	2.057	2.253	26.7	52.46	545.6	6
$^{63}Cu/Zn(nap)_2(4\text{-}Mepy)_2$ [2-N(*trans*) + 4-O]	10.3	1.9	2.0882	2.0538	2.3014	119	65	437	7
$^{63}Cu/Zn(na)_2(py)_2$[2-N(*trans*) + 4-O]	8.5	1	2.0723	2.0658	2.3126	55.4	6.3	473.3	8
$^{63}Cu/Zn(dtc)_2$ 4-S	9	—	2.0236	2.0236	2.1085	67.15	67.15	427	9
$^{63}Cu/Ni(dtc)_2$ 4-S	2.0	—	—	—	—	—	—	—	3
$Cu/Zn(dtc)_2$ 4-S	9.4	1.6	—	—	—	—	—	—	10

$^{63}Cu/Ni(dtc)_2^{3-}$ 4-S	2.1	—	2.0197	2.0191	2.0805	127	124.7	491.3	11
$^{63}Cu/Ni(dtc)_2^{2-}$ 4-S	2.1	—	2.0227	2.0186	2.0856	115.4	103.4	468	11
$^{63}Cu/Ni(i\text{-mnt})_2^{3-}$ 4-S	2.1	—	2.023	2.019	2.086	115.4	105	467	11
$^{63}Cu/Ni(S_2P(OC_3H_5)_2)_2$ 4-S	5.4–5.7	—	2.0199	2.0230	2.0856	91.43	98.33	451.4	11
$^{63}Cu/Ni(mnt)_2^{2-}$ 4-S	3.6	−1.5	2.0210	2.0199	2.0837	118	117	481	11
$^{63}Cu/Pd(SdbmO)_2$ [2-S + 2-O]	3.0	1.2	2.0294	2.0313	2.1450	105.5	112.7	481	11
Powder									
Cu/Pd$(acac)_2$ 4-O	24	—	2.051	2.054	2.264	63	60	545.6	12
$Cu(acac)_2$ frozen ethanol	21.9	—	2.062	2.062	2.29	38.37	38.37	503.65	13
$Cu(acac)_2$ frozen pyridine	25.5	—	2.066	2.066	2.294	35.38	35.38	485.66	13
$Cu(acac)_2$ frozen $CHCl_2$	20.4	—	2.0611	2.0611	2.286	38.97	38.97	518.64	13
$^{63}Cu/Pd(acac)_3$	9	0.6	2.0490	2.0525	2.262	72	79	568	14
Cu in Stellacyanin (Type 1)									
[2-N + S + ?]	15	−1	2.018	2.075	2.282	167	87	96	15
Cu Bacitracin [2-N(*cis*) + 2-O]	15.9	0	2.0575	2.0469	2.261	60	53.4	534	16
Cu in Stellacyanin (Type 1)	22.5	−1.5	2.02	2.07	2.28	167	87	96	17
$Cu(dtc)_2$ frozen pyridine 4-S	21.5	0.25	2.030	2.033	2.120	43	76	423	18

[a] 1. Bleaney *et al.* (1955); 2. Maki and McGarvey (1958); 3. So and Belford (1969); 4. Bonomo and Pilbrow (1981); 5. Harrison and Assour (1965); 6. Toriyama and Iwasaki (1975); 7. Attanasio (1977); 8. Attanasio and Gardini (1978); 9. Reddy and Srinivasan (1965); 10. Keijzers *et al.* (1975b); 11. White and Belford (1976); 12. Rollmann and Chan (1969); 13. Belford and Duan (1978); 14. Rothenberger *et al.* (1986); 15. Roberts *et al.* (1980); 16. Seebauer *et al.* (1983); 17. Fan *et al.* (1988); 18. Liczwek *et al.* (1983).

Table 2 is intended to be a rather complete list of available copper magnetic parameters for those complexes for which the quadrupole interaction is reported. It is notable for its shortness; there is not much information on *QD*, and even less on *QE* (we adopt the notation for the quadrupole coupling used by White and Belford (1976)).

Many workers cited in Table 2 have built convincing arguments that the quadrupole tensor gives valuable information on ligation. The general tendency for *QD* to decrease on going from oxygen to sulphur ligation is noted, as is the exception in the last entry in the table. The extreme paucity of data on *QD* with nitrogen ligation is noted. The superhyperfine interaction makes the problem difficult. It is in principle possible to obtain hyperfine rhombicity in the same analysis that yields quadrupole couplings and *g*-tensor rhombicity (see Section 5.1).

Hyde and Froncisz (1982) considered the use of other microwave frequencies and suggested a new experiment: quadrupole interactions could be obtained at very low microwave frequencies by observing the spectrally clear $g_{\parallel}$ region with $H_1 \| H_0$. Based on this idea, an octave-bandwidth 1-2-GHz bridge was developed and the idea confirmed (Rothenberger *et al.*, 1986, and Fig. 8).

Rothenberger *et al.* (1986) showed that by simulating *L*-band powder spectra at both $H_1 \|$ and $H_1 \perp H_0$, essentially all parameters could be obtained in a system where there were no nitrogen ligands and no strong couplings to nearby protons. As *g*- and *A*-strain increase and lines broaden because of superhyperfine couplings, the problem increases in difficulty, and the information that can reliably be extracted is reduced. Determination of copper quadrupole couplings in frozen solution in the presence of nitrogen ligation has not yet been carried out. Most biologically significant compounds have nitrogen ligands, and an objective for the future can be to use multifrequency ESR to obtain *QD* (and even *QE*) on compounds of this class (Rothenberger, 1988).

5.3. Postacquisition Data Processing

In Section 3.4.2, the use of pseudomodulation for formation of second-harmonic displays and for resolution enhancement was discussed. Further development of postacquisition processing of EPR spectra can be anticipated in future years.

Pseudomodulation is described mathematically as the convolution of a spectrum $S(H)$ with a delta function that is sinusoidally modulated in position:

$$S_m(H, t) = S(H) * \delta\left(H - \frac{H_m}{2} \cos \omega t\right) \tag{9}$$

Here H_m is the p-p pseudomodulation amplitude and $S_m(H, t)$ is the resulting signal at the input to a lock-in amplifier. Harmonics are produced, as one would expect:

$$S_m(H, t) = \sum_{n=0}^{\infty} S_n(H) \cos n\omega t \tag{10}$$

Here $S_n(H)$ are the harmonics. It is an interesting observation that pseudomodulation partitions information into the n harmonics. As one possible extension of pseudomodulation, one could apply the algorithm to a simulation as well as to an experimental spectrum and test the quality of the simulation by requiring all simulated and experimental harmonics to match.

Another idea, suggested by Hyde *et al.* (1992), was to apply a Hilbert transform to a spectrum, converting it to a dispersion display, followed by a calculation of the first harmonic. This display seemed particularly appropriate for copper spectra.

One can write

$$S_m(H, t) = \sum_{n=0}^{\infty} \frac{\partial^n S(H)}{\partial H^n} * h_n(H, H_m) \cos n\omega t \tag{11}$$

Here, h_n are filter functions that are rather complicated sums of Bessel functions. Many alternative filter functions could be considered for convolution with derivatives of the original spectra, and evaluations could be made of their relative merits.

Some of these ideas are illustrated in Fig. 25. The starting spectrum, from azurin, is labeled (a). The second harmonic of this spectrum, shown in (b), strongly emphasizes the superhyperfine features. By subtracting a portion of (b) from (a), a resolution-enhanced spectrum (c) is obtained. The tradeoff between increased noise and increased resolution is apparent. Figure 25d is the dispersion spectrum produced by applying a Hilbert transform to the starting spectrum, and Fig. 25e is the first harmonic of (d). It is suggested that this display should be used more widely. Spectra (b) and (e) have the same symmetry properties, and both can be used for resolution enhancement. Figure 25f shows the result of subtracting a portion of (e) from (a). Comparing the resolution-enhanced spectra of Figs. 25c, 25f, one is again made aware of the tradeoff between signal-to-noise ratio and resolution.

Computer processing of spectroscopic data is certain to increase in future years.

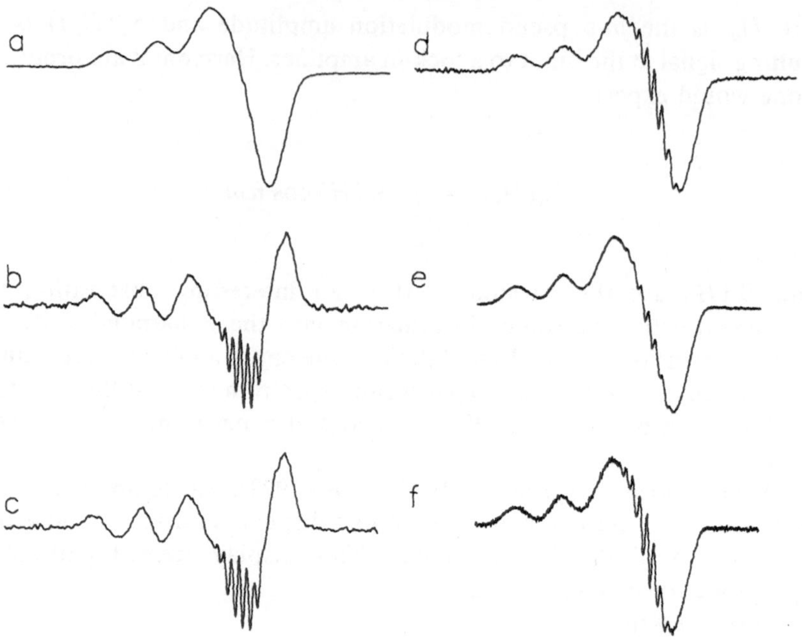

Figure 25. Pseudomodulation study of the nitrogen superhyperfine couplings of the blue copper protein azurin. (a) Original spectrum obtained at *S* band, 400-G sweep. (b) Second harmonic of (a) using 4-G pseudomodulation amplitude. If the peak-to-peak density of (a) is set to 1, then it is actually 0.016 in (b). It is convenient here to normalize second-harmonic peak-to-peak signal intensities. (c) 0.16 of (b) (normalized) subtracted from (a). (d) Hilbert transform of (a) to form the dispersion signal. (e) First harmonic of (d) using 2-G pseudo modulation amplitude. The peak-to-peak signal intensity of (e) is actually 0.027 of (a). (f) Resolution-enhanced spectrum obtained by subtracting 0.27 of (e) (normalized) from (a). From Hyde *et al.* (1992).

5.4. Computer Simulations of Copper Spectra under Slow-Motion Conditions

It took many years for J. H. Freed and his colleagues to develop efficient computer programs for simulation of nitroxide radical spectra in the slow-motion range and to verify them by precise comparison with experimental spectra. Extension to fluid-phase spectra of copper is desirable. Below is a list of possible problems, not necessarily complete.

- Copper quadrupole coupling should be included.
- Motional averaging of forbidden transitions must be considered.
- Hyperfine coupling to nitrogen ligands and the copper nucleus itself should be included.
- Motion might affect nitrogen and copper nuclear relaxation rates in such a manner as to affect the spectra.

- Dynamic second-order shifts in spectral position are an apparent aspect of slow-motion spectra. High-field approximations will fail at low frequencies.
- The spin–rotation interaction must be included.
- Slow-motion spectra with $H_1 \| H_0$ are of interest.

In addition to problems in chemical physics as suggested by this list, there are a number of uncertainties in the properties of the copper complexes:

- Vibrational and rotational modes of the complex may be important.
- Exchange rates with axial ligands may need to be considered.
- On the ESR time scale, some systems may be better represented as mixtures of slowly interchanging conformations.
- Solvent effects may be significant.

It is our thesis that theoretical development is now poised to go forward, depending on multifrequency ESR spectroscopy for verification using carefully selected model compounds.

REFERENCES

Aasa, R., Albracht, S. P. J., Falk, K. E., Lann, B., and Vänngård, T., 1976, *Biochim. Biophys. Acta* **422**:260.

Abragam, A., 1970, *The Principles of Nuclear Magnetism,* Oxford University Press, London.

Abragam, A., and Bleaney, B., 1970, *Electron Paramagnetic Resonance of Transition Ions,* Oxford University Press, London, p. 911.

Adman, E. T., 1991, in *Advances in Protein Chemistry: Metalloproteins: Structural Aspects,* Vol. 42 (C. R. Abfinsen, F. M. Richards, J. J. Edsall, and D. S. Eisenberg, eds.), Academic Press, San Diego, pp. 145–197.

Adman, E. T., Turley, S., Bramson, R. Petratosi, K., Banner, D., Tsernoglon, D., Beppu, T., and Watanabe, H., 1989, *J. Biol. Chem.* **264**: 87.

Alonso-Amigo, M. G., and Schlick, S., 1989, *Macromolecules* **22**:2628.

Antholine, W. E., Basosi, R., Hyde, J. S., Lyman, S., and Petering, D. H., 1984a, *Inorg. Chem.* **23**:3563.

Antholine, W. E., Hyde, J. S., Sealy, R. C., and Petering, D. H., 1984b, *J. Biol. Chem.* **259**:4437.

Antholine, W. E., Riedy, G., Hyde, J. S., Basosi, R., and Petering, D. H., 1984c, *J. Biomol. Struct. Dynam.*, **2**:469.

Atkins, D., and Kivelson, D., 1966, *J. Chem. Phys.* **44**:169.

Attanasio, D., 1977, *J. Magn. Reson.* **26**:81.

Attanasio, D., and Gardini, M., 1978, *J. Magn. Reson.* **32**:411.

Baffa, O., and Hanna, P. M., 1987, unpublished work during visits to the National Biomedical ESR Center in Milwaukee.

Baker, E. N., 1988, *J. Mol. Biol.* **203**:1071.

Basosi, R., 1988, *J. Phys. Chem.* **92**:992.

Basosi, R., Antholine, W. E., Froncisz, W., and Hyde, J. S., 1984, *J. Chem. Phys.* **81**:6869.

Basosi, R., Valensin, G., Gaggelli, E., Froncisz, W., Pasenkiewicz-Gierula, M., Antholine, W. E., and Hyde, J. S., 1986, *Inorg. Chem.* **25**:3006.

Bednarzek, J., and Schlick, S., 1991, *J. Am. Chem. Soc.* **113**:3303.
Belford, R. L., and Duan, D. C., 1978, *J. Magn. Reson.* **29**:293.
Bleaney, B., Bowers, K. D., and Ingram, D. J. E., 1955, *Proc. Royal Soc.* (*London*) **A228**:147.
Bloch, F., 1956, *Phys. Rev.* **102**:104.
Boas, J. F., Pilbrow, J. R., and Smith T. D., 1978, in *Biological Magnetic Resonance*, Vol. 1 (L. J. Berliner and J. Reuben, eds.), Plenum, New York, p. 345.
Bonomo, R. P., and Pilbrow, J. R., 1981, *J. Magn. Reson.* **45**:404.
Bränden, R., Malmström, B., and Vänngård, T., 1971, *Eur. J. Biochem.* **18**:238.
Brill, A. S., 1977, *Transition Metals in Biochemistry*, Springer-Verlag, Berlin, p. 186.
Bruno, G. V., Harrington, J. K., and Eastman, M. P., 1977, *J. Phys. Chem.* **81**:1111.
Budil, D. E., Earle, K. A., Lynch, W. B., and Freed, J. H., 1989, in *Advanced EPR: Applications in Biology and Chemistry* (A. J. Hoff, ed.) Elsevier, Amsterdam, pp. 307–340.
Buluggiu, E., and Vera, A., 1973, *J. Chem. Phys.* **59**:2886.
Buluggiu, E., Dascola, G., Giori, D. C., and Vera, A., 1971, *J. Chem. Phys.*, **54**:2191.
Buluggiu, E., Vera, A., and Tomlinson, A. A. G., 1972, *J. Chem. Phys.* **56**:5602.
Campbell, R. F. and Freed, J. H., 1980, *J. Phys. Chem.* **84**:2668.
Coffman, R. E., and Beuttner, G. R., 1979, *J. Phys. Chem.* **83**:2392.
Conesa, J. C., and Soria, J., 1979, *J. Magn. Res.* **33**:295.
Coyle, C. L., Zumft, W. G., Kroneck, P. M. H., Körner, H., and Jakob, H., 1985, *Eur. J. Biochem.* **153**:459.
Dougherty, G., Pilbrow, J. R., Skorobogaty, A., and Smith, T. D., 1985, *J. Chem. Soc.* **281**:1739.
Dunham, W. R., Sands, R. H., Shaw, W. R., and Beinert, H., 1983, *Biochim. Biophys. Acta*, **748**:73.
Eaton, G. R., and Eaton, S. S., 1988a, *Spectroscopy*, **3**:34.
Eaton, G. R., and Eaton, S. S., 1988b, *Bull. Magn. Reson.* **10**:3.
Eaton, S. S., and Eaton, G. R., 1978, *Coord. Chem. Rev.* **26**:207.
Eaton, S. S., and Eaton, G. R., 1988, *Coord. Chem. Rev.* **83**:29.
Eaton, S. S., More, K. R., Dubois, D. L., Boywel, P. M., and Eaton, G. R., 1980, *J. Magn. Reson.* **41**:150.
Eaton, S. S., More, K. R., Sawant, B. M., Boywel, P. M., and Eaton, G. R., 1983, *J. Magn. Reson.* **42**:435.
Falk, K.-E., Ivanova, E., Roos, B., and Vänngård, T., 1970, *Inorg. Chem.* **9**:556.
Fan, C., Taylor, H., Bank, J., and Scholes, C. P., 1988, *J. Magn. Reson.* **76**:74.
Francesconi, L. C., Corbin, D. R., Clauss, A. W., Hendrickson, D. N., and Stuky, G. D., 1981, *Inorg. Chem.* **20**:2059.
Fraenkel, G. K., 1965, *J. Chem. Phys.* **62**:6275.
Fraenkel, G. K., 1967, *J. Phys. Chem.* **71**:139.
Freed, J. H., 1987, in *Rotational Dynamics of Small and Macromolecules in Liquids* (T. Dorfmuller and R. Pecora, eds.), Springer-Verlag, New York, p. 89.
Froncisz, W., and Aisen, P., 1982, *Biochim. Biophys. Acta* **700**:55.
Froncisz, W., and Hyde, J. S., 1980, *J. Chem. Phys.* **73**:3123.
Froncisz, W., and Hyde, J. S., 1982, *J. Magn. Reson.* **47**:515.
Froncisz, W., Scholes, C. P., Hyde, J. S., Wei, Y.-H., King, T. E., Shaw, R. W., and Beinert, H., 1979, *J. Biol. Chem.* **254**:7482.
Gersmann, H., and Swalen, J., 1962, *J. Chem. Phys.* **36**:3221.
Goodman, B. A., and Raynor, B., 1970, *Adv. Inorg. Chem. Radiochem.* **13**:135.
Guss, J. M., Harrowell, P. R., Murata, M., Norris, V. A., and Freeman, H. C., 1986, *J. Mol. Biol.* **192**:361.
Haindl, E., Möbius, K., and Oloff, H., 1985, *Z. Naturforsch* **40A**:169.
Hanna, P. M., McMillin, D. R., Pasenkiewicz-Gierula, M., Antholine, W. E., and Reinhammer, B., 1988, *Biochem. J.* **253**:561.

Harrison, S. E., and Assour, J. M., 1965, *J. Chem. Phys.* **40**:365.
Herring, F. G., and Park, J. M., 1979, *J. Magn. Reson.* **36**:311.
Hoffman, B., Roberts, J. E., Swanson, M., Speck, S. H., and Margoliash, E., 1980, *Proc. Natl. Acad. Sci. USA* **77**:1452.
Hyde, J. S., and Froncisz, W., 1982, *Ann. Rev. Biophys. Bioeng.* **11**:391.
Hyde, J. S., and Froncisz, W., 1986, *Electron Spin Reson.* **10A**:175.
Hyde, J. S., and Froncisz, W., 1989, in *Advanced EPR: Applications in Biology and Biochemistry* (A. J. Hoff, ed.), Elsevier, Amsterdam, pp. 277–306.
Hyde, J. S., and Sarna, T., 1978, *J. Chem. Phys.* **68**:4439.
Hyde, J. S., Antholine, W. E., and Basosi, R., 1985, in *Biological and Inorganic Copper Chemistry*, Vol. 2 (K. D. Karlin and J. Zubieta, eds.), Adenine, Guilderland, NY, p. 239.
Hyde, J. S., Antholine, W. E., Froncisz, W., and Basosi, R., 1986, in *Advanced Magnetic Resonance Techniques in Systems of High Molecular Complexity* (N. Niccolai and G. Valensin, eds.), Birkhauser, Boston, pp. 363–384.
Hyde, J. S., Pasenkiewicz-Gierula, M., Basosi, R., Froncisz, W., and Antholine, W. E., 1989, *J. Magn. Reson.* **82**:63.
Hyde, J. S., Pasenkiewicz-Gierula, M., Jesmanowicz, A., and Antholine, W. E., 1990, *Appl. Magn. Reson.* **1**:483.
Hyde, J. S., Jesmanowicz, A., Ratke, J. J., and Antholine, W. E., 1992, *J. Magn. Reson.* **96**:1.
Iitaka, Y., Nakamura, H., Nakatani, T., Murasko, Y., Fujii, A., Takita, T., and Umezawa, H., 1978, *J. Antibiot. (Tokyo)* **31**:1070.
Keijzers, C. P., Paulussen, F. G. M., and de Boer, E., 1975a, *Mol. Phys.* **29**:973.
Keijzers, C. P., van der Meer, P. L. A. C. M., and de Boer, E., 1975b, *Mol. Phys.* **6**:1733.
Kivelson, D., 1960, *J. Chem. Phys.* **33**:1094.
Kivelson, D., 1972, in *Electron Spin Relaxation in Liquids* (L. T. Muss and P. W. Atkins, eds.) Plenum, New York, pp. 213–277.
Kivelson, D., and Neiman, R., 1961, *J. Chem. Phys.* **35**:149.
Kroneck, P. M. H., Antholine, W. E., Kastrau, D. H. W., Buse, G., Steffens, G. C. M., and Zumft, W. G., 1990, *FEBS Lett.* **268**:274.
Lebedev, Y. S., 1987, *Teor. Eksper. Khim.* **73**:66.
Lewis, W. B., Alei, M., Jr., and Morgan, L. O., 1966, *J. Chem. Phys.* **44**:2409.
Liczwek, D. L., Belford, R. L., Pilbrow, J. R., and Hyde, J. S., 1983, *J. Phys. Chem.* **87**:2509.
Low, W., 1960, *Paramagnetic Resonance in Solids*, Academic Press, New York.
Maki, A. H., and McGarvey, B. R., 1958, *J. Chem. Phys.* **29**:31.
Malkin, R., Malmström, B., and Vänngård, T., 1968, *FEBS Lett.* **1**:50.
McConnell, H. M., 1956, *J. Chem. Phys.* **25**:709.
McGarvey, B. R., 1956, *J. Phys. Chem.* **60**:71.
Moro, G., and Freed, J. H., 1981, *J. Chem. Phys.* **74**:3757.
Morpurgo, L., Calabrese, L., Desideri, A., and Rotilio, G., 1981, *Biochem. J.* **193**:639.
Noack, M., and Gordon, G., 1968, *J. Chem. Phys.* **48**:2689.
Orii, Y., and Morita, M., 1977, *J. Biochem. (Tokyo)* **81**:163.
Ovchinnikov, I. V., and Konstantinov, V. N., 1978, *J. Magn. Reson.* **32**:179.
Pasenkiewicz-Gierula, M., Antholine, W. E., Subczynski, W. K., Baffa, O., Hyde, J. S., and Petering, D. H., 1987a, *Inorg. Chem.* **26**:3945.
Pasenkiewicz-Gierula, M., Froncisz, W., Basosi, R., Antholine, W. E., and Hyde, J. S., 1987b, *Inorg. Chem.* **26**:801.
Peisach, J. and Blumberg, W. E., 1974, *Arch. Biochem. Biophys.* **165**:691.
Pember, S. O., Benkovic, J. J., Villafranca, J. J., Pasenkiewicz-Gierula, M., and Antholine, W. E., 1987, *Biochemistry* **26**:4477.
Pezeshk, A., and Coffman, R. E., 1986, *J. Phys. Chem.* **90**:6638.
Pilbrow, J. R., 1988, *Bull. Magn. Reson.* **10**:32.

Pilbrow, J. R., 1990, *Transition Ion Electron Paramagnetic Resonance*, Oxford University Press, London, pp. 211–259.
Pogni, R., Basosi, R., Antholine, W. E., and Hyde, J. S., 1990 (private communication).
Rakhit, G., Antholine, W. E., Froncisz, W., Hyde, J. S., Pilbrow, J. R., Sinclair, G. R., and Sarkar, B., 1985, *J. Inorg. Biochem.* **25**:217.
Rao, E. A., Saryan, L. A., Antholine, W. E., and Petering, D. H., 1980, *J. Med. Chem.* **23**:1310.
Reddy, T. R., and Srinivasan, R., 1965, *J. Chem. Phys.* **43**:1404.
Redfield, A. G., 1957, *IBM J. Res. Develop.* **1**:19.
Rex, G. C., and Schlick, S., 1987, *Am. Chem. Soc. Symp. Ser.* **350**:265.
Rist, G. H., and Hyde, J. S., 1969, *J. Chem. Phys.* **50**:4532.
Roberts, J. E., Brown, T. G., Hoffman, B. M., and Peisach, J., 1980, *J. Am. Chem. Soc.* **102**:825.
Rollmann, L. D., and Chan, S. I., 1969, *J. Chem. Phys.* **50**:3416.
Rothenberger, K. S., 1988, Determination of Quadrupole Couplings via *L*-Band, $B_0 \| B_1$ EPR, Ph.D. dissertation, U. of Illinois.
Rothenberger, K. S., Nilges, M. J., Altman, T. E., Glab, K., Belford, R. L., Froncisz, W., and Hyde, J. S., 1986, *Chem. Phys. Lett.* **124**:295.
Saryan, A., Mailer, K., Krishnaruti, C., Antholine, W. E., and Petering, D. H., 1981, *Biochem. Pharmacol.* **30**:1595.
Sausville, E. A., Peisach, J., and Horowitz, S. B., 1978a, *Biochemistry* **17**:2740.
Sausville, E. A., Stein, R. W., Peisach, J., and Horowitz, S. B., 1978b, *Biochemistry* **17**:2746.
Schlick, S., and Alonso-Amigo, M. G., 1987, *J. Chem. Soc., Faraday Trans. I* **83**:3575.
Schlick, S., and Alonso-Amigo, M. G., 1989, *Macromolecules* **22**:2634.
Schneider, D. J., and Freed, J. H., 1989a, *Adv. Chem. Phys.* **73**:387.
Schneider, D. J., and Freed, J. H., 1989b, in *Biological Magnetic Resonance Vol.* 8, *Spin Labeling Theory and Applications* (L. J. Berliner and J. Reuben, eds.), Plenum, New York, pp. 1–76.
Sealy, R. C., Hyde, J. S., and Antholine, W. E., 1985, in *Modern Physical Methods in Biochemistry* (A. Neuberger and L. L. M. Van Deenen, eds.), Elsevier, Amsterdam, pp. 69–148.
Seebauer, E. G., Duliba, E. P., Scogin, D. A., Dennis, R. B., and Belford, R. L., 1983, *J. Am. Chem. Soc.* **105**:4926.
So, H., and Belford, R. L., 1969, *J. Am. Chem. Soc.* **91**:2392.
Spiro, T. G., ed., 1981, *Copper Proteins: Metal Ion Biology, Vol. 3*, Wiley-Interscience, New York.
Stevens, T. H., Martin, C. T., Wang, H., Brudvig, G. W., Scholes, C. P., and Chan, S.-I., 1982, *J. Biol. Chem.* **257**:12,106.
Subczynski, W. K., Antholine, W. E., Hyde, J. S., and Petering, D. H., 1987, *J. Am. Chem. Soc.* **109**:46.
Subczynski, W. K., Antholine, W. E., Hyde, J. S., and Kusumi, A., 1991, *Biochemistry* **30**:8578.
Toriyama, K., and Iwasaki, M., 1975, *J. Am. Chem. Soc.* **97**:6456.
Umezawa, H., Meada, K., Takeuchi, Y., and Okami, Y., 1966, *J. Antibiot., Ser. A.* **19**:200.
Vänngård, T., 1972, in *Biological Applications of Electron Spin Resonance* (H. M. Swartz, J. R. Bolton, and D. C. Borg, eds.), Wiley-Interscience, New York, p. 411.
Werst, M. M., Davoust, C. E., and Hoffman, B. M., 1991, *J. Am. Chem. Soc.* **113**:1533.
White, L. K., and Belford, R. L., 1976, *J. Am. Chem. Soc.* **98**:4428.
Wilson, R., and Kivelson, D., 1966, *J. Chem. Phys.* **44**:4445.
Yen, T. F., 1969, *Electron Spin Resonance of Metal Complexes*, Plenum, New York.

4

Metalloenzyme Active-Site Structure and Function through Multifrequency CW and Pulsed ENDOR

Brian M. Hoffman, Victoria J. DeRose, Peter E. Doan, Ryszard J. Gurbiel, Andrew L. P. Houseman, and Joshua Telser

1. INTRODUCTION

Electron paramagnetic resonance (EPR) techniques have long been major tools in efforts to determine the structure and function of metalloenzyme active sites (Beinert *et al.*, 1962; Hoff, 1989). Much of the information EPR provides about the composition, structure, and bonding of a paramagnetic metal center is obtained by analysis of hyperfine coupling constants (Abragam and Bleaney, 1970; Atherton, 1973) that arise from interactions between the spin of the unpaired electron(s) and the spins of nuclei associated with the metal center, endogenous ligands, or bound substrate. At the most basic level, the observation of hyperfine coupling and its assignment to one or more nuclei (e.g., ^{1}H, ^{14}N) provide information about the chemical composition of the center. Detailed analysis of these couplings can provide information about its geometry or about substrate binding, as well as deep

---

Brian M. Hoffman, Victoria J. DeRose, Peter E. Doan, Ryszard J. Gurbiel, Andrew L. P. Houseman, and Joshua Telser • Department of Chemistry, Northwestern University, Evanston, Illinois 60208-3113.

Biological Magnetic Resonance, Volume 13: EMR of Paramagnetic Molecules, edited by Lawrence J. Berliner and Jacques Reuben. Plenum Press, New York, 1993.

insights into chemical bonding. In principle these coupling constants can be calculated from splittings in the EPR spectrum. However, as illustrated in Fig. 1, for most metalloproteins these splittings cannot be resolved, and thus the chemical information they carry is lost.

Electron–nuclear double resonance (ENDOR) spectroscopy recovers this information (Feher, 1959; Schweiger, 1982; Möbius and Lubitz, 1987; Hoffman, 1991). An ENDOR experiment provides an NMR spectrum of nuclei that interact with the electron spin of a paramagnetic center, and analysis of the observed ENDOR frequencies gives the desired electron–nuclear coupling constants. ENDOR is classified as a double-resonance technique because an NMR transition is detected not directly, but rather through a change in the EPR signal intensity at fixed magnetic field. The spectral resolution of ENDOR can be as much as three orders of magnitude better than that of conventional EPR. This increase permits the detection and characterization of electron–nuclear hyperfine interactions for systems whose EPR spectra show no hyperfine splittings. Furthermore, the technique is inherently broad-banded. It is comparably easy to detect ENDOR signals from every type of nucleus. We have examined signals from ^{1}H, ^{2}H, ^{13}C, $^{14,15}N$, ^{17}O, ^{33}S, ^{57}Fe, ^{61}Ni, $^{63,65}Cu$, and $^{95,97}Mo$ that are present either as constitutive components of a metalloenzyme active site or as part of a bound

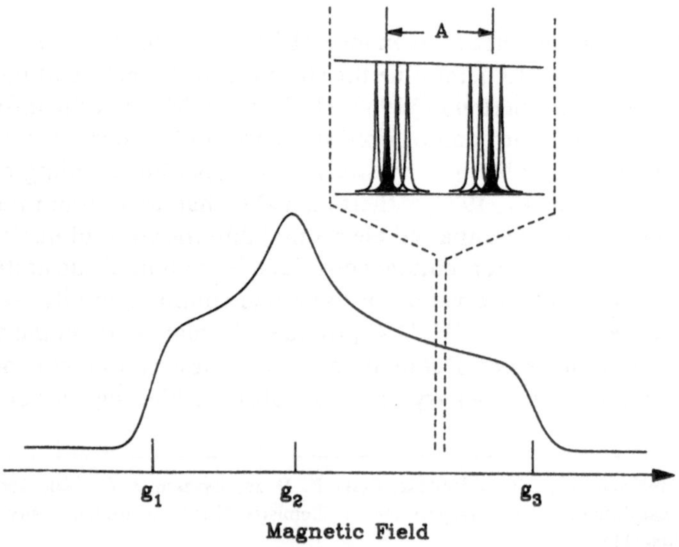

Figure 1. Rhombic EPR absorption envelope illustrating spin packets from individual centers that add together to give the inhomogeneously broadened powder EPR pattern. In particular, a pair of packets representing a hyperfine splitting of A for an $I = \frac{1}{2}$ nucleus is depicted.

ligand or substrate. As exemplified in the study of aconitase (Section 4.2), with proper isotopic labelling it is possible to characterize every type of atomic site associated with a paramagnetic center of a biomolecule. Finally, although measurements typically employ frozen-solution samples, an ENDOR spectrum generally arises from only a subset of molecules with defined, selected orientations, allowing us to deduce full hyperfine tensors including their orientations with respect to the *g* frame (see Appendix).

ENDOR offers two important additional forms of selectivity. One of these might be called *EPR selection.* Even if a sample exhibits EPR signals from multiple paramagnetic species, it is possible to use ENDOR to study each species individually, provided only that each EPR spectrum does not exactly overlap with others. Examples include studies of sulfite reductase (Cline *et al.*, 1986) and Ni-hydrogenase (Fan *et al.*, 1991). The second is *site specificity.* In contrast to NMR, the ENDOR spectrum reflects only those nuclei that have a hyperfine interaction with the electron spin system being observed. For example, it is possible to examine ^{57}Fe resonances from a specific metal cluster without interference from other clusters, as is seen in the case of the nitrogenase MoFe protein (True *et al.*, 1988).

ENDOR's breadth of applicability is illustrated by dividing metalloenzyme active sites into several nonexclusive pairs of categories:

- A site may be *well characterized,* perhaps with a known structure, as with Type 1 copper (Cu^{2+}) centers. Here, the goal of a study would be a more complete understanding of the site's electronic structure and a comparison among proteins of that structural type (Section 3.1). In contrast, the site might be a *black box* such as the "H-cluster" of Fe-hydrogenase (Section 3.2) or, until the recent determination of its crystal structure (Kim and Rees, 1992), the molybdenum-iron cofactor of the nitrogenase MoFe proteins (True *et al.*, 1988; McLean *et al.*, 1987).
- As a second distinction, one may use ENDOR to study the structure of a *metal center* itself, as in our examination of Rieske iron–sulfur centers (Section 4.1). On the other hand, detection of ENDOR signals from a *bound substrate* can be used to determine enzyme mechanism, as in our study of aconitase (Section 4.2).
- Finally, the *resting state* of a protein is often the most accessible to study, as with a variety of heme systems (Scholes, 1979). In favorable cases, however, ENDOR can provide detailed information about key *reaction intermediates* as well (Section 5.1; Roberts *et al.*, 1981).

Section 2 reviews the essentials of the ENDOR technique. It describes briefly first the three classes of ENDOR response (Section 2.1), then discusses in simple terms the spectra obtained from the most important of these classes (Section 2.2). This discussion includes a description of observed ENDOR patterns and frequencies, as well as an explanation of the correlation between molecular orientation and resonance magnetic field ("orientation selection") utilized to determine magnetic interaction tensors from studies of frozen-solution (polycrystalline) samples. Section 2.2 is amplified in an appendix that develops complete expressions for the ENDOR frequencies and presents in detail the theory and application of orientation selection (see also Hoffman *et al.*, 1989). Finally, Section 2.3 describes the several CW and pulsed ENDOR techniques currently employed.

The core of this chapter is to be found in Sections 3 and 4. Section 3 illustrates the necessity of using multiple approaches in ENDOR spectroscopy in order to extract the maximum possible information in an ENDOR experiment. This section in particular shows the importance of using frequencies above X band, but also critically discusses the relative advantages of CW and pulsed techniques. The issues are addressed by example with discussions of ENDOR studies of Type 1 (blue) Cu(II) proteins and of [nFe–mS] clusters. Section 4 is designed to show the profound insights that ENDOR studies of metalloproteins can provide. This is again done by example, with summaries of the studies of the [2Fe–2S] Rieske centers and of the [4Fe–4S] cluster of aconitase and its interactions with substrate.

We have not attempted a complete literature review of metallobiomolecule ENDOR, for that has been done admirably by Lowe (1992; see also Hoffman *et al.*, 1989; Hoffman, 1991). However, Section 5 addresses a topic not covered by Lowe, amino-acid free-radical sites in metalloenzymes.

2. THE ENDOR MEASUREMENT

In an ENDOR experiment, a paramagnetic center is placed in a static magnetic field $\mathbf{B}_0$ and subjected to a radio-frequency field whose frequency is swept as would be done in a CW NMR experiment. A change in the center's EPR signal intensity constitutes the ENDOR response.

2.1. Types of ENDOR Response

It is convenient to divide the types of ENDOR responses into three classes.

Distant ENDOR produces a signal at the nuclear Larmor frequency and involves nuclei in the bulk of the sample, which can include the protein matrix or the solvent (Lambe *et al.*, 1961; Wenckebach *et al.*, 1971; Boroske and Möbius, 1977; Pattison and Yim, 1981). It involves nuclei so remote from the paramagnetic center that the electron–nuclear interactions are smaller than nuclear–nuclear dipole couplings. A distinguishing experimental characteristic is that distant ENDOR signals relax on the relatively slow time scale of nuclear spin–lattice relaxation. Distant ENDOR has been used to perform high-resolution solid-state NMR on the diamagnetic host molecules in doped molecular crystals (Boroske and Möbius, 1977), but has had no application in the present context.

Local ENDOR and *matrix ENDOR* are both associated with nuclei that experience electron–nuclear interactions larger than nuclear–nuclear dipolar interactions. Normally, the ENDOR response varies on the time scale of electron spin–lattice relaxation. We denote by local ENDOR spectra those from nuclei that have spectrally resolved interactions with the unpaired spin of the paramagnetic center (Feher, 1959; Abragam and Bleaney, 1970). This operational definition roughly means that the linewidth of the resonances is less than their splitting. Local ENDOR is the primary focus of studies on metallobiomolecules and of this review. Such nuclei can be associated with constitutive elements of the center (e.g., ^{57}Fe, ^{33}S of an Fe–S cluster), with endogenous ligands to a metal ion (e.g., with coordinated or remote ^{14}N of a histidyl imidazole) or with exogenous ligands or substrate (e.g., CN^-, citrate). Typically this involves some form of bonding interaction that gives rise to an isotropic component to the hyperfine coupling, but the label would apply to nuclei in the immediate vicinity of a paramagnetic center that exhibit resolved, through-space dipolar coupling.

Matrix ENDOR also involves nuclei in the vicinity of the electron spin, but it is loosely distinguishable from local ENDOR in that the nuclei have unresolvably small, typically dipolar, couplings. Matrix ENDOR is seldom of direct interest in the present context, although it is of great interest in others (Kevan, 1987).

In both matrix and distant ENDOR, the observed signal is an unresolved line centered at the Larmor frequency of the nucleus. Operationally this has the consequence of obliterating local ENDOR lines from nuclei with small hyperfine couplings. It can be quite a handicap in CW ^2H ENDOR of D_2O-exchanged samples, because even the local couplings are small. This is less a problem in ^1H ENDOR, where much larger couplings are observed. The distant ENDOR signals can be eliminated in pulsed ENDOR. Surprisingly, ^{13}C matrix and distant ENDOR also can be seen in natural-abundance (1.1%) protein samples (Section 3.2).

2.2. Description of Local ENDOR Spectra

2.2.1. Resonance Frequencies

When a diamagnetic molecule is placed in a static magnetic field $\mathbf{B}_0$, a nucleus N exhibits a single NMR line centered at its Larmor frequency

ν_N, as determined by the magnitude of the external field $B_0 = |\mathbf{B}_0|$ and its nuclear g factor, $h\nu_N = g_N\beta_n B_0$, where β_n is the nuclear magneton. However, a hyperfine-coupled nucleus in a paramagnetic center feels a second magnetic field in addition to the external applied field (see Pilbrow, 1990). The hyperfine interaction with the electron spin appears to the nucleus as an internal field with magnitude $B^N \equiv h|A^N|/2g_n\beta_n$, where A^N is the hyperfine coupling constant. For example, a coupling of $A = 10$ MHz corresponds to internal fields of 0.117 T for ^1H, 1.63 T for ^{14}N, 0.87 T for ^{17}O, and 3.63 T for ^{57}Fe. As shown in Fig. 2, these two fields can be aligned parallel or antiparallel, and the magnitude of the net field experienced by a nucleus has one of two values $B^n_\pm = |B_0 \pm B^N|$. Correspondingly, for a nucleus with $I = \frac{1}{2}$, two ENDOR transitions are detected, with first-order resonance frequencies given by

$$\begin{aligned} h\nu^N_\pm &= g_N\beta_n|B_0 \pm B^N_i| \\ &= g_N\beta_n B^N_\pm \\ &= h\left|\nu_N \pm \frac{A^N}{2}\right| \end{aligned} \tag{1}$$

This equation corresponds to Eq. (A11b) of the Appendix.

For a proton, the external field is generally larger than the internal field ($\nu_H > A^H/2$); for example, if $A^H = 10$ MHz, then $B^N = 0.12$ T, whereas

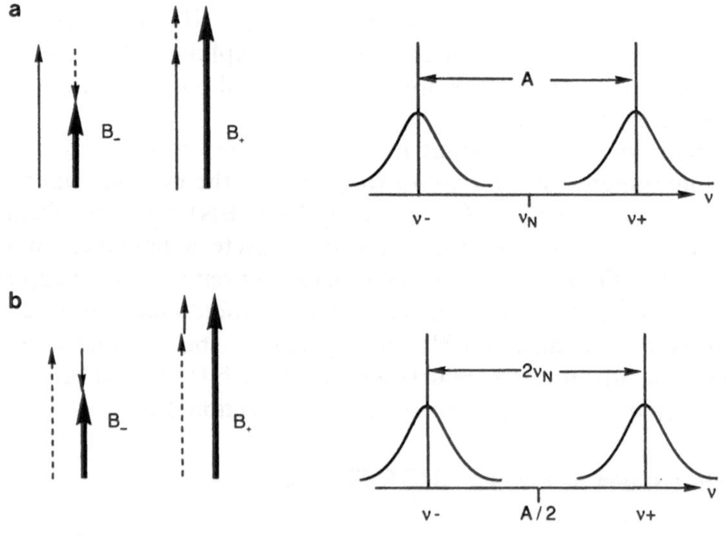

Figure 2. Schematic representation of the manner in which external field ($\mathbf{B}_0 \rightarrow$) and internal hyperfine field ($\mathbf{B}_i$ --→) add to give two possible resultant fields with magnitudes $B_\pm$ (→) and a two-line ENDOR spectrum [Eq. (1)]. Small internal field ($B_i < B_0$) as seen for protons; (b) large internal field ($B_i > B_0$), more typical for other nuclei.

at X band for $g = 2$, $B_0 \sim 0.35$ T. According to Eq. (1) the ^{1}H ENDOR spectrum then is a "hyperfine-split" doublet, centered at the free-proton Larmor frequency ν_H and split by A^H (Fig. 2a). For a given g value of observation, ν_H is proportional to the microwave frequency of the ENDOR spectrometer, so the center of the ^{1}H ENDOR pattern shifts correspondingly. For example, in a conventional X-band (~9 GHz) spectrometer the ^{1}H resonances are centered at $\nu_H \sim 14$ MHz for $g \sim 2$, whereas in a Q-band (35 GHz) spectrometer $\nu_H \sim 52$ MHz for $g \sim 2$. In contrast, for other nuclei, typically $A^N/2 > \nu_N$ and the internal field is the larger. For ^{14}N as an example, if A (^{14}N) = 10 MHz, then $B^N = 1.63$ T $> B_0 \sim 0.35$ T for X band, $g = 2$. In this case the ENDOR pattern is a "Larmor-split" doublet, centered at $A^N/2$ and split by $2\nu_N$ (Fig. 2b).

The difference between the ENDOR behavior of protons and that of other nuclei can be exploited to eliminate the biggest frustration in X-band ENDOR of metallobiomolecules. In many metalloenzymes ^{1}H signals in an X-band spectrometer overlap and obscure signals from other nuclei (e.g., ^{14}N, ^{57}Fe, ^{17}O). To illustrate, consider the simplified but realistic case (see Section 3.1) of a paramagnetic center that has a ^{14}N nucleus ($I = 1$) with hyperfine constant $A^N = 28$ MHz and protons with a range of couplings. The ^{14}N ENDOR pattern would be centered at $A^N/2 = 14$ MHz. It would consist of two lines split by $2\nu_N \approx 2$ MHz at an X-band field of 0.3300 T, with the possibility of an additional splitting by the quadrupole interaction. At this field, the proton ENDOR pattern would also be centered at 14 MHz, and signals from protons with $A^H \lesssim 5$ MHz would completely overlap the ^{14}N signals (Fig. 3A). This problem is eliminated entirely by shifting the proton pattern to a center frequency of $\nu_H \sim 50$ MHz in a Q-band spectrometer (Fig. 3B).

A nucleus with $I > \frac{1}{2}$ also possesses a quadrupole moment that interacts with nonspherical components of the total charge distribution of its surroundings (Abragam and Bleaney, 1970; Atherton, 1973). This interaction, if resolved, splits each of the two lines at $\nu_\pm^N$ [Eq. (1)] into $2I$ lines, with transitions at frequencies given (approximately) by

$$\nu_\pm(m) = \left| \pm \frac{A^N}{2} + \nu_N + (3P^N/2)(2m - 1) \right| \tag{2}$$

where P^N is the quadrupole coupling constant and $I \geq m > -I + 1$. This corresponds to Eq. (A19b) of the Appendix. The quadrupole coupling gives important information about bonding in the metal center (Brown and Hoffman, 1980; Scholes *et al.*, 1982) and is not manifest in ordinary EPR spectra even when hyperfine splittings are observed.

Thus, in both cases depicted in Fig. 2 the described hyperfine coupling constant can be read directly from the ENDOR spectrum. As will be seen

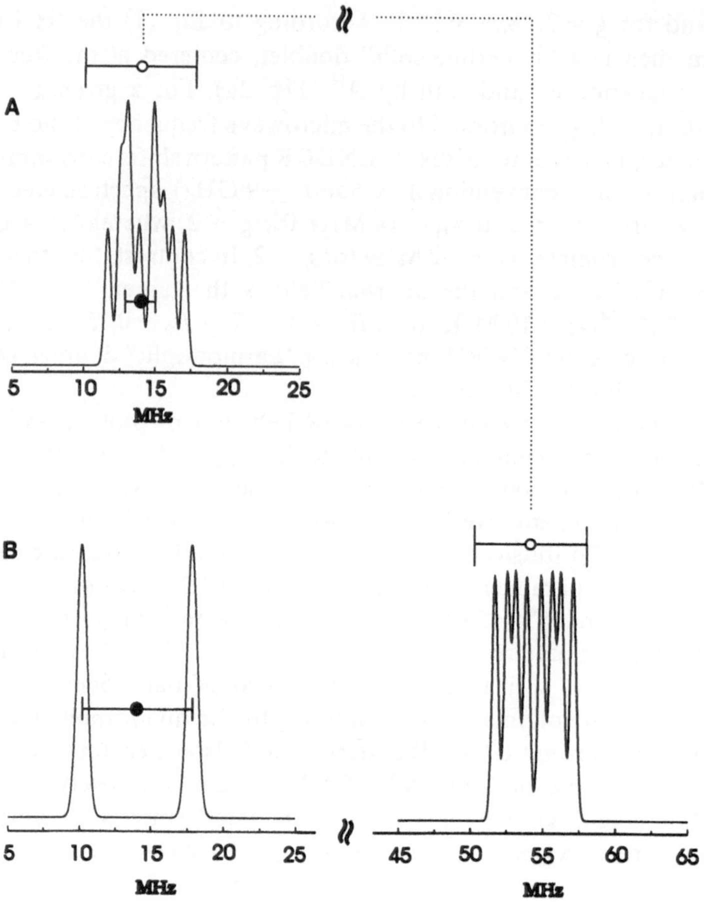

Figure 3. Computer-generated ENDOR spectra of an $S = \frac{1}{2}$ system containing one set of nitrogen couplings and four proton couplings, $1 \leq A^H \leq 7$ MHz. The open circles indicate ν_H and the attached connecting bars (⊢⊣) indicate the maximum value of A^H. The filled circles indicate $A^N/2 = 14$ MHz, and the attached connecting bars, $2\nu_N$. (A) *X*-band (~9 GHz) ENDOR spectrum showing overlap of ^1H and ^{14}N ENDOR signals: $\nu_H = 14.3$ MHz; $2\nu_N = 2.0$ MHz. (B) *Q*-band (35 GHz) ENDOR spectrum showing how ^1H and ^{14}N ENDOR signal overlap is alleviated: $\nu_H = 53$ MHz; $2\nu_N = 7.6$ MHz.

below, the ENDOR technique is able to measure couplings of as little as a few tenths of a megahertz, whereas resolved hyperfine splittings smaller than the tens of megahertz range are seldom seen in the EPR spectra of metallobiomolecules.

Both the ENDOR and EPR techniques in general give the magnitude but not the sign of a hyperfine coupling constant. The technique of double-ENDOR or TRIPLE-resonance employs two rf fields and can give the *relative* signs of two hyperfine constants. This technique can be applied both in CW (Möbius *et al.*, 1989) and in pulsed (Gemperle and Schweiger, 1991) ENDOR.

2.2.2. Orientation Selection

The samples employed in the ENDOR studies of metalloenzymes are almost always frozen solutions and thus contain a random distribution of all possible protein orientations with respect to the applied magnetic field $\mathbf{B}_0$. From the point of view of the molecules, this means that there is an equal probability for the field to have any orientation (angles θ, ϕ) with respect to the molecular framework. However, this does not mean that all information about the orientation of the hyperfine-interaction tensors is lost. If a molecule exhibits resolved anisotropic magnetic interactions, namely anisotropic g, hyperfine, or zero-field splitting tensors, then the EPR intensity at each fixed applied field arises only from a particular, restricted, and mathematically well defined set of orientations of the field relative to the molecular framework (Hoffman *et al.*, 1984, 1985; Hoffman and Gurbiel, 1989). This correlation of field and orientation ("orientation selection") provides a means of obtaining complete hyperfine tensors from ENDOR studies of polycrystalline (frozen-solution) samples.

Consider the EPR spectrum of a center with a rhombic g tensor ($g_1 > g_2 > g_3$), as shown in Fig. 4. ENDOR spectra taken with the magnetic field set at either of the extreme edges of the EPR spectrum arise only from those molecules for which the magnetic field happens to be directed along the corresponding g-tensor axis (g_1, Position D; g_3, Position A) (Rist and Hyde, 1970). However, the full description of a nuclear hyperfine tensor

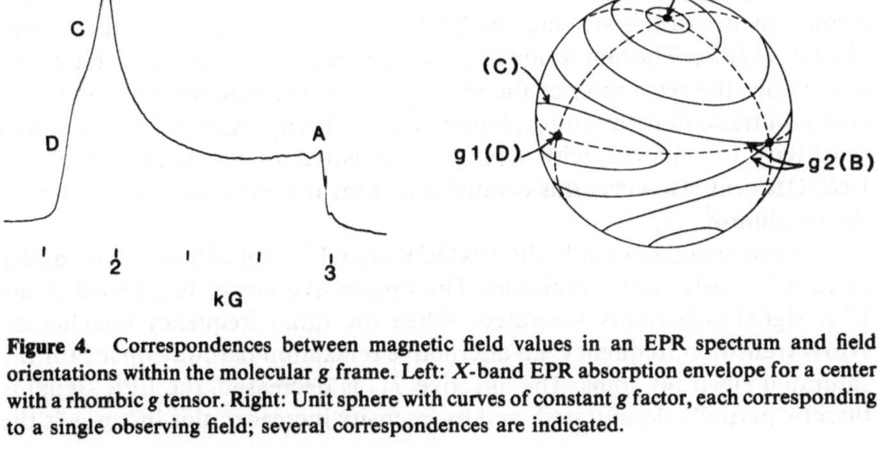

Figure 4. Correspondences between magnetic field values in an EPR spectrum and field orientations within the molecular g frame. Left: *X*-band EPR absorption envelope for a center with a rhombic g tensor. Right: Unit sphere with curves of constant g factor, each corresponding to a single observing field; several correspondences are indicated.

requires more information than is contained in these two single-crystal-like spectra. An ENDOR spectrum obtained using an intermediate g value (magnetic field) (e.g., Figure 4, positions B, C), is not associated with a single orientation, but it does arise only from those molecules for which the orientation of the field happens to satisfy the relation $g(\theta, \phi) = g$, where $g(\theta, \phi)$ is the angle-dependent g factor (Abragam and Bleaney, 1970; Pilbrow, 1990) given in Eq. (A3) of the Appendix. As indicated in Fig. 4, for each g value this set of orientations can be represented on the unit sphere by a curve s_g as described in the Appendix (Hoffman *et al.*, 1984). A series of ENDOR spectra collected at fields (g values) across the EPR envelope samples different sets of molecular orientations, by analogy with the way that rotating a single crystal in a field samples different individual molecular orientations. This fact serves as the basis of an analysis procedure we have developed (Hoffman *et al.*, 1989; True *et al.*, 1988; Gurbiel *et al.*, 1989) wherein such a series of orientation-selected ENDOR spectra can be simulated to determine the principal values of the nuclear hyperfine tensor and the orientation of the tensor relative to the g-tensor framework. Results of this procedure are presented in Section 4. A detailed description is given in the Appendix.

2.3. Detecting the ENDOR Response

2.3.1. CW ENDOR

Two modes of ENDOR detection are generally used (Kevan and Kispert, 1976). In the most common, embodied in the currently available Bruker, Inc. spectrometer, the ENDOR response is observed as changes in the EPR absorption-mode in-phase signal without field modulation, while sweeping an rf source. The signal is encoded by modulating the rf frequency, typically at $\nu_{mod} \sim 10$ kHz, and decoded by a phase-sensitive detector. As a consequence of this scheme, the ENDOR signal appears as the derivative display $dI/d\nu_{rf}$. The amplitude $\Delta\nu_{mod}$ of rf modulation in units of frequency determines the resolution of the spectrum, and the procedure involves the common tradeoff between resolution and sensitivity. As in EPR, maximum sensitivity is achieved with $\Delta\nu_{mod}$ about equal to the linewidth of the ENDOR peak. However this condition distorts the line shape and decreases the resolution.

In this detection mode the ENDOR signal is typically observed under so-called steady-state conditions. The microwave power is set so that the EPR signal is partially saturated. When the radio frequency matches an NMR transition frequency, an alternative relaxation pathway opens for the saturated electrons spins. The effective T_{1e} is decreased, the EPR signal is thereby partially desaturated, and its intensity increases: this increase is the

ENDOR response. The degree of desaturation depends on the complicated interplay of all relaxation processes within the spin system, and therefore the relative intensities of the ENDOR lines do not necessarily reflect the number of contributing nuclei, as is not the case in NMR or EPR.

In low-temperature studies ($T < 4.2$ K) of metalloproteins an alternate detection scheme that uses standard EPR field modulation, typically 100 kHz, often is advantageous. At these temperatures electron spin relaxation typically is very slow ($1/T_{1e} < 10^3\ s^{-1}$ at 2 K is common). Under saturating conditions the presence of the field modulation causes the electron spin system to exhibit rapid passage effects. To first approximation, the absorption-mode signal vanishes, and EPR must be detected with the microwave bridge tuned to the dispersion mode. The resulting dispersion-mode EPR signal as detected at 100 kHz takes the form the EPR absorption envelope would have if it were detected without field modulation. The ENDOR response is the change that occurs in this EPR signal when an unmodulated rf field is swept through a nuclear transition. In this scheme, an ENDOR spectrum appears as an absorption rather than a derivative display (Feher, 1959).

This approach involves a form of "transient" or "packet-shifting" (Kevan and Kispert, 1976), rather than steady-state ENDOR. The amplitude ΔB_{mod} of the field modulation defines a window in the broad EPR absorption envelope in which a set of electron spin packets contributes to the EPR signal. The spin packet with which a given molecule is associated resonates at a field determined by the orientation of its nuclear spins. When the rf field induces a nuclear spin transition in a molecule, the resonance field of its spin packet is shifted by an amount equal to the nuclear hyperfine coupling A. If $A \gtrsim \Delta B_{mod}$, this rf-induced shift of spin packets into and out of the observation window changes the EPR signal intensity; this change constitutes the ENDOR response. The phenomenon is more complex than it might appear and, depending on experimental conditions, the ENDOR response can be either an increase or a decrease in the EPR signal. (For example, see Duglav *et al.*, 1974.) In our experience a decrease is the norm. Again, one sees a tradeoff between sensitivity and resolution. As ΔB_{mod} is increased, the sensitivity increases but the resolution, namely the minimal A that can be detected, decreases.

To our knowledge no systematic comparisons of the two detection procedures have been reported. However, one gets the impression that frequency modulation is preferred at temperatures above $T \sim 4$ K (where passage effects do not typically appear) and for studies of free radicals, whereas the field modulation scheme is superior at low temperatures. In part this is because the dispersion detection of rapid-passage signals gives larger EPR signals and better signal-to-noise ratios, provided that saturation conditions can be achieved at microwave powers below the level at which source noise becomes dominant. We note that the ENDOR response of a system where electron spins are in adiabatic rapid passage is conceptually related to the Davies pulsed-ENDOR scheme described immediately below. As in pulsed ENDOR, adiabatic packet-shifting ENDOR should not be as sensitive to competing relaxation processes as is steady-state ENDOR. However, anomalous patterns are often observed in *Q*-band ENDOR spectra. Both the ν_+ and ν_- branches commonly appear as decreases in EPR signal intensity, but with ν_+ much more intense than ν_- (see Section 3.1.3). Such a disparity in intensities was observed in a very early study of a cofactor radical in the enzyme methanol dehydrogenase (de Beer *et al.*, 1979). However, de Beer *et al.* found both ν_+ and ν_- as increases in EPR signal intensity, as did Feher (1959) in his seminal work. Occasionally, the ν_+ branch appears as a decrease in intensity, whereas the ν_- appears as an increase.

2.3.2. Pulsed ENDOR

There are two commonly used pulsed ENDOR techniques (Gemperle and Schweiger, 1991). For each the ENDOR response is a change in the intensity of an electron spin echo that arises because a nuclear-resonance transition shifts electron-spin packets ("polarization transfer"). Both are often referred to as electron spin echo ENDOR (ESE-ENDOR).

The first ESE-ENDOR technique to be devised (Mims, 1965) is eponymously called Mims ENDOR. In this experiment, three $\pi/2$ microwave pulses are used to generate a stimulated echo (Fig. 5a); the time between pulses 1 and 2 is called τ, that between pulses 2 and 3, T. An rf pulse, ideally a π pulse, is applied during time T (usually between 5 and 60 μs).

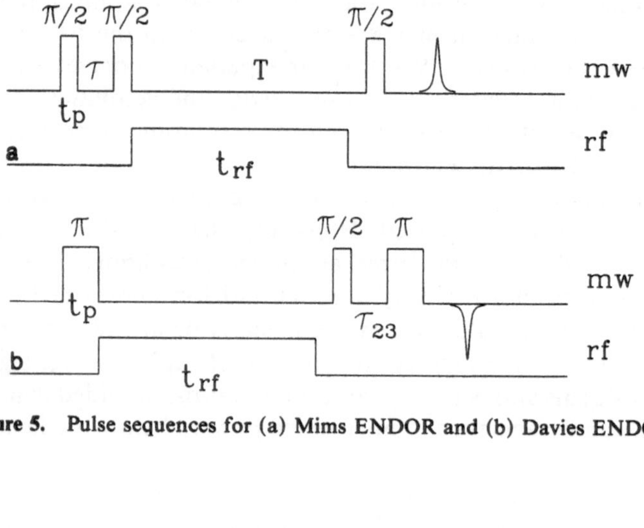

Figure 5. Pulse sequences for (a) Mims ENDOR and (b) Davies ENDOR.

Resonant NMR transitions shift spin packets and decrease the electron spin echo created by microwave pulse 3. This decrease is the ENDOR response.

The second technique (Davies, 1974) is similarly called Davies ENDOR. In this experiment (Fig. 5b), a preparation π microwave pulse of duration t_p inverts spin packets or burns a hole (of width $\sim 1/t_p$) in a broad line. The resulting magnetization is subsequently detected by a $\pi/2$–π two-pulse echo sequence. The application of an on-resonance rf pulse after the preparation phase increases the magnetization measured by the two-pulse detection sequence.

Both of these techniques are inherently hyperfine selective in that the strength of the ENDOR response is determined jointly by the experimental parameters and the hyperfine coupling. In a Mims ENDOR sequence, the ENDOR response I^N for a nucleus N varies with the time τ between the first and second pulses of the stimulated echo sequence according to the relation (Liao and Hartmann, 1973)

$$I^N \propto 1 - \cos(2\pi A^N \tau) \tag{3}$$

with A^N in MHz and τ in μs. When $A^N\tau = 1, 2, 3, \ldots$, the intensity falls to near zero, resulting in a "blind spot" in the spectrum. When $A^N\tau = 0.5$, $1.5, \ldots$, the ENDOR response reaches a maximum. If measurements are made as a function of τ, such a modulation of ENDOR intensities allows a resonance to be identified by its A^N value. Liao and Hartmann (1973) performed this experiment on a ruby single crystal. However, in a powder spectrum of a quadrupolar nucleus ($I \geq 1$) the technique cannot be applied in a transparent manner because molecules with different orientations can exhibit resonances with the same frequency but not associated with the same value of A. However, when $A^N\tau \lesssim 0.5$ the technique has good sensitivity without the complication of blind spots. Given that instrumental characteristics limit τ to $\gtrsim 0.1$ μs, this means that the Mims technique is appropriate if $A^N \lesssim 5$ MHz. This technique has proven to be highly successful in studies of ^2H and weakly coupled ^{14}N (see Sections 3.2.2 and 4.2.2.1).

In a Davies experiment, the ENDOR response is jointly dependent on A^N and the width t_p of the initial inverting microwave pulse. If we define a selectivity parameter $\eta^N = A^N t_p$, the absolute ENDOR response is determined by

$$I(\eta) = I_0\left(\frac{1.4\eta}{0.7^2 + \eta^2}\right) \tag{4}$$

where I_0 is the maximum ENDOR response. For a given value of A^N, the ENDOR response is optimized by selecting t_p so that $\eta = 0.7$ (Fan *et al.*, 1992a, b; Grupp and Mehring, 1990). For example, when $t_p \sim 30$ ns, a Davies experiment gives optimum signals from nuclei with A about 20 MHz, but signals from sites with $A < 5$ MHz are strongly attenuated. This characteristic can mitigate some of the problems described in Section 2.2 of spectral overlap of weakly coupled protons with strongly coupled nuclei. This form of hyperfine selection has been referred to as POSHE (Doan *et al.*, 1991) or self-ENDOR (Gemperle and Schweiger, 1991). For this type of selection to work, the I_0 values of the different nuclei must be similar in magnitude, or else selectivity is diminished.

Another pulsed-ENDOR technique that has been used in studies of metalloproteins employs an electron-nuclear-electron triple resonance scheme in a variant of Davies ENDOR called hyperfine selective ENDOR (HS-ENDOR). In one form, the initial (preparation) pulse of a Davies sequence is at one microwave frequency and the frequency of the $\pi/2$–π detection sequence is offset by Δ. Only those ENDOR resonances with hyperfine couplings close to Δ are observed (Thomann and Bernardo, 1990). A conceptually equivalent form employs a field jump of value $\Delta B_0 = h\Delta/gB$ after the rf pulse of the Davies sequence (Bühlmann *et al.*, 1989). As will be seen (Section 3.1), such procedures give an alternate approach to solving the proton–heteronuclear overlap problem and have the theoretical capability of unravelling complicated spectral overlap from systems with similar hyperfine couplings. However, HS-ENDOR can resolve only peaks that differ in A by more than $\sim 1/t_p$ and thus suffers the common problem of requiring a compromise between sensitivity and resolution.

3. MULTIPLE APPROACHES IN ENDOR

The Type 1 Cu(II) sites of blue copper proteins and the [nFe–mS] clusters of Fe–S proteins are ideal systems for illustrating and comparing the multiple approaches available in CW and pulsed ENDOR. Both exhibit strong overlap of ^{1}H and heteronuclear resonance in X-band spectrometers: in the former, it involves ^{1}H and ^{14}N; in the latter, ^{1}H and ^{57}Fe. Both have been studied by all available techniques.

3.1. ^{1}H and ^{14}N ENDOR of Blue Copper Proteins

The Type 1 sites of blue copper proteins have unusual optical and magnetic properties, which include intense ($\varepsilon = 3500$–$6000\ \mathrm{M^{-1}\,cm^{-1}}$) absorption bands in the visible region of the spectrum ($\lambda = 600$–625 nm), an approximately axial EPR spectrum with very low hyperfine splitting constants ($A_\parallel \approx 150$ MHz), and relatively high reduction potentials (Ainscough *et al.*, 1987; Gray and Solomon, 1981). Extensive efforts have been made to understand the spectral features and the associated electronic structure of these unusual sites. Crystal structures have been reported for the two single-site blue copper electron-transfer proteins, plastocyanin (Guss and Freeman, 1983) and azurin (Norris *et al.*, 1986). In both cases, the Cu^{2+} ion has three strongly bound ligands, the thiolate sulfur of a cysteine and the imidazole nitrogens of two histidines (N1, N2). The primary coordination geometry of Cu^{2+} in these two proteins is very similar, with inequivalent Cu–N1 and Cu–N2 bond distances and S-Cu–N1 and S-Cu–N2 bond angles, and can be schematically represented as follows:

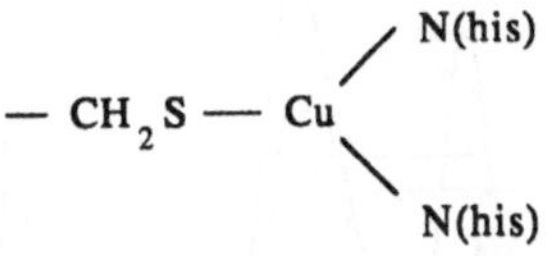

An earlier *X*-band ENDOR study of the ^{1}H, ^{14}N, and ^{63,65}Cu resonances of the Type 1 copper blue centers of azurin, plastocyanin, stellacyanin, and (Type 2-reduced) tree and fungal laccases showed that the coordination environment of Cu in all these centers is broadly similar (Roberts *et al.*, 1984). Each exhibits strongly coupled proton resonances ($A^H \sim$ 20–30 MHz) from methylene protons of the coordinated cysteinyl mercaptide, ^{14}N resonances from at least one nitrogen ligand, and resonances from ^{63,65}Cu in the range 60–190 MHz. However, this study was limited because of the extensive overlap between ^{14}N and ^{1}H signals.

We therefore reinvestigated the Type 1 copper centers from five different proteins with a *Q*-band (35 GHz) ENDOR spectrometer (Werst *et al.*, 1991). The increase of microwave frequency from *X* band (~9 GHz) to *Q* band (35 GHz) gave complete spectral separation of the ^{14}N and ^{1}H resonances. The use of the higher frequency also enhances the effect of *g* anisotropy in achieving orientation selection, as noted in early work by Becker and Kwiram (1976) and de Beer *et al.* (1979). These signals were analyzed in detail individually to give estimates of the spin densities on all three strongly coordinated ligands and of the dihedral angles of the cysteinyl β-protons relative to the C^{β}-S-Cu plane. These proteins were subsequently studied using pulsed ENDOR techniques.

3.1.1. Comparison of *X*- and *Q*-band ^{1}H and ^{14}N ENDOR Spectra

Figures 6A and 6B show the single-crystal-like ($g_{\|}$) ENDOR spectrum of the Type 1 Cu^{2+} center of azurin, taken at 9.6 GHz and 11.7 GHz, respectively (Roberts *et al.*, 1984). These *X*-band spectra represent a complicated overlap of resonances from protons and nitrogens, and their complete interpretation is impossible. However, the slight difference in ν_H at the two microwave frequencies was used in attempts to assign the ^{14}N and ^{1}H features. The proton Larmor frequency is $\nu_H \sim$ 14 MHz in Fig. 6A, but $\nu_H \sim$ 17 MHz in Fig. 6B. Comparison of the spectra clearly indicates that

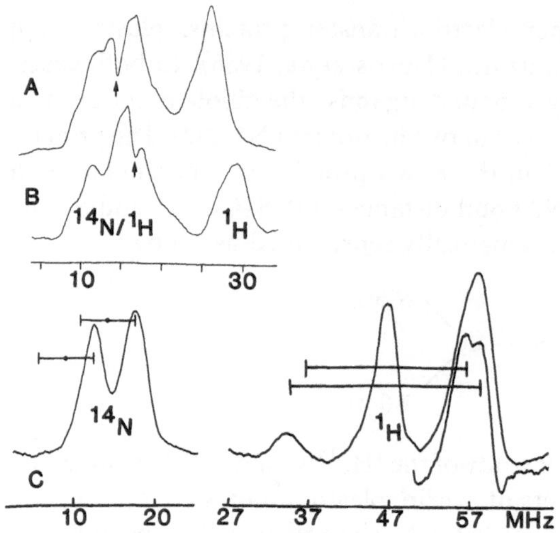

Figure 6. ^{1}H and ^{14}N ENDOR of azurin near $g_{\parallel}$ at (A) 9.6 GHz, (B) 11.7 GHz, and (C) 35.2 GHz. *Conditions*: temperature, 2 K; microwave power, 2 μW (A, B) and 50 μW (C); 100-kHz field modulation amplitude, 0.3 mT; rf scan rate, 2.5 MHz/s. The arrows indicate the proton Larmor frequency at each microwave frequency.

the well-resolved, intense high-frequency signal at ~26 MHz in Fig. 6A shifts to ~29 MHz at the higher microwave frequency of Fig. 6B and thus can be assigned to the ν_+ feature of a strongly coupled proton(s) with $A^H = 24$ MHz [Eq. (1)].

In contrast, ^{14}N resonances centered at $A^N/2$ do not shift between Fig. 6A (9.6 GHz) and Fig. 6B (11.7 GHz). Furthermore, at these two microwave frequencies $\nu_N = 1.0$ and 1.1 MHz, respectively, and hence the change in the splitting of the ν_- and ν_+ ^{14}N resonances ($2\nu_N \sim 2$ MHz) is undetectable. The signals in the range 7–22 MHz are poorly resolved, and it is difficult to distinguish them because of weakly coupled protons and those from nitrogen. However, it was argued that a poorly resolved feature centered at 9 MHz and ~2 MHz in width is seen in both Figs. 6A and 6B, and this was assigned to ^{14}N ($A^N \sim 18$ MHz). Any additional nitrogen resonances appearing at $A^N/2 \sim 13$–18 MHz would be wholly obscured by the signals from weakly coupled protons centered at ν_H. Such difficulties in making assignments arose for each blue copper protein. Furthermore, the ^{14}N signals assigned at *X* band could not be analyzed in detail by measuring their dependence on the observing field.

3.1.2. ^{1}H *Q*-Band ENDOR

Five Type 1 copper proteins were studied in a 35-GHz ENDOR spectrometer (Werst *et al.*, 1991). The ^{1}H ENDOR response in the *Q*-band, single-crystal-like spectrum of azurin taken near $g_{\parallel}$ (Fig. 6C) is centered at $\nu_H \sim 47$ MHz. This is representative of spectra from the five Type 1 centers examined; the ν_+ branch [Eq. (1)] of each of the ^{1}H spectra is considerably more intense than the ν_- branch. In each of the Type 1 proteins, strongly coupled protons are observed, with hyperfine splittings of $A^{H} \sim 15$–31 MHz (Table 1). These hyperfine couplings are effectively isotropic (10–20% anisotropy). In addition, all the proteins display signals associated with weakly coupled protons, $A^{H} \leq 6$ MHz. At least some of these weakly coupled protons are from the coordinating groups.

The strongly coupled proton signals $A^{H} \sim 15$–31 MHz for the Type 1 Cu proteins arise from the $—CH_2S^-$ side chain of the cysteine coordinated to Cu. We proposed that these β-methylene proton hyperfine coupling constants depend on two molecular parameters: the dihedral angle $\theta(H^{\beta i})$, defined by the $H^{\beta i}—C^{\beta i}—S—Cu$ bonds (see Fig. 7) and the total spin density ρ_S on sulfur. Equation (5) gives the resulting expression for $A^{H\beta i}$.

$$\begin{aligned} A^{H\beta i} &= [B \cos^2 \theta(H^{\beta i}) + C]\rho_S \\ &= \left[\cos^2 \theta(H^{\beta i}) + \frac{C}{B}\right] B\rho_S \qquad i = 1, 2 \end{aligned} \tag{5}$$

Here $\theta(H^{\beta i})$ is the dihedral angle of the ith β-proton and $\theta(H^{\beta 1}) \approx \theta(H^{\beta 2}) - 2\pi/3$; B and C reflect spin delocalization to H^{β} through hyperconjugation and σ-bond polarization ($C \ll B$) respectively. Analysis of the data for these blue copper proteins yields a quite narrow range of angles: $-58° \leq \theta(H^{\beta 2}) \leq -50°$ (Table 1). This clearly suggests that the geometry of the cysteinyl-Cu bond is remarkably well conserved, in agreement with resonance Raman and crystallographic data (Han *et al.*, 1991). In all these proteins ρ_S is also conserved. Using as a reasonable estimate $B \sim 100$ MHz

TABLE 1

^{1}H and ^{14}N Isotropic Hyperfine Coupling Constants (MHz) of Blue Copper Centers[a]

Substance	^{1}H[b] A^{H1} $(H^{\beta 2})$	A^{H2} $(H^{\beta 1})$	θ $(H^{\beta 2})$	ρ_s (MHz)	^{14}N A^{N1}	A^{N2}
Plastocyanin	27	16	−50	0.44	22	22[c]
Azurin	23	18	−54	0.43	27	17
Stellacyanin[d]	20	20	−58	0.42	33	17
Fungal laccase[e]	31	25	(−55)	(0.59)	36	23
Tree laccase	25	19	−54	0.46	40[f]	33

[a] Measured in the $g_{\parallel}$ region of the EPR spectrum. All hyperfine interactions with the exception of the ^{14}N interaction of N1 of tree laccase are roughly isotropic (see text). From Werst *et al.* (1991).

[b] The quantities ρ_s and $\theta(H^{\beta 2})$ are defined in the text and calculated using Eq. (5) as parameterized from the $\theta(H^{\beta 2})$ and the A^{Hi} of plastocyanin, taking $\theta(H^{\beta 1}) = \theta(H^{\beta 2}) + 116°$, and using $B = 100$ MHz.

[c] See text for the assignment of equivalent nitrogens.

[d] See text for assignment of equivalent protons.

[e] Parentheses around derived values reflect a probability that the plastocyanin parametrization of Eq. (5) may not apply.

[f] Tensor components $A_{3,2,1} = (32, 43, 47)$ MHz.

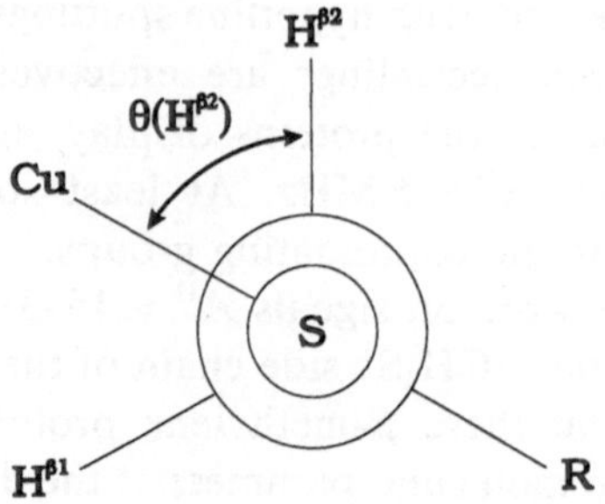

Figure 7. Newman projection for β-methylene proton hyperfine coupling constants in Type 1 Cu.

(Gordy, 1980), then the ^{1}H ENDOR data gives $\rho_S \sim 0.45$, in good agreement with theoretical calculations (Penfield *et al.*, 1985; Gewirth and Solomon, 1988). The quite different behavior of fungal laccase (Table 1) indicates it is distinct from the other blue copper proteins.

3.1.3. ^{14}N *Q*-Band ENDOR

3.1.3.1. Plastocyanin, Azurin, and Stellacyanin. The low-frequency portion ($\nu < 30$ MHz) of the *Q*-band ENDOR spectra taken at $g_{\parallel}$ for these single-site copper proteins (Fig. 6C) shows intense resonances that were assigned to ν_+ for nitrogen coordinated to Cu (Werst *et al.*, 1991). The ENDOR spectra of azurin (Fig. 6C) show two intense peaks that we assigned to the ν_+ resonances of two inequivalent ^{14}N ligands. The ν_- partners at lower frequency are too weak to be detected; this low relative intensity parallels that for the ν_- peaks of the strongly coupled protons of all the Type 1 centers. None of these *Q*-band spectra shows resolved quadrupolar splittings for ^{14}N ($I = 1$). Likewise, ^{14}N quadrupolar splittings were not resolved for the nitrogen that is directly coordinated to copper in $[Cu(Im)_4]^{2+}$ (VanCamp *et al.*, 1981) or in the Cu_A site of cytochrome oxidase (Gurbiel *et al.*, 1983a).

The ^{14}N *Q*-band ENDOR spectra taken at selected *g* values across the EPR envelope indicate that the hyperfine splittings of N1 and N2 are roughly isotropic (~10% anisotropy), with $A^{N1} \approx 27$ MHz, $A^{N2} \approx 17$ MHz. The ^{14}N *Q*-band ENDOR spectrum of stellacyanin resembles that of azurin, with resonances from two inequivalent ^{14}N ligands, although the hyperfine coupling of N1 is slightly larger in the former protein. In contrast, for poplar plastocyanin the observation of a single ^{14}N resonance implies that the nitrogens are magnetically equivalent, $A^{N1} \sim A^{N2}$ (Table 1). Analysis of the ^{14}N hyperfine couplings was shown to give spin densities on nitrogen that are in excellent agreement with theory (see Gewirth and Solomon, 1988; Solomon *et al.*, 1989).

3.1.3.2. Fungal and Tree Laccase. The Type 1 site of fungal laccase is like those of the single-site proteins azurin and stellacyanin in that it has distinctly inequivalent ^{14}N ligands with roughly isotopic hyperfine interactions (Table 1). However, the Type 1 center of tree laccase is unique among these blue copper proteins. First, the ^{14}N isotropic hyperfine couplings are larger in tree laccase. The hyperfine coupling for N1 is essentially the same as the coupling observed for the nitrogen ligand in $[Cu(imidazole)_4]^{2+}$; the hyperfine coupling for N2 is only slightly smaller. In addition, the hyperfine tensor of N1 has substantial anisotropy, as in a Cu^{2+}-doped histidine monodeuterohydrate (McDowell *et al.*, 1989), whereas this is not true for any of the other Type 1 sites or for $[Cu(imidazole)_4]^{2+}$. It has been proposed that a nearly isotropic ^{14}N coupling can be taken as a signature of imidazole coordination to Cu(II) (Yokoi, 1982). The data for the Type 1 centers in azurin, etc., support this idea, but the results for tree laccase (Surerus *et al.*, 1992) (and also for Cu_B of cytochrome oxidase) suggest that this proposed correlation is not universal.

3.1.4. ^{14}N Pulsed ENDOR

Both Davies POSHE and HS-ENDOR have been applied to obtain ^{14}N spectra from blue copper proteins at *X* band. Using the former technique with azurin, an ENDOR spectrum was taken at $g_{\perp}$, with t_p adjusted to suppress proton resonances with small couplings. It revealed resonances from the two types of nitrogen resonances seen in the *Q*-band studies; signals from the strongly coupled methylene protons ($A^H = 18, 25$ MHz) also were observed (Fig. 8) (Fan *et al.*, 1992a).

HS-ENDOR was used to study a partially denatured high-pH form of stellacyanin (Thomann *et al.*, 1991b). It showed that the resonances from the two coordinated nitrogens ($A^N = 19, 35$ MHz) and from the cysteine methylene protons change little in going from the native to high-pH forms of the enzyme. However, the high-pH form shows an additional resonance, which appears at 21 MHz and is assigned to a nitrogen with $A^N = 42$ MHz. The intensity of this new peak varies when the HS-ENDOR spectrum is recorded at different *g* values as the result of hyperfine anisotropy. These data suggest that this new signal is from an amide nitrogen rather than an imidazole nitrogen.

3.1.5. Comparison of Results from Alternate ENDOR Techniques

The blue copper proteins provide the most extensive illustration of the use of various ENDOR techniques to overcome proton–nitrogen spectral overlap. The results obtained from 35-GHz CW ENDOR are certainly the

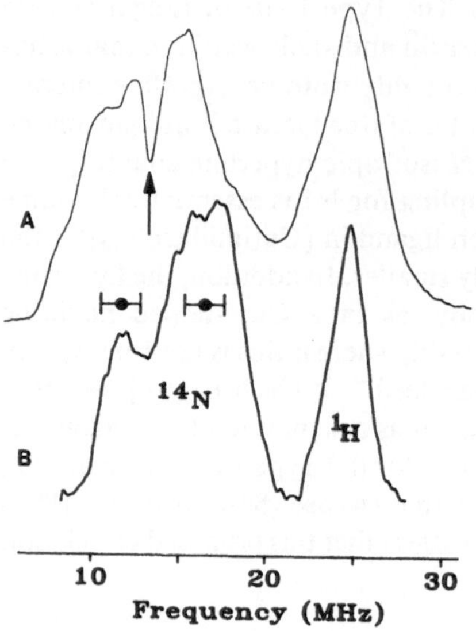

Figure 8. *X*-band ENDOR spectra of *Pseudomonas aeruginosa* azurin. (A) CW spectrum; (B) POSHE-ENDOR spectrum. The proton Larmor frequency ν_H is shown as ↑. The assignment for the ^{14}N resonances is shown on top of spectrum (B), with ● representing $A^N/2$. Experimental conditions for (B) were t_p = 32 ns, t_{rf} = 10 μs; B_0 = 3155 G; ν_e = 9.15 GHz; temperature 4 K; 600 scans.

most complete. By separating the proton and nitrogen resonances into different frequency regions, each type of pattern could be studied in detail, yielding the ^1H and ^{14}N couplings summarized in Table 1. Obviously, this advantage would also exist for a 35-GHz pulsed spectrometer.

The use of two microwave frequencies in the *X*-band range (Roberts *et al.*, 1984) was moderately successful in distinguishing the severely overlapped ^1H and ^{14}N features (Fig. 6). In each case the proton hyperfine coupling for H1 (Table 1) was correctly identified. Of the four proteins with an H2 resonance, only those for the two laccases were resolved; the *X*-band H2 resonance of plastocyanin was incorrectly attributed to a very strongly coupled nitrogen ($A^N \sim 49$ MHz) and the H2 resonance of azurin was not detected. The two ^{14}N resonances for stellacyanin and fungal laccase, the single resonance for plastocyanin, and the N2 resonance for azurin were correctly assigned and their coupling constants were reliably obtained. At *X* band, the ENDOR signal for N1 of azurin is centered at ~14 MHz and is totally obscured by the proton signal (Figs. 6A, 6B). The ^{14}N resonances for tree laccase at *X* band were correctly assigned, but the complex ^{14}N interaction described above could not possibly have been detected or fully analyzed.

When ^1H resonances severely overlap those of heteronuclei (^{14}N, ^{17}O, etc.), attempts to interpret spectra taken at a single *X*-band frequency are not reliable, as shown by erroneous assignments in several studies (e.g., Desideri *et al.*, 1985; Cline *et al.* 1985a) that have since been corrected by studies at 35 GHz (e.g., Werst *et al.*, 1991; Gurbiel *et al.*, 1989). The present discussion makes it clear that appreciable success can be achieved by use of two *X*-band frequencies but that such studies cannot be viewed as definitive, and a wider range in microwave frequencies is required.

The applications to blue copper proteins of the two *X*-band Davies pulsed ENDOR techniques, POSHE and HS-ENDOR, show that such advanced techniques can assist appreciably if one is limited to experiments at *X* band. However, the use of multiple frequencies clearly would be superior in pulse as it is in CW ENDOR.

The Davies POSHE spectrum of azurin gives clean ^{14}N patterns that underlie the weak proton envelope. There is little distortion of the ENDOR orientation-selected line shape from this simple technique of hyperfine selection. This leaves open the possibility of spectral simulation to obtain principal values (see below). However, the strongly coupled protons still are seen with intensity similar to those of the nitrogen signals and in the same spectral region. Attempts to study these proton resonances using POSHE at *X* band would be difficult due to ambiguities in the lower-field portion ($g_{\|}$), as these proton resonances would overlap the more strongly coupled ^{14}N resonance.

HS-ENDOR gives the benefit of directly measuring the *A* values of resonances. The different ^{14}N peaks can be observed independently and the assignments are then made trivially. However, when the peak shows hyperfine anisotropy this narrow-banded selection technique destroys information available from the orientation-selected powder spectrum. HS-ENDOR can be used most effectively at single-crystal-like orientations and in systems with only slight hyperfine anisotropy.

3.2. ENDOR of [*n*Fe–*m*S] Clusters

The pioneering *X*-band ^{57}Fe and ^{1}H ENDOR studies of [2Fe–2S] and [4Fe–4S] cluster proteins by Sands and co-workers have been extensively reviewed (Sands, 1979). More recently we have performed *X*-band ENDOR studies on a number of [*n*Fe–*m*S] cluster proteins (Hoffman, 1991; Lowe, 1992).

The problems that can arise in such experiments are illustrated in the *X*-band ENDOR studies of the H-cluster of the bidirectional and uptake Fe-hydrogenases isolated from *Clostridium pasteurianum W5* (Wang *et al.*, 1984; Telser *et al.*, 1986b). Study of the CO-bound enzymes was limited by overlap between ^{1}H ENDOR signals and those from ^{57}Fe from cluster iron and ^{13}C from cluster-bound ^{13}CO. This difficulty was partially overcome by subtraction of a given natural isotopic abundance ENDOR spectrum from the corresponding ^{57}Fe-enriched spectrum. This is shown in Fig. 9. However, despite careful attempts at spectral normalization, such procedures are subject to large uncertainties. Not least, cross-relaxation can alter the

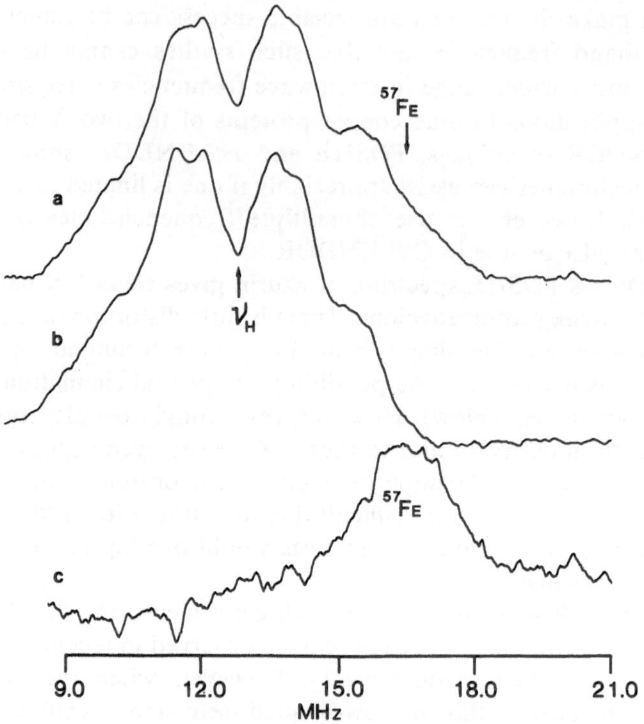

Figure 9. ENDOR scan of CO-bound oxidized Fe-hydrogenase I: (a) ^{57}Fe enriched, (b) natural abundance, (c) computer subtraction of (b) from (a) (× 2 gain). *Conditions*: temperature, 2 K; microwave frequency, 8.705 GHz; microwave power, 0.6 μW; magnetic field, 0.3050 T, $g = 2.04$; 100-kHz field modulation amplitude, 0.25 mT; time constant, 0.008 s; rf scan rate, 10 MHz/s; 2450 scans.

intensities of overlapping ^{1}H or ^{57}Fe (Scholes *et al.*, 1982), and such phenomena can cause an ^{57}Fe signal to be missed in such a subtraction.

Such problems are eliminated in experiments performed at 35 GHz. The investigations of the [2Fe–2S]$^{1+}$ cluster of the Rieske proteins and of the [4Fe–4S]$^{1+}$ cluster of aconitase are discussed in detail in Section 4. In this section we illustrate the use of the several ENDOR techniques discussed above through studies on a number of [nFe–mS] proteins with natural isotopic abundances (Houseman *et al.*, 1992). Figure 10 presents *Q*-band (35 GHz) ENDOR spectra of four proteins all in natural isotopic abundance, with [4Fe–4S] clusters. The spectra cover roughly the same rf range as those shown in Fig. 9, and all were recorded at the field of greatest intensity in the EPR envelope, near $g = 2$. Here the Larmor frequency of ^{1}H is approximately 53 MHz, well above the frequency region of the ENDOR spectrum in which most other nuclei usually resonate. This leaves an unobscured view of resonances from ^{13}C, ^{14}N, ^{15}N, ^{17}O, ^{57}Fe, etc.

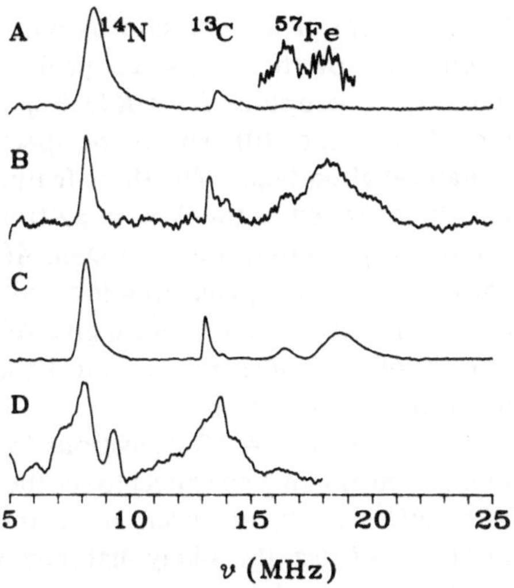

Figure 10. 35-GHz ENDOR of several natural-abundance Fe–S proteins: (A) *D. gigas* Fd II, (B) *V. vinosum* HiPIP, (C) *E. halophila* HiPIP, (D) *Anabaena 7120* Fd. *Conditions*: temperature, 2 K; rf scan rate, (A–C) 1, (D) 0.5 MHz/s; time constant, (A–C) 0.032, (D) 0.128 s; microwave power, (A) 7.9, (B) 50, (C) 7.9, (D) 0.5 mW; microwave frequency, (A, B) 35.33, (C) 34.70, (D) 34.68 GHz; spectra were taken at g_2 with magnetic field B_0 of (A) 1.260, (B) 1.234, (C) 1.219, (D) 1.308 Y; 100-kHz field modulation amplitude, (A–C) 0.8, (D) 4 mT; number of scans, (A) 50, (B, C) 100, (D) 20. Spectrum (D) taken with negative rf sweep.

Each of these spectra shows peaks at ~9 and ~13 MHz. The appearance of such peaks is surprising, for in these proteins the ligands to the cluster Fe are sulfur, which has no significant percentage of an isotope with nonzero nuclear spin. However, at the magnetic field of observation the Larmor frequency for ^{13}C is $\nu(^{13}\text{C}) \approx 13$ MHz and the resonances near 13 MHz were identified as distant or matrix ENDOR (Section 2.1) from ^{13}C in natural abundance. ENDOR studies of *Anabaena* Fd with ^{13}C enrichment confirmed this assignment. They also confirmed that resolved features at ~13 MHz in the spectra of Fig. 10 (particularly D) can be assigned to local ENDOR peaks with couplings $A(^{13}\text{C}) \approx 1$ MHz. These undoubtedly are associated with the β carbons of cysteine coordinated to the cluster; there are two resolved couplings, which likely are associated with ligands to different iron ions.

The strongest peak in each of the spectra is assigned to a $\Delta m_I = \pm 2$ transition of ^{14}N. Such nominally forbidden transitions are allowed through state mixing by the $^{14}\text{N}(I = 1)$ quadrupole interaction. They are not usually seen at X band, the most likely reason being that for $g = 2$ the $\Delta m_I = \pm 2$ transitions occur at $2\nu(^{14}\text{N}) = 2$ MHz and are difficult to detect. However, they have been seen in several ESEEM studies of weakly coupled ^{14}N (Mims and Peisach, 1989). In a powder sample the maximum ENDOR intensity of the $\Delta m = \pm 2$ transition is expected at frequencies

$$\nu_{\pm 2} = 2[(\nu_{\text{N}} \pm A^{\text{N}})^2 + K^2(3 + \eta^2)]^{1/2} \qquad (6)$$

where $K = e^2qQ/4h = P_{\text{max}}/2$ is the quadrupole coupling constant and η is the asymmetry parameter (Dikanov and Astashkin, 1989; Flanagan and

Singel, 1987). This equation assumes A^{N}, $K \ll 2\nu_N$. Thus, the resonant frequencies of the $\Delta m_I = \pm 2$ peaks can be used to derive the effective quadrupole coupling $K' \equiv K(3 + \eta^2)^{1/2}$ and the hyperfine coupling of a local ^{14}N. Figure 10D shows the spectrum of *Anabaena* Fd with all nuclei in natural abundance. The three features centered near $2\nu(^{14}\text{N}) \approx 8.0$ MHz may be assigned to the $\Delta m_I = \pm 2$ transitions of nitrogen. The center peak can be assigned to distant nitrogen, $A(^{14}\text{N})_{\text{dist}} \approx 0$, and the flanking doublet corresponds to $\nu_{\pm 2}$ peaks of a local nitrogen. The positions of the local ^{14}N doublet peaks in *Anabaena* Fd give $A(^{14}\text{N1}) = 1.0$ MHz and $K' = 1.5$ MHz. Spectra of *Anabaena* Fd isotopically enriched with ^{15}N confirm the assignment of these resonances.

Because $\Delta m_I = \pm 2$ transitions for distant ^{14}N are clearly observed in Fig. 10, one might also expect to see the corresponding $\Delta m_I = \pm 1$ transitions. With $A(^{14}\text{N}) \approx 0$, however, these transitions should be in resonance at $\nu(^{14}\text{N}) \approx 4$ MHz. It is likely that they typically do not appear in the 35-GHz spectra because they are spread over a broader frequency range and consequently are less intense because they are more sensitive to anisotropy.

The appearance of strong ^{14}N $\Delta m_I = \pm 2$ transitions and natural-abundance ^{13}C resonances is not a simple blessing. Most importantly, these peaks can interfere with other ENDOR signals. Also, they might possibly be misinterpreted as ^{14}N $\Delta m_I = \pm 1$ resonances. For example, the peaks in Fig. 10 could be erroneously assigned to ν_+ resonances from two different ^{14}N sites, with coupling of 18 and 10 MHz, respectively.

Finally, every spectrum in Fig. 10A–C shows resonances near 18 MHz that arise from natural-abundance ^{57}Fe (2.2%). Consequently, these Fe–S clusters must all have Fe sites with hyperfine couplings $A(^{57}\text{Fe}) \approx 36$ MHz [Eq. (3)]. Comparable ^{57}Fe couplings have been seen with Mössbauer spectroscopy (Surerus *et al.*, 1989) and in ENDOR studies of ^{57}Fe-enriched aconitase (see Section 4.2.1.1; Werst *et al.*, 1990b), *Chromatium vinosum* HiPIP, and *Clostridium pasteurianum* Fd (Anderson *et al.*, 1975; Sands, 1979). In favorable cases, such as *C. vinosum* (Fig. 10B) and *E. halophila* (Fig. 10C) HiPIPs, the natural-abundance ^{57}Fe ENDOR intensity is adequate to perform complete analysis of the ^{57}Fe hyperfine tensor(s). This opens up the possibility of studying natural-abundance proteins when iron enrichment is impractical. The comparison of these spectra with the *X*-band spectra in Fig. 9 shows the benefit of eliminating the overlap of ^{1}H signal, regardless of whether ^{57}Fe enrichment needs to be employed.

H bonding to the [*m*Fe–*n*S] clusters in Fe–S proteins is of interest because this interaction may well modulate cluster function (Backes *et al.*, 1991). We found that Mims-pulsed deuteron ENDOR spectroscopy of H–D exchanged Fe–S proteins provides significant new opportunities both for probing H bonding to metal clusters and for examining exogenous ligands to such clusters (Fan *et al.*, 1992a).

A ^{2}H *Q*-band CW ENDOR spectrum of the $[2Fe\text{-}2S]^{+}$ cluster of D_2O-exchanged *Anabaena* Fd (Fig. 11A) shows only a featureless signal at the deuterium Larmor frequency ν_D (7.9 MHz at $H_0 = 12{,}000$ G), and one cannot determine to what extent this signal is due to distant ENDOR. In contrast, the Mims pulsed-ENDOR technique yields highly resolved local deuterium ENDOR spectra from which hyperfine and quadrupole splittings can be determined directly. Figure 11B is a single-crystal-like ^{2}H Mims ENDOR spectrum taken at the high-field ($g_3 = 1.92$) edge of the *Anabaena* Fd EPR absorption envelope, using the same small sample (~10 μL) used at *Q* band. It shows a pair of peaks centered at ν^D and separated by 0.60 MHz. This splitting, far too large to be associated with the quadrupolar interaction, represents a deuteron hyperfine coupling $A^D \approx$ 0.60 MHz (Edmonds, 1977). The peaks further show a partially resolved quadrupole splitting of $3P^D \approx 0.1$ MHz. As no exogenous ligands are associated with the cluster, this local deuteron can be assigned to a N–H · · · S hydrogen bond. This is presumably one of the two putative strong H bonds seen in the crystal structure of the oxidized protein: arginine-42 H-bonded to a cysteinyl mercaptide sulfur bound to iron or arginine-258 H-bonded to a bridging cluster S^{2-} (Rypniewski *et al.*, 1991).

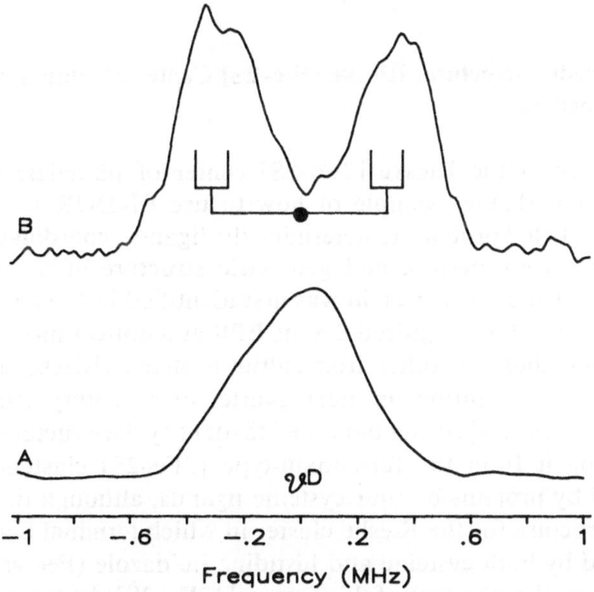

Figure 11. ^{2}H *Q*-band (A) CW ENDOR and (B) Mims ENDOR spectra of the $[2Fe\text{-}2S]^{+}$ cluster of *Anabaena* ferredoxin in D_2O solvent. For comparison, spectra are plotted as $\delta\nu = \nu - \nu_D$. *Conditions*: (A) microwave frequency, 34.54 GHz; B_0, 1.2670 T; ν_D, 8.3 MHz; rf scan rate, 0.25 MHz/s; 200 scans; (B) microwave frequency, 9.15 GHz; B_0, 0.3357 T; microwave pulse width, 16 ns; t_{rf}, 40 μs; τ_{12}, 420 ns; ν_D, 2.2 MHz; 64 scans.

The explanation for the difference in resolution of the CW and pulsed deuterium signals is likely as follows: In the former, distant ENDOR signals (Section 2.1) from noninteracting deuterons overwhelm the local ENDOR of H-bonded deuterons. In pulsed ENDOR, an individual pulse sequence ($\leq 50\ \mu s$) is shorter than the time for the spin diffusion processes that are the basis of the distant CW-ENDOR response; thus the distant ENDOR response is "quenched", and local ENDOR signals become visible.

Finally, weakly coupled nitrogens in the Fe–S cluster of Fe-hydrogenase I from *C. pasteurium* have been studied using Mims pulsed ENDOR (Thomann *et al.*, 1991a). Two different ^{14}N sites were reported with different hyperfine and quadrupole couplings. Both allowed ($\Delta m_I = \pm 1$) and forbidden ($\Delta m_I = \pm 2$) transitions were observed in the ENDOR spectrum between 2 and 6 MHz. The ENDOR study was used to augment multifrequency ESEEM spectra taken at the low-field (high-g) edge of the spectrum. The spin parameters of one of these two sites were analyzed in detail with $|A_{iso}| = 1.2$ MHz and $K = 1.21$ MHz.

4. ENDOR AS THE SOLUTION TO PROBLEMS ABOUT METALLOBIOMOLECULES

4.1. Active-Site Structure: Rieske [2Fe–2S] Center of Phthalate Dioxygenase

The study of the Rieske [2Fe-2S] center of phthalate dioxygenase provides a remarkable example of how to use ENDOR spectroscopy of isotopically labeled proteins to determine the ligands, coordination environment, electronic properties, and geometric structure of a protein-bound metal cluster. The Rieske protein was first identified in 1964 by J. S. Rieske and co-workers, who recognized that its EPR and optical spectra were quite different from those of other iron–sulfur proteins (Rieske *et al.*, 1964). Subsequent studies during the next quarter of a century established the presence of a [2Fe–2S] cluster but failed to identify the structural differences distinguishing it from the ferredoxin-type [2Fe–2S] clusters, which are coordinated by proteins by four cysteine ligands, although the studies did suggest a structure for the Rieske cluster in which terminal S and N atoms are provided by both cysteine and histidine imidazole (Fee *et al.*, 1986).

To resolve the structure of this unusual [2Fe–2S] cluster, we performed in collaboration with Dr. J. Fee and Professors D. Ballou and T. Ohnishi 9- and 35-GHz ENDOR studies on the Rieske-type center of globally and selectively ^{15}N-enriched phthalate dioxygenase (PDO) prepared from *Pseudomonas cepacia* (Gurbiel *et al.*, 1989) and on the Rieske center of globally ^{15}N-enriched cytochrome bc_1 from *Rhodobactor capsulatus* (BC1) (Gurbiel *et al.*, 1991). The results confirm that two imidazole nitrogens from histidine are coordinated to the [2Fe–2S] cluster. This is in contrast to the classical ferredoxin-type [2Fe–2S] centers, in which all ligation is by sulfur of cysteine residues. Through analysis of the polycrystalline ENDOR spectra of the $[2Fe–2S]^{1+}$ state, we further established that the two histidines are coordinated to the Fe^{2+} site of the (Fe^{2+}, Fe^{3+}) spin-coupled pair, described the coordination to this ion, and determined molecular orbital parameters describing the bonding of the two nitrogen ligands.

4.1.1. Ligand Identification

Five different samples of *P. cepacia* PDO were prepared each with a specific labelling pattern. The *X*-band ENDOR spectrum at $g_2 = 1.92$ of natural-abundance PDO shows a broad pattern of ^{14}N resonances that by itself is not amenable to analysis (Fig. 12). In contrast, spectra of globally ^{15}N-enriched PDO taken at g_2 exhibit two sharp Larmor-split doublets from ^{15}N centered at $A(^{15}N)/2 = 2.55$ and 3.35 MHz and show the expected splitting [Eq. (1)] of $2\nu(^{15}N) = 3.1$ MHz (Fig. 12). Assignment of these doublets to two magnetically distinct ^{15}N-labelled nitrogenous ligands coordinated to the [2Fe–2S] cluster is confirmed by single-crystal-like spectra

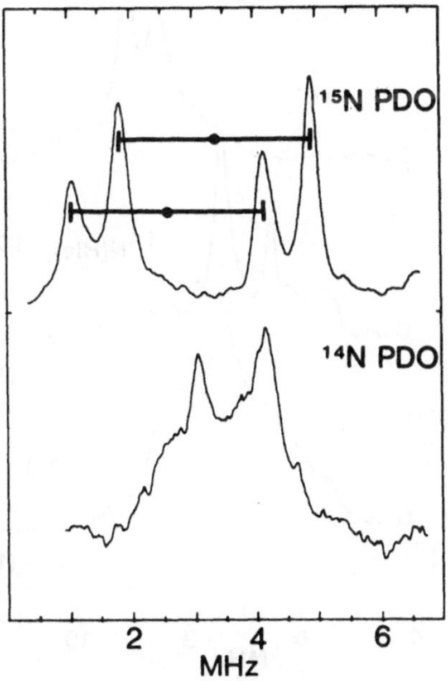

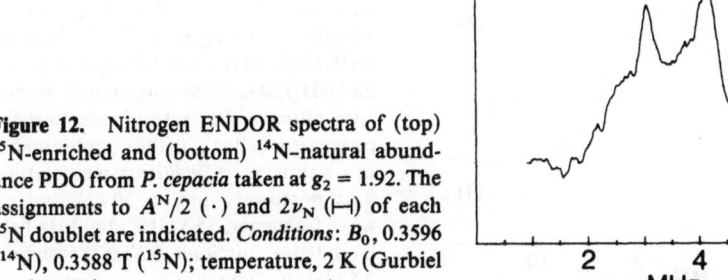

Figure 12. Nitrogen ENDOR spectra of (top) ^{15}N-enriched and (bottom) ^{14}N-natural abundance PDO from *P. cepacia* taken at $g_2 = 1.92$. The assignments to $A^N/2$ (·) and $2\nu_N$ (⊢⊣) of each ^{15}N doublet are indicated. *Conditions*: B_0, 0.3596 (^{14}N), 0.3588 T (^{15}N); temperature, 2 K (Gurbiel *et al.*, 1989).

taken at $g_1 = 2.01$. Equivalent data for natural-abundance and uniformly ^{15}N-labelled BC1 confirm that the same applies to this mitochondrial Rieske center.

These two ligands are further identified as histidines by comparison of ENDOR measurements on PDO *uniformly* labeled with ^{15}N (Fig. 13B), on PDO *specifically* labeled with ^{15}N-histidine in a ^{14}N background (Fig. 13A), and on PDO with ^{14}N-histidine in a ^{15}N background (Fig. 13C). The absence of additional ^{15}N signals in (C), where all nonhistidyl nitrogens have been labeled with ^{15}N, indicates that the nitrogen ligands to the [2Fe–2S] center are only from histidine. The ^{14}N resonances of (C) are more resolved than those of (D). This appears to result from unidentified relaxation processes associated with the bulk ^{14}N nuclei of the protein, and it greatly facilitated the analysis of the ^{14}N quadrupole tensors (see below).

The results of the above experiments do not address the problem of which nitrogen of a histidine imidazole ring is coordinated to the iron, N(δ) or N(ε):

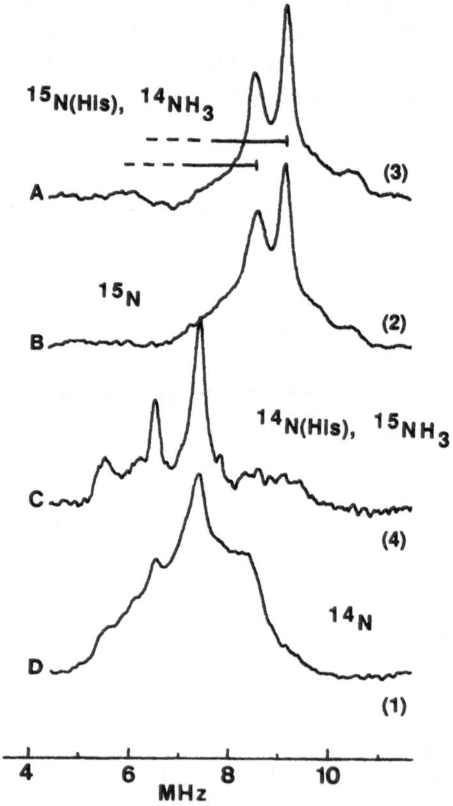

Figure 13. Nitrogen ENDOR spectra at $g_2 = 1.92$ and *Q*-band frequency of PDO extracted from (A) auxotroph of *P. cepacia* grown on a medium of ^{15}N histidine and NH_3 ($(NH_4)_2SO_4$) containing natural-abundance nitrogen; (B) normal *P. cepacia* grown on ^{15}N-labelled $(NH_4)_2SO_4$; (C) auxotroph of *P. cepacia* grown on a medium of histidine containing natural-abundance nitrogen and ^{15}N-labelled $(NH_4)_2SO_4$; (D) normal *P. cepacia* grown on $(NH_4)_2SO_4$ containing natural-abundance nitrogen. For ease of reference the traces are labelled with the sample number in parentheses. *Conditions*: temperature, 2 K; microwave power, 50 μW; microwave frequency, 35.3 GHz; 100-kHz field modulation amplitude, 0.32 mT; scan rate, 2.5 MHz/s, 300 scans.

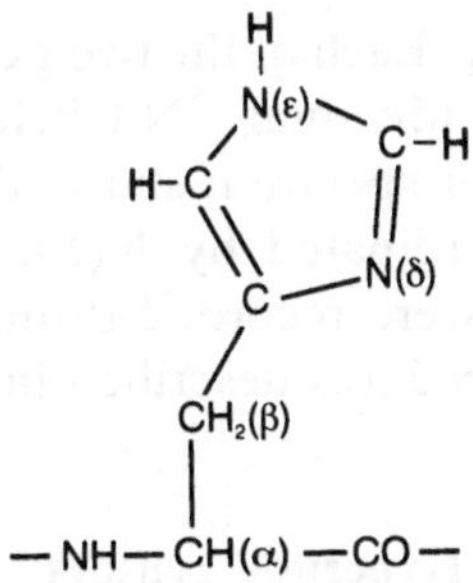

This structure suggests (deliberately) that coordination to a metal ion would occur through N(δ); in fact, both modes of binding commonly occur. We have determined the mode of histidine binding to the Rieske center by examining the ENDOR spectra of phthalate dioxygenase specifically labeled with ^{15}N-δ-histidine (Gurbiel *et al.*, 1993b).

Figure 14 shows an excellent match between corresponding pairs of ENDOR spectra for uniformly ^{15}N-labeled (upper traces) and ^{15}N-δ-histidine (lower traces) PDO taken at three g values across the EPR envelope. In particular, the spectra at g = 2.01 were taken at the low-field edge of the EPR envelope, where signals arise from a single-crystal-like

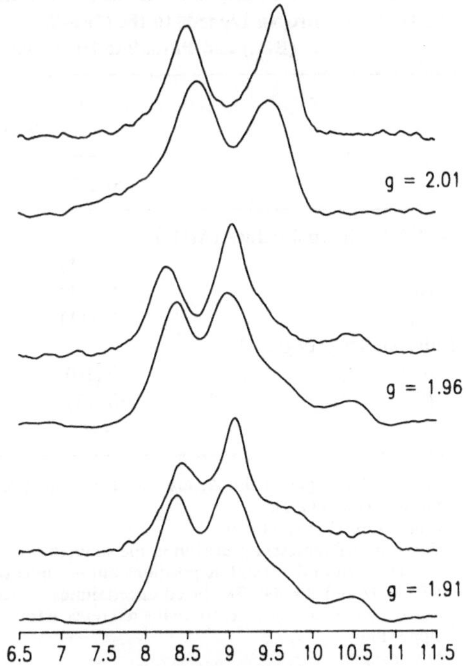

Figure 14. ENDOR spectra of PDO taken at three different g values. In each case the upper trace is from PDO uniformly labelled with ^{15}N and the lower trace is from PDO labelled specifically with ^{15}N-δ histidine.

orientation associated with g_1. Each of the two peaks in the upper spectrum corresponds to ν_+ from an individual ^{15}N-labeled histidine bound to the Fe^{2+} of the cluster. The lower spectrum shows these same peaks, proving that *both* histidines are coordinated by N(δ). As further confirmation, spectra of the two samples were recorded throughout the EPR envelope and simulated using the procedures described in the Appendix; these also match.

4.1.2. Determination of ^{15}N Hyperfine Tensors

^{15}N *X*-band ENDOR spectra of enriched PDO and of BC1 were measured across the EPR envelope, and the ^{15}N hyperfine tensors for the two coordinated histidyl nitrogens were determined using the analysis procedures mentioned above. Simulations using the two hyperfine tensors reported in Table 2 fully reproduce not only the positions of the two sharp resonances but also the breadth of the pattern to higher frequency and other resolved features. The hyperfine tensors were determined by simulating spectra at all observing fields; the full set of experimental and theoretical ENDOR frequencies is summarized in the field(g)–frequency plots for PDO and BC1 (Fig. 15).

TABLE 2
Hyperfine Tensor Principal Values and Orientations Relative to *g*-Tensor Principal Axes for the Histidyl-Nitrogen Ligands to the [2Fe–2S] Cluster of Cytochrome *bc*$_1$ from *R. capsulatus*[a] (BC1) and Phthalate Dioxygenase from *P. cepacia*[b] (PDO)

	Substance			
	BC1		PDO	
	Site 1[c]	Site 2[d]	Site 1	Site 2
A (^{15}N) principal values (MHz)				
A_1	5.0 (1)	6.0 (2)	4.6 (2)	6.4 (1)
A_2	5.6 (1)	7.4 (5)	5.4 (1)	7.0 (1)
A_3	8.4 (1)	9.7 (2)	8.1 (1)	9.8 (2)
Euler angles[e] (degrees)				
α	0 (10)	0 (10)	0 (30)	0 (10)
β	38[f] (3)	40[g] (4)	35 (5)	50 (5)
γ	—[h]	—[h]	0 (10)	0 (30)

[a] The absolute signs of the A_i are not determined; for an individual site all have the same sign. From Gurbiel *et al.* (1991).
[b] From Gurbiel *et al.* (1989).
[c] Uncertainties represent precision in the simulations, not necessarily the accuracy.
[d] As described in the text, line positions but not intensities are better fitted with principal values (5.8, 7.9, 9.9) MHz and β = 44°. The listed uncertainties reflect this discrepancy.
[e] Euler angles relate the A tensor to the reference g frame; they are discussed in the text and in the Appendix.
[f] Upper limit.
[g] Lower limit.
[h] This angle cannot be determined solely from data on the ^{15}N protein.

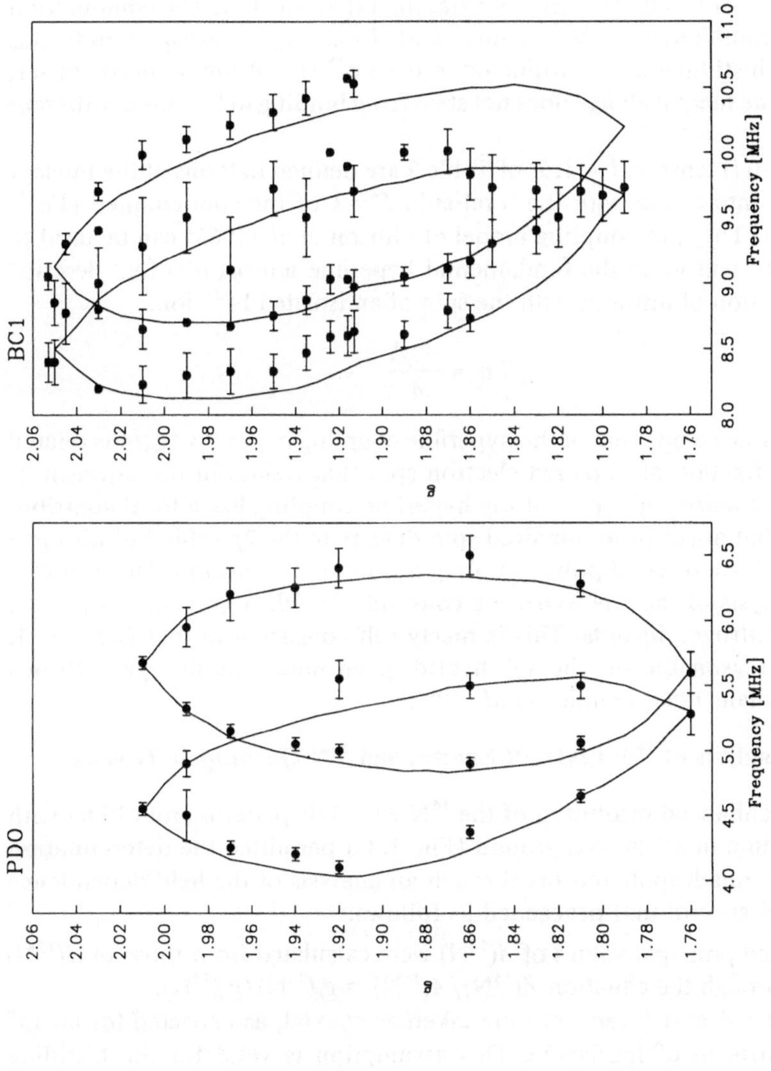

Figure 15. Plot of the frequency of the $\nu_+(^{15}N)$ ENDOR features for PDO and BC_1 versus the observing g value (static field). The theoretical values, calculated with the hyperfine tensor parameters given in Table 2, are indicated as solid lines. Left: PDO recorded at ~9.5 GHz; right: BC_1 recorded at ~35.4 GHz.

Analysis of the ^{15}N hyperfine tensors provides a means of characterizing the Fe–N bonding. The hyperfine tensors, and thus the two bonds, are quite inequivalent, with largest tensor components for PDO of $A_3(^{15}N1) = 8.1$ MHz and $A_3(^{15}N2) = 9.8$ MHz. The principal values for $A(1)$ are indistinguishable for BC1 and PDO. Those for $A(2)$ differ from those for $A(1)$, but again are quite similar for the two proteins; however, it does appear that $A(2)$ for PDO is more nearly axial than that for BC1. The combination of data from ENDOR, Mössbauer, and Raman spectroscopies indicates that both histidines are coordinated to the Fe^{2+} site of the reduced cluster, and thus the inequivalence does not stem from binding to Fe ions of different valency.

The $A(i)$ tensors, $i = 1, 2$, of Table 2 are defined in terms of the nuclear hyperfine interaction with the total spin $S = \frac{1}{2}$ of the spin-coupled [Fe^{3+}, Fe^{2+}] pair. The spin-coupling model of Gibson *et al.* (1966) can be used to relate these tensors to the fundamental hyperfine tensors $a(i)$ that describe the interaction of nitrogen with the spin of an isolated Fe^{2+} ion:

$$a = \frac{-3A}{4}$$

The isotropic component of the hyperfine coupling $a_{\text{iso}} = -3A_{\text{iso}}/4$ is related to f_s, the fraction of unpaired electron spin that resides in the nitrogen $2s$ orbital; the anisotropic part of the hyperfine coupling has a local contribution a_{2p} that arises from unpaired spin density in the $2p$ orbital of nitrogen (f_{2p}) and from direct dipolar coupling a_{3d} to the ferrous ion. The approximate analysis of the ^{15}N hyperfine constants for PDO gave $f_{2s} \sim f_{2p} \sim 2\%$ for both nitrogen ligands. This is nicely self-consistent in that $f_{2p}/f_{2s} \sim 1$, which is reasonable for the sp^n hybrid of an imidazole nitrogen (Brown and Hoffman, 1980; Scholes *et al.*, 1982).

4.1.3. Analysis of ^{14}N ENDOR Spectra and ^{14}N Quadrupole Tensors

The enhanced resolution of the ^{14}N ENDOR patterns from PDO with ^{14}N-histidine in a ^{15}N background (Fig. 13C) permitted the determination of the ^{14}N quadrupole tensors through an analysis of the field dependence of the ^{14}N spectra that proceeded as follows:

1. The principal values of $A(^{14}N)$ were calculated from those of $A(^{15}N)$ through the equation $A(^{15}N)/A(^{14}N) = g_n(^{15}N)/g_n(^{14}N)$.
2. The A and P tensors were taken as coaxial, as expected for an sp^n nitrogen of imidazole. This assumption is valid for the histidine coordinated to Fe^{3+} in metmyoglobin (Scholes *et al.*, 1982).
3. The Euler angles α and β given in Table 2 were used.
4. The principal values of the quadrupole tensor were taken to be the same for N(1) and N(2).

With these restrictions the ^{14}N simulations depend on four unknown parameters: two principal values of P, namely P_1 and P_2 (the quadrupole tensor is traceless; $P_1 + P_2 + P_3 = 0$), and $\gamma(1)$ and $\gamma(2)$, the rotations of the two tensors about their Fe–N bond. However, the constraint that P_1 and P_2 be chosen to reproduce the single-crystal-like spectrum at g_1 eliminated one of these as a variable. The parameters best reproducing the experiments are listed in Table 2.

It is noteworthy that the tensor values independently derived are almost identical to those for the axial histidine bound to the heme of aquomet-Mb (Scholes *et al.*, 1982). This agreement permits use of the Mb results to deduce the physical orientation of the imidazole planes in PDO relative to the quadrupole axes. In Mb, the tensor component perpendicular to this plane is $P_2 = 0.31$ (80); we assign the same value to PDO. Thus, for $\alpha = 0$ as found here, the parameter γ corresponds to the dihedral angle between the imidazole and the g_1–g_3 planes, with the latter corresponding to the N(1)–Fe–N(2) plane.

4.1.4. Structure of the Cluster

The process of determining the hyperfine and quadrupole coupling tensors gives information not only about the orbitals involved in bonding, but also about the geometry of the cluster directly. The analysis of the ^{15}N hyperfine tensors supports a geometric model (Fig. 16) in which the four protein-donated ligands and two iron ions lie in the g_1–g_3 plane; recent ^{57}Fe ENDOR data confirm that the Fe–Fe vector corresponds to g_1 (to be published). The geometry of histidine coordination in the two clusters is qualitatively the same, but shows distinct differences. The analysis for PDO gave $\bar{\beta}(N1) \equiv \pi - \beta(N1) \sim 35°$ and $\bar{\beta}(N2) \sim 55°$ for the angles between Fe–N and Fe–Fe vectors, whereas the BC1 center seems more symmetric, with $\bar{\beta}(N1) \sim \bar{\beta}(N2)$. The N–Fe–N bite angle is $\bar{\beta}(1) + \bar{\beta}(2) \sim 90(10)°$ in both clusters, roughly consistent with tetrahedral coordination at Fe^{2+}. The analysis of the ^{14}N quadrupole tensors further gave an estimate of the dihedral twist of the two imidazole rings out of the N–Fe–N plane (γ in Fig. 16). It is relatively small for N1 ($\gamma_1 \to 0$) and larger for N2; molecular modeling shows that steric constraints prevent both rings from lying flat in the g_1–g_3 plane ($\gamma_1 = \gamma_2 = 0$). Finally, because the ^{14}N and ^1H ENDOR patterns for PDO and BC1 are virtually indistinguishable, we inferred that the structure presented here is generally applicable to Rieske-type centers.

4.2. Substrate Interactions and Enzyme Mechanism: Aconitase

Aconitase catalyzes the stereospecific interconversion of citrate and isocitrate *via* the dehydrated intermediate *cis*-aconitate:

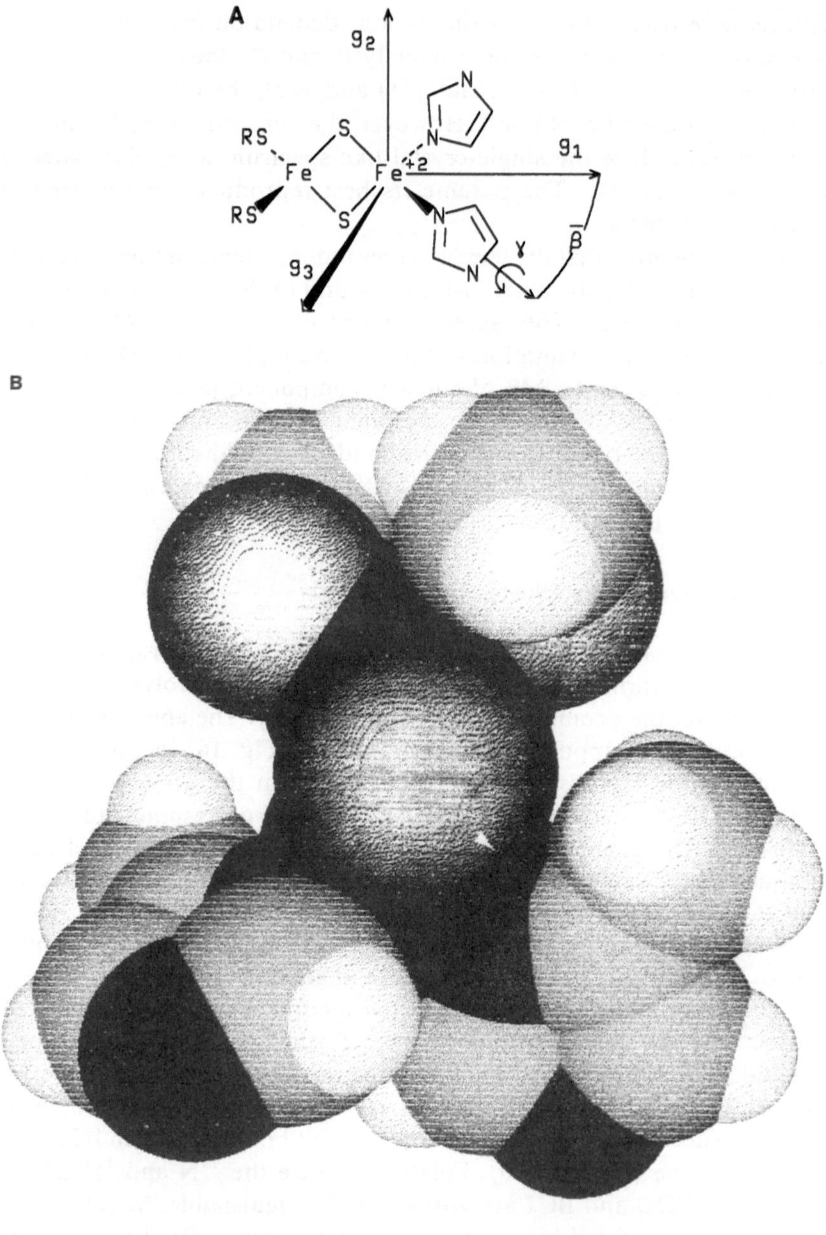

Figure 16. Structure of the Rieske-type [2Fe–2S] cluster of PDO as determined by ENDOR spectroscopy. (A) The [2Fe–2S] core and the $(g_1–g_2)$ plane lie in the paper; the bite angles β and dihedral angles γ are discussed in the text. (B) Space-filling model using two (CH_3-5) and two 2-methyl imidazoles as ligands. The view is down onto the g_1–g_3 plane, with the two Fe and four ligand atoms lying in the plane.

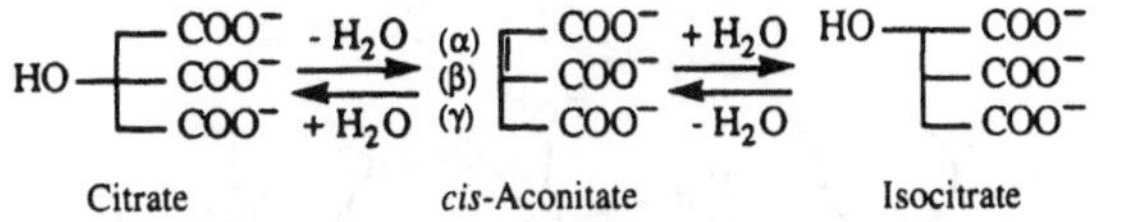

The active site contains a $[4Fe\text{–}4S]^{2+}$ cluster, which does not act in electron transport but rather performs a catalytic function through interaction of substrate at a specific iron site of the cluster (Fe_a) (Kent *et al.*, 1985). The cluster can be reduced to the EPR-active $[4Fe\text{–}4S]^{+}$ state with substantial (~30%) retention of activity.

In collaboration with Professors H. Beinert and M. C. Kennedy we individually characterized the 4Fe and the 4S of the cluster both in the enzyme's substrate-free (E) and substrate-bound (ES) forms by ^{57}Fe and ^{33}S CW ENDOR at *Q* band (Werst *et al.*, 1990b). We used $^{1,2}H$, ^{17}O, and ^{13}C CW ENDOR (Telser *et al.*, 1986a; Kennedy *et al.*, 1987; Werst *et al.*, 1990a) and $^{1,2}H$ pulsed ENDOR (Fan *et al.*, 1992a) to characterize the binding of solvent-derived species, substrate, and substrate analogues. The results, which have been confirmed by X-ray diffraction studies (Lauble *et al.*, 1992), led to a major revision in the accepted enzyme mechanism (Beinert and Kennedy, 1989; Emptage, 1987).

4.2.1. Cluster Properties and Structure of the Enzyme–Substrate Complex

The ^{57}Fe ENDOR data confirm the Mössbauer result (Kent *et al.*, 1985) showing that the four iron sites fall into two classes, a single-site Fe_a and three inequivalent Fe_b sites, and that only Fe_a responds significantly to substrate binding. ENDOR further shows that the three Fe_b sites are themselves inequivalent, and that this is true both in the substrate-free enzyme (E) and the enzyme–substrate complex (ES). Analysis of ^{33}S resonances from the $[4Fe\text{–}4S]^{+}$ cluster of the enzyme-substrate complex suggests that the sulfur sites occur as two pairs ($S_{\alpha 1}, S_{\alpha 2}; S_{\beta 1}, S_{\beta 2}$) with remarkably small spin density on sulfur, and it further discloses their spatial relation to the Fe sites (Werst *et al.*, 1990b). Figure 17 summarizes this information about the four Fe ions and four inorganic sulfides within the context of the X-ray diffraction structure (Robbins and Stout, 1989b; Lauble *et al.*, 1992), which shows that cysteines are bound to the three iron ions classified spectroscopically as Fe_b.

4.2.1.1. ^{57}Fe ENDOR. Activation and exchange reactions were used to label iron subsites of the $[4Fe\text{–}4S]^{+}$ cluster isotopically with ^{57}Fe ($I = \frac{1}{2}$) in either the site of substrate binding, Fe_a, or the three Fe_b sites (Werst *et al.*, 1990b). Figure 18A–C shows ^{57}Fe-ENDOR spectra of $E(^{57}Fe_a)$ obtained by monitoring the EPR signal at positions corresponding to $g_{1,2,3} = 2.06$, 1.93, and 1.86, respectively. In each case, the ^{57}Fe nucleus gives a doublet

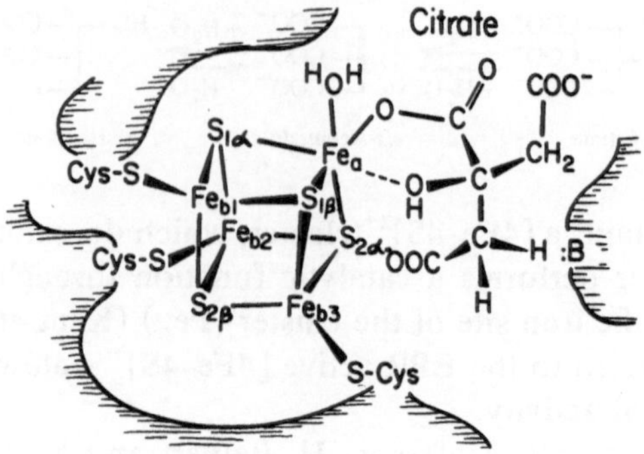

Figure 17. Representation of ENDOR-derived information about the $[4Fe\text{-}4S]^+$ cluster of the aconitase enzyme–substrate complex, showing the two pairs of sulfur detected by ENDOR, S1α, S1β and S2α, S2β in relationship to the four inequivalent iron sites Fe_a, Fe_{b1}, Fe_{b2} and Fe_{b3}, along with the bound substrate.

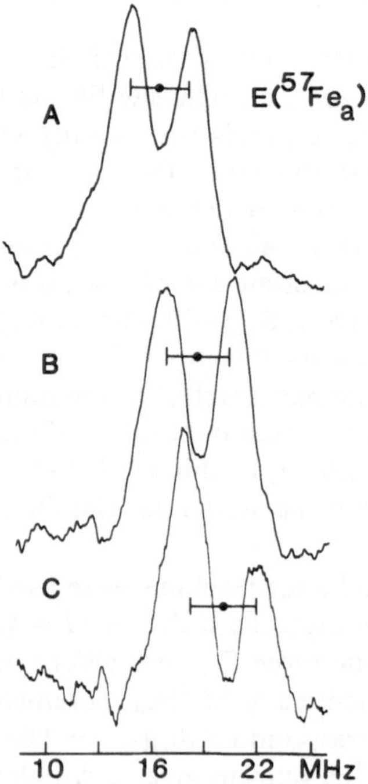

Figure 18. ^{57}Fe *Q*-band (34.61 GHz) ENDOR of Fe_a site in substrate-free aconitase at (A) $g = 2.06$, $B_0 = 1.21$ T; (B) $g = 1.93$, $B_0 = 1.281$ T; (C) $g = 1.86$, $B_0 = 1.33$ T. The assignments to $A/2$ (·) and twice the Larmor frequency (|—|) are indicated. *Conditions*: temperature, 2 K; microwave power, 0.08 mW; 100-kHz field modulation amplitude, 0.8 mT; rf scan rate, 6 MHz/s; time constant, 0.032 s.

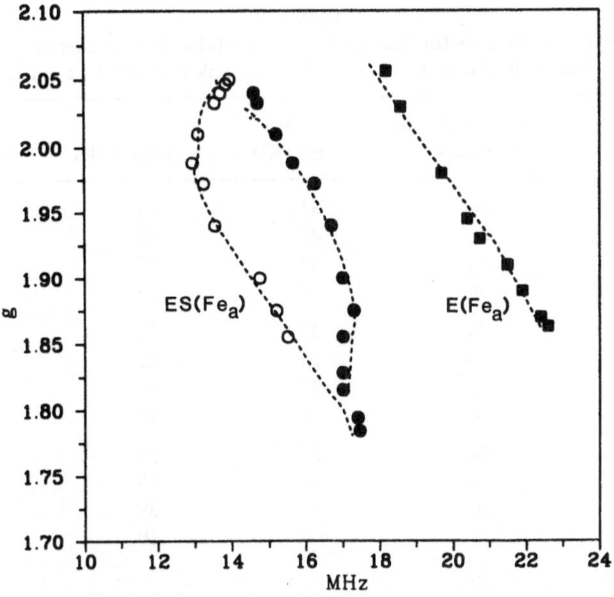

Figure 19. ^{57}Fe ENDOR peak position for the ν_+ feature(s) for Fe_a in substrate-free aconitase enzyme (squares) and substrate-bound aconitase (circles) versus the observing g value. Theoretical values calculated using parameters in Table 3 are indicated as dashed lines.

centered at $A^{Fe}/2$. As the observing g value is moved from g_1 to g_3, the center of the doublet progressively shifts to increasing rf, with the hyperfine coupling increasing from $A(g_1) = 32.8$ MHz to $A(g_3) = 40$ MHz. Figure 19 shows a plot of the ν_+ ENDOR frequencies as a function of g value for both E($^{57}Fe_a$) and ES($^{57}Fe_a$). The points represent the experimentally obtained frequencies and the dashed lines are from computer simulations. For E($^{57}Fe_a$) the smooth variation of ν_+ with g value can be simulated with a nearly axial A tensor coaxial with the g tensor. When substrate binds to the enzyme, the A tensor changes dramatically. Each of the principal values for E($^{57}Fe_a$) is reduced by ~10 MHz in ES($^{57}Fe_a$), and A is no longer coaxial with g. The principal values of A and the Euler angles that describe its orientation are listed in Table 3 for both states.

The Fe_b sites of the $[4Fe\text{-}4S]^+$ cluster have also been studied both in the presence and the absence of substrate. In both cases, resonances from three inequivalent iron ions have been observed. In contrast to Fe_a, the hyperfine couplings of the Fe_b sites change very little upon substrate binding (Table 3).

4.2.1.2. *^{33}S ENDOR.* Through exchange reactions analogous to those used for ^{57}Fe, aconitase was reconstituted with ^{33}S ($I = \frac{3}{2}$) in the inorganic sulfide sites. Figure 20 shows the ν_+ features of the resulting ^{33}S ENDOR

TABLE 3
^{57}Fe Hyperfine Tensors for Four Fe Sites of the $[4Fe–4S]^+$ Cluster of Aconitase, Substrate-Free (E) and in the Enzyme–Substrate (ES) Complex

Site	Principal value or Euler angle[a]	E (ENDOR)	ES (ENDOR)	ES (Mössbauer)[b]
Fe_a	A_1	33	23	26
	A_2	41	32	34
	A_3	42	32	34
	α	0	25	—
Fe_{b3}	A_1	35	35	−32[c]
	A_2	34	42	−39
	A_3	42	43	−39
	α	0	10	—
Fe_{b2}	A_1	27	31	−32[c]
	A_2	34	38	−39
	A_3	37	39	−39
	α	0	10	—
Fe_{b1}	A_1	—	—	12
	A_2	25[d] (18)	24[d] (17)	22
	A_3	23[d] (15)	24[d] (17)	12

[a] If we begin with **A** and **g** coaxial, α refers to the rotation of the A tensor about its A_3 axis. From Werst *et al.* (1990b).
[b] From Kent *et al.* (1985).
[c] Mössbauer does not distinguish between the Fe_{b2} and Fe_{b3} sites.
[d] There are two possible interpretations for the Fe_{b1} site; the less favored is placed in parentheses.

spectra of $E(^{33}S)$ taken at g_1, g_2, and g_3. In the single-crystal-like spectrum at g_1, $\nu_+ \sim 8.5$ MHz, which corresponds to $A(S) \sim 6.5$ MHz. The spectrum at g_3 consists of two peaks also centered at ~6.5 MHz and is interpretable as arising from at least two types of S^{2-} ions, one with $A(S) \sim 8.5$ MHz and one with $A(S) \sim 6.3$ MHz. The spectrum at g_2 has four discernible peaks (Fig. 20B). Three of these have remained near $\nu \sim 8.5$ MHz; the fourth has moved to higher frequency and gives $A(g_2) \sim 12$ MHz. The full field dependence of the ENDOR spectra (Fig. 20 plus other data) can be simulated by assuming that there are two pairs of equivalent S^{2-} ions that have no resolved quadrupole splittings; one pair has a hyperfine tensor $A(S1) \sim [6.8, 10.1, 6.5]$ MHz and the other has the tensor $A(S2) \sim [6.3, 12.4, 8.7]$ MHz. A simulation for the spectrum at g_2 is presented in Fig. 20B. The analogous experiment for ES (^{33}S) again showed two pairs of bridging sulfide ions, one with $A(S1) \lesssim 1$ MHz, the other with $A(S2) \sim$ 6–12 MHz.

Considerations of spin coupling within the cluster also indicate that the four bridging sulfide ions within the cluster should divide into two pairs, in accordance with the measurement, and gave the spatial relationships of

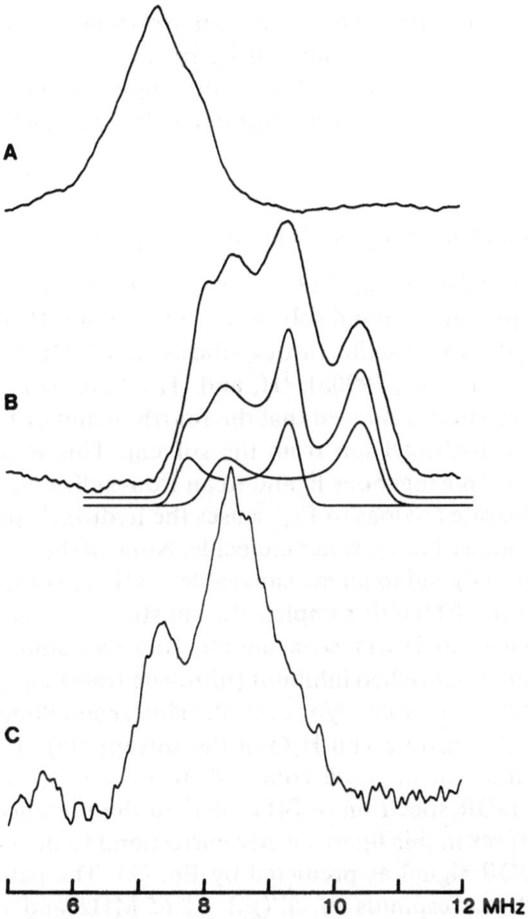

Figure 20. ^{33}S ENDOR at *Q* band (34.55 GHz) of the $[4Fe\text{-}4S]^+$ form of aconitase E(^{33}S): (A) $g = 2.04$, $B_0 = 1.21$ T; (B) $g = 1.93$; $B_0 = 1.28$ T; (C) $g = 1.85$, $B_0 = 1.335$ T. The simulation below spectrum (B) shows individual contributions from the two types of S^{2-} ions, along with their sum. Hyperfine tensors employed are given in the text; linewidth and intensities were adjusted visually. This simulation employed Eq. (A31) rather than (A29) and the intensities match the experimental results better than in the original simulation. *Conditions*: temperature, 2 K; microwave power, 0.05 mW; 100-kHz field modulation amplitude, 0.8 mT; rf scan rate, 0.75 MHz/s; time constant, 0.064 s. Spectrum (A) was scanned from high to low rf; spectra (B) and (C) were scanned from low to high. Consequently, for direct comparisons, (A) should be shifted ~1 MHz toward higher frequency.

Fe and S ions shown in Fig. 17. The analysis indicates that the intrinsic isotropic hyperfine interaction constant for S^{2-} bonded to an iron ion in the ES cluster is extremely small, $a(S) \sim 2$–4 MHz. The ^{33}S ENDOR results for the enzyme without substrate (E) can be interpreted in terms of two types of sites with $a(S) \lesssim 17$ MHz. For comparison, an electron in a 3*s* orbital on sulfur is predicted to give an isotropic coupling of $a_0(S) \sim$ 3500 MHz, and thus the *net* spin density in sulfur 3*s* is less than 0.5% in either form. The sulfides also exhibit anisotropic couplings equivalent to $\lesssim$1% *net* asphericity in the spin delocalization to S^{2-} 3*p* orbitals per neighboring iron ion.

4.2.2. Structure of the Enzyme–Substrate Complex

4.2.2.1. H_xO Binding by ^{17}O and $^{1,2}H$ ENDOR. We investigated the possible participation of bound solvent H_xO (H_2O or OH^-) in the catalytic dehydration–hydration reaction through the use of ENDOR with ^{17}O (Telser *et al.*, 1986a; Werst *et al.*, 1990a), ^1H, and ^2H (Werst *et al.*, 1990a; Fan *et al.*, 1992a). These studies showed that the fourth ligand of Fe_a in substrate-free enzyme is a hydroxyl ion from the solvent. This represents the first demonstration of an exogenous ligand to an iron–sulfur cluster. Binding of substrate or substrate analogs to Fe_a causes the hydroxyl species to become protonated, forming a bound water molecule. None of the observed ENDOR signals could be assigned to an exchangeable —OH of substrate. This result suggests that in the ENDOR samples, the substrate is largely bound to the enzyme in its dehydrated form, *cis*-aconitate. However, studies of the enzyme complexed to an ^{17}O-enriched inhibitor (nitroisocitrate) suggested that when citrate is bound during the catalytic cycle the cluster simultaneously coordinates the —OH of substrate and H_2O of the solvent (Fig. 17).

As an example of the data collected in this study, Fig. 21 shows an ^{17}O ($I = \frac{5}{2}$) ENDOR spectrum of $[4Fe\text{–}4S]^+$ in the presence of citrate and $H_2{}^{17}O$. The features in this figure clearly correspond to the pattern expected for a ^{17}O ENDOR signal as predicted by Eq. (2). The pattern is centered at $A^O/2$, which corresponds to $A^O(g_2) = 8.65$ MHz and consists of two groups of $2I = 5$ lines each, centered at $A^O/2$ and separated by $2\nu_O$. Within each group, the lines are separated by $3P^O$. As is usually the case, the higher-frequency group is more intense. The separation of the lines within each group gives an effective value of $P^O = 0.28$ MHz. Similarly, we performed ENDOR on aconitase in $H_2{}^{17}O$ without substrate and observed an ^{17}O signal with $A^O(g_2) = 10.4$ MHz. The observation of an ^{17}O signal with such substantial hyperfine interaction, which was found to be essentially isotropic, shows that the ^{17}O-containing species coordinates directly to the $[4Fe\text{–}4S]^+$ cluster. This is consistent with the crystal structure of the [4Fe–4S] cluster, indicating that the fourth ligand of Fe_a in the substrate-free enzyme is a solvent species, H_2O or OH^-. The ^{17}O hyperfine and quadrupole tensor parameters for $H_x{}^{17}O$ bound to E, ES, and the nitroisocitrate-bound enzyme EN are summarized in Table 4. The parameters for $E[H_2{}^{17}O]$ differ from those for the other two, which agree with one another.

^1H and ^2H CW and pulsed ENDOR were used to characterize these solvent species in greater detail. Comparison of the $g_3 = 1.86$ single-crystal-

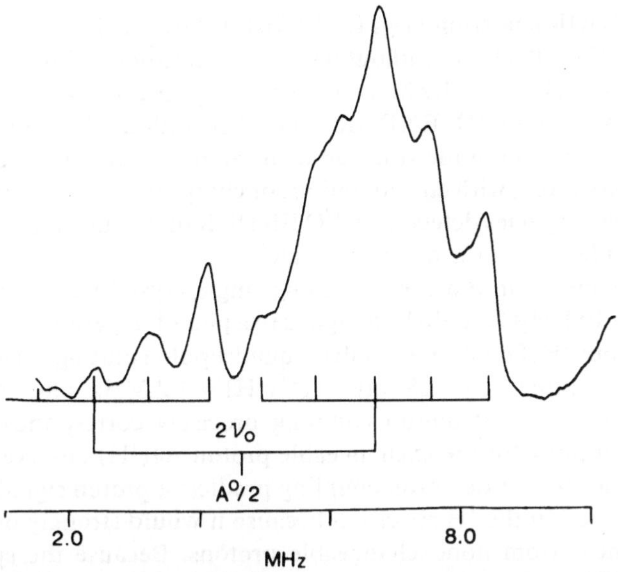

Figure 21. ^{17}O ENDOR spectrum at $g = 1.88$ for $[4Fe\text{-}4S]^{1+}$ substrate-bound aconitase (ES) in $H_2\,^{17}O$ solution. The pattern expected for ^{17}O ENDOR is indicated; the smaller splittings correspond to $3P^0$. *Conditions*: temperature, 2 K; B_0, 0.3650 T; 100-kHz field modulation amplitude, ~0.3 mT; microwave frequency, 9.53 GHz; microwave power, 1 μW; time constant, 0.032 s; rf scan rate, 3.0 MHz/s; 3432 scans.

like 35-GHz ^1H CW ENDOR spectra of $E[^1H_2O]$ and $E[^2H_2O]$ reveal at least one ^1H Larmor doublet associated with the H_xO species detected by ^{17}O ENDOR (Werst *et al.*, 1990a).

The ENDOR intensity associated with the exchangeable proton is poorly resolved because of overlapping resonances from the many nonexchangeable protons. However, there is no interference in the corresponding ^2H ($I = 1$) CW ENDOR measurements because they show only deuterons that replace protons upon solvent exchange. The single-crystal-like 35-GHz

TABLE 4
^{17}O Hyperfine (A) and Quadrupole (P) Tensors for $H_x\,^{17}O$ Bound to the $[4Fe\text{-}4S]^{1+}$ Cluster of Aconitase[a]

	Species	$A_{1,2,3}$			$P_{1,2,3}$		
$E[H_2\,^{17}O]^b$	OH^-	9	12	10		—	
$ES[H_2\,^{17}O]^c$	H_2O	9.0	8.5	8.4	−0.5	0.35	0.15
$EN[H_2\,^{17}O]^d$	H_2O	9.0	8.5	8.4	−0.5	0.35	0.15

[a] Assignments of the protonation states of cluster-bound H_xO are discussed in the text. From Telser *et al.* (1986a).
[b] Euler angles: $\alpha = \beta = \gamma = 0$. Resolved quadrupole splitting is not observed for this ^{17}O species. For E, ES, EN, see text.
[c] Euler angles: $\alpha \sim 35°$, $\beta = \gamma = 0$.
[d] Euler angles: $\alpha \sim 35°$, $\beta = \gamma = 0$. The hydroxyl group of nitroisocitrate also binds to the cluster.

^2H CW ENDOR spectrum at g_3 for $E[^2H_2O]$ shows a single hyperfine-split doublet $A(^2H) = 0.6$ MHz without resolved quadrupole splitting. This result corresponds to $A(^1H) = 3.9$ MHz, the coupling observed for the exchangeable proton seen in ^1H ENDOR. The observation of a single type of exchangeable proton in the resolved deuterium spectrum (as well as in the ^1H spectrum), along with considerations of charge balance, led us to suggest that the solvent species detected by ^{17}O ENDOR in the substrate-free enzyme is a hydroxide ion, not a water molecule.

The signals from the corresponding single-crystal-like g_3 ^2H ENDOR spectrum of $ES[^2H_2O]$ could be assigned to a pair of hyperfine-split deuteron doublets from ^2H_xO without resolved quadrupole splittings. The resulting hyperfine couplings for ES are $A1^{ES}(^2H) = 1.2$ MHz and $A2^{ES}(^2H) = 0.5$ MHz. The larger deuteron coupling precisely corresponds to the ^1H hyperfine coupling for the exchangeable proton $A1(^1H)$ observed in the ^1H ENDOR; the smaller deuteron coupling predicts a proton signal that is not readily detected in the ^1H spectrum because it would strongly overlap with the resonances from nonexchangeable protons. Because the spectrum of the $[4Fe\text{–}4S]^+$ cluster of ES reveals two distinct exchangeable protons, we assigned the cluster-bound H_xO as an asymmetrically bound water molecule. The ^{17}O hyperfine tensors of OH^- bound to the cluster of E and of H_2O bound to ES (Table 4) show significant differences, supporting the assignment of different protonation states of H_xO in the two cases.

This experiment was repeated and confirmed with ^2H Mims ENDOR at *X* band, which shows much improved resolution (Fan *et al.*, 1992b). Figure 22A is a single-crystal-like deuterium ENDOR spectrum of $E[^2H_2O]$ taken at the high-field edge ($g = 1.86$) of the EPR absorption envelope. It shows one pair of peaks from ^2H_xO split by the hyperfine interaction $A^D \approx 0.50$ ($A^H \approx 3.25$) MHz, with further quadrupole splitting $3P^D \approx 0.12$ MHz. The pulsed ENDOR spectrum taken at $g_3 = 1.78$ for the enzyme with citrate in ^2H_2O ($ES[^2H_2O]$) (Fig. 22B) shows a more complex pattern, with four deuterium peaks from two inequivalent nuclei that have $A^D = 1.22$, 0.46 ($A^H = 7.9$, 3.0) MHz and no observable quadrupole splitting at this field. Comparison with the ^1H resonances lost on H–D exchange (Werst *et al.*, 1990a) confirms the latter assignment.

4.2.2.2. Substrate Binding by ^{17}O ENDOR. To complete the picture of substrate binding to the cluster, we investigated whether one or more carboxyls of substrate bind to the cluster (Kennedy *et al.*, 1987; Werst *et al.*, 1990a). Figure 23 presents ENDOR spectra of $[4Fe\text{–}4S]^+$ aconitase in the presence of citrate whose carboxyl groups have been individually labeled with ^{17}O. The ENDOR measurements with substrate whose central (β)-carboxyl group is ^{17}O labeled show a strong ^{17}O pattern (Fig. 23, middle spectrum), but no ^{17}O ENDOR signal was observed when either of the terminal carboxyl groups (α or γ) was ^{17}O labeled (Fig. 23, upper and lower spectra). Thus, under the experimental conditions the central carboxyl

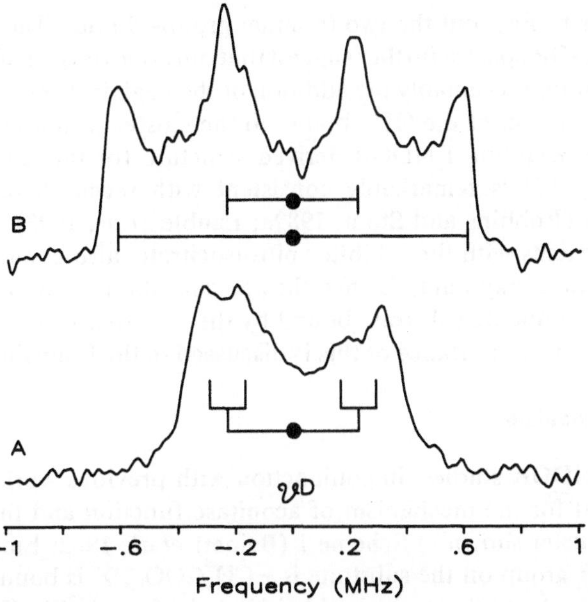

Figure 22. ^2H Mims ENDOR spectra of the $[4Fe\text{-}4S]^+$ cluster of aconitase (A) without substrate and (B) with substrate. Spectra are plotted as $\delta\nu = \nu - \nu_D$. *Conditions*: temperature, 2 K; microwave frequency, (A) 9.14 GHz, (B) 9.09 GHz; B_0, (A) 0.3525 T, (B) 0.3645 T; t_p, 16 ns; t_{rf}, 50 μs; τ_{12}, (A) 380 ns, (B) 440 ns; scans (A) 168, (B) 880.

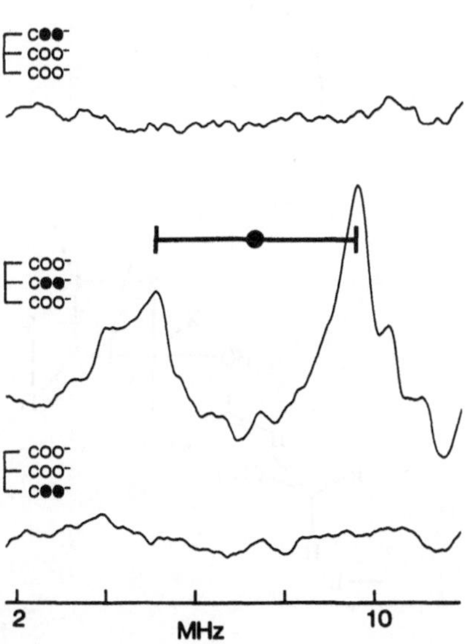

Figure 23. ^{17}O ENDOR spectra at $g = 1.85$ of reduced aconitase in the presence of substrate whose —COO^- groups have been individually labeled with ^{17}O. Upper: label at α-carboxyl, middle: at β-carboxyl, lower: at γ-carboxyl; (·) indicates $A^O/2$, and (⊢⊣) indicates $2\nu_O$. The hydroxyl is not shown in the cartoon substrate because aconitase converts it to an equilibrium mixture. *Conditions*: temperature, 2 K; B_0, 0.3815 T; microwave frequency, 9.92 GHz.

group binds to Fe_a, but the two terminal groups do not. The simplicity of the ^{17}O ENDOR spectra further suggest that only one oxygen of the carboxyl group is bound. Presumably the addition of the negatively charged carboxyl causes protonation of the OH^- bound to the cluster without substrate (see above). The resulting ENDOR-derived structure for the substrate-bound cluster (Fig. 17) is remarkably consistent with recent X-ray diffraction information (Robbins and Stout, 1989a; Lauble *et al.*, 1992).

Experiments with the inhibitor nitroisocitrate, also confirmed and extended by the X-ray data, further showed that the enzyme nonetheless is able to accommodate substrate bound by the α-carboxyl, as well as by the β-carboxyl. The importance of this is discussed in the following subsection.

4.2.3. Mechanism

The ENDOR studies, in conjunction with previous work, have led to the proposal for the mechanism of aconitase function and the role of the [4Fe–4S] cluster shown in Scheme 1 (Beinert *et al.*, 1962; Emptage, 1987). Here the $-R$ group on the substrate is $-CH_2COO^-$, X is bound hydroxide, and $-B$: represents the amino acid side chain from ^{642}Ser (Lauble *et al.*, 1992) that stereospecifically transfers a proton between citrate and isocitrate.

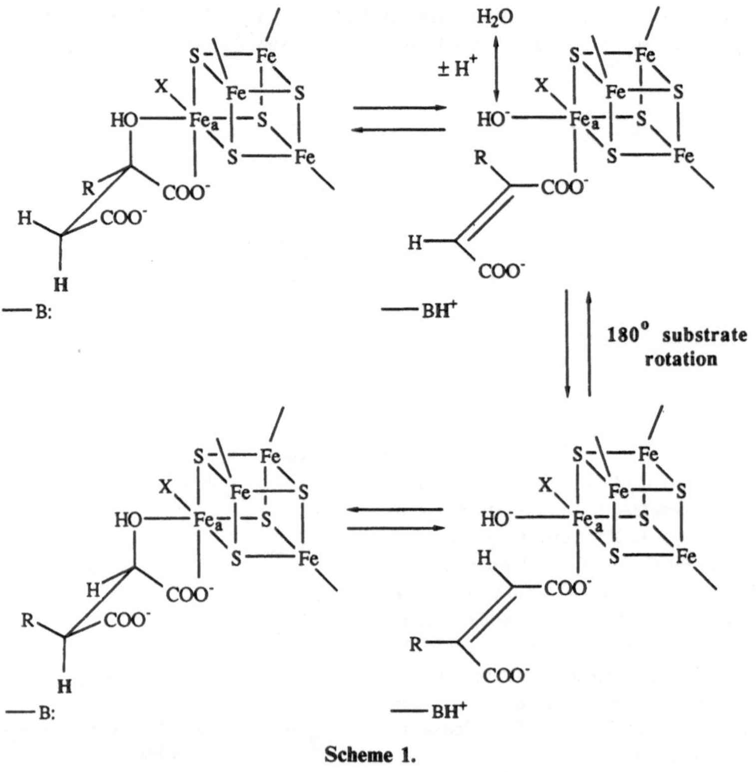

Scheme 1.

The cluster appears to function as follows:

1. It helps to position the substrate through the binding of one carboxyl.
2. It coordinates and accepts the hydroxyl of substrate during the dehydration of citrate and isocitrate.
3. It donates a bound hydroxyl during the rehydration of *cis*-aconitate.

The stereochemistry of the reaction, the specific transfer of the proton from citrate to isocitrate, and the coordination of a single carboxyl and adjacent hydroxyl group of substrate together require that *cis*-aconitate disengage from the active site, rotate 180°, and switch from binding the α-carboxyl to the β-carboxyl during the catalytic cycle.

5. METALLOENZYMES WITH AMINO ACID RADICALS

The previous sections of this review have described the use of ENDOR in studying metal centers in proteins. Many other enzymes employ organic radicals in their catalytic cycles. Most such stable radicals characterized to date are found on protein cofactors, and ENDOR has been applied quite successfully to identify and to investigate the electronic structures of the π-based radicals of porphyrin macrocyles, quinone species, and chlorophyll and bacteriochlorophyll species in photosynthetic reaction centers. These studies have been reviewed (Möbius *et al.*, 1989; Scholes, 1979; Prince *et al.*, 1988). In addition to these cofactor radicals, an increasing number of enzymes has been shown to use amino acids to form radicals important in catalysis. ENDOR has been used to study three metalloenzymes with such radicals, to identify the amino acid species that hosts the radical, and to determine structural and electronic parameters that relate to their chemical properties. Unlike the studies of paramagnetic metal centers, proton ENDOR alone is often capable of giving much insight into the properties of amino acid radicals.

5.1. Cytochrome *c* Peroxidase

Cytochrome *c* peroxidase (C*c*P) catalyzes the H_2O_2-dependent oxidation of ferrocytochrome *c* and is capable of storing two oxidizing equivalents when treated with peroxide. As in peroxidases, such as horseradish peroxidase and chloroperoxidase, and in catalase, this intermediate state of C*c*P stores one oxidizing equivalent as an oxy-ferryl species (Dawson, 1988). In these other peroxidases, the second equivalent is stored as the π-cation radical of the porphyrin macrocycle. A variety of physical evidence indicates, however, that C*c*P stores the second oxidizing equivalent as a stable, EPR-active amino acid radical. From the crystal structure of C*c*P two tryptophan residues, ^{51}Trp and ^{191}Trp, were identified as being near the Fe porphyrin and thus candidates for the radical (Finzel *et al.*, 1984; Poulos, 1988). However, the highly unusual EPR signal from this radical, which appears axial with $g_{\parallel} = 2.04$ and $g_{\perp} = 2.01$, and the proton ENDOR (see below) had prompted proposals that the signal was from a peroxy radical (Wittenberg *et al.*, 1968) or was derived from methionine (Hoffman *et al.*, 1981).

The identity of the radical in C*c*P was determined in collaboration with Professor D. Goodin through a combination of ENDOR and site-directed mutagenesis (Sivaraja *et al.*, 1989). Proton ENDOR of the native oxidized C*c*P radical shows sharp resonances from protons with hyperfine couplings ranging from 2–20 MHz (Hoffman *et al.*, 1979, 1981; Fishel *et al.*, 1991). The complicated proton ENDOR signal was reduced to a single doublet with $A^{H} \sim 17$ MHz when the organism was grown on deuterated ($d8$) tryptophan. That signal was further shown to be exchangeable in D_2O. These experiments proved that the signal is associated with a radical on ^{51}Trp or ^{191}Trp. Substitution of ^{51}Trp for Phe by site-directed mutagenesis leaves the EPR and proton ENDOR spectra virtually unchanged (Fishel *et al.*, 1987, 1991; Sivaraja *et al.*, 1989). Thus the amino acid radical in C*c*P is identified as ^{191}Trp. This residue lies approximately 5.1 Å from the heme active site and is oriented with its ring roughly perpendicular to the heme and parallel to the distal imidazole ligand to the heme.

In recent work (Houseman *et al.*, 1993), we have found that simulations of the EPR spectrum and of the ENDOR spectra at different field positions support a model in which the radical and the Fe(IV)=O in C*c*P are exchange-coupled with a distribution of coupling constants. Unfortunately, this distribution greatly complicates analysis of ENDOR spectra.

5.2. Tyrosine Radicals

Several stable amino acid radicals in metalloenzymes have now been identified with tyrosine. Ribonucleotide reductase (RR) and prostaglandin H synthase both contain tyrosine radicals (Stubbe, 1989; Karthein *et al.*, 1988), and the Photosystem II (PSII) reaction center contains two redox-active tyrosines with distinguishable electron-transfer properties (Debus *et al.*, 1988). The radical species in galactose oxidase was also recently determined to be derived from tyrosine (Whittaker and Whittaker, 1990).

5.2.1. Photosystem II and Ribonucleotide Reductase

In a series of studies on the PSII and RR radicals, Babcock and co-workers (Barry *et al.*, 1990; Bender *et al.*, 1989) have used both EPR and ENDOR of specifically deuterated tyrosine to assign the proton hyperfine couplings to specific positions on the tyrosine ring, as well as the contribution from the methylene protons at the β-carbon position. Using these coupling parameters, the spin density at each position in the tyrosine ring could be derived. The tyrosine radical follows an odd-alternate spin distribution model for ring-based radicals, in which the spin distribution alternates between positive and negative neighbors around the ring. Large positive spin density is found on carbons C_1 (Fig. 24) and $C_{3,5}$ and on the phenoxy O. The $C_{3,5}$ ring protons are characterized by highly anisotropic, nearly dipolar couplings ($A \sim$ 8–30 MHz). In contrast, the methylene protons at the β-carbon position exhibit very strong, relatively isotropic proton hyperfine interactions ($A \sim$ 40–60 MHz) due to hyperconjugation between the β-proton and the ring π-system and can be analyzed (see Section 3.1) in terms of the dihedral angle between the C_1 p_z orbital perpendicular to the plane of the tyrosine ring and the C_1–C_β–H plane. This strong hyperfine interaction is responsible, along with the $C_{3,5}$ α-protons, for the splittings of the radical EPR signals. As in the case of the cysteine ligands to the Cu proteins discussed above (Section 3.1.2), this dihedral angle was found to be highly constrained in the cases studied. The presence of hydrogen bonding interactions and conclusions about the charge on the radical could also be obtained using proton ENDOR on these systems.

5.2.2. Galactose Oxidase

In a similar fashion, the proton hyperfine couplings of the protein-based radical of galactose oxidase were recently analyzed using ENDOR (Babcock *et al.*, 1992). Galactose oxidase contains a single Cu atom and catalyzes the $2e^-$ oxidation of galactose to its aldehyde, with reduction of O_2 to H_2O_2. The highest oxidation state of galactose oxidase is an EPR-silent form in

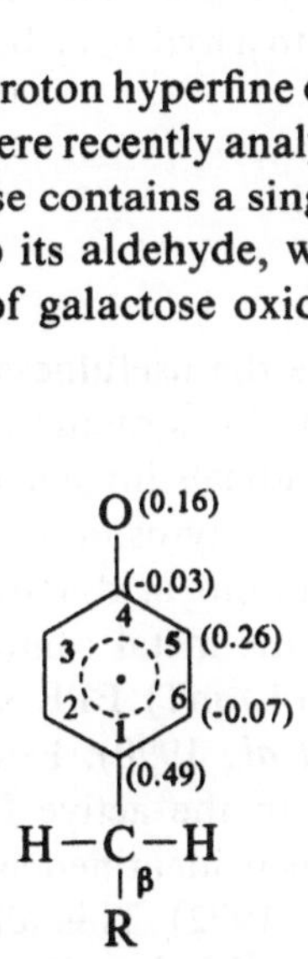

Figure 24. Numbering scheme for the tyrosyl radical. ENDOR-derived spin-density distributions in the tyrosyl radical of ribonucleotide reductase are shown in parentheses. Adapted from Bender *et al.* (1989).

which the Cu(II) is antiferromagnetically coupled to the radical $S = \frac{1}{2}$ species, proposed to be an amino acid radical. Upon removal of Cu and oxidation of the apoprotein a distinctive three-line EPR signal from the radical can be formed in modest yield. The EPR is perturbed upon incorporation of β-methylene deuterated tyrosine, confirming tyrosine as its source (Whittaker and Whittaker, 1990). Proton ENDOR disclosed a strong, relatively isotropic ($A_{iso} = 41$ MHz) β-proton coupling and resonances consistent with weaker, highly anisotropic proton hyperfine interactions with the ring $C_{3,5}$ protons. However, the structure of the radical EPR signal could not be simulated with parameters consistent with a normal tyrosine radical. Rather, a successful EPR simulation using the parameters derived from proton ENDOR required just one α-carbon proton and the β-protons. This implied the surprising existence of a derivatized form of tyrosine. In fact, the recent X-ray crystal structure of galactose oxidase showed that ^{272}Tyr has a thioether linkage at C_3 (Ito *et al.*, 1991).

Further ENDOR studies using alkylthio-substituted phenol model compounds confirmed this result and showed that the addition of a sulfur ligand did not substantially change the odd-alternate ring spin density characteristics by delocalizing the spin onto the sulfur. In particular, monitoring of the proton ENDOR of a methyl group substituent *para* to the ring phenol showed that *o*-methylthio substitution resulted in only ~25% reduction of spin density at the *para* (C_1) carbon. Finally, in this study it was concluded that the weak and almost purely dipolar coupling to an exchangeable proton in the enzyme radical was due to a hydrogen bond from the species replacing Cu in the apoenzyme.

5.3. Discussion

Such studies demonstrate the usefulness of detailed proton ENDOR studies in determining very specific structural and spin density information concerning these ring-based radicals involved in protein catalytic mechanisms. Other amino acids besides tyrosine and tryptophan have been suggested as sources of stable amino acid radicals in proteins. A histidine radical, exchange-coupled to the metal center in PSII has been suggested as responsible for a broad (14.7 mT) EPR signal detected in specifically inhibited centers (Boussac *et al.*, 1990). Perhaps most unexpectedly, the persistent free radical found in the active form of the enzyme pyruvate formate–lyase has recently been identified with ^{13}C labeling as a glycine radical (Volker-Wagner *et al.*, 1992). This alkyl-based radical presumably is stabilized through resonance with the adjacent peptide bond. This finding demonstrates the potential for biochemically important radical species from *any* of the amino acids, not merely the aromatic, π-based systems.

6. CONCLUSION

The examples presented above have been selected among many to illustrate how the ENDOR technique can answer questions about metalloenzyme structure and function that have resisted other approaches. They make clear the importance of the development of procedures for the analysis of frozen-solution ENDOR spectra, the utilization of multifrequency CW and time-domain capabilities, and the availability of isotopic labeling. It is easy to predict further expansion of the use of ENDOR spectroscopy in the study of metallobiomolecules.

ACKNOWLEDGMENTS. We wish to acknowledge the collaborators on the projects mentioned above, others associated with projects that could not be mentioned, and the graduate students and postdoctoral fellows involved in this work. These studies could not have been performed without the technical expertise of Mr. Clark E. Davoust or the support of the NIH, NSF, and USDA.

APPENDIX: ENDOR ANALYSIS

This Appendix begins by recalling the formulation of the orientation-dependent g factor (Section A.1). It then derives the orientation-dependent ENDOR frequencies for a nucleus where g, A, and P tensors are of arbitrary symmetry and orientation (Section A.2). [For more extended discussions, see Pilbrow (1990).] It then presents expressions of particular transparency for analyzing ENDOR spectra of systems with axial resolved EPR spectra (Section A.3). Next comes a presentation of the general theoretical treatment of orientation selection for ENDOR of polycrystalline (powder) samples, along with a brief discussion of systems with resolved axial EPR spectra (Section A.4). A discussion of the actual implementation of this theory is then presented (Section A.5).

A.1. EPR Spectra

We consider a paramagnetic center whose ground state is an isolated Kramers doublet at the temperature of observation. This might be a system with a true spin $S = \frac{1}{2}$ or an odd-electron high-spin system that has a zero-field splitting large compared to the temperature, as well as one lowest doublet.

The EPR signal from a single Kramers doublet can always be described in a representation based on a fictitious spin $S' = \frac{1}{2}$. In this representation the spectrum is characterized by a g' tensor coaxial with the fine-structure interaction. We choose to work in a right-handed coordinate system in which $g_1' > g_2' > g_3'$.

$$\mathscr{H}'_{ze} = \hat{\mathbf{S}} \cdot g' \cdot \mathbf{B} \tag{A1}$$

For $S = \frac{1}{2}$, then, $S = S'$ and $g' = g$. When the sample is subjected to a magnetic field **B** whose direction within the molecular g'-tensor (fine-structure) reference frame is described by the polar and azimuthal angles (θ, ϕ), then the EPR resonance field for that orientation is described by the angle-dependent g' value (Abragam and Bleaney, 1970)

$$h\nu = g'\beta B \tag{A2}$$

and

$$\begin{aligned} g'^2 &= g'^2(\theta, \phi) \\ &= \tilde{l} \cdot \tilde{g}' \cdot g' \cdot l \\ &= \sum g_i'^2 l_i^2 \end{aligned} \tag{A3}$$

where $\mathbf{B} = B\mathbf{l}$ and $\mathbf{l} = (l_1, l_2, l_3)$ is a unit vector parallel to $\mathbf{B}$: $\mathbf{l} = (\sin\theta\cos\phi, \sin\theta\sin\phi, \cos\theta)$.

The EPR spectrum of a frozen-solution (polycrystalline) protein sample is a superposition of the signals from molecules randomly oriented with respect to the field. The shapes for such powder patterns are well known (Atherton, 1973). Figure 1 presents the absorption envelope for a nitrogenase MoFe protein, *A*v1, which is representative of a center with g' tensor of rhombic symmetry $g_1' \neq g_2' \neq g_3'$. When considering ENDOR spectra of polycrystalline samples and their simulations, the observing field is best denoted by the corresponding g' value. In discussing experimental results, it is common to omit the prime on g' unless confusion can ensue.

A.2. ENDOR Frequencies: *g* and *A* of Arbitrary Symmetry

The interaction between the electron spin S of a paramagnetic center and the nuclear spin I of an individual site within the center is described by adding a hyperfine interaction term to the Hamiltonian (Abragam and Bleaney, 1970)

$$\mathscr{H}_{\mathrm{hf,loc}} = hI \cdot A \cdot S = hI \cdot \begin{pmatrix} A_1 & & \\ & A_2 & \\ & & A_3 \end{pmatrix} \cdot S \tag{A4}$$

The A for each site is diagonal within its own local principal-axis coordinate frame, and the orientation of that frame with respect to the g tensor reference frame can be expressed in terms of the three Euler angles (α, β, γ). (Our treatment of coordination transformations follows Mathews and Walker, 1965.) Normally in ENDOR spectroscopy each site can be treated separately and spectra from distinct sites are additive. However, when resonances from different nuclei overlap, cross-relaxation can have observable effects.

The first step in obtaining the ENDOR transition frequencies involves expressing the hyperfine interaction tensor in the g-tensor reference frame, using the rotation matrix $M(\alpha, \beta, \gamma)$ (Atherton, 1973):

$$^{g}A = \tilde{M} \cdot A \cdot M \tag{A5}$$

This leads to a hyperfine Hamiltonian in the $S = \frac{1}{2}$ representation,

$$\mathcal{H}_{\mathrm{hf}} = hI \cdot {}^{g}A \cdot S = h \sum {}^{g}A_{ij} I_i S_j \tag{A6}$$

Through use of the Wigner–Eckart theorem (Abragam and Bleaney, 1970), this can be transformed to the effective-spin $(S' = \frac{1}{2})$ representation

$$\mathcal{H}_{\mathrm{hf}} = hI \cdot A' \cdot S'; \qquad A'_{ij} = \frac{{}^{g}A_{ij} g'_j}{g_j} \tag{A7a}$$

and can be written in matrix form:

$$A = {}^{g}A \cdot \overline{g'}; \qquad \overline{g'_{ij}} = \frac{\delta_{ij} g'_j}{g_j} \tag{A7b}$$

Often the spin-S g values deviate negligibly from $g = 2.0$, and then for all but the most exacting work, $\mathbf{A}'$ may be expressed in terms of the g' tensor

$$A = \frac{{}^{g}A \cdot g'}{2}; \qquad g'_{ij} = \delta_{ij} g'_j \tag{A7c}$$

The final term needed to describe an ENDOR measurement of an $I = \frac{1}{2}$ nucleus is the nuclear-Zeeman interaction. In most cases this can be taken as a scalar coupling

$$\mathcal{H}_{\mathrm{nZ}} = \beta_{\mathrm{N}} g_{\mathrm{N}} I \cdot \mathbf{B} \tag{A8a}$$

and is characterized by a nuclear Larmor frequency $\nu_N = g_N\beta_N B$, where g_N is the g factor for the free nucleus. However, low-lying electron-spin excited states can give rise to a large, anisotropic pseudo-nuclear-Zeeman effect (Abragam and Bleaney, 1970). Because of coupling to these states, the nuclear-Zeeman interaction is described by an effective nuclear g^N tensor (True *et al.*, 1988; Hoffman *et al.*, 1989) and takes the form

$$\mathcal{H}_{nZ} = \beta_N I \cdot g^N \cdot \mathbf{B} \tag{A8b}$$

The deviations of the effective nuclear g value from g_N are inversely proportional to the energy difference between the ground-state Kramers doublet and low-lying excited spin states. Thus, these energy differences can be determined to high precision from a measurement of the pseudo-nuclear-Zeeman effect (True *et al.*, 1988).

For a nucleus with $I = \frac{1}{2}$, the transition frequencies measured in an ENDOR experiment are determined by the nuclear interaction terms, Eqs. (A7) and (A8). These can be derived (Thuomas and Lund, 1975) for a single orientation of the external field (θ, ϕ) from a nuclear interaction Hamiltonian

$$\mathcal{H}_{int} = hI \cdot \mathbf{V}_{\pm} \tag{A9}$$

where $\mathbf{V}_{\pm}$ is the vector sum of the effective hyperfine and nuclear-Zeeman fields

$$\begin{aligned} \mathbf{V}_{\pm} &= \left[\left(\pm\frac{1}{2}\right)\frac{1}{g}A' \cdot g' - \frac{\beta_N}{h}\mathbf{B} \cdot g^N\right] \cdot \mathbf{l} \\ &\equiv K_{\pm} \cdot \mathbf{l} \end{aligned} \tag{A10}$$

and the subscript $\pm$ refers to the electronic quantum number $m_s' = \pm\frac{1}{2}$. Without regard to the relative magnitude of hyperfine and nuclear-Zeeman terms, the first-order, orientation-dependent ENDOR transition frequencies for $I = \frac{1}{2}$ then are

$$\begin{aligned} \nu_{\pm} &= K_{\pm} = [K_{\pm}^2]^{1/2} \\ \mathrm{K}_{\pm}^2 &= \tilde{\mathbf{l}} \cdot \tilde{K}_{\pm} \cdot K_{\pm}\mathbf{l} \equiv K_{\pm}^2 \cdot \mathbf{l} \end{aligned} \tag{A11a}$$

When either the hyperfine or nuclear-Zeeman term is appreciably larger than the other, it is appropriate to discuss spectra in terms of the approximate equation

$$\nu_{\pm} = \left| \pm\frac{A'}{2} + \nu_N \right| \tag{A11b}$$

where A', the angle-dependent hyperfine splitting parameter, is given by

$$(g')^2(A')^2 = \tilde{\mathbf{l}} \cdot \tilde{g} \cdot (A')^2 \cdot g \cdot \mathbf{l} \tag{A12}$$

and the nuclear-Zeeman splitting is written

$$\nu_N = g^N \beta_N B \tag{A13a}$$

where g^N, the angle-dependent nuclear g factor, is defined by

$$g'A'g^N = \tilde{\mathbf{l}} \cdot \tilde{g} \cdot \tilde{A}' \cdot g^N \cdot \mathbf{l} \tag{A13b}$$

which reduces to ν_N and g_N when the pseudo–nuclear-Zeeman effect is unimportant. For protons in biological systems, $A/2 < \nu_H$ and Eq. (A11b) becomes

$$\nu_\pm = \nu_H \pm \frac{A'}{2} \tag{A11c}$$

whereas for ^{57}Fe, ^{13}C, and ^{14}N, often $A'/2 > \nu_N$, and Eq. (A11b) gives

$$\nu_\pm = \frac{A'}{2} \pm \nu_N \tag{A11d}$$

It is useful to note that the lower Kramers doublet of an $S > \frac{1}{2}$ spin system with large zero-field splitting has a strongly anisotropic magnetic moment, which forces the effective hyperfine interaction tensor A' [Eq. (A7)] to be highly anisotropic even when the intrinsic site hyperfine interaction is isotropic, $A_1 = A_2 = A_3 = a$ [Eq. (A4)]. For example, for a high-spin system with axial g tensor, by Eq. (A7) the ENDOR frequencies for a nucleus with isotropic A [Eq. (A4)] would be determined by a hyperfine tensor that is nonetheless highly anisotropic in the $S' = \frac{1}{2}$ representation:

$$A'_{1,2} = a\frac{g'_{1,2}}{g_\perp} \sim a\left(\frac{2S+1}{2}\right); \qquad A'_3 = a\frac{g'_3}{g_\parallel} \sim a \tag{A14a}$$

This effect spreads out the ENDOR spectra taken at low observing fields and greatly enhances resolution.

When the nuclear spin is $I > \frac{1}{2}$ the ENDOR transition frequencies are modified by the nuclear quadrupole interaction (Abragam and Bleaney, 1970; Atherton, 1973):

$$\mathscr{H}_Q = hI \cdot P \cdot I$$

The quadrupole splitting tensor P is diagonal in an axis frame that need not be coaxial with the hyperfine tensor frame. If the quadrupole frame is considered as being rotated away from the corresponding hyperfine frame by Euler angles $\alpha_p, \beta_p, \gamma_p$, then it is related to the g tensor reference frame by the rotation matrix $M_P = M(\alpha_p, \beta_p, \gamma_p) \cdot M(\alpha, \beta, \gamma)$. Of the three diagonal elements of **P**, only two are independent because the tensor is traceless:

$$\mathcal{H}_Q = \sum_{i=1} P_i I_i^2$$
$$P_1 + P_2 + P_3 = 0; \qquad P_i = \frac{he^2 q_i Q}{h2I(2I-1)} \tag{A15}$$

Often, the two independent parameters are taken to be the component of largest magnitude, P_3, and the anisotropy parameter η, where $0 \leq \eta = (P_2 - P_1)/P_3 \leq 1$, with $P_2 \geq P_1$.

Because we include the possible use of the fictitious spin S', the computation of the quadrupole contribution involves expression of this tensor in the g-tensor reference frame using the rotation matrix M_P defined above:

$$g_P = \tilde{M}_P \cdot P \cdot M_P \tag{A16}$$

With this transformation, the nuclear-hyperfine, Zeeman, and quadrupole-interaction tensors all are expressed in the g' coordinate frame. The quadrupole coupling contribution to the Hamiltonian has the form

$$\mathcal{H}_p = hI \cdot {}^g P \cdot I \tag{A17}$$

We consider first its first-order contributions to ENDOR frequencies. These involve the angle-dependent and electron-spin projection-dependent splitting constant $P_\pm$, given by

$$K_\pm^2 P_\pm = \tilde{1} \cdot \tilde{K}_\pm \cdot {}^g P \cdot K_\pm \cdot 1 \tag{A18a}$$

The result of these considerations for the ENDOR spectrum is as follows: In a single-crystal ENDOR measurement one expects a set of equivalent nuclei of spin I to give $4I$ ENDOR transitions (selection rules, $\Delta m_s = 0$ and $\Delta m = \pm 1$). The first-order single-crystal transition frequencies given in Eq. (A2) for $I = \frac{1}{2}$ are augmented for $I > \frac{1}{2}$ by an m-dependent quadrupole term and become

$$\nu_\pm(m) = \nu_\pm + (2m-1)\left(\frac{3P_\pm}{2}\right) P_\pm, \qquad -I+1 \leq m \leq I \tag{A19a}$$

an equation that is valid for $P_\pm < \nu_\pm$. The extension of the more approximate equation for transition frequencies, Eq. (A11a), to nuclei of arbitrary value of I is

$$\nu_\pm(m) = \left| \pm\frac{A'}{2} + \nu_N + \frac{3P}{2}(2m-1) \right| \qquad \text{(A19b)}$$

where the angle-dependent quadrupole-splitting constant P is given by

$$(g')^2(A')^2P = \tilde{\mathbf{l}} \cdot \tilde{g}' \cdot \tilde{A}' \cdot {}^gP \cdot A' \cdot g' \cdot \mathbf{l} \qquad \text{(A18b)}$$

The validity of a first-order treatment of the quadrupole coupling term requires that it be small compared to the other nuclear terms, namely $P_\pm < \nu_\pm$. When this is not so, for nuclei with arbitrary $I > \frac{1}{2}$ one must include higher-order terms (Kevan and Kispert, 1976). However, the nuclear Hamiltonian $\mathcal{H}_{nuc}$, that governs the ENDOR frequencies for $I > \frac{1}{2}$ is the sum of Eqs. (A9) and (A8a), and Eq. (A9) corresponds to the scalar interaction of I with the vector $\mathbf{V}_\pm$ defined in the g frame, Eq. (A10). This vector may be expressed in the quadrupole principle-axis frame as ${}^P\mathbf{V}_\pm = M_p \cdot \mathbf{V}_\pm$, which permits us to write

$$\begin{aligned} \mathcal{H}_{nuc} &= \mathcal{H}_{int} + \mathcal{H}_{nZ} \\ &= h^P\mathbf{V}_\pm \cdot I + I \cdot P \cdot I \end{aligned} \qquad \text{(A20)}$$

which represents an extension of the work of Bowman and Massoth (1987) for an isotropic g tensor. Muha has presented a general, explicit, and closed-form solution for this eigenvalue problem for $I = 1$ and $I = \frac{3}{2}$ that is valid for arbitrary values of the hyperfine and quadrupole couplings (Muha, 1982, 1983). This solution diagonalizes the individual m_s submanifolds, and thus the condition for applicability is merely that the nuclear terms be much smaller than the electron-Zeeman term, which is inevitably the case for unresolved hyperfine interactions in metallobiomolecules.

We summarize here only the results for $I = 1$ because of their wide applicability in ENDOR studies of ^{14}N and ^2H. The eigenvalues of Eq. (A20) given by Muha can be written in a modified notation as

$$h\nu_{\pm,n} = \left(\frac{4}{3}|p_\pm|\right)^{1/2} \cos\left[\theta_0^\pm + \frac{2n\pi}{3}\right] \qquad (n = 0, 1, 2) \qquad \text{(A21)}$$

where

$$|p_\pm| = {}^PV_\pm^2 - (P_1P_2 + P_1P_3 + P_2P_3) \qquad q_\pm = \sum ({}^PV_{\pm i}^2 P_i) - P_1P_2P_3$$

$$C_\pm = \left(\frac{3}{|p_\pm|}\right)^{3/2} \frac{q_\pm}{2}; \qquad \theta_0^\pm = \frac{1}{3}\cos^{-1} C_\pm; \qquad {}^PV_\pm^2 = \sum_i {}^PV_{\pm i}^2 \tag{A22}$$

and we emphasize that the components of ${}^P\mathbf{V}_\pm$ [Eqs. (A9, A20)] are functions of the field-orientation angles (θ, ϕ) through the unit vector **l** [Eq. (A10)]. The three differences between the $\nu_{+,n}$ correspond to the frequencies for the two $\Delta m = 1$ and the $\Delta m = 2$ transitions within the $m_s = +\frac{1}{2}$ electron-spin manifold, and similarly for the three $\nu_{-,n}$ differences. The corresponding eigenvectors also were presented, which makes possible a ready calculation of transition probabilities.

A.3. Endor Frequencies: Centers With Axial EPR Spectra and Analytic Expressions for Nuclei with Dipolar Couplings

As a special case of the above approach we presented a formulation for rapid simulations of polycrystalline ENDOR patterns for systems whose EPR spectra exhibit axial resolved magnetic interactions (g tensor and "central"-nucleus hyperfine couplings) (Hoffman and Gurbiel, 1989). Such centers include nitroxide free radicals, $(VO)^{2+}$, and $(Cu)^{2+}$ complexes (e.g., Mustafi *et al.*, 1990). The ENDOR response of interest is that from a nucleus with *un*resolved splittings and possibly nonaxial hyperfine couplings. The equations presented above simplify when the hyperfine tensor of the nucleus being observed in ENDOR has axial symmetry in its principal axis frame, as would occur for an $I = \frac{1}{2}$ nucleus (^1H or ^{19}F) interacting by a through-space electron–nuclear dipolar interaction. In such cases the field dependence of the ENDOR frequencies can be described by simple analytic functions, well suited to least-squares determination of metrical parameters that define the nuclear position. Sample calculations were given for a proton interacting with a nitroxide radical.

Although procedures were presented for treating ENDOR from a nucleus with unresolved hyperfine interactions (e.g., ^{14}N, ^1H, ^{19}F) of arbitrary symmetry, it is instructive to consider how geometric information can be obtained from ENDOR frequencies when the anisotropic hyperfine coupling arises from the classic dipolar interaction between a point-dipole electron spin and a nucleus separated by the vector $\mathbf{r} = r\mathbf{n}$ ($\mathbf{n}$ a unit vector):

$$A = a + \left(\frac{g_e g_N \beta \beta_N}{r^3}\right)(3\cos^2\omega - 1)$$

$$\equiv a + A_D(3\cos^2\omega - 1), \tag{A23}$$

where ω is the angle between **r** and the external field, which lies along the unit vector $\mathbf{l} = (\sin\theta\cos\phi, \sin\theta\sin\phi, \cos\theta)$ (Fig. 25); the existence of a weak contact interaction is provided for by the constant a, although this would be zero for an ideal geometric probe. Again, because the resolved magnetic interactions have axial symmetry, the nucleus may be taken to be in the x–z plane of the g frame (Fig. 25), and the orientation of **n** is defined by the polar angle β: $\mathbf{n} = (\sin\beta, 0, \cos\beta)$. It then follows that

$$\cos\omega = \mathbf{n}\cdot\mathbf{l}$$

$$= (\sin\beta\sin\theta)\cos\phi + \cos\beta\cos\theta \tag{A24}$$

The analysis started by considering the polycrystalline ENDOR spectrum taken at observing field **B**, in which the EPR intensity arises from molecules such that the field lies at a unique polar angle for each m_I of the central nucleus $\theta(H, m_I)$. For each m_I with EPR intensity at field of magnitude B, the polycrystalline ENDOR spectrum that arises from the interaction of Eq. (A23) has as many as three pairs of distinct features centered at ν_{N}, with their frequencies given by the formula

$$\nu_{\pm,i}(m_I) = \nu_{\mathrm{N}} \pm \frac{A_i(m_I)}{2} \tag{A25}$$

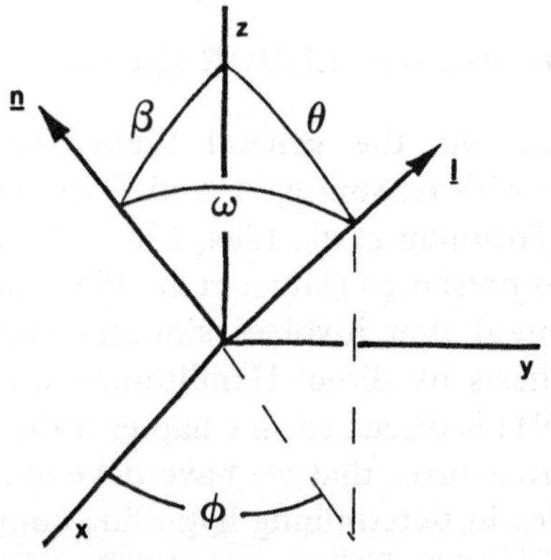

Figure 25. Definition of the angles θ, ϕ determining the unit vector along the magnetic field **l** and of the angle β determining the orientation of **n**, the elctron-spin–nuclear-spin unit vector; ω is the angle between **l** and **n**.

where

$$A_{1,2}(m_I) = a + A_D[3\cos^2(\beta \pm \theta(B, m_I)) - 1]$$
$$A_3(m_I) = a - A_D, \qquad \pi \geq \theta(B, m_I) \geq \left(\frac{\pi}{2} - \beta\right) \tag{A26}$$

Thus, through use of an explicit equation given by Hoffman and Gurbiel that relates θ and B, Eq. (A26) gives simple analytic expressions for the ENDOR doublet splitting $\Delta\nu_i(m_I) = A_i(m_I)$, in terms of three parameters that define A, namely, a, A_D, and β; these can be used to relate frequencies of features on polycrystalline ENDOR patterns to the value of the observing field in precise equivalence to the relation between frequency and orientation in single-crystal experiments (e.g., Abragam and Bleaney, 1970). The use of Eq. (A26) or a more general form to analyze the field dependence of observed ENDOR splittings from dipolar-coupled ^1H or ^{19}F gives the metrical parameters (r, β) needed to define the nuclear position r within the molecular g frame.

Several features implicit in Eqs. (A25) and (A26) should be noted:

1. Hyperfine couplings that arise from a molecule with dipolar interactions where $\beta > 0$ exhibit one ENDOR doublet whose splitting is independent of the observing field [Eq. (A26)] and thus might be mistaken as arising from isotropic coupling.
2. The maximum ENDOR splitting is $\Delta\nu_{max} = 2(a + 2A_D)$; it occurs for the doublet associated with A_2 and the observing field gives the value for β: $\theta(B, m_I) = \beta$.
3. Although a least-squares fit of the ENDOR splittings measured over the full range of observing field gives the most accurate metrical results, observation and assignment of three doublets allows direct analytical solutions for the three parameters desired.

A.4. Simulating Polycrystalline ENDOR Spectra

This section presents the general theory of orientation-selective ENDOR for systems with tensors g, A, and P of arbitrary symmetry and relative orientation (Hoffman *et al.*, 1984, 1985). Theory restricted to axial systems also has been presented (Hurst *et al.*, 1985; Section A.3 above). An elegant recent treatment that involves simultaneous EPR and ENDOR powder-spectra synthesis by direct Hamiltonian diagonalization (Kreiter and Hüttermann, 1991) is discussed in Chapter 5 this volume. This section then describes the procedures that we have developed for the application of the general theories in determining hyperfine coupling tensors (True *et al.*, 1988; Gurbiel *et al.*, 1989; Telser *et al.*, 1986a; Cline *et al.*, 1985b, 1986). Analysis of quadrupole tensors from polycrystalline spectra was illustrated in the work on the Rieske center of phthalate dioxygenase (Section 4.1.3).

Consider the common situation of a frozen solution of a protein with a metal center whose EPR spectrum is described by a g' tensor that has rhombic (or axial) symmetry and is without resolved hyperfine structure. ENDOR spectra are taken with the external field fixed within the polycrystalline EPR envelope at a selected value B which corresponds to a g value determined by the spectrometer frequency, $g = h\nu/\beta B$. As recognized by Hyde and his co-workers (Rist and Hyde, 1970), ENDOR spectra taken with the magnetic field set at the extreme edges of the frozen-solution EPR envelope (positions A and D in Fig. 4), near the maximal or minimal g values, give single-crystal-like patterns from the subset of molecules for which the magnetic field happens to be directed along a g-tensor axis. Although an ENDOR spectrum obtained using an intermediate field and g value (e.g., position C) does not arise from a single orientation, nonetheless it is associated with a well-defined subset of molecular orientations. Ignoring for the moment the existence of a nonzero component EPR linewidth (employ the EPR envelope of a δ-function component line), the EPR signal intensity at field of magnitude B and thus the endor spectrum arise from those selected molecular orientations associated with the curve on the unit sphere s_g composed of points for which the orientation-dependent spectroscopic splitting factor [Eq. (A.3)] satisfies the condition $g'(\theta, \phi) = g$. This is illustrated in Fig. 4. However, although g is constant along the curve s_g, the orientation-dependent ENDOR frequencies $\nu_{\pm}(m, \theta, \phi)$ [Eq. (A11), (A19), or (A21)] are not. Thus, the ENDOR intensity in a spectrum taken at g occurs in a range of frequencies spanning the values of $\nu_{\pm}$ associated with the selected subset of orientations associated with s_g. The calculation of this intensity requires a proper parametrization. The element of arc for a general curve on a sphere is given by

$$ds^2 = (d\theta)^2 + \sin^2\theta(d\phi)^2$$

For a curve of constant g, (ϕ, g) is the natural choice for independent variables and one can solve Eq. (A3) to give

$$\sin^2\theta(\phi, g) = \frac{g^2 - g_3^2}{(g_2^2 - g_3^2) + (g_1^2 - g_2^2)\cos^2\phi} \tag{A27}$$

The result for the element of arc (corrected for a typographical error in Hoffman *et al.*, 1984) is

$$\begin{aligned} ds_g &= \left(\frac{\partial s}{\partial \phi}\right)_g d\phi \\ &= \left[\frac{(g_1^2 - g_2^2)^2}{(g^2 - g_3^2)^2}\frac{\sin^6\theta(\phi, g)}{\cos^2\theta(\phi, g)}\cos^2\phi\sin^2\phi + \sin^2\theta\right]^{1/2} d\phi \end{aligned} \tag{A28}$$

where Eq. (A27) is employed.

At any observing g value within the EPR envelope of a polycrystalline (frozen-solution) sample, the intensity of a superposition ENDOR spectrum at radiofrequency, ν, can be written as a sum of convolutions over the ENDOR frequencies [Eq. (A11)] that arise on the curve s_g; our original formulation was

$$^{\delta}I(\nu, g) = \sum_m \sum_{\pm} \int_{s_g} L(\nu - \nu_{\pm}(m))e(\nu_{\pm}(m))\, ds \qquad \text{(A29a)}$$

$$= \sum_m \sum_{\pm} \int_{s_g} L(\nu - \nu_{\pm}(m))e(\nu_{\pm}(m)) \left(\frac{\partial s}{\partial \phi}\right)_g d\phi \qquad \text{(A29b)}$$

$L(x)$ is an ENDOR line-shape function, and $e(\nu)$ is the hyperfine enhancement factor (Abragam and Bleaney, 1970; Atherton, 1973). For $I = \frac{1}{2}$ the nuclear transition observed is $(m = +\frac{1}{2} \leftrightarrow m = -\frac{1}{2})$ and the summation involves only the electron-spin quantum number $m_s' = \pm\frac{1}{2}$; for $I > \frac{1}{2}$, quadrupole terms must be included [Eqs. (A19), (A21)] and the sum extended over the additional nuclear transitions. In the case of an EPR spectrum that shows resolved hyperfine couplings with a central metal ion (e.g., Cu^{2+}), the extension of this approach involves an additional sum over nuclear spin projections of that nucleus. This occasional complication becomes increasingly unimportant as the microwave frequency is increased.

Recently, we have concluded that it is more appropriate to consider the area element $d\sigma$ associated with point (ϕ, g) rather than the element of arc. The superposition ENDOR spectrum associated with a particular g value along the EPR envelope is obtained by an integration of the area element along the curve s_g. We find that

$$d\sigma = \left(\frac{g}{g^2 - g_3^2}\right) \frac{\sin^2 \theta(\phi, g)}{\cos \theta(\phi, g)}\, d\phi\, dg$$

$$\equiv w(\phi, g)\, d\phi\, dg \qquad \text{(A30)}$$

where $\sin^2 \theta(\phi, g)$ again is given by Eq. (A27). The integral along the curve of constant g then becomes

$$^{\delta}I(\nu, g) = \sum_m \sum_{\pm} \int_{s_g} L(\nu - \nu_{\pm}(m)e(\nu_{\pm}(m))w(\phi, g)\, d\phi \qquad \text{(A31)}$$

Calculations indicate that simulations with Eq. (A31) better reproduce the observed ENDOR intensities, as is noted in the legend to Fig. 20.

One should recognize here that the conceptual utility of this approach, as illustrated in Fig. 4, is coupled with the computational advantage that it reduces the simulation procedure to a single integral [Eq. (A29)]. This may be contrasted with the restricted (to axial symmetry) theory of Hurst *et al.* (1985), which requires a double integral, as does the general theory of Kreiter and Hütterman (1991).

The above equations in general are adequate for determining hyperfine and quadrupole tensors. However, to achieve final simulations of experimental ENDOR spectra, the restriction in Eqs. (A29) and (A31) to a δ-function EPR pattern must be relaxed. The complete expression for the relative ENDOR intensity at frequency ν for an applied field set to a particular g value involves the convolution of ${}^{\delta}I(\nu, g)$, the EPR envelope function derived by Kneubühl (1960), $S(g)$, and a component EPR line-shape function $R(g, g')$:

$$I(\nu, g) = \int_{g_{\min}}^{g_{\max}} dg'\, S(g')\,{}^{\delta}I(\nu, g')R(g, g') \tag{A32}$$

This theory has been implemented as a QUICKBASIC™ program. The component line-shape functions, L and R, both were taken as Gaussians and the linewidths as isotropic. In applications such as those discussed below, the hyperfine enhancement factor usually could be ignored [$e(\nu) = 1$; Eqs. (A29), (A31)]. The above-mentioned simulation program is now available from the authors upon request.

It is convenient to represent the overall ENDOR response of a particular site in terms of a $(g-\nu)$ plot as in Fig. 26 and in the sections on Rieske centers and on aconitase, where each feature in an ENDOR spectrum at a given g is represented by its frequency; for clarity in Fig. 26, a Larmor-split doublet has been given a single point at the frequency corresponding to its center ($A'/2$). For the case represented there, with $A'_1 > A'_2 > A'_3$, and still considering coaxial tensors, the higher-frequency doublet, which we label D_{13}, moves smoothly between the frequencies associated with the two extremal, single-crystal-like fields at g_1 and g_3. The other, lower-frequency doublet can best be labelled D_{12} at fields such that $g'_1 > g > g'_2$ and D_{23} when $g'_2 > g > g'_3$. It shifts to a center at $A'_2/2$ as the field approaches g'_2 from either above or below.

If the tensors g' and A' are not coaxial, then the ENDOR spectra and their field variation become more complex. For example, if the A' principal-axis frame is rotated from the g' frame by a rotation about a single g' axis, say g'_i, then the D_{jk} doublet tends to split into two doublets for fields between g'_j and g'_k. In particular, for a rotation about the g'_3 axis as illustrated in Fig. 26, the D_{12} step-doublet splits into two such steps in the low-field portion of the EPR spectrum, where $g'_1 > g > g'_2$. In coaxial cases where

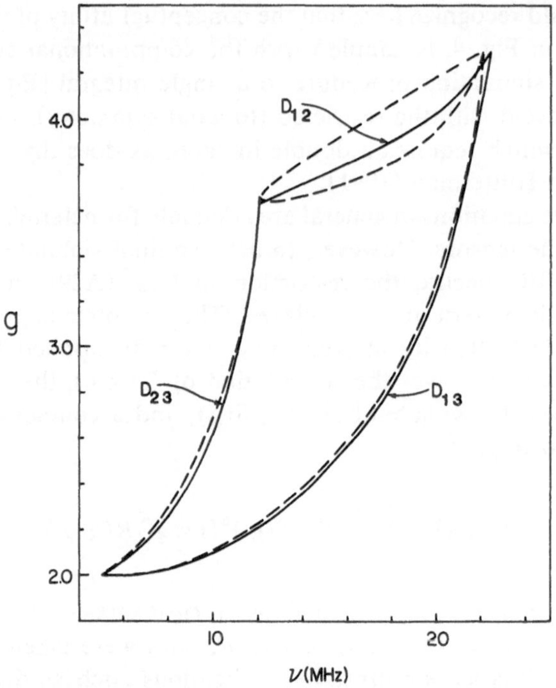

Figure 26. Calculated polycrystalline ENDOR patterns versus observing g values for a rhombic g tensor (4.32, 3.68, 2.01) and an $I = \frac{1}{2}$ hyperfine tensor, (−20.85, −13, −9.8) MHz. The center frequency of the resolved features of calculated spectra are given by the intersection(s) of a horizontal with the displayed curves; Larmor splittings are not indicated. Collinear A and g tensors (——); noncollinear tensors, ($\alpha = 15°$, $\beta = \gamma = 0$) (---).

the edge of the ENDOR spectrum displays one doublet of peaks, on rotation the edge is expected to split into two peak-doublets. Note also that when g and hyperfine tensors are not coaxial the hyperfine values measured at the edges of the EPR spectrum in general do not correspond to a principal-axis value.

More complicated orientations of A' and g' frames, described by two or three Euler angles, can give correspondingly more complex patterns and field variations. Conversely, as shown above, a set of spectra taken at multiple fields generates a field–frequency pattern that is interpretable in terms of the relative orientation of tensors g' and A' as well as of the principal values of the tensors A'.

Finally, as an illustration of the formulas for centers of axial symmetry, consider a frozen solution in which a nitroxide free-radical is incorporated as a paramagnetic probe of structure. The magnetic anisotropy of a nitroxide is dominated by hyperfine interaction with ^{14}N of the NO moiety; this can be idealized as having axial symmetry with principal values, $A_{\parallel}^{N} = 100$ MHz,

$A_\perp^N = 20$ MHz. In addition, we include the weak g anisotropy, $g_\parallel = 2.002$, $g_\perp = 2.007$.

Figure 27 plots the field dependence of the ENDOR frequencies for a proton at a distance of 6 Å from the N—O moiety considered as a point dipole. Plots are given for several relative orientations in the molecular frame, $\beta = 0$, 20°, 70°, 90°; for simplicity, the calculations have assumed $a = 0$. The theory discussed here was used to obtain metrical parameters for a nitroxide spin-labelled amino acid (Mustafi *et al.*, 1990).

A.5. Analysis Procedure

The process of obtaining hyperfine tensor principal values and orientations begins with the accumulation and indexing of ENDOR spectra at

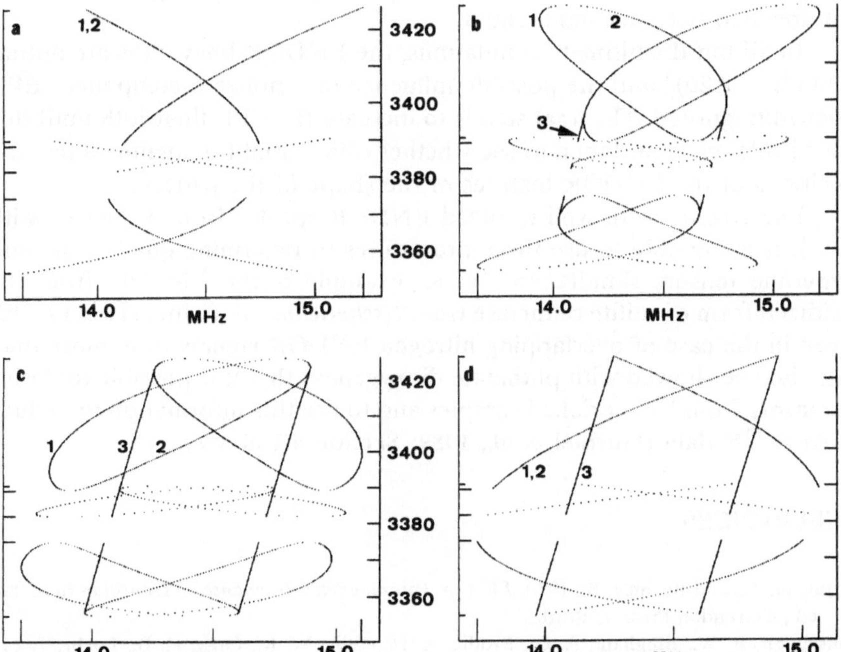

Figure 27. Field dependence of the ENDOR frequencies for a proton interacting according to Eqs. (A25) and (A26) ($a = 0$) with a nitroxide whose magnetic parameters are $g_\parallel = 2.002$, $g_\perp = 2.007$, $A_\parallel^N = 100$ MHz, $A_\perp^N = 20$ MHz. The abscissa on the right side of each plot gives the value of the magnetic field, while that on the left side marks the resonance positions of the EPR spectrum for $\theta = 0$ ($g_\parallel$, $A_\parallel^N$) and $\theta = \pi/2$ ($g_\perp$, $A_\perp^N$). The numbers (i = 1-3) label doublets according to Eqs. (A25) and (A26). The proton is taken to be 6.0 Å away from the point-dipole nitroxide with orientation in the molecular frame of (a) $\beta = 0$, (b) $\beta = 20°$, (c) $\beta = 70°$, (d) $\beta = 90°$.

multiple fields across the EPR envelope. Next, a first approximation to the hyperfine principal values A_3 and A_1 is obtained from the ENDOR frequencies measured in the single-crystal-like spectra obtained, respectively, at the high- and low-field edges of the EPR envelope; A_2 is estimated from the spread of frequencies (when observable) in the spectrum taken at $g_{mid} = g'_2$. Then, the nature of the relative orientation of the g' and A tensors is inferred from the development of the ENDOR pattern as the field increases from the low-field (g'_1) edge of the EPR spectrum. At this stage, many simulations of selected spectra typically are performed by varying the spin-S hyperfine interaction principal values and the relative orientation of g' tensor and nuclear coordinate frames. These variations are constrained such that the parameters correctly predict the resonance frequencies measured in the single-crystal-like spectra. The analysis begins with the simplest assumption—namely, the minimal noncoaxiality of tensors g' and A'—and is terminated when the entire accessible field–frequency range of ENDOR features had been accommodated.

In all but the ultimate simulations, the ENDOR linewidths are optimized [Eq. (A20)] and the possible influence of a nonzero component EPR linewidth ignored. The final step is to increase the EPR linewidth until this has visible effect and then to see whether differential broadening alters the positions of the ENDOR features or the shape of the pattern.

For systems with well-resolved ENDOR spectra from a nucleus with $I > \frac{1}{2}$, it is possible to use these procedures to determine quadrupole and hyperfine tensors simultaneously; an example is the ^{14}N data from the oxidized form of sulfite reductase from *Escherichia coli* (Cline *et al.*, 1985b). Even in the case of overlapping nitrogen ENDOR signals from more than one site, we showed with phthalate dioxygenase that it is possible to derive A tensors from ^{15}N-enriched samples and to use this information to deduce P from ^{14}N data (Gurbiel *et al.*, 1989; Section 4.1 above).

REFERENCES

Abragam, A., and Bleaney, B., 1970, *Electron Paramagnetic Resonance of Transition Ions*, 2nd ed., Clarendon Press, Oxford.

Ainscough, E. W., Bingham, A. G., Brodie, A. M., Ellis, W. R., Gray, H. B., Loehr, T. M., Plowman, J. E., Norris, G. E., and Baker, E. N., 1987, *Biochemistry* **26**:71.

Anderson, R. E., Anger, G., Petersson, L., Ehrenberg, A., Cammack, R., Hall, D. O., Mullinger, R., and Rao, K. K., 1975, *Biochim. Biophys. Acta* **376**:63.

Atherton, N. M., 1973, *Electron Spin Resonance: Theory and Applications*, Halstead Press, New York.

Babcock, G. T., El-Deeb, M. K., Sandusky, P. O., Whittaker, M. M., and Whittaker, J. W., 1992, *J. Am. Chem. Soc.* **114**:3727.

Backes, G., Mino, Y., Loehr, T. M., Meyer, T. E., Cusanovich, M. A., Sweeney, W. V., Adman, E. T., and Sanders-Loehr, J., 1991, *J. Am. Chem. Soc.* **113**:2055.

Barry, B. A., El-Deeb, M. K., Sandusky, P. O., and Babcock, G. T., 1990, *J. Biol. Chem.* **265**:20,139.

Becker, D., and Kwiram, A. L., 1976, *Chem. Phys. Lett.* **39**:180.

Beinert, H., and Kennedy, M. C., 1989, *Eur. J. Biochem.* **186**:5.

Beinert, H., Griffiths, D. E., Wharton, D. C., and Sands, R. H., 1962, *J. Biol. Chem.* **237**:2337.

Bender, C. J., Sahlin, M., Babcock, G. T., Barry, B. A., Chandrashekar, T. K., Salowe, S. P., Stubbe, J. A., Lindström, B., Petersson, L., Ehrenberg, A., and Sjöberg, B.-M., 1989, *J. Am. Chem. Soc.* **111**:8076.

Boroske, E., and Möbius, K., 1977, *J. Magn. Reson.* **28**:325.

Boussac, A., Zimmerman, J.-L., Rutherford, A. W., and Lavergne, J., 1990, *Nature (London)* **347**:303.

Bowman, M. K., and Massoth, R. J., 1987, Nuclear spin eigenvalues and eigenvectors in electron spin echo modulation, in *Electronic Magnetic Resonance of the Solid State* (J. A. Weil, ed.), Canadian Society for Chemistry, Ottawa, pp. 99–110.

Brown, T. G., and Hoffman, B. M., 1980, *Mol. Phys.* **39**:1073.

Bühlmann, C., Schweiger, A., and Ernst, R. R., 1989, *Chem. Phys. Lett.* **154**:285.

Cline, J. F., Hoffman, B. M., Mims, W. B., LaHaie, E., Ballou, D. P., and Fee, J. A., 1985a, *J. Biol. Chem.* **260**:3251.

Cline, J. F., Janick, P. A., Siegel, L. M., and Hoffman, B. M., 1985b, *Biochemistry* **24**:7942.

Cline, J. F., Janick, P. A., Siegel, L. M., and Hoffman, B. M., 1986, *Biochemistry* **25**:4647.

Davies, E. R., 1974, *Phys. Lett.* **47A**:1.

Dawson, J. H., 1988, *Science* **240**:433.

de Beer, R., van Ormondt, D., van Ast, M. A., Banen, R., Duine, J. A., and Frank, J., 1979, *J. Chem. Phys.* **70**:4491.

Debus, R. J., Barry, B. A., Babcock, G. T., and McIntosh, L., 1988, *Proc. Natl. Acad. Sci. USA* **85**:427.

Desideri, A., Morpurgo, L., Agostinelli, E., Baker, G. J., and Raynor, J. B., 1985, *Biochim. Biophys. Acta* **831**:8.

Dikanov, S. A., and Astashkin, A. V., 1989, ESEEM of disordered systems: Theory and applications, in *Advanced EPR: Applications in Biology and Biochemistry* (A. J. Hoff, ed.), Elsevier, Amsterdam, pp. 59–117.

Doan, P. E., Fan, C., Davoust, C. E., and Hoffman, B. M., 1991, *J. Magn. Reson.* **95**:196.

Duglav, A. V., Mitrofanov, Yu. F., and Pol'skii, Yu. E., 1974, *Sov. Phys. Solid State* **16**:111.

Edmonds, D. T., 1977, *Phys. Reports* **29**:233.

Emptage, M. H., 1987, Aconitase, in *Metal Clusters in Proteins* (L. Que, ed.), American Chemical Society, Washington, D.C., pp. 343–371.

Fan, C., Teixeira, M, Moura, J., Moura, I., Huynh, B.-H., le Gall, J., Peck, H. D., Jr., and Hoffman, B. M., 1991, *J. Am. Chem. Soc.* **113**:20.

Fan, C., Doan, P. E., Davoust, C. E., and Hoffman, B. M., 1992a, *J. Magn. Reson.* **98**:62.

Fan, C., Kennedy, M. C., Beinert, H., and Hoffman, B. M., 1992b, *J. Am. Chem. Soc.* **114**:374.

Fee, J. A., Kuila, D., Mather, M. W., and Yoshida, T., 1986, *Biochim. Biophys. Acta* **853**:153.

Feher, G., 1959, *Phys. Rev.* **114**:1219.

Finzel, B. C., Poulos, T. L., and Kraut, J., 1984, *J. Biol. Chem.* **259**:13,027.

Fishel, L. A., Villafranca, J. E., Mauro, J. M., and Kraut, J., 1987, *Biochemistry* **26**:351.

Fishel, L. A., Farnum, M. F., Mauro, J. M., Miller, M. A., Kraut, J., Liu, Y., Tan, X., and Scholes, C. P., 1991, *Biochemistry* **30**:1986.

Flanagan, H. L., and Singel, D. J., 1987, *J. Chem. Phys.* **87**:5606.

Gemperle, C., and Schweiger, A., 1991, *Chem. Rev.* **91**:1481.

Gewirth, A. A., and Solomon, E. I., 1988, *J. Am. Chem. Soc.* **110**:3811.

Gray, H. B., and Solomon, E. I., 1981, in *Copper Proteins.* Vol. 3 (T. G. Spiro, ed.), Wiley, New York, pp. 1–39.

Grupp, A., and Mehring, M., 1990, Pulsed ENDOR spectroscopy in solids, in *Modern Pulsed and Continuous-Wave Electron Spin Resonance* (L. Kevan and M. K. Bowman, eds.), Wiley, New York, pp. 195–229.

Gurbiel, R. J., Batie, C. J., Sivaraja, M., True, A. E., Fee, J. A., Hoffman, B. M., and Ballou, D. P., 1989, *Biochemistry* **28**:4861.

Gurbiel, R. J., Ohnishi, T., Robertson, D. E., Daldal, F., and Hoffman, B. M., 1991, *Biochemistry* **30**:11579.

Guss, J. M., and Freeman, H. C., 1983, *J. Mol. Biol.* **169**:521.

Han, J., Adman, E. T., Beppu, T., Codd, R., Freeman, H. C., Huq, L., Loehr, T. M., and Sanders-Loehr, J., 1991, *Biochemistry* **30**:10,904.

Henderson, T. A., Hurst, G. C., and Kreilick, R. W., 1985, *J. Am. Chem. Soc.* **107**:7299.

Hoff, A. J., ed., 1989, *Advanced EPR: Applications in Biology and Biochemistry*, Elsevier, Amsterdam.

Hoffman, B. M., 1991, *Acc. Chem. Res.* **24**:164.

Hoffman, B. M., and Gurbiel, R. J., 1989, *J. Magn. Reson.* **82**:309.

Hoffman, B. M., Roberts, J. E., Brown, T. G., Kang, C. H., and Margoliash, E., 1979, *Proc. Natl. Acad. Sci. USA* **76**:6132.

Hoffman, B. M., Roberts, J. E., Kang, C. K., and Margoliash, E., 1981, *J. Biol. Chem.* **256**:6556.

Hoffman, B. M., Martinsen, J., and Venters, R. A., 1984, *J. Magn. Reson.* **59**:110.

Hoffman, B. M., Venters, R. A., and Martinsen, J., 1985, *J. Magn. Reson.* **62**:537.

Hoffman, B. M., Gurbiel, R. J., Werst, M. M., and Sivaraja, M., 1989, Electron nuclear double resonance (ENDOR) of metalloenzymes, in *Advanced EPR: Applications in Biology and Biochemistry* (A. J. Hoff, ed.), Elsevier, Amsterdam, pp. 541–591.

Houseman, A. L. P., Oh, Byung-H., Kennedy, M. C., Fan, C., Werst, M. M., Beinert, H., Markley, J. L., and Hoffman, B. M., 1992, *Biochemistry* **31**:2073.

Houseman, A. L. P., Doan, P. E., Goodin, D. B., and Hoffman, B. M., 1993, *Biochemistry* (in press).

Hurst, G. C., Henderson, T. A., and Kreilick, R. W., 1985, *J. Am. Chem. Soc.* **107**:7294.

Ito, N., Phillips, S. E., Stevens, C., Ogel, Z. B., McPherson, M. J., Kean, J. N., Yadav, K. D. S., and Knowles, A. P. F., 1991, *Nature* **350**:87.

Karthein, R., Dietz, R., Nastainczyk, W., and Ruf, H. H., 1988, *Eur. J. Biochem.* **171**:313.

Kennedy, M. C., Werst, M. M., Telser, J., Emptage, M. H., Beinert, H., and Hoffman, B. M., 1987, *Proc. Natl. Acad. Sci. USA* **84**:8854.

Kent, T. A., Emptage, M. H., Merkle, H., Kennedy, M. C., Beinert, H., and Münck, E., 1985, *J. Biol. Chem.* **260**:6871.

Kevan, L., 1987, *Acc. Chem. Res.* **20**:1.

Kevan, L., and Kispert, L. D., 1976, *Electron Spin Double Resonance Spectroscopy*, Wiley, New York.

Kim, J., and Rees, D. C., 1992, *Science* **257**:1677.

Kneubühl, F. K., 1960, *J. Chem. Phys.* **33**:1074.

Kreiter, A., and Hüttermann, J., 1991, *J. Magn. Reson.* **93**:12.

Lambe, J., Laurance, N., McIrvine, E. C., and Terhune, R. W., 1961, *Phys. Rev.* **122**:1161.

Lauble, H., Kennedy, M. C., Beinert, H., and Stout, C. D., 1992, *Biochemistry* **31**:2735.

Liao, P. F., and Hartmann, S. R., 1973, *Phys. Rev.* **8**:69.

Lowe, D. J., 1992, *Prog. Biophys. Molec. Biol.* **57**:1.

Mathews, J., and Walker, R. L., 1965, *Mathematical Methods of Physics*, W. A. Benjamin, Inc., New York.

McDowell, C. A., Naito, A., Sastry, D. L., Cui, Y. U., Sha, K., and Yu, S. X., 1989, *J. Mol. Struc.* **195**:361.

McLean, P. A., True, A. E., Nelson, M. J., Chapman, S., Godfrey, M. R., Teo, B.-K., Orme-Johnson, W. H., and Hoffman, B. M., 1987, *J. Am. Chem. Soc.* **109**:943.

Mims, W. B., 1965, *Proc. Roy. Soc. A* (*London*) **283**:452.

Mims, W. B., and Peisach, J., 1989, ESEEM and LEFE of metalloproteins and model compounds, in *Advanced EPR: Applications in Biology and Biochemistry* (A. J. Hoff, ed.), Elsevier, Amsterdam, pp. 1-57.

Möbius, K., and Lubitz, W., 1987, ENDOR spectroscopy in photobiology and biochemistry, in *Biological Magnetic Resonance.* Vol. 7 (L. J. Berliner and J. Reuben, eds.), Plenum, New York, pp. 129-247.

Möbius, K., Lubitz, W., and Plato, M., 1989, Liquid-state ENDOR and triple resonance, in *Advanced EPR: Applications in Biology and Biochemistry* (A. J. Hoff, ed.), Elsevier, Amsterdam, pp. 441-499.

Muha, G. M., 1982, *J. Magn. Reson.* **49**:431.

Muha, G. M., 1983, *J. Magn. Reson.* **53**:85.

Mustafi, D., Sachleben, J. R., Wells, G. B., and Makinen, M. W., 1990, *J. Am. Chem. Soc.* **112**:2558.

Norris, G. E., Anderson, B. F., and Baker, E. N., 1986, *J. Am. Chem. Soc.* **108**:2784.

Pattison, M. R., and Yim, Y. W., 1981, *J. Magn. Reson.* **43**:193.

Penfield, K. W., Gewirth, A. A., and Solomon, E. I., 1985, *J. Am. Chem. Soc.* **107**:4519.

Pilbrow, J. R., 1990, *Transition Ion Electron Paramagnetic Resonance*, Clarendon Press, Oxford.

Poulos, T. L., 1988, *Adv. Inorg. Biochem.* **7**:1.

Prince, R. C., Halbert, T. R., and Upton, T. H., 1988, in *Advances in Membrane Biochemistry and Bioenergetics* (C. H. Kim *et al.*, eds), Plenum, New York, pp. 469-477.

Rieske, J. S., MacLennan, D. H., and Coleman, R., 1964, *Biochem. Biophys. Res. Commun.* **15**:338.

Rist, G. H., and Hyde, J. S., 1970, *J. Chem. Phys.* **52**:4633.

Robbins, A. H., and Stout, C. D., 1989a, *Proteins: Struct., Funct., Genet.* **5**:289.

Robbins, A. H., and Stout, C. D., 1989b, *Proc. Natl. Acad. Sci. USA* **86**:3639.

Roberts, J. E., Hoffman, B. M., Rutter, R., and Hager, L. P., 1981, *J. Am. Chem. Soc.* **103**:7654.

Roberts, J. E., Cline, J. F., Lum, V., Freeman, H., Gray, H. B., Peisach, J., Reinhammar, B., and Hoffman, B. M., 1984, *J. Am. Chem. Soc.* **106**:5324.

Rypniewski, W. R., Breiter, D. R., Benning, M. M., Wesenberg, G., Oh, B.-H., Markley, J. L., Rayment, I., and Holden, H. M., 1991, *Biochemistry* **30**:4126.

Sands, R. H., 1979, ENDOR and ELDOR on iron-sulfur proteins, in *Multiple Electron Resonance Spectroscopy* (M. M. Dorio and J. H. Freed, eds.), Plenum, New York, pp. 331-374.

Scholes, C. P., 1979, ENDOR on hemes and hemoproteins, in *Multiple Electron Resonance Spectroscopy* (M. M. Dorio and J. H. Freed, eds.), Plenum, New York, pp. 297-329.

Scholes, C. P., Lapidot, A., Mascarenhas, R., Inubushi, T., Isaacson, R. A., and Feher, G., 1982, *J. Am. Chem. Soc.* **104**:2724.

Schweiger, A., 1982, *Structure and Bonding* **51**:1-121.

Sivaraja, M., Goodin, D. B., Smith, M., and Hoffman, B. M., 1989, *Science* **245**:738.

Solomon, E. I., Gewirth, A. A., and Westmoreland, T. D., 1989, EPR spectra of active sites in copper proteins, in *Advanced EPR: Applications in Biology and Biochemistry* (A. J. Hoff, ed.), Elsevier, Amsterdam, pp. 865-907.

Stubbe, J., 1989, *Ann. Rev. Biochem.* **58**:257.

Surerus, K. K., Kennedy, M. C., Beinert, H., and Münck, E., 1989, *Proc. Natl. Acad. Sci. USA* **86**:9846.

Surerus, K. K., Oertling, W. A., Fan, C., Gurbiel, R. J., Einarsdóttir, O., Antholine, W. E., Dyer, R. B., Hoffman, B. M., Woodruff, W. H., and Fee, J., 1992, *Proc. Natl. Acad. Sci. USA* **89**:3195.

Telser, J., Emptage, M. H., Merkle, H., Kennedy, M. C., Beinert, H., and Hoffman, B. M., 1986a, *J. Biol. Chem.* **261**:4840.

Telser, J., Benecky, M. J., Adams, M. W. W., Mortenson, L. E., and Hoffman, B. M., 1986b, *J. Biol. Chem.* **261**:13,536.

Thomann, H., and Bernardo, M., 1990, *Chem. Phys. Lett.* **169**:5.

Thomann, H., Bernardo, M., and Adams, M. W. W., 1991a, *J. Am. Chem. Soc.* **113**:7044.

Thomann, H., Bernardo, M., Baldwin, M. J., Lowery, M. D., and Solomon, E. I., 1991b, *J. Am. Chem. Soc.* **113**:5911.

Thuomas, K.-Å., and Lund, A., 1975, *J. Magn. Reson.* **18**:12.

True, A. E., Nelson, M. J., Venters, R. A., Orme-Johnson, W. H., and Hoffman, B. M., 1988, *J. Am. Chem. Soc.* **110**:1935.

VanCamp, H. L., Sands, R. H., and Fee, J. A., 1981, *J. Chem. Phys.* **75**:2098.

Volker-Wagner, A. F., Frey, M., Neugebauer, F. A., Schäfer, W., and Knappe, J., 1992, *Proc. Natl. Acad. Sci. USA* **89**:996.

Wang, G., Benecky, M. J., Huynh, B. H., Cline, J. F., Adams, M. W. W., Mortenson, L. E., Hoffman, B. M., and Münck, E., 1984, *J. Biol. Chem.* **259**:14,328.

Wenckebach, W. T., Schreurs, L. A. H., Hoogstraate, H., and Swanenburg, T. J. B., 1971, *Physica* **52**:455.

Werst, M. M., Kennedy, M. C., Beinert, H., and Hoffman, B. M., 1990a, *Biochemistry* **29**:10,526.

Werst, M. M., Kennedy, M. C., Houseman, A. L. P., Beinert, H., and Hoffman, B. M., 1990b, *Biochemistry* **29**:10,533.

Werst, M. M., Davoust, C. E., and Hoffman, B. M., 1991, *J. Am. Chem. Soc.* **113**:1533.

Whittaker, M. M., and Whittaker, J. W., 1990, *J. Biol. Chem.* **265**:9610.

Wittenberg, B. A., Kampa, L., Wittenberg, J. B., Blumberg, W. E., and Peisach, J., 1968, *J. Biol. Chem.* **243**:1863.

Yokoi, H., 1982, *Biochem. Biophys. Res. Commun.* **108**:1278.

5

ENDOR of Randomly Oriented Mononuclear Metalloproteins

Toward Structural Determinations of the Prosthetic Group

Jürgen Hüttermann

1. INTRODUCTION

ENDOR (*e*lectron *n*uclear *do*uble *r*esonance) spectroscopy has matured considerably since its early applications involving color centers and phosphorous donors more than 30 yr ago (Feher, 1956; Seidel, 1961). Although it basically involves EPR (*e*lectron *p*aramagnetic *r*esonance) detection of NMR transitions in paramagnetic systems, many schemes for its application have appeared. A major line can be drawn between transient and stationary ENDOR, a differentiation which dates back to the two initial investigations. Transient phenomena are best covered today under the term "pulsed ENDOR," schemes and applications for which are dealt with in Chapter 7 of this volume. This chapter deals with stationary transitions.

Of the many aspects rendering ENDOR a highly interesting tool in spectroscopy of paramagnetic systems (for recent reviews see Schweiger, 1982; Hüttermann and Kappl, 1987; Möbius and Lubitz, 1987; Hoffman *et al.*, 1989), the one that has a strong impact on metalloprotein studies is the

Jürgen Hüttermann • Fachrichtung Biophysik und Physikalische Grundlagen der Medizin, Universität des Saarlandes, 6650 Homburg/Saar, Germany.

Biological Magnetic Resonance, Volume 13: EMR of Paramagnetic Molecules, edited by Lawrence J. Berliner and Jacques Reuben. Plenum Press, New York, 1993.

increase in spectral resolution compared to EPR, due to linewidth reduction. Although this is accompanied by a considerable loss of sensitivity, the present state of the experimental ENDOR facilities, commercial or home-built, is sufficiently advanced to give good signal-to-noise ratios for many applications. It is the gain in information achieved by probing the weakly coupled proton and the nitrogen hyperfine interactions that will be explored here in some detail. Using mononuclear metalloproteins from our own work as examples, we wish to show that in combination with the orientation selectivity inherent in powder-type EPR spectra of strongly anisotropic electron Zeeman or other interaction terms, the spectral resolution provides ENDOR with the potential of delivering structural information about the prosthetic group with single-crystal-type resolution but from a randomly oriented specimen. In order to utilize this capacity fully, several prerequisites have to be fulfilled; these will be discussed together with other limitations of the method.

2. THEORY

The basic theory of ENDOR energy levels and corresponding transition frequencies is well etablished. In the present context we consider a spin Hamiltonian $\mathcal{H}$ which comprises metal-ion (M) and ligand (L) terms

$$\mathcal{H} = (\beta\tilde{\mathbf{S}}\cdot\mathbf{g}\cdot\mathbf{H} + \tilde{\mathbf{S}}\cdot\mathbf{A}\cdot\mathbf{I} + \tilde{\mathbf{I}}\cdot\mathbf{P}\cdot\mathbf{I})_M + \sum_L (\tilde{\mathbf{S}}\cdot\mathbf{A}\cdot\mathbf{I} + \tilde{\mathbf{I}}\cdot\mathbf{P}\cdot\mathbf{I} + \beta' g_n \mathbf{H}\cdot\mathbf{I}) \tag{1}$$

with the usual meaning of the symbols: β, β' the electronic and nuclear magnetic moments; **S** and **I** the electron and nuclear spin operators or their transpose ($\tilde{\ }$); **g**, **A**, and **P** the electronic g-, the metal-ion and ligand hyperfine and the quadrupolar interaction tensors, respectively. No fine structure and no metal-ion nuclear Zeeman term is included in Eq. (1) since for the mononuclear $S = \frac{1}{2}$ systems with metal ions of small nuclear moments treated here, these do not contribute significantly. Eigenvalue solutions to the Hamiltonian have been given frequently in the literature assuming different symmetries of tensors and coordinate systems as well as various orders of perturbation theory. Clear formulations with little restriction other than the high-field approximation have been given, e.g., by Iwasaki (1974), for up to second-order perturbation theory. Our own strategy involves two kinds of programs, which provide different degrees of accuracy.

Consider the typical situation of a mononuclear metal ion embedded in an environment of organic ligands, as depicted schematically in Fig. 1,

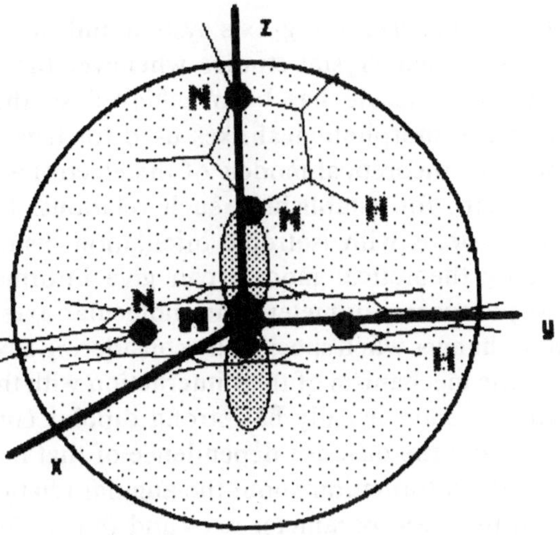

Figure 1. Schematic depiction of a metal-ion ligand complex and associated axis system.

using a porphyrin skeleton together with an axial imidazole nitrogen ligand. Due to the ligand field acting on the, say, $3d$ metal orbitals, there exists a relation between the molecular axis system and that of the g tensor. In our example the axes fall together and are denoted x, y, and z. Omitting lineshapes and linewidths, the EPR spectrum at a certain direction of the magnetic field $\mathbf{H} = H \cdot (h_1, h_2, h_3)$ given by the polar angles Θ and Φ with respect to (x, y, z) at a specific, fixed microwave frequency ν can be written as resonance positions H_{res} on the magnetic field scale as

$$H_{\text{res}} = \frac{h\nu - M_I \cdot A(\Theta, \Phi) - M_I^N \cdot A^N(\Theta, \Phi)}{\beta \cdot g(\Theta, \Phi)} \tag{2}$$

in which M_I and M_I^N are the metal-ion and the ligand nuclear spin states and A the respective hyperfine couplings. The first-order values for these and the effective g values are calculated as

$$g(\Theta, \Phi) = \left[\sum_i (g_i \cdot h_i)^2\right]^{1/2}, \qquad i = 1, 2, 3 \tag{3}$$

and

$$A(\Theta, \Phi) = g(\Theta, \Phi)^{-1} \cdot \left[\sum_i \left(\sum_j A_{ji} \cdot g_j \cdot h_j\right)^2\right]^{1/2}, \qquad i, j = 1, 2, 3 \tag{4}$$

The angular relations between the g-axis system and the molecular frame can be derived from single-crystal studies whenever the crystal-structure determination of the molecule has been achieved so that the direction cosines with respect to morphological axes used for tensor determination are known. Otherwise, theoretical models of the electronic structure of the complex are necessary for establishing such relations. For powder-type EPR-spectra from systems with isotropic orientation distribution of paramagnetic sites, only theoretical models can give information about the connection between tensor parameters and molecular axes.

This situation changes when specifically considering ENDOR of weakly coupled protons. The mechanism of their interaction with the unpaired spin of the paramagnetic center usually has strong dipolar contributions. The ensuing anisotropy and the distance dependence of that interaction can be utilized to retrieve the information about the angular relationships between the molecular and the EPR-parameter axes and thus eventually to reconstruct the geometry of the complex within a coordination shell of between about 4 to 6 Å, as indicated in Fig. 1. The spatial resolution is distance dependent; for practical purposes one can estimate 0.1 to 0.5 Å. This method goes back to the work of Rist and Hyde (1970), who introduced the term "orientation selection" and realized that ENDOR intensity of anisotropic proton couplings from powder EPR would give discernible lines only at certain cumulation points. These authors proposed to use the g-turning points in EPR as working positions for ENDOR in order to extract, in favorable cases, single-crystal-like information. More recently Yordanow and co-workers (1986) worked out the angular relations between unpaired spin and the interacting proton in systems of axial symmetry, and Henderson *et al.* (1985) as well as Hurst *et al.* (1985) have exploited both the angular and the distance part of the interaction in detail in order to determine the position of a proton from powder ENDOR analysis.

The first-order perturbation theory gives proton ENDOR resonance frequencies ν, which are dependent on the magnetic field $\mathbf{H}$ and the spin state M_S as follows:

$$\nu(\mathbf{H}, M_S) = \sqrt{\sum_i \left[\frac{M_S}{g(\Theta, \Phi)} \left(\sum_j g_j h_j A_{ji}^{\mathrm{N}} \right) - h_i \nu_0 \right]^2} \tag{5}$$

with the nuclear hyperfine tensor $\mathbf{A}^N$ written in MHz as elements

$$hA_{ij}^N = r^{-3} \cdot (-\beta g_N \beta') \cdot g_i (3 r_i r_j - \delta_{ij}) + \mathbf{A}_{\mathrm{iso}}^N \cdot \delta_{ij} \tag{6}$$

comprising the dipolar and the isotropic interaction. The value of ν_0 in Eq. (5) is given by $\nu_0 = g_N \beta' H_{\mathrm{res}} / h$ with the Planck constant h and the nuclear g factor g_N. The hierarchy of interaction $\mathbf{A}^N$ in Eqs. (2) and (5) is such that in the former it influences H_{res}, whereas in the latter it is discernible only by ENDOR. An example illustrating Eq. (2) is a coordinated ^{14}N nucleus in a CuN_4 complex, to be discussed below.

The foregoing formulas hold for unit spin density concentrated in one point. Often, a more realistic situation is a distributed spin-density. In such cases, either a center of gravity with respect to one proton is an approximation (Kappl, 1988) or a computer program can produce a distributed calculation according to Eqs. (5) and (6) with subsequent addition and normalization.

Another factor that can be of considerable importance in the reconstruction of the spatial relationships of atoms in the prosthetic group is the nitrogen (^{14}N, $I = 1$) interaction evaluation. Although it may appear that the large width of lines obtained (often 1 MHz or more) should impede extraction of structural information, their analysis frequently yields valuable information about symmetry between tensors and, moreover, provides an important means of estimating distributed spin density. One approximation amenable to data analysis is the assumption that the hyperfine interaction is larger than the quadrupolar term. In the case of coaxial **A** and **P** tensors one can calculate the resulting energy correction in first order in a straightforward manner using formulas of Iwasaki (1974). The more general matrix formulation yields as additive term

$$-\tfrac{1}{2}(\tilde{\mathbf{k}} \cdot \mathbf{P} \cdot \mathbf{k})[I(I+1) - 3M_I^2] \tag{7}$$

with

$$(\tilde{\mathbf{k}} \cdot \mathbf{P} \cdot \mathbf{k}) = \frac{\tilde{\mathbf{h}}\tilde{\mathbf{K}}(M_S) \cdot \mathbf{P} \cdot \mathbf{K}(M_S) \cdot \mathbf{h}}{K^2(M_S)} \tag{8}$$

and

$$\mathbf{K}(S_u) = \left(\frac{\mathbf{A} \cdot \mathbf{g}}{g}\right)(S_u) - g_N\beta' H \cdot \mathbf{E} \qquad (\mathbf{E} = \text{unit tensor}). \tag{9}$$

We have written a computer program called HEROS (*h*igh-resolution-*E*NDOR of *r*andomly *o*riented *s*ystems). For proton interactions it follows the formalism outlined by Henderson, Hurst, and Kreilick (1985a, b) but is amended by a multicenter spin-density option and an extension for rhombic EPR symmetry. For the nitrogen nuclei the Iwasaki formulas, Eqs. (7)–(9) are employed. There are no restrictions in terms of collinearity of tensors. The storage and representation of calculated spectra complies with the normalization of our EPR-spectra manipulation and simulation program HOMER (Ohlmann, 1992). Thus, calculated spectra can be compared on screen with experimental spectra, and the latter can be manipulated (baseline correction, expansion between cursors, etc.).

To visualize "orientation selection," consider for an axial case (Cu, $I = \frac{3}{2}$, one isotope only), the ESR spectrum for a powder and its angular development between the "parallel" (z) and the "perpendicular" (x, y) directions together with the proton ENDOR resonances for two magnetic field positions in this framework, as shown in Fig. 2. The proton is situated 3 Å away along the x axis and is considered for its dipolar interaction only. Setting the ENDOR excitation window on a true single-crystal-like orientation (left part) as on the $m_I = +\frac{3}{2}$ sublevel (often called "$g_\parallel$ extreme" in the literature) only those orientations involved contribute, yielding a symmetrical two-line group. The situation is very much different at the second working point. Not only are all Cu nuclear sublevels exited simultaneously, but each comprises a different angular window of orientations

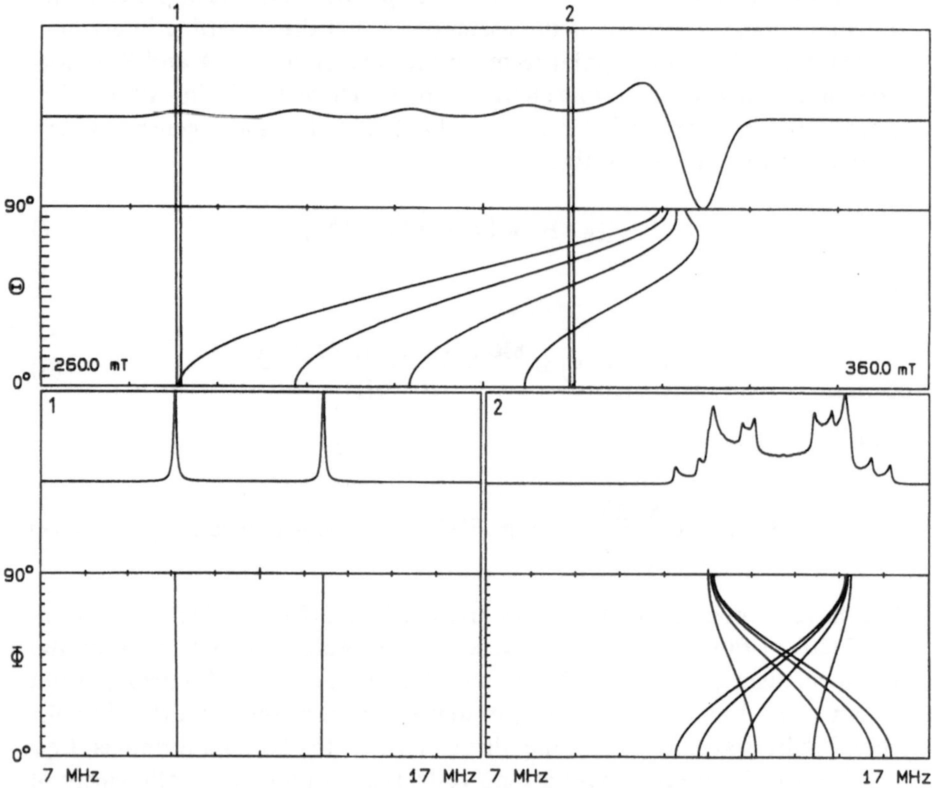

Figure 2. Simulated orientational variation and resulting powder EPR spectrum for an axially symmetrical ($S = \frac{1}{2}$, $I = \frac{3}{2}$ system (top); simulated orientational variation and resulting powder ENDOR spectrum for a proton at different EPR positions (bottom).

Θ. Therefore, although the proton is situated at a highly symmetrical position with respect to the g tensor, a total of five to six line pairs is observed. Patterns of similar complexity evolve from protons at less symmetrical sites at less complex parts of the EPR spectrum. An example is shown in Fig. 3 for the same axial case but using a proton in an off-axis direction.

Even more complex relations are found for rhombic EPR symmetry. Again it may be helpful to visualize, in Fig. 4, the angular development of the EPR powder spectrum of copper, although the quartet lines from one g factor are followed only to the two other g values, omitting the lines in between. In ENDOR, a proton again situated 3 Å away but along the z axis now gives a very intricate powder pattern at any other working point than $g_{\parallel}$ extreme (left), and the graph Φ versus ν has no closed form (right). Still, plots of those values of (Φ, Θ) pairs for which resonance is observed can be obtained. Again it becomes clear that ENDOR intensity in a powder-type EPR situation builds up at points with little angular variation of hyperfine interaction. Although it is obvious that for such cases there is no straightforward way of extrapolating backwards from a given ENDOR spectrum to the position of the proton in question, especially if more than one interacting proton is present, it is also clear that unfolding the "ENDOR space" by moving the excitation window over the powder EPR spectrum step by step gives systematic information that, if fully analyzed, contains unique information about proton positions with respect to the metal-ion center.

As an example, which can be related to Fig. 1, we compare, for the same axial g system and other parameters as in Fig. 2, the ENDOR response for a proton not situated along one axis but still $|\mathbf{r}| = 3$ Å away from the metal ion with polar angles $\Phi = 30°$ and $\Theta = 60°$. As is shown in Fig. 3, even at $g_{\parallel}$ extreme a four-line pattern is obtained now, whereas at a midfield position the situation becomes much more complex. At any rate, the proton position is, at least in principle, fully reflected by the spectra if they can be analyzed.

The analysis requires an array of additional experimental techniques. In the context of numerical calculations, we mention that there may be cases in which partial knowledge of the association of an ENDOR resonance line group with a certain proton exists, say, from one specific field position in the EPR spectrum. It then can be helpful to calculate coordinates and their range of variation with assumed ENDOR couplings at other working points (Kappl, 1988). Together with "forward" simulations with variations in spin-density distributions, in an iterative way an approximate picture of the spatial arrangement of the protons around the metal ion can be obtained.

Examples of simulations of ^{14}N interactions are given in Fig. 5, using an axially symmetrical $S = \frac{1}{2}$, $I = \frac{7}{2}$ system, which can be regarded as an appropriate example for low-spin cobaltous complexes. The nitrogen (^{14}N,

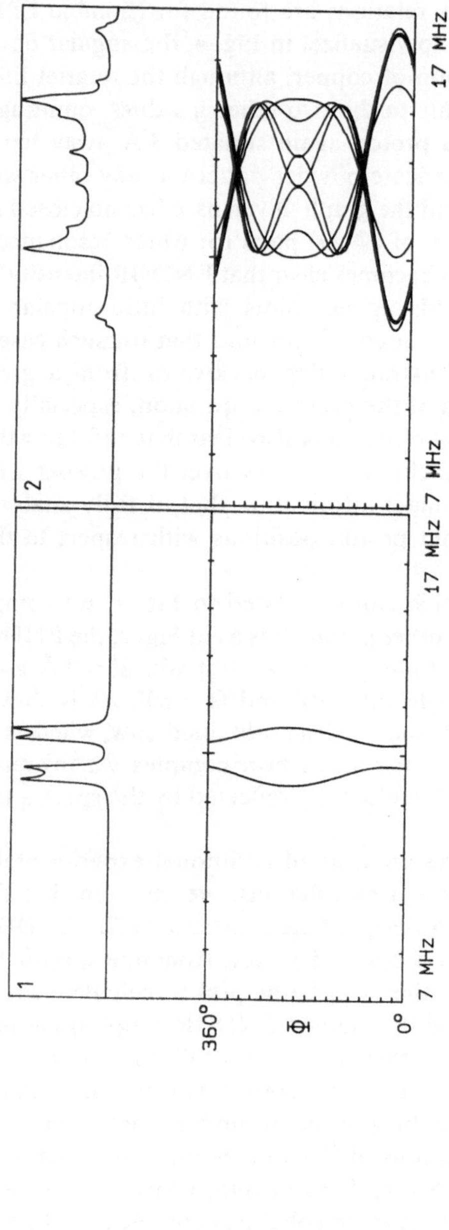

Figure 3. Angular variation and resulting power ENDOR spectrum for the proton of Fig. 2 but tilted to $\Theta = 60°$ and $\Phi = 30°$.

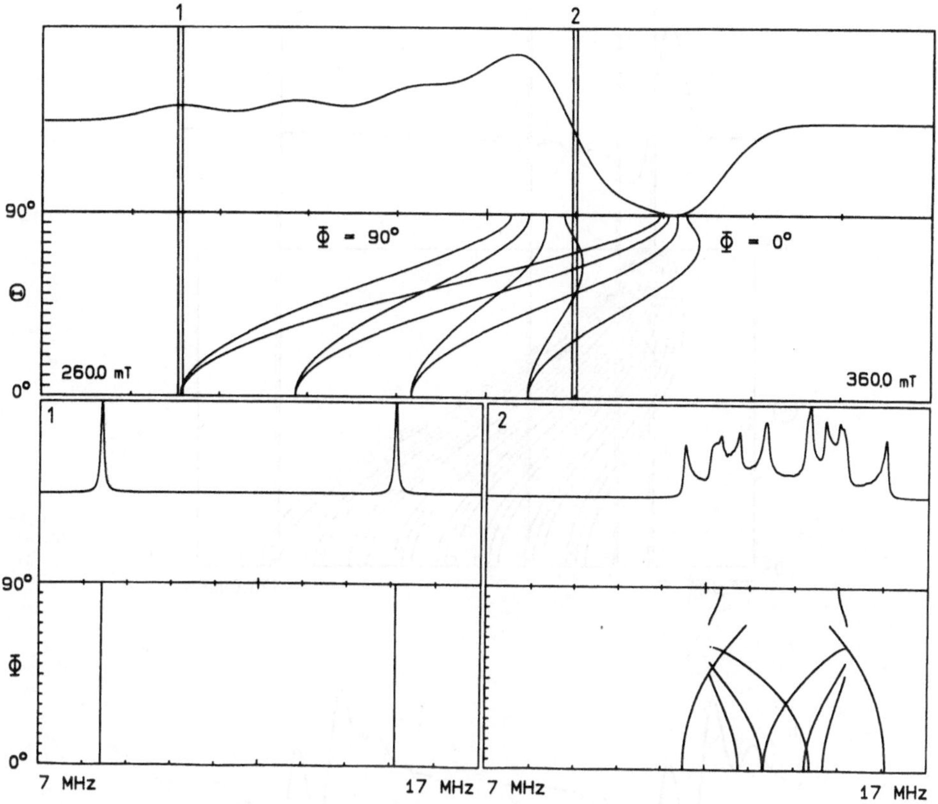

Figure 4. As in Fig. 2, but for a rhombic EPR symmetry.

$I = 1$) hyperfine tensor is collinear with the g and major ($I = \frac{7}{2}$) hyperfine tensors. The simulation data resemble those discussed below for cobaltous porphyrin complexes with imidazole as axial ligand, modelling the Co-substituted hemo- and myoglobins.

The resolution of the spectra and thus the line-widths recorded determines the precision with which one can reconstruct the prosthetic group by analyzing nitrogen interactions and the weakly coupled proton. Therefore, frequency modulation to yield a first-derivative display is our experimental mode of ENDOR detection although loss of intensity compared to amplitude modulation for broader lines must also be considered. Another factor is the accuracy of simulation, for which we show both absorption and first derivatives in the examples. In this respect it can be desirable to have a program with fewer restrictions than the first-order perturbation theory discussed above. For this purpose we have made use of a core of a former QCPE (*q*uantum *c*hemistry *p*rogram *e*xchange, University of Indiana, Indiana) program denoted Magnspec 3. This program, described

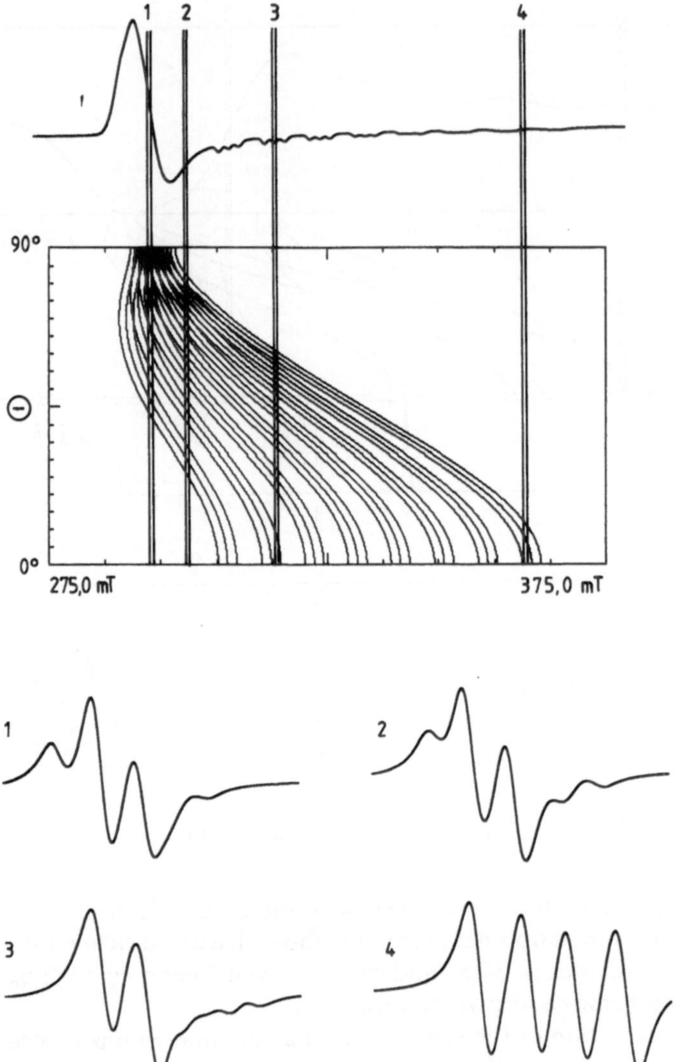

Figure 5. Simulated EPR (top) and ^{14}N ENDOR spectra (bottom) for an axial ($S = \frac{1}{2}$, $I = \frac{7}{2}$) spin system showing the development of the latter with variation of the excitation window over different orientations and nuclear sublevels.

by Mackey *et al.* (1960) among others, calculates energy levels and transition probabilities by setting up a matrix for the spin Hamiltonian and diagonalizing it by the Jacobi algorithm without any restrictions other than the high-field approximation. Having previously used this program for obtaining single-crystal simulations for organic free radicals involving halogen nuclear interaction (e.g., Haindl and Hüttermann, 1978), we have since then added a powder-orientation generator and, more recently, an ENDOR part (Kreiter and Hüttermann, 1991). Although the handling of the program is somewhat sophisticated and the computing time can be enormous, it can often be a useful strategy to simultaneously calculate EPR and ENDOR spectra through the same simulation using the first-order values as starting parameters. Care was taken to construct and use an orientation generator adapted to the symmetry of the problem in order to minimize computing time. The problem of ENDOR line intensities is approximated by multiplying the calculated transition probabilities of ESR with those of matching NMR intensities without including any relaxational term or adjusting for hyperfine enhancement. For most applications, this approach has so far been sufficient.

Since in practical studies with proteins there are usually quite a few protons and several nitrogen nuclei interacting with the unpaired spin in its immediate neighborhood, the problem of assignment by simulation can be enormous, since the situation of multiple line overlap often occurs, preventing a straightforward and unique solution. Moreover, the spin density in the complex may not be known with the necessary accuracy, nor is the distance estimate always sufficiently well established, and these two factors determine the reliability of the data base for simulation. In order to provide a sound base for the reconstruction of the prosthetic group, several additional techniques and measures should be applied. These involve not only isotope exchange, especially exchangeable protons versus deuterons, but also the use of ^{15}N, ^{57}Fe, ^{17}O, ^{63}Cu, etc. whenever possible. In the same context, the study of model compounds is indispensable in order to secure as much interaction as possible in a non-protein environment, allowing, among other results, for the chemical substitution of different protons by, e.g., carbon-bound deuterons or methyl groups. In addition single-crystal studies should amend the investigations of the powder-type specimen in order to extract as much angular relations as possible. Often, EPR studies are sufficient to derive the directions of the g tensor with respect to the molecular axes. Another, very important tool is the use of modified proteins, i.e., natural or genetically produced mutants.

3. EXAMPLES

3.1. Hemoglobin and Model Compounds

The structure-function relationship in hemoglobin (Hb) has been a topic of extensive research over the past 30 years or more (for a recent summary, see Perutz *et al.*, 1987). The contribution of EPR spectroscopy to the key issue of the control of the globin part on the oxygen affinity of the heme group in the tetrameric protein has, however, been limited, largely because the native molecule is EPR-inaccessible in both oxygenation conditions (deoxy Hb, $S = 2$; oxyHb, $S = 0$). Therefore, many studies have dealt with the Fe^{3+} metheme; these, however, do not lead to conclusions about phenomena associated with the conformational changes induced in the native protein upon oxygenation (T–R transition). This problem can be overcome either by using paramagnetic ligands to the Fe^{2+}-ion, which do not impede the structural changes or by exchanging Fe with another transition metal ion that is EPR-active in both oxygenation states. So far, research along these lines has yielded only two tools of potential relevance: Nitrosyl($NO^{\bullet}$) radicals can act as suitable paramagnetic ligands to Fe^{2+}, and Co^{2+} is a substitute for iron which adds oxygen in a comparable way. Following this rationale we have studied both systems.

3.1.1. NO Ligation

Most of the structural information we have obtained so far from heme systems comes from $NO^{\bullet}$ ligation, which was applied to hemoglobin (HbNO), myoglobin (MbNO), isolated α and β chains (αNO, βNO), and model componds such as NO-ligated Fe-tetraphenylporphyrin-imidazole (Im-FeTPP-NO). The ultimate aim of the studies is characterizing the hyperfine properties of the nitrogen nuclei and of the protons in the immediate heme environment and understanding their effects on the conformation transition induced in HbNO by pH and allosteric effectors (Hüttermann and Kappl, 1987; Höhn *et al.*, 1983; Kappl and Hüttermann, 1989a, b). For MbNO not only an EPR (Nitschke, 1982) but also an ENDOR analysis was performed using single crystals (Kappl and Hüttermann, 1989a). The combination of all of these methods and specimens led to a current understanding, depicted in Fig. 6, of the geometry of the heme and its environment. The configuration of the NO ligand and the proximal histidine (F8) shown mostly rests on the EPR single-crystal analysis of the $^{15}N(NO)$ (Nitschke, 1982) and the ENDOR derivation of the $^{14}N(N_{\varepsilon}$-His F8) interaction tensors (Kappl and Hüttermann, 1989a). The resulting geometry is embedded in that derived from the X-ray structure of metMb with H_2O as ligand at low temperature (Hartmann *et al.*, 1982). It is normalized to the heme plane, which is assumed not to change its position on cooling; support for this assumption comes from early work on metMb (Benett *et al.*, 1957). It is probable that the very low temperature structure derived from ENDOR data reflects the structural transitions occurring on cooling of the crystal.

The structural representation shown is the result of an ongoing effort of many years. Two severe obstacles had to be overcome, which exemplify some specific problems in protein ENDOR work and the necessity of

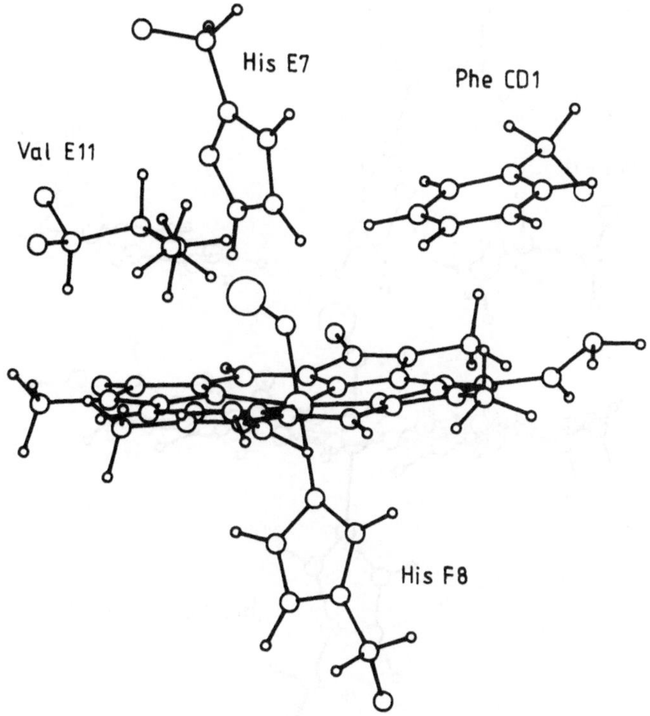

Figure 6. Bonding geometry of ENDOR-reconstructed heme NO group (5 K).

applying as many of the aforementioned additional techniques as possible in order to gain a coherent picture. One problem involved the NO ligation for which two main bond geometries can occur in 6-coordinated NO-heme complexes. These geometries show different EPR spectra and give varying contributions to the spectra between 77 K and room temperature. For myoglobin, this situation was first studied in frozen aqueous solutions at *X*-band frequencies (9.5 GHz) by Morse and Chan (1980), who also investigated other NO-heme complexes. These authors discussed an equilibrium between two positions of iron with respect to the heme plane, below and above, leading to differences in the bond length to N_ε of His(F8). However, Hori, Ikeda-Sato, and Yonetani (1981) in single-crystal EPR work found that there are two different Fe—N—O bond angles and tilts with respect to the heme normal. One configuration dominating at 300 K shows a nearly linear Fe—N—O bond (bond angle 153°), with only a small deviation (≈20°) between the heme normal and the Fe—N bond. This transforms into a considerably more bent one (Fe—N tilt ≈35° and Fe—N—O angle ≈110°), which is dominating the spectra at 77 K and is shown in Fig. 7. The other problem derives from the protein backbone and its flexibility

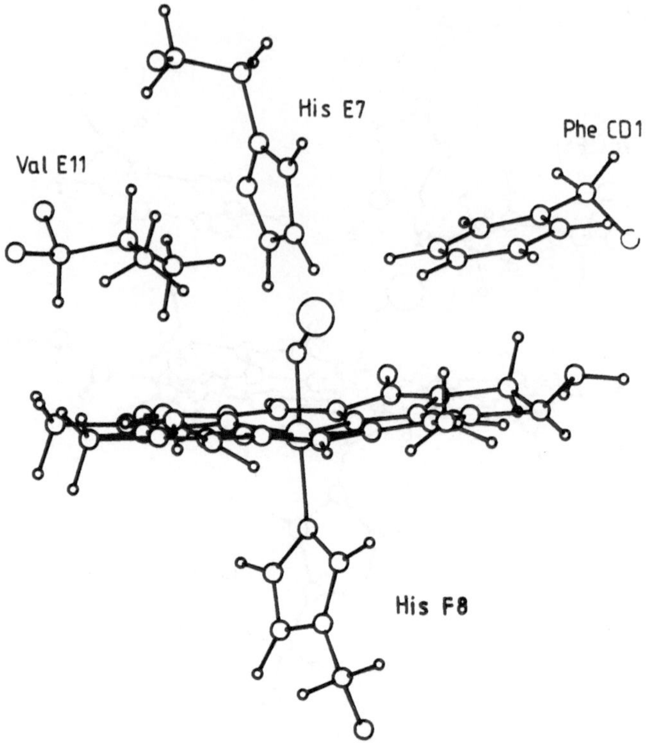

Figure 7. Bonding geometry of the Fe—N—O bonds in MbNO at 300 K. From Hori *et al.* (1981).

with temperature. Figure 8 shows an overlay of the X-ray-derived structures at 300 K and 100 K, indicating fairly large shifts of amino acid residues in the heme environment. The resulting proton positions must be established and taken into account when trying to assign, e.g., proton ENDOR spectra.

The data given by Hori *et al.* (1981) for the g and A[N(NO)] tensors for MbNO were supported by an independent EPR single-crystal study in our laboratory, although an error in the signs of direction cosines was detected (Nitschke, 1982). The EPR determination of the A tensor could, however, give only an estimate for the minimum value and its direction with an accuracy limited by the linewidth. In a subsequent ENDOR analysis the enhanced precision gave data that required the O(NO) to bend over in the direction shown in Fig. 6. Another achievement of the ENDOR resolution enhancement is the determination of the position of the N_ε atom of the proximal histidine (F8). EPR spectroscopy alone could deal only with the strong N(NO) interaction. It is very satisfying that the independent determination of the N_ε hyperfine and quadrupolar interaction tensors positioned this atom roughly along the linear extension of the N(NO)—Fe

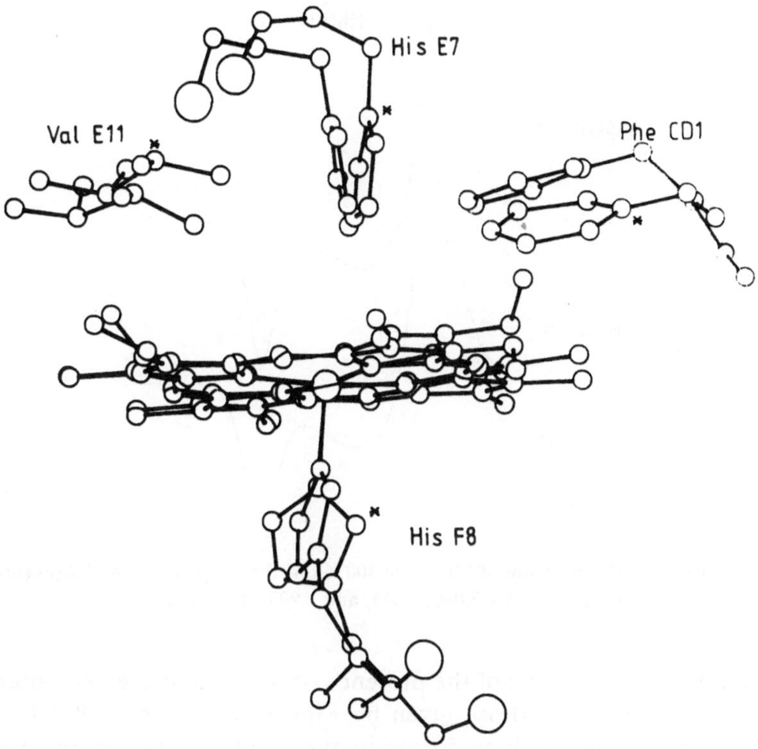

Figure 8. Overlay of prosthetic groups of Mb as seen by X rays at 300 (*) and 120 K.

direction. The whole imidazole plane could be fixed by assuming that the minimum value of the quadrupolar interaction tensor points along the normal of imidazole plane. This makes it coplanar with the Fe—N—O plane (Kappl and Hüttermann, 1989a).

Even once one knows quite well the majority species at 77 K, the problem exists that the configuration dominating the spectra at 300 K is not as well characterized. There is a question as to whether this is the configuration Morse and Chan (1980) address in their equilibrium model. In their study of solutions, the major species at 300 K was found to have axial symmetry and to contribute as a minor form at low temperatures, as is manifest from one of its g values ($g = 2.04$). The g tensor given by Hori *et al.* (1981) and by Nitschke (1982) for the room temperature configuration in MbNO is, however, rhombic. Recent measurements in our laboratory of the evolution of powder EPR from MbNO with temperature at *Q*-band frequencies gave clear evidence that the species at 300 K has mostly axial symmetry with only slight rhombic distortion, as shown in Fig. 9 together with a simulation (Kleber, 1989). Especially, the signal at $g = 2.04$, which

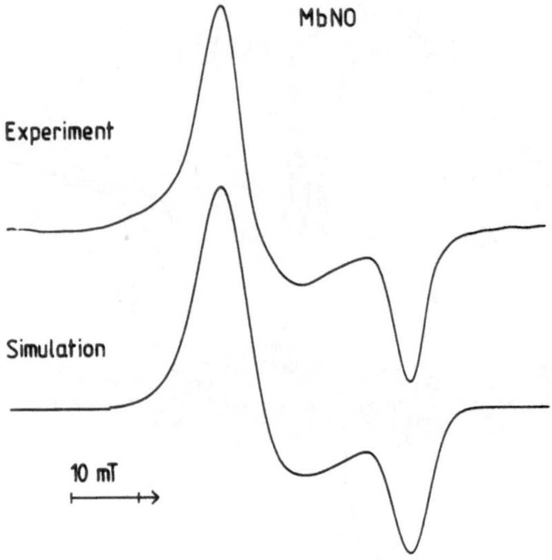

Figure 9. *Q*-band EPR spectrum of MbNO at 300 K together (top) with simulation (bottom) using *g* factors 2.040, 2.035, and 1.999, respectively.

is often taken as indicative of the presence of the second species, coincides with our data and with those given by Morse and Chan (1980) for that species. Perhaps the packing forces in the single crystals provide for a slightly different geometry, although Hori *et al.* (1981) do not favor the idea of an equilibrium with two components. The situation is even more complex in MbNO since there are signs of a second low-temperature species, which, however, is a minority with unknown geometry.

Although the very intricate interplay between energetic and geometrical factors—which, especially for Mb, has also been observed in the context of the FeO_2^- complex formed on low-temperature irradiation of MbO_2 (Kappl *et al.* 1985)—appears to demand a cautious approach to simplifying models, we nevertheless suggest that, *grosso modo*, the group of NO-heme complexes exhibits mainly two groups of bonding stereochemistries, one represented by the stretched, nearly linear form and the other by the strongly bent one. MbNO as well as αNO in isolated chains and in the tetrameric molecule HbNO both undergo, with decreasing temperature, the transformation between the two geometries. The β chains (βNO), however, show little variation in their temperature profile of spectra but rather retain their 300 K appearance. Since 6-coordinated model compounds like Im—Fe(TPP)—NO also take on the bent form at low temperature, the conclusion suggests itself that the protein stress does not force this form in MbNO or αNO but rather prevents the occupation of the energetically obviously more favorable bent configuration in the β chains. This is in line with X-ray data which indicate that the pocket is usually much smaller in the β chains than in the α chains and, of course, in Mb. Differences in α and β chain conformations were also observed in a single-crystal EPR study of HbNO (Doetschman and Utterback, 1981), but the tensors given in that work do not align with the ones presented for either species here. In our single crystal ENDOR study of MbNO (Kappl and Hüttermann, 1989a) we ascertained that, whenever the data permitted the source to be determined, the data obtained came from the majority species (the one given in Fig. 6).

It is interesting to note that only the α, not the β chains, isolated or in the tetrameric protein molecule undergo a conformational change which can be considered as an r–t tertiary structure transition. It has long been speculated that a certain EPR specrum first reported by Hille *et al.* (1979) should involve a 5-coordinated NO–heme group. In order to exemplify the differences from the 6-coordinate complexes we show two examples of procedures to obtain them in Fig. 10 (Kleber, 1989). One utilizes the difference in EPR spectra between HbNO in R and in T state and the other brings the α chains directly into their t state. In comparison we show the spectra of the α chains in their r state and of the β chains. The obvious difference in spin-density distribution between αNO(r) and αNO(t) leading to three sharp ^{14}N hyperfine lines in the latter was interpreted as resulting from the loss of the imidazole as the sixth ligand. This proposal is in line with EPR findings from 5-coordinated model compounds (Scheidt and Frisse, 1975) and with our ENDOR results from αNO in its t state (Höhn *et al.*, 1983). We could clearly show that the bond to the proximal

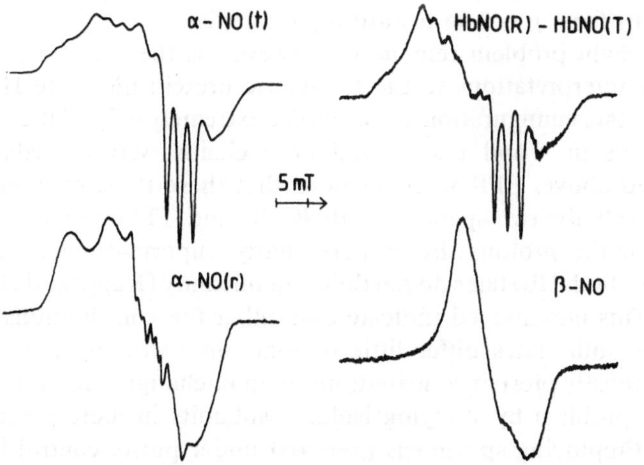

Figure 10. EPR spectra of difference between HbNO(R) and HbNO(T) (top right), α-NO(t) (top left), αNO(r) (bottom left), and β-NO hemoglobin (bottom right) subunits.

imidazole ligand opens upon decreasing the pH and/or adding allosteric effectors. One would like to think that the strong distortion of the NO bond on cooling, together with the corresponding twist in the proximal histidine in the α chains, reflects the large degree of flexibility of the O—N—Fe—N_ε bonding situation, which may be related to the possibility of the loss of the proximal histidine. It is known from EPR studies that the T-state properties of HbNO are connected with α chain changes (cf. Fig. 10, top right), whereas the β chains remain unchanged (Henry and Banerjee, 1973).

With a distorted N_ε—Fe—N—O geometry and position changes in the amino acids at low temperatures, one may perhaps not expect to achieve the reconstruction of the MbNO prosthetic group protons to be derived simply from ENDOR measurement of frozen solutions. It was, however, unexpected that proton ENDOR data from MbNO single crystals also gave only limited information. For the very weakly coupled protons ($\leq$1 MHz) in the immediate vicinity of ν_n, the anisotropy and number of lines turned out to prevent an unambiguous pursuit of resonances through the three planes of measurement; for the larger couplings, only some windows of crystal orientations were amenable to direct data evaluation. As a consequence, even the single-crystal analysis could not be performed without simulations using adopted tensors and proton assignments (Kappl and Hüttermann, 1989a). Nevertheless, in conjunction with the other techniques mentioned it was possible to arrive at a reconstruction (cf. Fig. 6), which we believe is very close to the actual situation. This is reflected in the good agreement between experimental spectra and their simulation, as shown in Fig. 11 at two g turning points (g_1 and g_2) of MbNO for both protons and nitrogens (N_ε). We note that there is an uncertainty of assignments inherently unavoidable due to the overlap of the second species, especially in the regions from $g \sim 2.04$ toward higher fields.

The main problem remaining, however, is the transfer of these data and their interpretations to the tetrameric protein molecule HbNO in its R- and T-state configurations. It would be extremely helpful if a clear picture of α chains in r and t state and of β chains were available since, as mentioned above, EPR work suggests that these three components make up additively the two spectra of HbNO(R) and -(T) (Henry and Banerjee, 1973). For the protons this is particularly important since the ENDOR patterns of both Hb states do not differ significantly (Kappl and Hüttermann, 1989b). This may indeed indicate that either the contributions of the two chains in both states differ little or that some working positions of the tetramer mostly present contributions of an unchanging subunit. We looked into that problem by studying isolated subunits in more detail than previously. Employing specimens prepared under purity control by gel electrophoresis, a study of α chains in their t state was repeated. The previous finding (Höhn *et al.*, 1983) from the study of nitrogen interaction, which

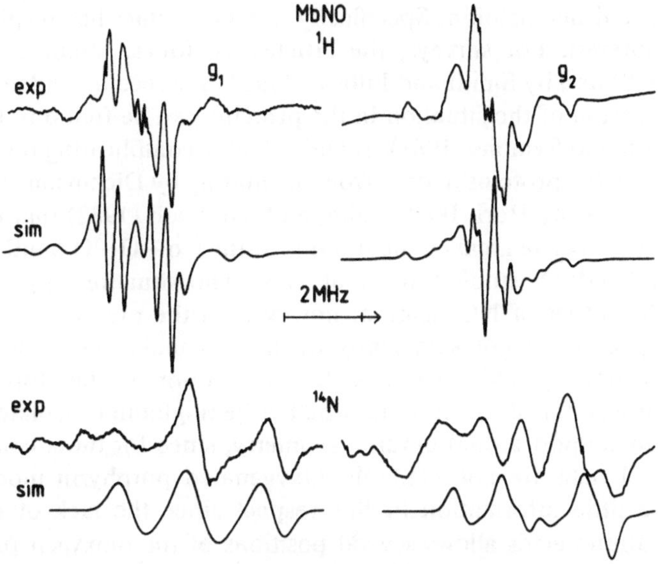

Figure 11. Experimental (top curve in each set) and simulated proton (top) and nitrogen (bottom) powder ENDOR in MbNO.

indicated loss of the proximal histidine in the t state, could be supported again. Moreover, for the protons, a comparison between αNO(r), αNO(t) and βNO gave differences between the three. Changes along g_1 and g_2 between αNO(r) and αNO(t) involve protons of the proximal histidine, which are lost in the latter specimen. Comparison with the β chains gives only small differences to the r-state α chain, which we have not yet assigned (Kleber, 1989). On the whole it appears, however, that only hybrid studies in HbNO with one subunit muted will bring about the real differences necessary to explain the R–T transition. These are under way in our laboratory, together with a more detailed simulation for the subunit spectra.

3.1.2. Co Substitution

Cobaltous hemoglobin as prepared first by Hoffman and Petering (1970) is a functional analog of the native protein. It binds dioxygen reversibly and cooperatively but with a reduced affinity (Chien and Dickinson, 1973). The major advantage for its use in EPR-related studies is its paramagnetism, both as oxy- and as deoxyhemoglobin (CoHb, $CoHbO_2$). In its oxygenated state, the molecule exhibits rhombic symmetry in the g and A tensors, whereas in the deoxygenated form the square-pyramidal configuration of the Co–porphyrin core gives an axially symmetrical spectrum.

There have been quite extensive EPR studies of both the oxy- and deoxy-state of Co–porphyrin complexes, modelling the situation in myoglobin and hemoglobin. Specifically, the O_2 adduct has inspired considerable interest. For surveys, the articles by Jones, Summerville, and Basolo (1979) and by Smith and Pilbrow (1981) are recommended. A more recent discussion of the situation in the proteins can be found in the work of Dickinson and Symons (1983). A major factor complicating the analysis of the oxygenated proteins derives from the finding by Dickinson and Chien (1980) as well as by Hori, Ikeda-Saito, and Yonetani (1982) that there are two distinct stereochemical configurations of the Co—O—O bond in single crystals of $CoMbO_2$ studied at about 77 K. The situation appears to be comparable to that of NO ligation; however, in the present case the two resulting spectra are not sufficiently distinct to make their contribution apparent in polycrystalline systems (frozen solutions). Therefore there is no possibility so far of ascertaining whether hemoglobin or isolated α and β chains do support a dual binding geometry, since for these compounds single-crystal data are not available. Oxygenated porphyrin model complexes give little information in this respect since the lack of directing distal-side amino acids allows several positions of the dioxygen projection over the porphyrin plane to be attained. Apart from the uncertainty of the number of configurations, there is also a problem with identifying the correct stereochemistry beforehand. Smith and Pilbrow (1981) discuss in some detail the influence of the Co—O—O bond angle and the resulting non-coaxialities on the shape of the powder-type EPR spectrum, which can partially be disentangled by proper simulation. The situation becomes more complex with more than one configuration to be considered.

There are fewer degrees of variabilities in the deoxygenated cobaltous porphyrin model compounds and in the proteins. Fairly precise values for the g- and A-tensor components of CoMb have been determined by X- and Q-band EPR spectroscopy (Inubushi and Yonetani, 1983). For tetrameric hemoglobin, hybrids containing Fe and Co subunits have shown that the a chains in the tetrameric protein are sensitive to properties representing R and T states, respectively. It was also reported that the isolated α chains respond in a similar way to changes in pH, a finding which we have so far been unable to reproduce (Klein, 1989). The β chains, on the other hand, were found to be fairly insensitive.

We have studied the ENDOR response of cobaltous hemoglobin, myoglobin, and model compounds for some time now. In a first report we could show that in oxygenated $CoHbO_2$ there was a proton interaction of about 5.5 MHz coupling, which was lost when changing the buffer conditions from H_2O to D_2O. We assigned this coupling to a hydrogen bond between the dioxygen ligand and the proton of the distal histidine (N_ε-His E7) (Höhn and Hüttermann, 1982). This was direct spectroscopic evidence of a proposal which had been under discussion for some time (e.g., Kitagawa *et al.*, 1982). In a later study, Walker and Bowen (1985) concentrated on hydrogen donation to O_2 in $CoMbO_2$ and model compounds from dynamical considerations in EPR spectra at room temperature. In a more detailed account of ENDOR results from deoxy and oxy CoMb, CoHb, and octaethyl- and tetraphenylporphyrin model compounds (OEP, TPP) ligated with pyridine as models of the proximal histidine we could show recently, among other findings, the existence of an exchangeable proton in $CoMbO_2$ (Hüttermann and Stabler, 1989). Its linewidth, however, suggested either a large variation in coupling values or the existence of orientation-inequivalent sites. Another interesting feature was the surprisingly large value of interactions coming from the proximal side although the spin density is nearly fully situated on the dioxygen. For the deoxy compounds the spectra of CoHb and CoMb were found to be fairly similar. Tentative assignments were achieved by comparison with the pyridine-ligated model compounds.

Aiming at modelling the proximal histidine more closely and at achieving better resolved data, Däges and Hüttermann (1992) recently completed another series of studies concentrating on deoxy model complexes (OEP, TPP) only. As axial ligand imidazole was used, for which some of the protons were selectively replaced by methyl couplings. This together with the application of newly developed simulation tools and an elaborate extraction of the EPR spectral parameters from *X* and *Q* band (9.5 GHz and 34 GHz) allowed us to arrive at a very detailed picture of the Co-porphyrin-imidazole core which could be reconstructed with very high precision and be compared with crystal-structure data (Dwyer *et al.*, 1974). It was intriguing to find that the proton positions deriving from simulations at $g_{\parallel}$ differed from those at $g_{\perp}$ when we used the distance parameters from the room-temperature crystal structure. On the other hand, for both positions one unique set could be obtained only by employing low-temperature crystal-structure distances from Mb (Hartmann *et al.*, 1982). This implies that the distance reductions in the porphyrin plane and in the axial directions upon cooling are different ("anisotropic shrinking").

Rather than the proton spectra themselves, the final geometry reconstructed from ENDOR at about 5 K is shown for CoTPP, in Fig. 12, in comparison with the room-temperature structure derived from X-ray diffraction data. The imidazole in the low-temperature reconstruction is drawn closer to Co and somewhat tilted toward a linear Co–N bond. With this knowledge we now understand the ENDOR response of CoMb which, as mentioned above, is very much like that CoHb (Hüttermann and Stabler, 1989). It turns out that in the previous investigation the larger proton interactions leading to coupling values of 5.6 to 5.9 MHz were overlooked along $g_{\perp}$. Their assignment to CE1 and CD2 protons of the proximal histidine based on $g_{\parallel}$ patterns is, on the other hand, supported. These larger interactions are usually smaller in intensity than are those of the inner part, since their spectral density is spread over a much larger range of frequencies.

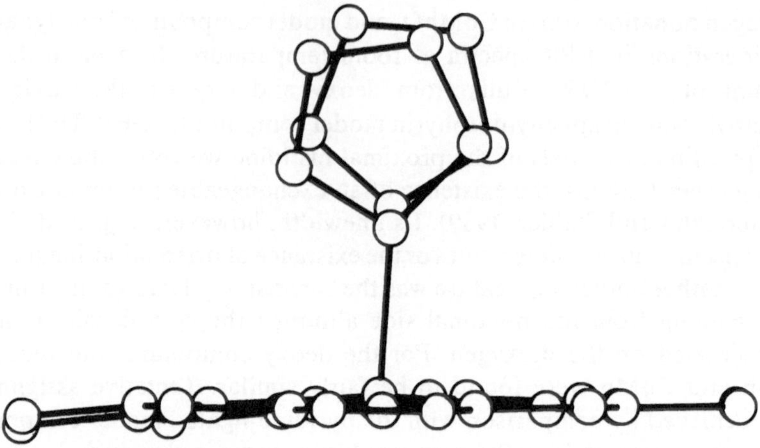

Figure 12. Overlay of room-temperature crystal structure of CoTPP–imidazole with structure deriving from ENDOR at about 5 K normalized to the Co position.

Otherwise the couplings are fairly similar to those of the model compound. The precise assignment of the latter requires changes in their placement in the protein. Figure 13 gives two ENDOR spectra for CoMb, indicating the field positions on the EPR spectrum. The assignment to be used because of the model-compound studies is also shown. As is demonstrated, all couplings can be accounted for by interactions of porphyrin-core and proximal histidine couplings. The distal side is too far away to contribute.

Another question that needs to be settled in more detail because of the new results is related to the ^{14}N interaction. Our previous preliminary analysis of CoMb and CoHb (Hüttermann and Stabler, 1989) suggested that a quartet pattern should be obtained for both proteins along $g_\perp$ and a triplet along $g_\parallel$, the corresponding ^{14}N hyperfine couplings reading about 42 MHz and 44.5 MHz. In contrast, the recent model-compound data give a triplet along $g_\perp$ and a quartet along $g_\parallel$, thus reversing the association of the two quadrupolar-interaction values with the ^{14}N hyperfine couplings. Also, the hyperfine coupling is appreciably more anisotropic, with values, for imidazole, of 40.2 MHz and 48.2 MHz for the perpendicular and the parallel directions, respectively. These differences, especially the directions of the quadrupolar-interaction values, are sufficiently large to suggest a reinvestigation of the proteins in order to assess the weight of the data. Establishing the connection between hyperfine and quadrupolar elements is important in relating the proximal histidine ring plane to the porphyrin core.

Turning now again to the oxy complexes, we recall that due to a spin density on the dioxygen ligand of 90% or more, sizable contributions of

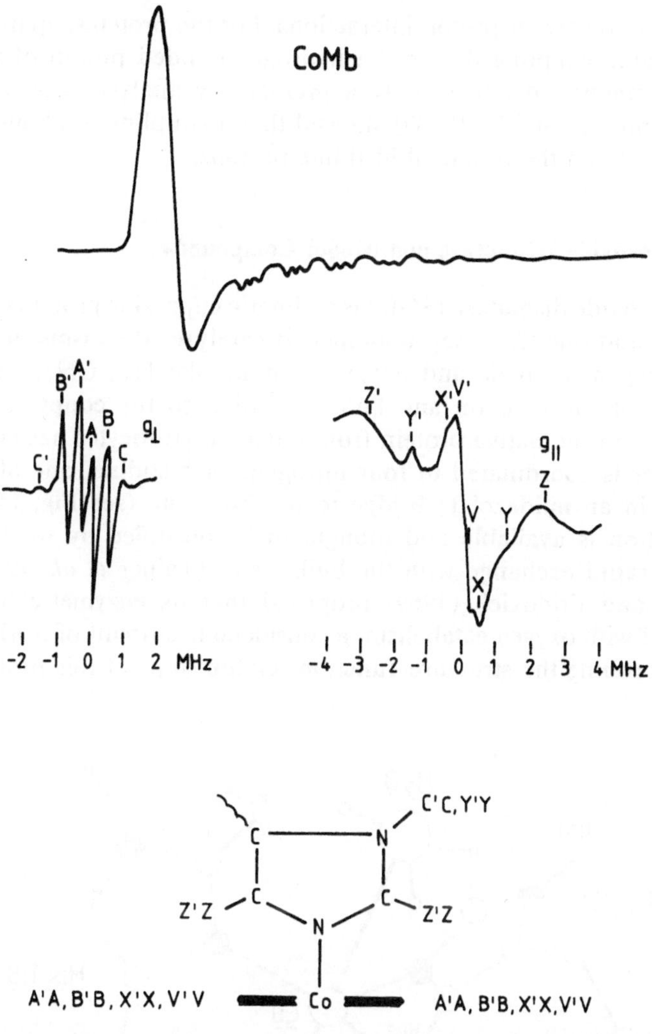

Figure 13. EPR (top) and ENDOR (middle) spectra for CoMb. ENDOR spectra were taken at $g_{\perp}$ (left) and $g_{\parallel}$ (right) positions. Line assignment to protons in porphyrin and imidazole parts are shown in the structure (bottom).

distal-side amino acids are to be expected. In our model-compound studies using pyridine as axial ligand, as mentioned above, it was surprising to find fairly large couplings ($\approx$2 MHz) resulting from the proximal protons (Hüttermann and Stabler, 1989). No solvent interactions were detected in toluene. Couplings of similar size were observed in most recent studies, still under way in our laboratory, in which imidazole is employed as axial ligand to OEP and TPP. These compounds also gave larger couplings, which should be ascribed to solvent-proton interactions. For the proteins, apart from the exchangeable coupling due to the hydrogen-bonded proton of the distal histidine mentioned above, only a preliminary analysis was performed. Hüttermann and Stabler (1989) showed that a coupling of about 2.5 MHz should reside on the proximal histidine protons.

3.2. Superoxide Dismutase and Model Compounds

Superoxide dismutase (SOD) is a dimeric cuprozinc protein containing one Zn^{2+} and one Cu^{2+} per monomer. It catalyzes the dismutation of O_2^- into hydrogen peroxide and oxygen. Anions like N_3^-, CN^-, and F^- are inhibitors of the reaction and known to bind to the copper site. X-ray diffraction on the native protein from bovine erythrocytes has shown that the copper is coordinated to four nitrogens of histidine, one of which is involved in an imidazolate bridge to the Zn^{2+} site (see Fig. 14). A fifth coordination is available and thought to be occupied by weakly bound water in rapid exchange with the bulk water (Tainer *et al.*, 1982). Since McCord and Fridovich (1969) proposed that its enzymatic function is connected with oxygen catabolism, a considerable amount of work has been done in probing the structure–function relationships of this protein. Most

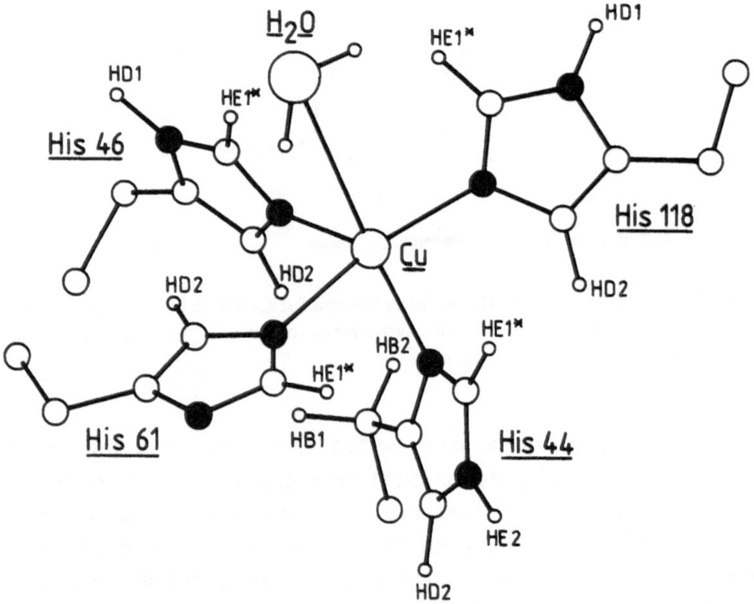

Figure 14. The structure of the Cu site of superoxide dismutase from bovine erythrocytes.

of the discussion nowadays is centered around the question of how the binding of the anions disturbs the coordination properties of the Cu–Zn sites.

EPR spectroscopy had already been applied to the protein before its function was clear (Malmström and Vanngard, 1960). The native protein was found to give rise to a spectrum with rhombic symmetry in the g tensor. Single-crystal EPR showed that the g_z component was approximately normal to a plane through all four nitrogens (Lieberman *et al.*, 1982). Binding of CN^- produced axial symmetry with resolvable hyperfine coupling assigned to three equivalent nitrogens. The use of $^{13}CN^-$ gave evidence for cyanide binding to copper through the carbon. Less strong but still discernible disturbances tending to axial symmetry were observed upon binding of F^- and N_3^-, as was ascertained by spectroscopy at the high microwave frequency of 34 GHz (Q band rather than the conventional 9.5 GHz X band) (Rotilio *et al.*, 1972).

Only a few investigations involving high-resolution EPR techniques like ENDOR or ESEEM (*e*lectron *s*pin *e*cho *e*nvelope *m*odulation) spectroscopy have been performed probing the weak nitrogen interactions. Subsequent to ESEEM of complexes of Cu^{2+} with imidazoles (Mims and Peisach, 1978) the same technique was applied to SOD by Fee, Mims, and Peisach (1981). In the former study, no difference was found for the couplings of the remote nitrogen in the diethylene-triamine-imidazole or the tetraimidazole complex. Also, for both imidazoles protonation at that nitrogen was postulated. In the protein, the authors studied the influence of N_3^- and CN^- binding on the echo from the complete and from the Zn^{2+}-free protein. Drastic changes were observed to occur when cyanide bound, whereas the influence of azide was not detectable.

Along the same lines, Van Camp, Sands, and Fee (1981) applied ENDOR to study the tetraimidazole complex in order to apply this knowledge to the protein (Van Camp *et al.*, 1982). The former study included selective imidazole deuteration in order to give ranges for proton interactions. Further emphasis was placed on establishing the hyperfine and quadrupolar parameters for the near and the remote nitrogens. Most of the work with the protein was performed on the CN^- complex, but some data were also given for azide binding. The spectra from the native protein, though considered too complex to interpret, were thought to be indicative of an inequivalence of ^{14}N interactions of the near nitrogen. The authors provided two sets of histidine nitrogen couplings for the CN^- complex, together with the finding that the anion should bind through its carbon site, yielding for two of the remaining nitrogens a larger, equivalent interaction of about 47 MHz and one smaller coupling of about 37 MHz because of the third nitrogen. This result is in contrast with Rotilio's conclusions from EPR simulations, which required all three nitrogens to have one coupling value (Rotilio *et al.*, 1972). Both groups agree, on the other hand, that CN^-

binding should form a square planar complex. As in EPR, ENDOR also showed differences between the cyanide and the azide, although these were not analyzed in detail.

Our own ENDOR work on SOD emphasized proton couplings together with nitrogen interactions from several specimens, native SOD from bovine and human erythrocytes, and proteins with the Zn^{2+} site depleted, as well as SOD modified by the binding of CN^- and N_3^- (Hüttermann *et al.*, 1988). Although the study provided a large body of data, it was insufficient for answering the main questions under discussion, connected with anion binding. These concerned the equivalence between CN^- and N_3^- binding in replacing the same histidine and the weakly coordinated water. NMR data seemed to suggest that both anions modify the copper center in the same way (Bertini *et al.*, 1985). Qualitatively we found that the reaction with N_3^- caused more subtle changes than CN^- binding. Nevertheless, we were unable to assign all protons unequivocally in the native protein and to follow the changes with modification. Also, for the nitrogen interactions, approximate values could be given only at the $g_{\parallel}$ position, but the perpendicular response was not understood. We therefore started a series of model-compound investigations in which both the EPR and the ENDOR parameters were studied with the aim of providing a sound base for interpretation of protein data (Scholl and Hüttermann, 1992). In addition, we concentrated our efforts on simulation routines. The following paragraph sums up our present understanding of the topic, emphasizing copper tetraimidazole, which was found to give the results most relevant to the protein.

Figure 15 shows the molecular structure of the tetraimidazole complex together with a typical *X*-band EPR spectrum displaying axial symmetry. ENDOR spectra were taken at the positions marked *a* through *h* on the EPR spectrum. Representative spectra emphasizing the nitrogen interaction are shown in Fig. 16 for fields *c* and *h*, giving the values for parallel and perpendicular couplings. Specifically, the perdeuterated compound shows the nitrogen region nearly free of overlap from protons. The resulting coupling values for ^{14}N as given by stick diagrams should be taken as indicative of the simulation parameters used. The interaction is at least close to being axially symmetrical, so that along $g_{\perp}$ both coupling values contribute. For field positions *e* and *g* we show the proton part of the ENDOR spectra. The top curve is experimental, the bottom is a simulation involving all 12 protons HC2, HC5, HC4 on the four imidazole rings (HN1 is left out in the deuterated specimen) using the crystal-structure distances and optimizing the isotropic term for best match. We believe that the agreement between simulation and experiment is encouraging, keeping in mind the enormous number of proton signals to be accounted for. Comparison with the values of Van Camp, Sands, and Fee (1982) shows that

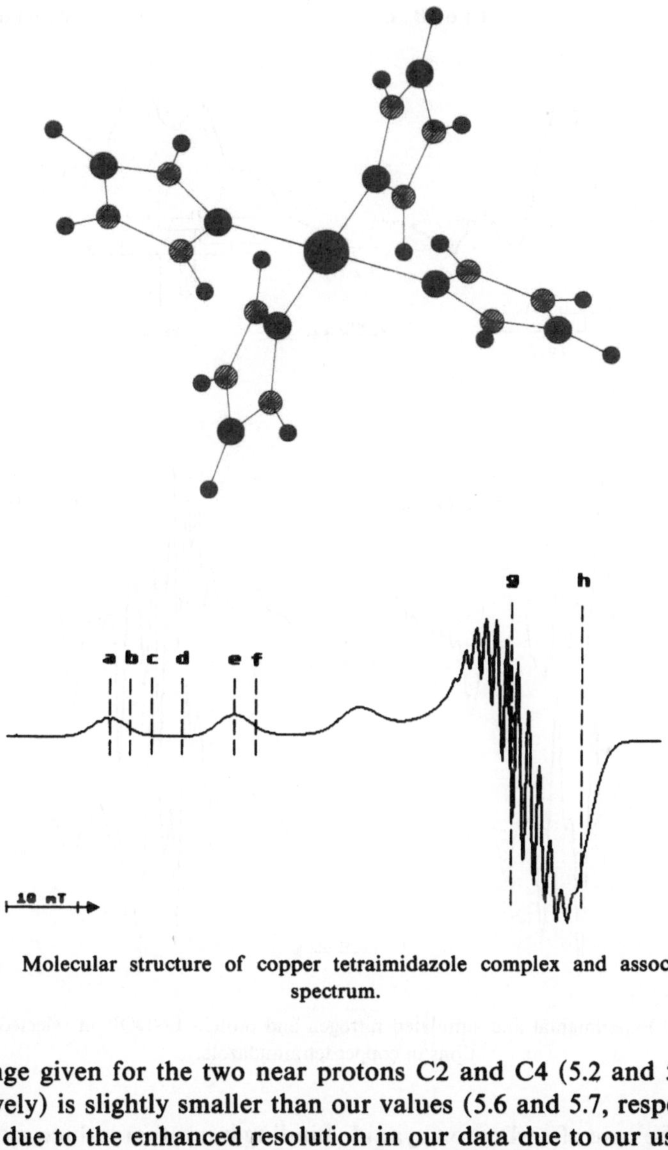

Figure 15. Molecular structure of copper tetraimidazole complex and associated EPR spectrum.

their range given for the two near protons C2 and C4 (5.2 and 5.3 MHz, respectively) is slightly smaller than our values (5.6 and 5.7, respectively), perhaps due to the enhanced resolution in our data due to our use of FM rather than AM. However, the C5 proton of the 1.7 MHz coupling in the former study is less by far than our value of 2.8 MHz. We also give, as an additional coupling, the exchangeable HN1 at 3.4 MHz.

Transfer of these results to the protein helps to clarify some of the problems we encountered previously. From comparison of the nitrogen coupling region of the native protein, its azide derivative, and its ^{15}N-substituted azide derivative we concluded previously that N_3 binding to the protein would not lead to coordination of the azide nitrogen the way it

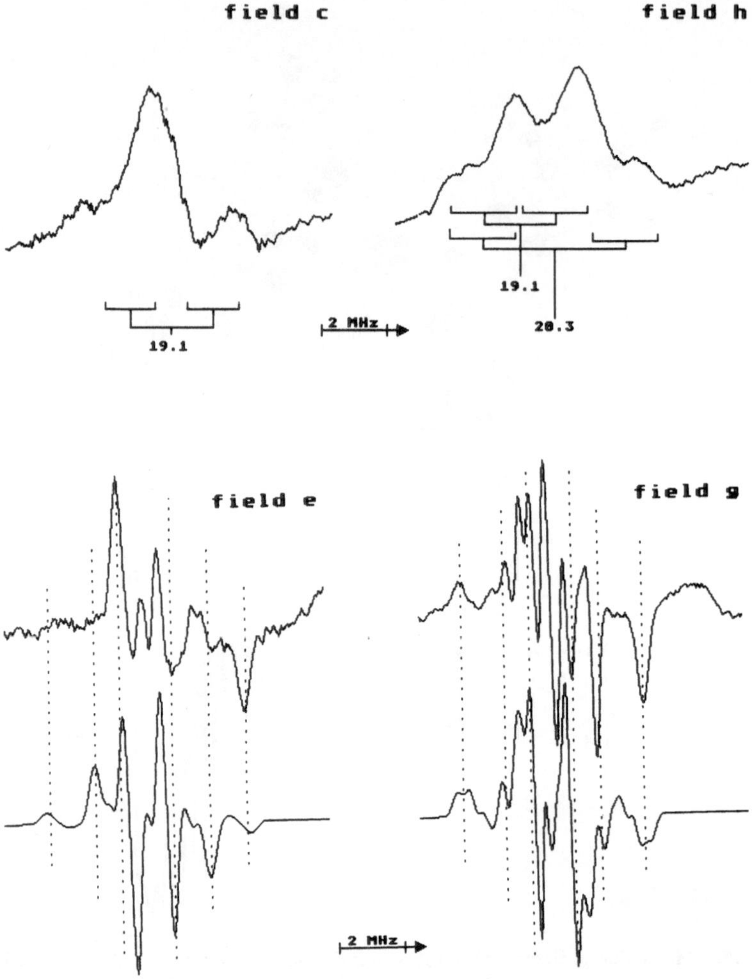

Figure 16. Experimental and simulated nitrogen and protein ENDOR at selected field positions in copper tetraimidazole.

does in the case of carbon on cyanide binding but rather induce more subtle changes. We noted, furthermore, that coupling values from the $g_\perp$ position could not be extracted, since we did not understand the pattern produced sufficiently well. The detailed knowledge from axial model compounds (tetraimidazole, tetrapyridine) at their $g_\perp$ positions allows us to sustain that main conclusion about azide binding and to give 48 MHz to 50 MHz as a reasonable estimate for the histidine-N coupling at the high-field position of the spectrum of both the native and the azide-substituted protein. However, since both have rhombic symmetry, the native protein exhibiting a stronger disturbance than the azide derivative, one cannot extract any principal values at that position.

Regarding proton interactions, one must note that the tetraimidazole model compound gives, as tetrapyridine does not, signals for several of the stronger interactions in which the high-frequency branch ($\nu(+)$) of the spectrum is of larger intensity than its counterpart in the range $\nu < \nu_n$ ($\nu(-)$ branch) (cf. Fig. 16). This is a feature which we had to invoke for some lines of the protein, too. Consider the situation depicted in Fig. 17. The top recording is an overlay of two proton simulations at the high-field position (denoted *d* in Hüttermann *et al.*, 1988). Comparison of the two top experimental patterns for bovine SOD (BESOD) in H_2O and D_2O buffer shows that one line is exchanged in the $\nu(+)$ part, whereas the expected "partner" in the $\nu(-)$ branch exhibits an unchanged intensity. At the same field position, the azide derivative shows a line in $\nu(+)$ for which there is no

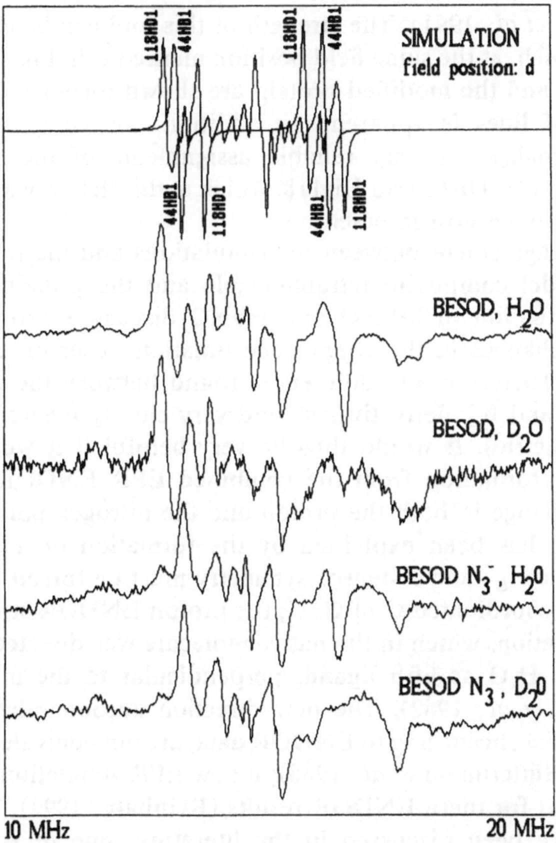

Figure 17. Simulated and experimental behavior of selected proton lines in SOD.

apparent partner with equal intensity in the $\nu(-)$ range. Exchange of the specimen in D_2O (bottom recording) leads to loss in the $\nu(+)$ branch. This behavior can be explained by the simulations of protons His44-HB1 and His118-HD1 (using the nomenclature for the bovine isoenzyme). The latter is exchangeable, is present in both BESOD and its N_3 derivative, and has high-frequency branch line significantly more intense than its $\nu(-)$ counterpart, in complete agreement with the behavior of the corresponding proton in tetraimidazole. The other proton, His44-HB1, is not exchangeable in D_2O, has inverted intensity relations between $\nu(+)$ and $\nu(-)$, and changes its coupling value upon azide binding.

This situation of overlap of lines from two protons combined with effects of inverted asymmetry in intensity relations exemplifies the need to reduce the number of proton interactions in order to make firm assignments possible. One such measure is the use of modified proteins. We have most recently given a preliminary account of measurements on SOD in which all HE1 protons (marked by asterisks in Fig. 14) were replaced by deuterons (Hüttermann *et al.*, 1991). The strength of this tool can be estimated from Fig. 18, in which, at the same field position marked *d* in Fig. 17, the spectra of the native and the modified protein are shown together. The reduction in number of lines is apparent. In addition, the comparison with the simulations makes for very reliable assignments of the HE1 protons, especially those of His61 and His118, which at this field position contribute largely to the outermost interactions.

The fair agreement between the simulations and the proton lines for both the model compound tetraimidazole and the protein implies that, unlike the situation in MbNO and HbNO discussed above, few if any geometrical changes in the copper coordination occur on cooling of the compounds. Therefore, the differences found between the native protein and its CN^- and N_3^- derivatives should very closely reflect their different binding geometries. It would thus be very helpful if it were possible to extract this information from the combined EPR–ENDOR information. The strong change in both the proton and the nitrogen parts obtained by CN^- binding has been explained by the formation of a square planar complex. A new *g*- and *A*-tensor symmetry must be forced on the copper by the newly coordinated C of CN^-; the proton ENDOR strongly asks for a new $g_{\parallel}$ direction, which in the native molecule was directed nearly along the vector to H_2O as fifth ligand, perpendicular to the mean N_4 plane (Liebermann *et al.*, 1982). The new direction accommodates only three nitrogens, which, according to ENDOR data, are not equivalent (Van Camp *et al.*, 1982; Hüttermann *et al.*, 1988); a new EPR simulation recently gave strong support for these ENDOR results (Reinhard, 1992). Two positions for CN^- have been discussed in the literature, one in which His44 is displaced by the cyanide coming from the so-called H_2O side and one in

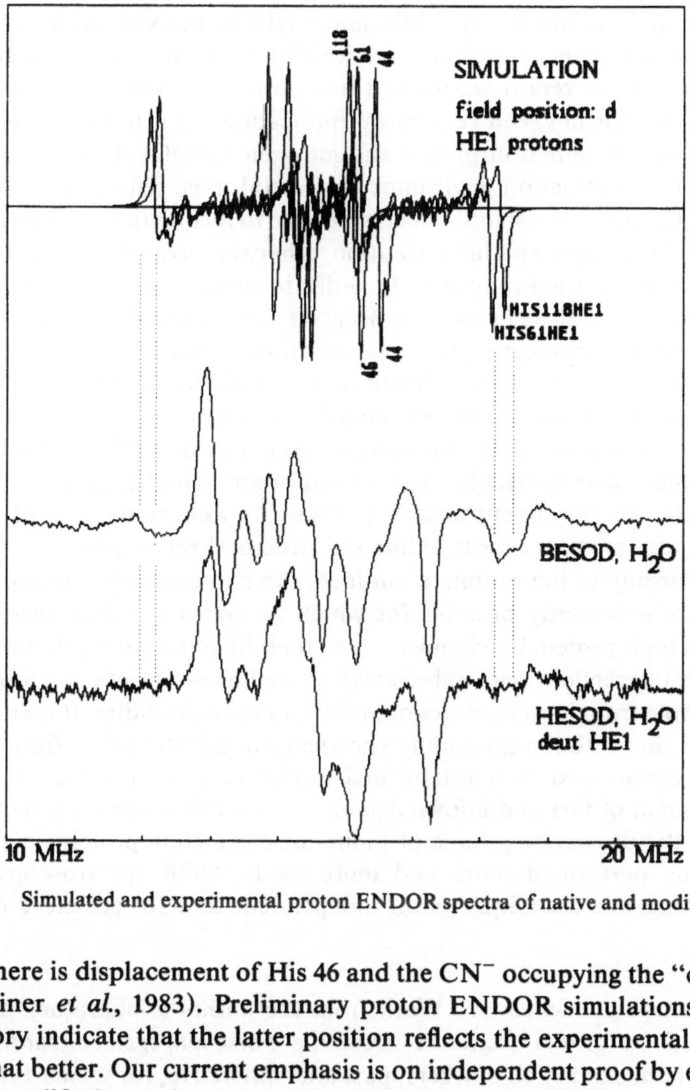

Figure 18. Simulated and experimental proton ENDOR spectra of native and modified SOD.

which there is displacement of His 46 and the CN^- occupying the "copper" site (Tainer *et al.*, 1983). Preliminary proton ENDOR simulations in our laboratory indicate that the latter position reflects the experimental spectra somewhat better. Our current emphasis is on independent proof by employing the modified protein to solve that question more reliably.

4. CONCLUSION

ENDOR studies on randomly oriented metalloproteins can in general be mainly classified into two groups. One concerns basic questions like identification of ligand nuclei (e.g., proving the absence or existence of histidine coordination) or titrating exchangeable proton interactions and documenting their change on, e.g., protein–substrate interaction. This group of investigations needs very little non-ENDOR-derived information and has a strong impact on proteins for which a structural picture is only just emerging. In this report we wanted to highlight a second class of studies, which so far has not seen very many applications. It is built on the implicit information contained in proton and nitrogen ENDOR from nuclei in the immediate coordination environment of metal ions, reflecting their spatial relationship to the metal ion. This can be used to reconstruct the coordination sphere with a high spatial resolution otherwise available only in X-ray diffraction from single crystals. In order to utilize this potential fully, an array of additional techniques is necessary to reduce the uncertainty due to the nonlinear connection between coordinates and spectra. These involve techniques associated with ENDOR itself, simulation of EPR and ENDOR spectra, and measurements of model compounds and of isotopically exchanged compounds. In addition, information from X-ray diffraction—e.g., of the native molecule—is very important. Finally, powerful genetic engineering to selectively modify proteins as well as the use of suitable mutants are important tools aiding the structural reconstitution.

According to the rationale outlined, the two examples chosen for this report are necessarily proteins for which an enormous data base already exists on their protein biochemistry (i.e., their ligand-binding characteristics and their interactions with substrates), as well as on structural properties—for example from X-ray diffraction, NMR, or optical studies. It is envisaged, however, that with the expanding knowledge of ENDOR data from suitable model systems and well established model proteins together with, e.g., development of fast and knowledge-based simulation routines, reconstructions of the prosthetic groups of mononuclear metalloproteins can, in the future, be performed more and more by ENDOR spectroscopy alone, establishing this technique as an independent tool for structure determination.

ACKNOWLEDGMENTS. Work from the author's laboratory has been supported by grants from the *Deutsche Forschungsgemeinschaft.* I owe sincere thanks to my co-workers, past and still active, for which I name the latter as representatives: Dr. R. Kappl, G. P. Däges, H. Reinhard, and H. J. Scholl. The work on SOD has become possible through cooperation with the group in Florence, I. Bertini and L. Banci.

REFERENCES

Bennett, J. E., Gibson, J. F., and Ingram, D. J. E., 1957, *Proc. R. Soc., Ser. A* (London) **240**:67.
Bertini, I., Briganti, F., Luchinat, C., Mancini, M., and Spina, G., 1985, *J. Magn. Reson.* **63**:41.
Chien, J. C. W., and Dickinson, L. C., 1973, *J. Biol. Chem.* **248**:5005.

Däges, G., and Hüttermann, J., 1992, *J. Phys. Chem.*, **96**:4787.
Dickinson, L. C., and Symons, M. C. R., 1983, *Chem. Soc. Rev.* **12**:387.
Dickinson, L. C., and Chien, J. C. W., 1980, *Proc. Natl. Acad. Sci. USA* **77**:1235.
Doetschman, D. C., and Utterback, S. G., 1981, *J. Am. Chem. Soc.* **103**:2847.
Dwyer, N. P., Madura, P., and Scheidt, W. R., 1974, *J. Am. Chem. Soc.* **96**:4815.
Fee, J. A., Peisach, J., and Mims, W. B., 1981, *J. Biol. Chem.* **256**:1910.
Feher, G., 1956, *Phys. Rev.* **103**:834.
Haindl, E., and Hüttermann, J., 1978, *J. Magn. Reson.* **30**:13.
Hartmann, H., Parak, F., Steigemann, W., Petsko, G. A., Ringe-Ponzi, D., and Frauenfelder, H., 1982, *Proc. Natl. Acad. Sci. USA* **79**:4967.
Henderson, T. A., Hurst, G. C., and Kreilick, R. W., 1985, *J. Am. Chem. Soc.* **107**:7299.
Henry, Y., and Banerjee, R., 1973, *J. Mol. Biol.* **73**:469.
Hille, R., Olson, J. S., and Palmer, G., 1979, *J. Biol. Chem.* **254**:12110.
Höhn, M., and Hüttermann, J., 1982, *J. Biol. Chem.* **257**:10,554.
Höhn, M., Hüttermann, J., Chien, J. C. W., and Dickinson, L. C., 1983, *J. Am. Chem. Soc.* **105**:109.
Hoffman, B. M., and Petering, D. H., 1970, *Proc. Natl. Acad. Sci. USA* **67**:637.
Hoffman, B. M., Gurbriel, R. J., Werst, M. M., and Sivaraja, M., 1989, in *Advanced EPR: Applications in Biology and Biochemistry* (A. J. Hoff, ed.), Elsevier, Amsterdam, pp. 541–591.
Hori, H., Ikeda-Saito, M., and Yonetani, T., 1981, *J. Biol. Chem.* **256**:7849.
Hori, H., Ikeda-Saito, M., and Yonetani, T., 1982, *J. Biol. Chem.* **257**:3636.
Hurst, G. C., Henderson, T. A., and Kreilick, R. W., 1985, *J. Am. Chem. Soc.* **107**:7294.
Hüttermann, J., and Kappl, R., 1987, in *Metal Ions in Biological Systems*, Vol. 22 (H. Sigel, ed.), Marcel Dekker, New York, pp. 1–80.
Hüttermann, J., and Stabler, R., 1989, in *Electron Magnetic Resonance of Disordered Systems* (N. D. Yordanov, ed.), World Scientific, Singapore, pp. 127–148.
Hüttermann, J., Kappl, R., Banci, L., and Bertini, I., 1988, *Biochim. Biophys. Acta* **956**:173.
Hüttermann, J., Kappl, R., Reinhard, H., Banci, L., Viezzoli, M. S., and Bertini, I., 1991, *Inorganic Biochemistry* **43**:224.
Inubushi, T., and Yonetani, T., 1983, *Biochemistry* **22**:1894.
Iwasaki, M., 1974, *J. Magn. Reson.* **16**:417.
Jones, R. D., Summerville, D. A., and Basolo, F., 1979, *Chem. Rev.* **79**:139.
Kappl, R., 1988, Ph.D. Thesis, U. of Regensburg, Germany.
Kappl, R., and Hüttermann, J., 1989a, *Israel J. Chem.* **29**:73.
Kappl, R., and Hüttermann, J., 1989b, in *Advanced EPR: Applications in Biology and Biochemistry* (A. J. Hoff, ed.), Elsevier, Amsterdam, pp. 501–540.
Kappl, R., Höhn-Berlage, M., Hüttermann, J., Bartlett, N., and Symons, M. C. R., 1985, *Biochim. Biophys. Acta* **827**:327.
Kitagawa, T., Ondrias, M. R., Rousseau, D. L., Ikeda-Saito, M., and Yonetani, T., 1982, *Nature* **298**:869.
Kleber, F., 1989, Diploma Thesis, Universität des Saarlandes, Germany.
Klein, C., 1989, Diploma Thesis, Universität des Saarlandes, Germany.
Kreiter, A., and Hüttermann, J., 1991, *J. Magn. Reson.* **93**:12.
Lieberman, R. A., Sands, R. H., and Fee, J. A., 1982, *J. Biol. Chem.* **257**:336.
Mackey, J. H., Kopp, M., Tynan, E. C., and Yen, T. F., 1960, in *Electron Spin Resonance of Metal Complexes* (T. F. Yen, ed.), Plenum, New York, pp. 33ff.
Malmström, B. G., and Vänngard, T., 1960, *J. Mol. Biol.* **2**:118.
McCord, J. M., and Fridovich, L., 1969, *J. Biol. Chem.* **244**:6049.
Mims, W. B., and Peisach, J., 1978, *J. Chem. Phys.* **69**:4921.
Möbius, K., and Lubitz, W., 1987, in *Biological Magnetic Resonance*, Vol. 7 (L. J. Berliner and J. Reuben, eds.), Plenum, New York, pp. 129–247.

Morse, H. R., and Chen, S. I., 1980, *J. Biol. Chem.* **225**:7876.
Nitschke, W., 1982, Diploma Thesis, University of Regensburg, Germany.
Ohlmann, J., 1992, Diploma Thesis, Universität des Saarlandes, Germany.
Perutz, M. F., Fermi, G., Luisi, B., Shaanan, B., and Liddington, R. C., 1987, in *Cold Spring Harbor Symposia on Quantitative Biology*, Vol. 52, p. 555.
Reinhard, H., 1992, Diploma thesis, Universität des Saarlandes, Germany.
Rist, G., and Hyde, J., 1970, *J. Chem. Phys.* **52**:4633.
Rotilio, G., Morpurgo, L., Giovagnoli, C., Calabrese, L., and Mondovi, B., 1972, *Biochemistry* **11**:2187.
Scheidt, W. R., and Frisse, M. J., 1975, *J. Am. Chem. Soc.* **97**:17.
Scholl, H.-J., and Hüttermann, J., 1992, *J. Phys. Chem.* **96**:9684.
Schweiger, A., 1982, *Struct. Bonding (Berlin)* **51**.
Seidel, H., 1961, *Z. Physik* **165**:218.
Smith, T. D., and Pilbrow, J. R., 1981, *Coordination Chem. Rev.* **39**:295.
Tainer, J. A., Getzoff, E. D., Beem, K. M., Richardson, J. S., and Richardson, D. C., 1982, *J. Mol. Biol.* **160**:181.
Tainer, J. A., Getzoff, E. D., Richardson, J. S., and Richardson, D. C., 1983, *Nature* **306**:284.
Van Camp, H. L., Sands, R. H., and Fee, J. A., 1981, *J. Chem. Phys.* **75**:2098.
Van Camp, H. L., Sands, R. H., and Fee, J. A., 1982, *Biochim. Biophys. Acta* **704**:75.
Walker, F. A., and Bowen, J., 1985, *J. Am. Chem. Soc.* **107**:7632.
Yordanov, N. D., Zdravkova, M., and Shopov, D., 1986, *Chem. Phys. Lett.* **127**:487.

6

High-Field EPR and ENDOR on Bioorganic Systems

Klaus Möbius

1. INTRODUCTION

Over the last 10 to 20 years NMR spectroscopy has advanced tremendously by employing sophisticated pulsed radio-frequency irradiation schemes and by exposing the sample to the strong magnetic fields attainable in superconducting magnets. Pulsing the rf field led to Fourier transform and two-dimensional correlation NMR; increasing B_0 and the rf frequency led to improved spectral resolution and sensitivity. Both developments were of decisive importance for successful applications of NMR to biological systems (Ernst *et al.*, 1987; Wüthrich, 1986).

Thanks to more recent improvements in microwave technology, similar trends to pulsed irradiation and high fields and frequencies can also be observed in EPR spectroscopy. In spite of these trends, most commercial EPR spectrometers are expected to adhere to *X*-band microwave frequencies (9.5 GHz, corresponding to a wavelength of 3 cm and, for $g = 2$, to a B_0 field of 0.34 T), since they represent a reasonable compromise between spectroscopic considerations (resolution, sensitivity) and practical ones (equipment costs, sample handling). Lower microwave frequencies (e.g., *S* band, 2–4 GHz) or higher ones (e.g., *Q* band, 34–36 GHz) are much less common in EPR spectrometers. For many bioorganic systems, however,

Klaus Möbius • Department of Physics, Free University of Berlin, D-1000 Berlin 33, Germany.

Biological Magnetic Resonance, Volume 13: EMR of Paramagnetic Molecules, edited by Lawrence J. Berliner and Jacques Reuben. Plenum Press, New York, 1993.

X-band EPR often runs into problems with spectral resolution because, for instance, the sample contains several radical species or different magnetic "sites" of rather similar *g* values or because a small *g*-factor anisotropy prevents clear features of the powder spectrum of disordered samples from being observed.

By analogy with modern NMR spectroscopy, these problems can, in principle, be overcome by applying high magnetic fields. For many systems, the increased Zeeman interaction ultimately leads to a complete separation of spectral features belonging to different *g* values. It is primarily this improved Zeeman magnetoselection which makes high-field EPR—and, even more, high-field ENDOR—so attractive when studying bioorganic systems. For such systems it may be an additional incentive that high-field EPR at correspondingly high frequencies is distinguished by the large filling factor of the microwave resonator. This leads to excellent detection sensitivity even for small single crystals.

High-field EPR can even become compulsory when high-spin transition metal complexes with large zero-field splittings are investigated. *X*- or *Q*-band frequencies are often too low for driving EPR transitions in such systems. Examples of such situations include some transition metal complexes of biological interest.

Beyond this, there are other, more physically or chemically motivated applications deriving benefit from high-field EPR. Rather than trying to be encyclopedic in coverage, this Chapter is primarily intended to report on the high-field EPR/ENDOR activities in Berlin. Recent work by other groups of the high-field/high-frequency community will only be reviewed, without going into detail. In the next section, the basic considerations in favor of high-field EPR/ENDOR are summarized, and an overview of applications by various groups is given.

2. BASIC PRINCIPLES AND OVERVIEW

The magnitude of the magnetic field at the center of an EPR spectrum is given by $B_0 = h\nu/g\mu_B$ (ν is the microwave frequency, g the g factor, h and μ_B the Planck constant and Bohr magneton, respectively). Consequently, if paramagnetic species with different g factors or with g-factor anisotropy are present, the difference in resonance field positions ΔB_0 is proportional to the frequency of electromagnetic irradiation:

$$\Delta B_0 = \frac{h\nu}{\mu_B}\left(\frac{1}{g_1} - \frac{1}{g_2}\right)$$

which, in principle, shows that higher frequencies and B_0 should lead to enhanced spectral resolution.

Except for transition metal complexes, for many bioorganic systems $g = 2$ and relative g variations $\Delta g/g$ rarely exceed 10^{-4}–10^{-3}. At X-band frequencies, therefore, the line separation due to g-value differences is only $\Delta B_0 = 0.03$–0.3 mT, which can easily be masked in solid-state samples with typical linewidths around 1 mT. An increase of microwave frequency by, for instance, a factor of 10 (W-band, 95 GHz) improves the spectral resolution accordingly, provided the linewidths—both their homogeneous and their inhomogeneous contributions—do not increase considerably with increasing B_0. As an example, Fig. 1 compares the EPR spectra of the benzosemiquinone radical anion in frozen solution at X band and W band. In contrast to the unresolved spectrum at X band, from which the g and hyperfine parameters can be extracted only by extensive computer deconvolution techniques, the W-band spectrum exhibits well-separated g-tensor components with resolved proton hyperfine structure on the g_{xx} and g_{zz} peaks. In this example, the linewidth does not significantly change between the two frequency bands.

The situation might be different in other systems, and a brief consideration of the sources of line broadening is appropriate. The homogeneous linewidth is determined by spin–lattice and phase relaxation mechanisms originating in a modulation of the relevant interaction terms in the spin Hamiltonian. The respective contributions to the relaxation rates are either field-independent (e.g., spin-rotation and anisotropic hyperfine interactions) or field-dependent [e.g., Zeeman interaction with anisotropic g tensor ($\propto B_0^2$), combined g and hyperfine anisotropic interaction ($\propto B_0$), a direct process of spin–lattice relaxation by spin–phonon interactions modulating the ligand field ($\propto B_0^4$) (Abragam and Bleaney, 1970)]. Also, the inhomogeneous linewidth, which is determined by unresolved substructures, can be either field-independent (e.g., fine structure for $S > \frac{1}{2}$ systems, site

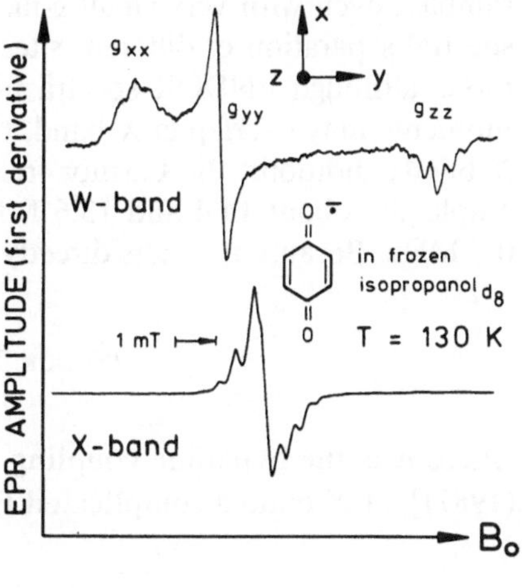

Figure 1. Comparison of the resolution between W-band (top) and X-band (bottom) EPR spectra of the benzosemiquinone anion radical in frozen perdeuterated isopropanol solution (Burghaus, 1991).

variations of the hyperfine tensors or unresolved isotropic and anisotropic hyperfine couplings) or field-dependent (e.g., distribution of g values arising from inhomogeneities in the local environments of individual spin probes (g strain, $\propto B_0$). In particular, for metalloproteins, such as low-spin ferric hemoprotein cytochrome c and iron–sulfur protein ferredoxin, the g strain broadening already dominates at X-band frequencies (Hagen, 1989).

Generally, the field dependence of the overall linewidth is difficult to predict theoretically and, for a specific project, experiments may be required to settle the dispute as to whether or not high-field EPR should be employed. Fortunately, with respect to resolving power, no prohibitive line broadening was observed up to 250 GHz for many classes of radicals both in liquid and rigid media, e.g., nitroxide and peroxide radicals (Dubinskii *et al.*, 1981; Tsvetkov *et al.*, 1987; Budil *et al.*, 1989), semiquinone radical anions (Burghaus, 1991; Gulin *et al.*, 1991a; Burghaus *et al.*, 1992b), and porphyrin- and chlorophyll-type radical cations (Burghaus, 1991; Burghaus *et al.*, 1991; Prisner *et al.*, 1992a). And even for metalloproteins, where g-strain line broadening might prevent high resolution at high fields, the onset of the slow-motion regime at high microwave frequencies can be very informative about molecular dynamics.

An important aspect of high-field/high-frequency EPR is the option to transform motionally narrowed spectra ($\omega_0 \tau_R \ll 1$) into the slow-motion regime ($\omega_0 \tau_R \approx 1$), with its increased sensitivity to line-shape effects by motional dynamics (ω_0 is the Zeeman angular frequency and τ_R is the correlation time of the motion). Nitroxide radicals are typical examples for this option when used as spin probes for the structure and dynamics of active sites in enzymes.

Double resonance extension to high-field ENDOR, for which the pioneering work was done by the Berlin group, has the additional advantage of allowing single crystal-like spectra to be taken from disordered solid samples, even with very small g-factor anisotropies. It also allows complete spectral separation of different sites with small g-value differences. Furthermore, although ENDOR spectra from nuclei with rather similar magnetic moments may overlap at X band, they can be separated at W band. Under X-band conditions, the Larmor frequencies ν_n of ^1H and ^{19}F nuclei, for example, are about 14.4 and 13.6 MHz, respectively; i.e., they differ by only 0.8 MHz. Because ν_n enters directly the ENDOR resonance condition

$$\nu_{\mathrm{ENDOR}}^{\pm} = |\nu_n \pm a/2|$$

where a is the hyperfine coupling constant (hfc) [see Möbius and Lubitz (1987)], at X band a complicated ENDOR pattern with intermixed ^1H and

^{19}F lines may occur. At *W* band, however, the corresponding difference in Larmor frequencies increases to 8 MHz, and in most cases the hyperfine lines of the two sets of nuclei become disentangled. At any rate, complex ENDOR spectra with large and small hfc's tend to become easier to analyze at high fields because, even for large hfc's, the lines remain symmetrically displayed around ν_n.

With these advantages in mind, continuous-wave (cw) high-field EPR spectrometers have been constructed in several laboratories, among which that of Lebedev in Moscow has to be acknowledged first. These spectrometers operate, for instance, at 70 GHz ($\lambda \approx 4$ mm) (Box *et al.*, 1979), at 95 GHz ($\lambda \approx 3$ mm) (Haindl *et al.*, 1985; Burghaus *et al.*, 1992a; Weber *et al.*, 1989; Clarkson *et al.*, 1990), at 150 GHz ($\lambda \approx 2$ mm) (Grinberg *et al.*, 1976; Lebedev, 1990), at 250 GHz ($\lambda \approx 1$ mm) (Lynch *et al.*, 1988, Budil *et al.*, 1989) and, more recently, even in the submillimeter region (Muller *et al.*, 1989; Barra *et al.*, 1990; Tarasov and Shakurov, 1991). So far, the extension to high-field ENDOR has been established up to the 3-mm band in the cw mode of operation (Burghaus *et al.*, 1988, 1992a). Most recently, high-field EPR spectrometers with pulsed microwave irradiation have also been described, operating at 3-mm (Allgeier *et al.*, 1990) and 2-mm wavelengths [Prisner *et al.* (1992b) and Bresgunov *et al.* (1991), respectively].

The community of millimeter-wave EPR and ENDOR spectroscopists is—fortunately—still very compact, consisting of fewer than ten groups, which till the ground in this promising research field. Recent overview articles from these groups stress quite different applications of high-field spectroscopy: General aspects of high-field EPR together with representative examples are presented by Lebedev (1990, 1991), Belford *et al.* (1987), and Clarkson *et al.* (1990). Instrumental details of their millimeter-wave spectrometers are given, together with the physical considerations behind their construction strategy, for instance by the groups of Lebedev in Moscow (Grinberg *et al.*, 1976, 1983; Lebedev, 1990), Freed in Ithaca (Lynch *et al.*, 1988; Budil *et al.*, 1989), Schmidt in Leiden (Weber *et al.*, 1989; Allgeier *et al.*, 1990), Möbius in Berlin (Burghaus *et al.*, 1988, 1992a), and Brunel in Grenoble (Muller *et al.*, 1989). Specific applications of high-field EPR to bioorganic systems are described in considerable detail, for example by Lebedev (1990), Budil *et al.* (1989), Tsvetkov (1989); Tsvetkov *et al.* (1987), Krinichnyi (1991), and Burghaus *et al.* (1992b).

Lebedev (1990), after summarizing the physical and technical aspects of cw 2-mm EPR, addresses structural studies on mono- and biradical systems. Systematic measurements of the principal values of the *g* and hyperfine (*A*) tensors of about a hundred nitroxide radicals in solid solutions have been carried out. In cyclic nitroxides, in which the nitrogen atom is bonded to two carbon ring atoms, the unpaired electron is mainly localized on the $>\dot{N}—O$ fragment of the radical. Such nitroxides are most commonly used as "spin labels" (Berliner, 1976, 1979) probing the immediate environment of the site of a biological system to which they are attached (Likhtenstein, 1988). For this purpose, the *g*- and *A*-tensor parameters of the isolated nitroxide radicals are needed as *a priori* knowledge. Variations in radical structure appear to affect mostly the g_{xx} tensor component. This result was interpreted semiquantitatively on the basis of accepted theoretical *g*-factor models for organic radicals. The sensitivity of g_{xx} to environmental effects was exploited to study solvent-solute intermolecular interactions of a large number of nitroxide radicals in various media.

The series of nitroxide radicals characterized by 2-mm EPR was considerably extended by Tsvetkov and co-workers (Tsvetkov *et al.*, 1987; Gulin *et al.*, 1991b), who measured *g*- and hyperfine (^{14}N, ^{19}F) tensor principal values of many imidazoline nitroxide radicals and their derivatives in frozen solution.

Stepping momentarily aside from biosystems, an interesting new application of high-field EPR to spin-labeled organic polymers should also be mentioned: Pelekh *et al.* (1991) studied semiconductive polypyrrole material modified by nitroxide radicals as spin probes in order to elucidate structural, conformational, and electrodynamic characteristics. Thanks to the high spectral resolution of 2-mm EPR, the distance between the nitroxide radicals and spinless charge carriers on the polymer chain could be determined, as could the effective dipole moment and size of the charge carrier and their mobility.

Peroxide radicals of biological importance were also studied by 2-mm EPR (Lebedev, 1990). Peroxide radicals are formed as both intermediate and final products in metabolic and radiation-induced reactions in biological systems. The unpaired electron in such centers is mainly localized on the $—O—O^{\bullet}$ fragment; i.e., there is only very weak hyperfine interaction in these centers and their *g* factors are very close together. Consequently, the identification of peroxide radicals by *X*-band EPR is difficult, but it could be achieved for many cases at 150 GHz.

Lebedev (1990) also reports on dynamic studies of molecular motion, such as rotation, reorientation, and libration. He stresses the point that by varying the Zeeman frequency from centimeter to millimeter bands, broad ranges of correlation times become accessible in both the slow- and fast-motion regimes; i.e., large temperature ranges can be studied. Due to the enhanced Zeeman magnetoselection, high-field EPR can be used to study slow-motion line-broadening effects of individual canonical features in the spectrum. In this way, the three principal components of the reorientation tensor for anisotropic rotation of nonspherical spin probes could be determined, whereas at *X*-band, often, only an average correlation time can be measured.

In continuation of earlier work (Lebedev, 1990) slow-motion 2-mm EPR spectra of nitroxide spin labels have recently been reanalyzed theoretically in terms of anisotropic rotations of the spin probes (Antsiferova and Lyubashevskaya, 1991). This work shows how the rotational anisotropy can be extracted from the relaxation behavior of the three canonical components of the powder spectrum, i.e., how to evaluate the anisotropic correlation times on the basis of the spectral parameters.

The increased spectral sensitivity of high-field EPR to motional dynamics is also stressed by Budil *et al.* (1989). They studied, for example, nitroxide spin probes in fluid isotropic solvents and liquid crystals by 1-mm EPR. In the case of toluene as the solvent, their spectra show characteristic features of the fast- to slow-motion regimes when varying the temperature between 290 and 190 K, thereby covering a range of correlation times τ_R from about 1 to 100 ps. The 1-mm spectra for the fast-motion regime ($\tau_R < 100$ ps) are found to be much more sensitive to motion effects than the corresponding 3-cm spectra. Already for $\tau_R \geq 100$ ps the slow-motion regime sets in in the 1-mm spectra. This is an order of magnitude faster than in the corresponding 3-cm spectra. Thus, comparing EPR spectra at different Zeeman frequencies allows motional spin relaxation studies to be pursued over wide ranges of temperatures.

This overview section is concluded by alluding to the review article by Krinichnyi (1991), which covers 2-mm EPR investigations on a variety of biological systems. It summarizes the advantages of high-field EPR when studying structural, conformational, and dynamic characteristics of spin probes. Spin-labeled human serum albumin, egg lysozyme, liposome membranes, inverted micelles, α-chymotrypsin, and natural biopolymers are given as examples. Besides tabulating the magnetic parameters of the immobilized spin probes, special attention is given to the determination of their microenvironment and molecular mobility.

3. HIGH-FIELD EPR/ENDOR IN PHOTOSYNTHESIS RESEARCH

Photosynthetic donor and acceptor pigments and their biomimetic model systems are anticipated to represent a particularly promising area of high-field EPR and ENDOR spectroscopy. Their molecular structure is so complex that *X*-band EPR/ENDOR spectra often suffer from line congestions with severe overlap. To enhance the resolution capability of our magnetic resonance instrumentation, we have built a 3-mm (*W*-band) EPR/ENDOR spectrometer over the last 10 years (Haindl *et al.*, 1985; Burghaus, 1991; Burghaus *et al.*, 1992a). The performance specifications achieved to date are well suited for studies of systems related to photosynthesis both in fluid or frozen solutions and in single crystals. In this section, the 3-mm spectrometer at Berlin will be described briefly and examples of recent applications to acceptor- and donor-ion radicals in photosynthesis will be presented.

3.1. The Berlin 3-mm EPR/ENDOR Spectrometer

Details of the *W*-band spectrometer are given by Burghaus *et al.* (1992a). To illustrate the operating principles, Fig. 2 shows a simplified block diagram of the computer-controlled EPR/ENDOR system. The microwave source is either a low-noise Gunn oscillator (6 mW output power) or a klystron (300 mW). The static magnetic field is generated by a cryomagnet (maximum field 6 T) with a room-temperature bore of 88-mm diameter. This bore is large enough to accommodate the EPR probe head together with additional components, such as field modulation and ENDOR coils, light pipe, and goniometer. The field homogeneity over a sample capillary of 1-mm diameter and 10-mm length is better than 0.01 mT, the long-term field stability is better than 0.001 mT/min, and the short-term stability is 10^{-4} mT at a detection time constant of 1 s. The field can be linearly varied with time at a sweep rate from 0 to 50 mT/min full range (linearity better than 10^{-3}). Field modulation up to 4 mT (peak-to-peak) at 1 kHz can be applied; for ENDOR, additional frequency modulation (20 kHz) and field-frequency locking are employed. To minimize attenuation losses at the *W*-band frequencies, oversized waveguides are used between the microwave bridge and the resonator. The resonator incorporating the sample is either of the Fabry-Perot type, with two concave mirrors, or a cylindrical TE_{011} cavity. Both resonators operate in the reflection mode with variable iris coupling to the waveguide (see Fig. 3). The microwave bridge is completed by a circulator and a helium-cooled hot-electron InSb bolometer as the detector. A convection-type cooling system provides stable sample temperatures between 30 and 300 K. Liquid or solid samples are studied in thin-walled quartz capillaries; single crystals can be rotated about three orthogonal axes. The EPR sensitivity of the spectrometer is estimated to be 2×10^7 spins for nonlossy samples with an unsaturated linewidth of 0.1 mT at 30 mW incident microwave power and 1 s time constant of the detection channel.

For ENDOR spectroscopy an external NMR coil is added with two loops if the Fabry–Perot resonator is used or four if the TE_{011} cavity is used (see Fig. 3); the coil is energized by a 500-W rf amplifier. With a tuned rf resonance circuit, a field of up to 2.5 mT B_2 (rotating frame) can be achieved at the sample site. The probe head is arranged in such a way that the external field $\mathbf{B}_0$, microwave field $\mathbf{B}_1$, and rf field $\mathbf{B}_2$ are mutually orthogonal.

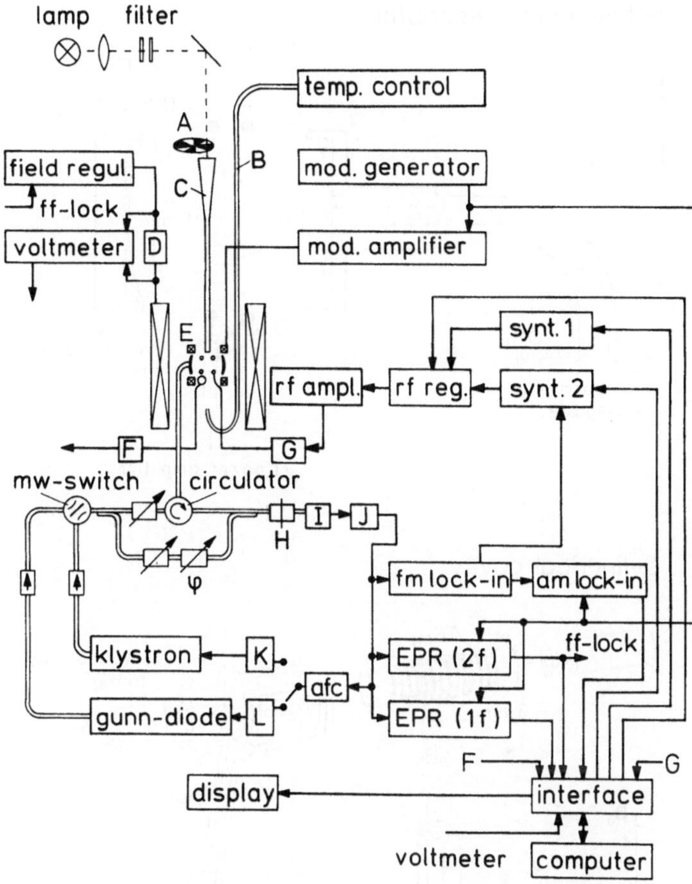

Figure 2. Simplified block diagram of the Berlin *W*-band EPR spectrometer with ENDOR and TRIPLE resonance capability (Möbius and Biehl, 1979). Essential components not obvious from the inscriptions are A, light chopper blade; B, Dewar system for sample cooling; C, quartz light pipe; D, normal resistor for field regulation of the cryomagnet; E, Fabry–Perot resonator with ENDOR and field-modulation coils; F, rf rectifier unit connected to the pickup coil for measuring and regulating the rf field strengths in ENDOR/TRIPLE experiments; G, rf power meter with interface; H, dc block; I, InSb bolometer for microwave detection; J, low-noise preamplifier; K and L, klystron and Gunn-diode power supply with regulator, respectively; rf reg, digital attenuator and rf mixer for computer-controlled ENDOR (synthesizer 2) or (synthesizers 1 and 2) TRIPLE resonance experiments. (Adapted from Burghaus *et al.*, 1992a.)

The *g* factors and hyperfine coupling constants are measured by simultaneously recording the EPR spectrum of a powder sample of Mn^{2+} (0.02% in MgO) whose magnetic parameters were previously determined at *X* band with high precision. By this method, field positions can be measured with an accuracy of better than ±0.005 mT, and the absolute *g*-factor error is not larger than $\pm 1 \times 10^{-5}$.

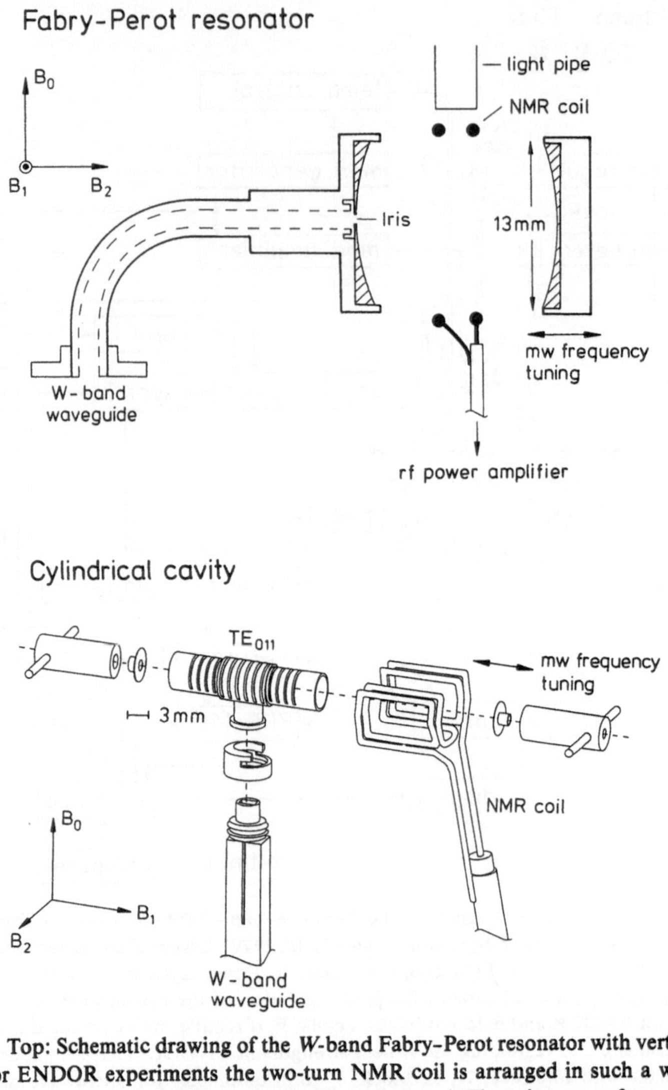

Figure 3. Top: Schematic drawing of the *W*-band Fabry-Perot resonator with vertical sample access. For ENDOR experiments the two-turn NMR coil is arranged in such a way that the Zeeman, microwave, and rf fields are mutually orthogonal. The microwave frequency tuning is achieved by changing the distance of the two mirrors. The microwave coupling to the *W*-band waveguide is achieved by moving a ceramic wedge immediately behind the iris of the mirror. For more details, see Burghaus *et al.* (1992a). Bottom: Exploded view of the ENDOR version of the TE_{011} cylindrical *W*-band cavity with horizontal sample access. The four-turn NMR coil provides the $\mathbf{B}_2$ perpendicular to $\mathbf{B}_0$ and $\mathbf{B}_1$. The microwave frequency can be varied by changing the cavity length with tuning pistons. To avoid rf-induced eddy currents, the pistons are made of ceramic with gold-plated end faces and the cavity wall is slotted. The microwave coupling is achieved by moving a teflon needle with a silver-plated top vertically in the waveguide toward an iris in the cavity wall. For more details, see Burghaus *et al.* (1992a).

To provide an impression of the overall performance of the spectrometer, Fig. 4 shows 3-mm EPR and ENDOR spectra of a single crystal of DANO (di-*p*-anisyl-nitroxide 1% in 4,4-di-methoxybenzophenone) with well-resolved structure from ^1H and ^{14}N dipolar and quadrupolar hyperfine interactions. There are several magnetically inequivalent sites in the crystal and, by varying the B_0 resonance field among various parts of the EPR spectrum at a given crystal orientation, selection of different sites with different hyperfine interactions can be achieved. From the ENDOR line positions ν_i^{ENDOR}, the dipolar and quadrupolar hyperfine couplings A_i and Q_i can be deduced according to

$$\nu_i^{\mathrm{ENDOR}} = \left| \tfrac{1}{2}A_i \pm \nu_n \pm Q_i \right|$$

Both hyperfine interaction parameters provide detailed information about the electronic structure and bond characteristics of the radical.

3.2. Semiquinone Anion Radicals as Models for Acceptor Pigments in Photosynthesis

Quinones play an important role as acceptor molecules in fundamental biological processes such as photosynthesis, where they are involved in electron transfer and proton translocation. To understand these processes

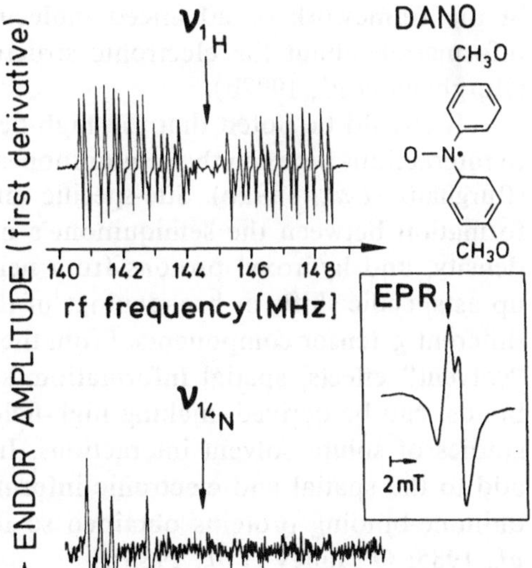

Figure 4. *W*-band EPR and ENDOR spectra of ^1H (top) and ^{14}N (bottom) nuclei in a magnetically dilute single crystal of DANO (2% di-*p*-anisyl-nitroxide in 4,4-dimethoxy-benzophenone) with crystal axis $a \| \mathbf{B}_0$. The Zeeman field at which ENDOR is performed is indicated by an arrow at the EPR spectrum (Burghaus, 1991).

from first principles, a knowledge of both the spatial and electronic structure of the donor (see Section 3.3) and acceptor reactants is essential. In specific transfer steps, the paramagnetic semiquinone form gets stabilized and becomes detectable by EPR and ENDOR methods. Examples include the ubisemiquinone in reaction centers of photosynthetic bacteria (Feher *et al.*, 1985), the plastosemiquinone in plant photosystem II (Klimov *et al.*, 1980) and the phyllosemiquinone (Vitamin K_1) in photosystem I (Thurnauer and Gast, 1985; Thurnauer *et al.*, 1987, 1989). Since these systems are largely immobilized in the biopolymer, their EPR spectra suffer from severe line-broadening effects. Therefore, in most cases (Hales, 1975, 1976a, b; Hales and Case, 1981) only broad Gaussian envelope lines were obtained at *X*- and even *Q*-band frequencies.

In an attempt to separate the spectral regions from different *g*-factor contributions, 3-mm high-field EPR and ENDOR spectra of frozen solutions of biologically important semiquinone radical anions and their simpler model systems have recently been recorded (Burghaus, 1991; Burghaus *et al.*, 1992b). Examples of the quinones studied are given in Fig. 5. As a typical representative, the *W*-band EPR spectrum of the phyllosemiquinone radical anion in perdeuterated isopropanol is shown in Fig. 6. The spectral features from the *g*-tensor components are, indeed, well resolved, and from least-squares fitting to a theoretical spectrum, the principal values of the *g* tensor could be determined with remarkably high precision [$g_{xx} = 2.00564\,(5)$, $g_{yy} = 2.00494\,(5)$, $g_{zz} = 2.00217\,(5)$]. Interpretation of the data in the framework of advanced molecular orbital theory gives valuable information about the electronic structure of these important molecules (Burghaus *et al.*, 1992b).

It should be noted that the high-field EPR spectra are very sensitive to interactions between the semiquinones and their immediate environment (Burghaus *et al.*, 1992b). Site-specific interactions, such as hydrogen-bond formation between the semiquinone's oxygen, which carries a large spin density, and hydroxyl protons from neighboring matrix molecules, show up as specific shifts or broadenings of the spectral peaks belonging to the different *g*-tensor components. From the anisotropic contributions to these "solvent" effects, spatial information, such as the direction of hydrogen bonds, can be derived, making high-field EPR a useful tool for structural studies of solute–solvent interactions. In this respect, high-field EPR can add to the spatial and electronic information about hydrogen bonding to quinone-binding proteins obtained so far by *X*-band ENDOR (Feher *et al.*, 1985; O'Malley *et al.*, 1985).

In a recent paper, Tsvetkov and co-workers (Gulin *et al.*, 1991a) reported the use of 2-mm EPR spectroscopy to study the adsorption of *p*-quinones on the activated surface of zeolites. The adsorption of *p*-benzoquinone, chloranil, and fluoranil is found to result in the formation

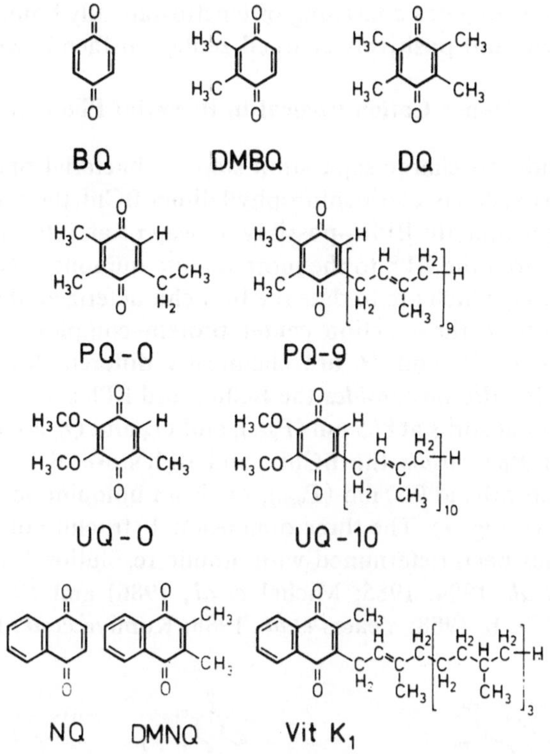

Figure 5. Molecular structures of quinones related to photosynthesis and their model quinones as studied by *W*-band EPR (Burghaus, 1991; Burghaus *et al.*, 1992b).

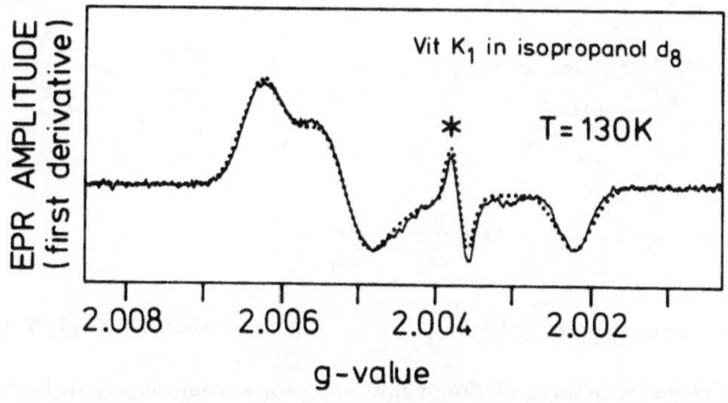

Figure 6. *W*-band EPR spectrum of a frozen solution of phyllosemiquinone (Vitamin K_1) anion radical in perdeuterated isopropanol (* signal from Mn^{2+}/MgO reference sample). The dotted line represents the least-squares fit of a theoretical spectrum to the experimental one (Burghaus, 1991; Burghaus *et al.*, 1992b).

of paramagnetic aggregates consisting of electrostatically bound semiquinone anion radicals and positively charged zeolite surface Lewis sites.

3.3. The Primary Donor Cation Radical in Bacterial Photosynthesis

The light-induced charge-separation steps in bacterial photosynthesis proceed from an excited bacteriochlorophyll dimer BChl, the primary donor *P*, to a bacteriopheophytin BPh—possibly by way of an intermediate BChl monomer—and from the BPh to the more remote quinones Q_A and Q_B. In the purple bacteria, which are to date the best characterized, the donor and acceptor pigments in the reaction center protein complex, consisting of protein subunits *L*, *M*, and *H*, are chemically different for the various organisms; e.g., for *Rb. sphaeroides* the BChl's and BPh's are of the *a* type, the primary donor absorbs at 865 nm (P_{865}), and Q_A and Q_B are ubiquinones (see Fig. 7); for *Rps. viridis* the BChl's and BPh's are of the *b* type, the primary donor absorbs at 960 nm (P_{960}), Q_A is an ubiquinone, and Q_B is a menaquinone (see Fig. 7). The three-dimensional structure of the reaction centers (RCs) has been determined with atomic resolution for *Rps. viridis* (Deisenhofer *et al.*, 1984, 1985; Michel *et al.*, 1986) and *Rb. sphaeroides* (Allen *et al.*, 1987a, b, 1988; Yeates *et al.*, 1988; Komiya *et al.*, 1988; Chang

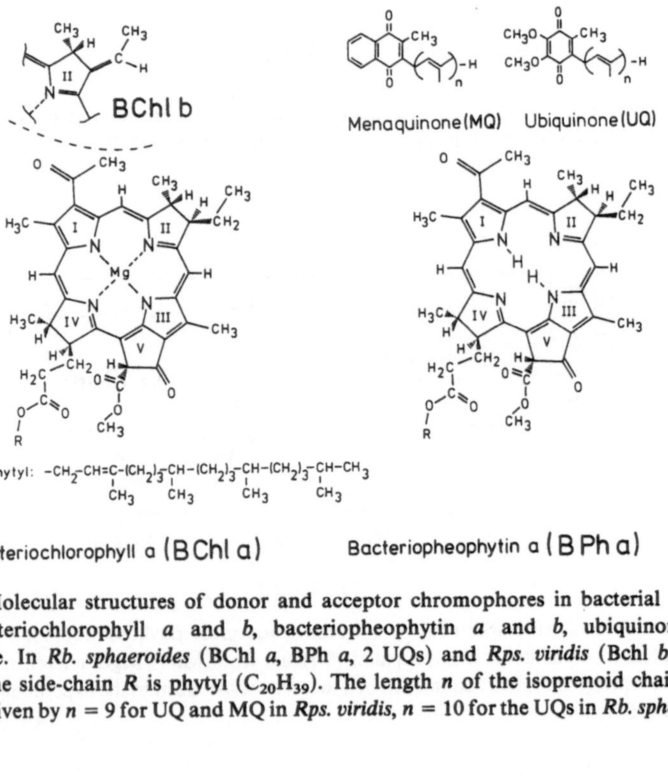

Figure 7. Molecular structures of donor and acceptor chromophores in bacterial reaction centers: bacteriochlorophyll *a* and *b*, bacteriopheophytin *a* and *b*, ubiquinone, and menaquinone. In *Rb. sphaeroides* (BChl *a*, BPh *a*, 2 UQs) and *Rps. viridis* (Bchl *b*, BPh *b*, UQ, MQ), the side-chain *R* is phytyl ($C_{20}H_{39}$). The length *n* of the isoprenoid chain of the quinones is given by $n = 9$ for UQ and MQ in *Rps. viridis*, $n = 10$ for the UQs in *Rb. sphaeroides*.

et al., 1991; El-Kabbani *et al.*, 1991) by *X*-ray diffraction. The cofactor arrangement in the RCs of both organisms is distinguished by an approximate C_2 symmetry axis perpendicular to the membrane plane. The local symmetry of this arrangement can be estimated from Fig. 8 for the RC of *Rb. sphaeroides R*-26 (a mutant lacking carotenoid). This figure, however, already indicates a broken symmetry for the electron transfer time constants, which are much shorter along the *L*-protein branch than along the *M* branch (see below).

High-resolution EPR and ENDOR spectroscopy together with advanced MO methods can supplement the structural information about the heavy-atom coordinates obtained from *X*-ray diffraction, since the positions of light atoms, such as hydrogens, become available and, furthermore, the distribution of the frontier electrons in the doublet ground and triplet excited states of the reactants can be determined from hyperfine

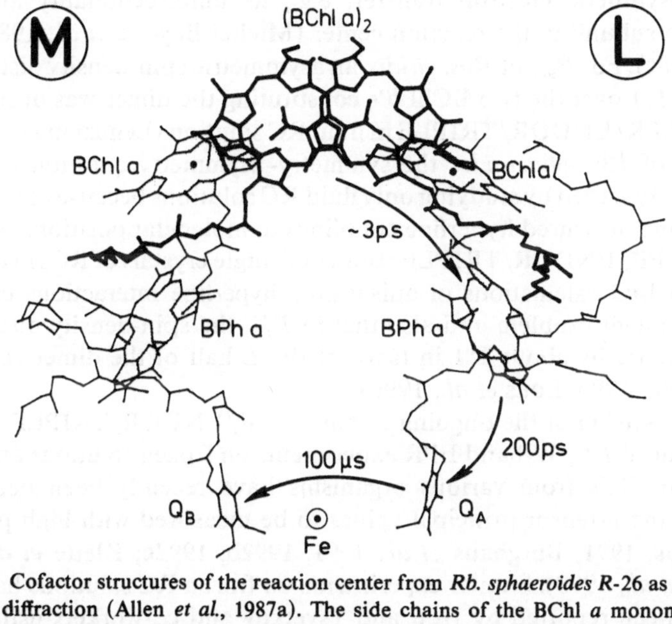

Figure 8. Cofactor structures of the reaction center from *Rb. sphaeroides R*-26 as determined by *X*-ray diffraction (Allen *et al.*, 1987a). The side chains of the BChl *a* monomers and of the quinones Q_A and Q_B have been partly truncated in the figure for better clarity. An approximate two-fold symmetry axis is aligned vertically in the plane of the paper and runs through the primary donor (BChl $a)_2$ dimer *P* near the periplasmic side of the membrane at the top and the iron atom Fe near the cytoplasmic side at the bottom of the membrane. The protein subunits *L* and *M* (not shown) each contain five transmembrane helices and several other helices that do not span the membrane (Allen *et al.*, 1987b). The *H* subunit (not shown) is located mainly near the cytoplasmic side. The light-induced electron transfer proceeds preferentially along the pigments of the *L* branch; i.e., the electron transfer along the *M* branch is much slower, with an *L*/*M* ratio for the time constants of more than 10 (Michel-Beyerle *et al.*, 1988).

interactions. All these pieces of information are needed to improve our understanding of the major processes in photosynthesis on the molecular level.

The spatial and electronic structures of the respective primary donors P_{865} and P_{960} are of particular interest because of their specialized function in initiating the vector process of primary charge separation. Not surprisingly, therefore, *X*-band EPR, ENDOR, and electron-nuclear-nuclear TRIPLE resonance (Möbius and Biehl, 1979) were—and still are—intensively used to determine the unpaired electron spin density distribution in the primary donor cation radicals $P_{865}^{+\cdot}$ and $P_{960}^{+\cdot}$, in remarkable detail (Lubitz *et al.*, 1985; Möbius and Lubitz, 1987; Lendzian *et al.*, 1988, 1990, 1992; Plato *et al.*, 1988a,b, 1991; Lous *et al.*, 1990; Lubitz, 1991).

The main motivation for our studies is to find an answer to the question of whether the symmetry, or asymmetry, of the electron distribution in the dimer halves of *P* is an important factor in controlling the vector properties of photosynthetic electron transfer, e.g., its unidirectionality along the *L*-protein subunit in the reaction center (Michel-Beyerle *et al.*, 1988; Plato *et al.*, 1988c). For $P_{960}^{+\cdot}$ of ***Rps. viridis***, an asymmetric spin-density distribution of about 2:1 over the two BChl *b*'s constituting the dimer was observed by *X*-band EPR/ENDOR/TRIPLE in fluid RC solution (Lendzian *et al.*, 1988). For $P_{865}^{+\cdot}$ of ***Rb. sphaeroides*** the symmetry-asymmetry question could not be definitely settled by studying only fluid RC solutions because of ambiguity in assigning measured hyperfine couplings to molecular positions. It turned out that EPR/ENDOR/TRIPLE studies of single crystals of RCs in conjunction with MO calculations of anisotropic hyperfine interactions can solve the assignment problem and also that in $P_{865}^{+\cdot}$ the spin-density distribution is asymmetric by about 2:1 in favor of the *L* half of the dimer (Lendzian *et al.*, 1990, 1993; Lous *et al.*, 1990).

In extension of the ongoing *X*-band EPR/ENDOR/TRIPLE work on light-induced $P^{+\cdot}$, *W*-band EPR experiments on frozen solutions and single crystals of RCs from various organisms have recently been performed, allowing the *g*-tensor principal values to be measured with high precision (Burghaus, 1991; Burghaus *et al.*, 1991, 1992b, 1992c; Klette *et al.*, 1992; Törring *et al.*, 1993). Similar experiments on frozen RC solutions have most recently been reported by Hoff and Tsvetkov and co-workers using 2-mm EPR (Gulin *et al.*, 1991c).

As an example of our *W*-band EPR work, the illuminated RC preparation of ***Rb. sphaeroides R***-26 with the Fe^{2+} replaced by Zn^{2+} was studied (Burghaus *et al.*, 1992c). The Zeeman magnetoselection at B_0 of 3.4 T was found to be high enough to allow simultaneous detection of the EPR spectra of the donor and acceptor ion radicals, $P_{865}^{+\cdot}$ and $Q_A^{-\cdot}$.

In another application of *W*-band EPR, the light-induced signal of $P_{865}^{+\cdot}$ in single-crystal RCs from ***Rb. sphaeroides R***-26 was studied with

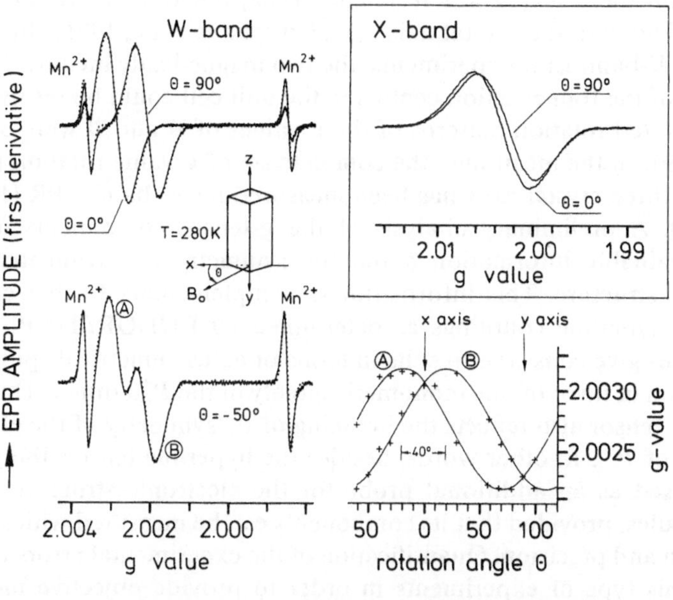

Figure 9. *W*-band and *X*-band (inset) EPR spectra of $P_{865}^{+\cdot}$ in a reaction center (RC) single crystal of *Rb. sphaeroides R*-26 for different orientations of $\mathbf{B}_0$ in the *x*–*y* crystal plane. Whereas at *X* band the line positions shift only slightly even between the extreme angular settings ($\theta = 0°$, $\theta = 90°$), the much larger shifts at *W* band can be precisely measured over the full angular range. The additional outer lines in the *W*-band spectra are hyperfine components of a Mn^{2+}/MgO powder sample serving as *g* factor and B_0 calibration standard. The experimental spectra were computer-simulated by first-derivative Gaussian lines (dotted) and the *g* values were obtained from the least-squares fits. (The differences between experimental and theoretical spectra are hardly discernible in the figure.) Top left: The settings $\theta = 0°$ and 90° correspond to the *x* and *y* C_2 crystal axes, along which all four RCs of the unit cell are magnetically equivalent (Allen *et al.*, 1987a). Bottom left: At $\theta = -50°$, the two magnetically inequivalent $P_{865}^{+\cdot}$ pairs (sites *A* and *B*) are best resolved in their EPR spectra. Bottom right: Rotation pattern of the *g* factor of the $P_{865}^{+\cdot}$ sites *A* and *B* in the *x*–*y* plane of the RC single crystal. The solid lines represent a least-squares fit of the experimental *g* values (+) to the theoretical function

$$g(\theta) = \tfrac{1}{2}(g_{xx} + g_{yy}) + [\tfrac{1}{4}(g_{xx} - g_{yy})^2 + g_{xy}^2]^{1/2} \cdot \cos 2(\theta - \Delta\theta)$$

with

$$\tan 2\Delta\theta = \frac{2g_{xy}}{(g_{xx} - g_{yy})}$$

The phase shift between sites *A* and *B*, $2\Delta\theta = 40°$, is marked in the figure. For further details and numerical values of the *g*-tensor components, see Burghaus *et al.* (1991) and Klette *et al.* (1992).

respect to its angular variation when the crystal was rotated about its crystal axes within the Fabry–Perot resonator. First, the RC was rotated in the Zeeman field with the crystal axis $z \| \mathbf{B}_0$ (Burghaus *et al.*, 1991). In contrast to earlier X-band EPR experiments, the two magnetically inequivalent sites A and B of the four reaction centers in the unit cell could be resolved, and well-separated rotation patterns of the g values of A and B were observed (see Fig. 9). In the meantime, the complete set of g-value rotation patterns about all three crystal axes has been measured by W-band EPR (Klette *et al.*, 1992). A preliminary analysis of the g-tensor components obtained yielded valuable information about the symmetry or asymmetry of the electronic structure. This information is complementary to that extracted from the hyperfine couplings as determined by ENDOR/TRIPLE. Both approaches give consistent results in terms of an asymmetrical spin-density distribution in favor of one monomeric moiety of the $P_{865}^{+\cdot}$ dimer. This means that the g tensor also reflects the breaking of C_2 symmetry of the electronic structure of $P^{+\cdot}$. In other words, besides the hyperfine tensor, the g tensor can be used as an additional probe for the electronic structure of large biomolecules, provided that its components can be measured with sufficient resolution and precision. Quantification of the experimental errors is important in this type of experiments in order to provide objective means for confidence levels of the determined parameters, particularly the orientations of the principal axes of the interaction tensors. Extension of this work to 3-mm ENDOR on RCs is in progress in Berlin.

4. PERSPECTIVES

High-field EPR and ENDOR on bioorganic systems are attractive principally for at least three reasons: improved spectral resolution, high detection sensitivity, and enhanced dynamic range for studies of molecular motion. Aside from practical considerations such as component performance and costs, the choice of an optimum frequency and field depends on the details of the spectrum under study. If, for instance, resolving power is of primary interest, the highest available Zeeman field should be employed until field inhomogeneities and field-dependent line broadenings set an upper limit. For many bioorganic systems, g-strain contributions to the inhomogeneous linewidth will ultimately become dominant, and further increases in field will not enhance EPR resolution. In this regard, 1–3-mm EPR represents a compromise between aspiration and reality.

High-field EPR at a wavelength of a few millimeters also seems to be a realistic compromise with respect to detection sensitivity. In principle, the sensitivity should increase significantly with increasing Zeeman frequency ω_0; in practice, however, existing high-field EPR spectrometers still fail to reach the theoretical limit (Burghaus *et al.*, 1992a). The minimum detectable number of spins $N_{\min}$ depends on ω_0; the sample volume; the temperature; the Q value and microwave coupling of the resonator; its filling factor; the EPR linewidth and field-modulation amplitude; the noise figures and insertion losses of the microwave source, components, and detector; the noise figure of the preamplifier; and the effective bandwidth of the lock-in amplifier. Under the (unrealistic) assumption that these noise figures are not deteriorating at very high microwave frequencies, for small samples of fixed size such as reaction-center single crystals, $N_{\min} \propto \omega_0^{-9/2}$ under otherwise constant experimental conditions (Feher, 1957; Poole, 1983). In comparison with a state-of-the-art X-band EPR spectrometer, for millimeter-wave EPR, not to mention sub millimeter-wave EPR, the limiting value of $N_{\min}$ is not accessible with present-day microwave technology. A new generation of ultrahigh-frequency microwave components with considerably improved noise characteristics must become available before the adequate detection sensitivity inherent in high-field EPR/ENDOR can be achieved. It should be emphasized, however, that full exploitation of the resolution capability of high-field EPR on bioorganic systems requires adequate detection sensitivity, since biological magnetic resonance generally suffers from weak signals. This is particularly true for ENDOR, which normally attains only 1–10% of EPR signal strength.

By analogy with NMR spectroscopy, the extension of high-field cw EPR to time-resolved methods with pulsed microwave irradiation schemes will ultimately make this spectroscopic tool indispensable for serious studies of structure and dynamics of biological systems in their doublet and triplet states.

ACKNOWLEDGMENTS. It is a pleasure to acknowledge the many valuable contributions from the members of the Berlin group involved in early and present-day high-field EPR/ENDOR developments, in particular from Olaf Burghaus, Thomas Götzinger, Edmund Haindl, Robert Klette, Martin Rohrer, Jens Törring, and Anna Tòth-Kischkat. Several results on photosynthesis-related donor and acceptor radical ions, reported on in this chapter, were obtained together with Wolfgang Lubitz (now at the Technical University of Berlin) and his co-workers Birgit Bönigk, Uwe Fink, and Fraser MacMillan and with Martin Plato from this laboratory. Illuminating discussions with Martina Huber and Friedhelm Lendzian from this laboratory about photosynthetic reaction centers are gratefully acknowledged. This work was supported by the Deutsche Forschungsgemeinschaft (SFB 337).

REFERENCES

Abragam, A., and Bleaney, B., 1970, *Electron Paramagnetic Resonance of Transition Ions*, Clarendon Press, Oxford, pp. 541–583.

Allen, J. P., Feher, G., Yeates, T. O., Komiya, H., and Rees, D. C., 1987a, *Proc. Natl. Acad. Sci. USA* **84**:5730.

Allen, J. P., Feher, G., Yeates, T. O., Komiya, H., and Rees, D. C., 1987b, *Proc. Natl. Acad. Sci. USA* **84**:6162.

Allen, J. P., Feher, G., Yeates, T. O., Komiya, H., and Rees, D. C., 1988, *Proc. Natl. Acad. Sci. USA* **85**:8487.

Allgeier, J., Disselhorst, J. A. J. M., Weber, R. T., Wenckebach, W. Th., and Schmidt, J., 1990, High-frequency pulsed electron spin resonance, in *Modern Pulsed and Continuous-Wave Electron Spin Resonance* (L. Kevan and M. K. Bowman, eds.), Wiley, New York, pp. 267–283.

Antsiferova, L. I., and Lyabashevskaya, E. V., 1991, *Soviet J. Chem. Phys.* **7**:2970.

Barra, A. L., Brunel, L. C., and Robert, J. B., 1990, *Chem. Phys. Lett.* **165**:107.

Belford, R. L., Clarkson, R. B., Cornelius, J. B., Rothenberger, K. S., Nilges, M. J., and Timken, M. D., 1987, EPR over three decades of frequency: Radiofrequency to infrared, in *Electron Magnetic Resonance of the Solid State* (J. A. Weil, M. K. Bowman, J. R. Morton, and K. F. Preston, eds.), Can. Soc. Chem., Ottawa, pp. 21–43.

Berliner, L. J., ed. 1976, *Spin Labeling: Theory and Application*, Vol. 1, Academic Press, New York.

Berliner, L. J., ed., 1979, *Spin Labeling: Theory and Application*, Vol. 2, Academic Press, New York.

Box, H. C., Budzinski, E. E., Freund, H. G., and Potter, W. R., 1979, *J. Chem. Phys.* **70**:1320.

Bresgunov, A. Yu., Dubinskii, A. A., Krimov, V. N., Poluektov, O. G., and Lebedev, Ya. S., 1991, *Appl. Magn. Res.* **2**:715.

Budil, D. E., Earle, K. A., Lynch, W. B., and Freed, J. H., 1989, Electron paramagnetic resonance at 1 mm wavelength, in *Advanced EPR: Applications in Biology and Biochemistry* (A. J. Hoff, ed.), Elsevier, Amsterdam, pp. 307–340.

Burghaus, O., 1991, Ph.D. thesis, Department of Physics, Free University of Berlin, Germany.

Burghaus, O., Tòth-Kischkat, A., Klette, R., and Möbius, K., 1988, *J. Magn. Reson.* **80**:383.

Burghaus, O., Plato, M., Bumann, D., Neumann, B., Lubitz, W., and Möbius, K., 1991, *Chem. Phys. Lett.* **185**:381.

Burghaus, O., Rohrer, M., Götzinger, T., Plato, M., and Möbius, K., 1992a, *Measurement Sci. Tech.* **3**:765.

Burghaus, O., Plato, M., Rohrer, M., Möbius, K., MacMillan, F., and Lubitz, W., 1992b, *J. Phys. Chem.*, submitted.

Chang, C.-H., El-Kabbani, O., Tiede, D., Norris, J. R., and Schiffer, M., 1991, *Biochem.* **30**:5352.

Clarkson, R. B., Waug, W., Nilges, M. J., and Belford, R. L., 1990, Influence of organic sulfur in very high frequency EPR of coal, in *Processing and Utilization of High-Sulfur Coals III* (R. Markusewski and T. D. Wheelock, eds.), Elsevier, Amsterdam, pp. 67–80.

Deisenhofer, J., Epp, O., Miki, K., Huber, R., and Michel, H., 1984, *J. Mol. Biol.* **180**:385.

Deisenhofer, J., Epp, O., Miki, K., Huber, R., and Michel, H., 1985, *Nature (London)* **318**:618.

Dubinskii, A. A., Grinberg, O. Ya., Kurochkin, V. I., Oranskii, L. G., Poluektov, O. G., and Lebedev, Ya. S., 1981, *Theor. Exp. Chem. (Engl. Transl.)* **17**:180.

El-Kabbani, O., Chang, C.-H., Tiede, D., Norris, J. R., and Schiffer, M., 1991, *Biochem.* **30**:5361.

Ernst, R. R., Bodenhausen, G., and Wokaun, A., 1987, *Principles of Nuclear Magnetic Resonance in One and Two Dimensions*, Clarendon Press, Oxford.

Feher, G., 1957, *Bell Syst. Techn. J.* **36**:449.

Feher, G., Isaacson, R. A., Okamura, M. Y., and Lubitz, W., 1985, ENDOR of semiquinones in RC's from *Rhodopseudomonas sphaeroides*, in *Antennas and Reaction Centers of Photosynthetic Bacteria* (M. E. Michel-Beyerle, ed.), Springer-Verlag, Berlin, pp. 174–189.

Grinberg, O. Ya., Dubinskii, A. A., Shuvalov, V. F., Oranskii, L. G., Kurochkin, V. I., and Lebedev, Ya. S., 1976, *Dokl. Phys. Chem. (Engl. Transl.)* **230**:923.

Grinberg, O. Ya., Dubinskii, A. A., and Lebedev, Ya. S., 1983, *Russ. Chem. Rev. (Engl. Transl.)* **52**:850.

Gulin, V. I., Samoilova, R. I., Dikanov, S. A., and Tsvetkov, Yu. D., 1991a, *Appl. Magn. Reson.* **2**:425.

Gulin, V. I., Dikanov, S. A., and Tsvetkov, Yu. D., 1991b, Hyperfine interaction with β-fluorine in nitroxide radicals, in *Electron Magnetic Resonance of Disordered Systems* (N. D. Yordanov, ed.), World Scientific, Singapore, pp. 92–111.

Gulin, V. I., Dikanov, S. A., Tsvetkov, Yu. D., Evelo, R. G., and Hoff, A. J., 1991c, *Proc. Conf. on ESE Spectroscopy*, Novosibirsk, in press.

Hagen, W. R., 1989, *g*-Strain: Inhomogeneous broadening in metalloprotein EPR, in *Advanced EPR, Applications in Biology and Biochemistry* (A. J. Hoff, ed.), Elsevier, Amsterdam, pp. 785–812.

Haindl, E., Möbius, K., and Oloff, H., 1985, *Z. Naturforsch.* **40a**:169.

Hales, B. J., 1975, *J. Am. Chem. Soc.* **97**:5993.

Hales, B. J., 1976a, *J. Am. Chem. Soc.* **98**:7350.

Hales, B. J., 1976b, *J. Chem. Phys.* **65**:3767.

Hales, B. J., and Case, E. E., 1981, *Biochim. Biophys. Acta* **637**:291.

Klette, R., Törring, J. T., Plato, M., Möbius, K., Bönigk, B., and Lubitz, W., 1992, *J. Phys. Chem.*, in press.

Klimov, V. V., Dolan, E., Shaw, E. R., and Ke, B., 1980, *Proc. Natl. Acad. Sci. USA* **77**:7227.

Komiya, H., Yeates, T. O., Rees, D. C., Allen, J. P., and Feher, G., 1988, *Proc. Natl. Acad. Sci. USA* **85**:9012.

Krinichnyi, V. I., 1991, *J. Biochem. Biophys. Methods* **23**:1.

Lebedev, Ya. S., 1990, High-frequency continuous-wave electron spin resonance, in *Modern Pulsed and Continuous-Wave Electron Spin Resonance* (L. Kevan and M. K. Bowman, eds.), Wiley, New York, pp. 365–404.

Lebedev, Ya. S., 1991, Single-crystal-like spectra from disordered systems as obtained by high field EPR, in *Electron Magnetic Resonance of Disordered Systems* (N. D. Yordanov, ed.), World Scientific, Singapore, pp. 69–91.

Lendzian, F., Lubitz, W., Scheer, H., Hoff, A. J., Plato, M., Tränkle, E., and Möbius, K., 1988, *Chem. Phys. Lett.* **148**:377.

Lendzian, F., Endeward, B., Plato, M., Bumann, D., Lubitz, W., and Möbius, K., 1990, ENDOR and TRIPLE resonance investigation of the primary donor cation radical in single crystals of *Rb. sphaeroides* reaction centers, in *Reaction Centers of Photosynthetic Bacteria* (M. E. Michel-Beyerle, ed.), Springer-Verlag, Berlin, pp. 57–68.

Lendzian, F., Huber, M., Isaacson, R. A., Endeward, B., Plato, M., Bönigk, B., Möbius, K., Lubitz, W., and Feher, G., 1993, *Biochim. Biophys. Acta,* submitted.

Lous, E. J., Huber, M., Isaacson, R. A., and Feher, G., 1990, EPR and ENDOR studies of the oxidized primary donor in single crystals of reaction centers of *Rb. sphaeroides R*-26, in *Reaction Centers of Photosynthetic Bacteria* (M. E. Michel-Beyerle, ed.), Springer-Verlag, Berlin, pp. 45–55.

Likhtenstein, G. I., 1988, *Chemical Physics of Metalloenzyme Catalysis*, Springer-Verlag, Berlin.

Lubitz, W., 1991, EPR and ENDOR studies of chlorophyll cation and anion radicals, in *Chlorophylls* (H. Scheer, ed.), CRC Press, Boca Raton, pp. 903–944.

Lubitz, W., Lendzian, F., Plato, M., Möbius, K., and Tränkle, E., 1985, ENDOR studies of the primary donor in bacterial reaction centers, in *Antennas and Reaction Centers of Photosynthetic Bacteria* (M. E. Michel-Beyerle, ed.), Springer-Verlag, Berlin, pp. 164–173.

Lynch, W. B., Earle, K. A., and Freed, J. H., 1988, *Rev. Sci. Instrum.* **59**:1345.

Michel, H., Epp, O., and Deisenhofer, J., 1986, *EMBO J.* **5**:2445.

Michel-Beyerle, M. E., Plato, M., Deisenhofer, J., Michel, H., Bixon, M., and Jortner, J., 1988, *Biochim. Biophys. Acta* **932**:52.

Möbius, K., and Biehl, R., 1979, Electron-nuclear-nuclear TRIPLE resonance of radicals in solution, in *Multiple Electron Resonance Spectroscopy* (M. M. Dorio and J. H. Freed, eds.), Plenum Press, New York, pp. 475–507.

Möbius, K., and Lubitz, W., 1987, ENDOR spectroscopy in photobiology and biochemistry, in *Biological Magnetic Resonance* (L. J. Berliner and J. Reuben, eds.), Vol. 7, Plenum Press, New York, pp. 129–247.

Muller, F., Hopkins, M. A., Coron, N., Grynberg, M., Brunel, L. C., and Martinez, G., 1989, *Rev. Sci. Instrum.* **60**:3681.

O'Malley, P. J., Chandrashekar, T. K., and Babcock, G. T., 1985, ENDOR characterization of hydrogen bonding to immobilized quinone anion radicals, in *Antennas and Reaction Centers of Photosynthetic Bacteria* (M. E. Michel-Beyerle, ed.), Springer-Verlag, Berlin, pp. 339–344.

Pelekh, A. E., Goldenberg, L. M., and Krinichnyi, V. I., 1991, *Synthetic Metals* **44**:205.

Plato, M., Lubitz, W., Lendzian, F., and Möbius, K., 1988a, *Israel J. Chem.* **28**:109.

Plato, M., Lendzian, F., Lubitz, W., Tränkle, E., and Möbius, K., 1988b, Molecular orbital studies on the primary donor P_{960} in reaction centers of *Rps. viridis*, in *The Photosynthetic Bacterial Reaction Center* (J. Breton and A. Vermeglio, eds.), Plenum Press, New York, pp. 379–388.

Plato, M., Möbius, K., Michel-Beyerle, M. E., Bixon, M., and Jortner, J., 1988c, *J. Am. Chem. Soc.* **110**:7279.

Plato, M., Möbius, K., and Lubitz, W., 1991, Molecular orbital calculations on chlorophyll radical ions, in *Chlorophylls* (H. Scheer, ed.), CRC Press, Boca Raton, pp. 1015–1046.

Poole, Jr., C. P., 1983, *Electron Spin Resonance: A Comprehensive Treatise on Experimental Techniques*, Wiley, New York.

Prisner, T. F., McDermott, A. E., Un, S., Norris, J. R., Thurnauer, M. C., and Griffin, R. G., 1992a, *Proc. Natl. Acad. Sci. USA*, in press.

Prisner, T. F., Un, S., and Griffin, R. G., 1992b, *Israel J. Chem.*, in press.

Tarasov, V. F., and Shakurov, G. S., 1991, *Appl. Magn. Reson.* **2**:571.

Thurnauer, M. C., and Gast, P., 1985, *Photobiochem. Photobiophys.* **9**:29.

Thurnauer, M. C., Gast, P., Petersen, J., and Stehlik, D., 1987, EPR evidence that the photosystem I acceptor A_1 is a quinone molecule, in *Progress in Photosynthesis Research*, Vol. 1, Martinus Nijhoff Publ., Dordrecht, pp. 237–240.

Thurnauer, M. C., Brown, J. W., Gast, P., and Feezel, L. L., 1989, *Radiat. Phys. Chem.* **34**:647.

Törring, J., Rohrer, M., Klette, R., Huber, M., Lubitz, W., Lendzian, F., Feick, R., and Möbius, K., 1993, to be published.

Tsvetkov, Yu. D., 1989, 2 mm band ESR and its applications, in *Electron Magnetic Resonance of Disordered Systems* (N. D. Yordanov, ed.), World Scientific, Singapore, pp. 15–57.

Tsvetkov, Yu. D., Gulin, V. I., Dikanov, S. A., and Grigor'ev, 1987, Determination of magnetic resonance parameters of imidazoline nitroxide radicals by the 2-mm EPR method, in *Electron Magnetic Resonance of the Solid State* (J. A. Weil, M. K. Bowman, J. R. Morton, and K. F. Preston, eds), Can. Soc. Chem., Ottawa, pp. 45–55.

Weber, R. T., Disselhorst, J. A. J. M., Prevo, L. J., Schmidt, J., and Wenckebach, W. Th., 1989, *J. Magn. Reson.* **81**:129.

Wüthrich, K., 1986, *NMR of Proteins and Nucleic Acids*, Wiley, New York.

Yeates, T. O., Komiya, H., Chirino, A., Rees, D. C., Allen, J. P., and Feher, G., 1988, *Proc. Natl. Acad. Sci. USA* **85**:7993.

7

Pulsed Electron Nuclear Double and Multiple Resonance Spectroscopy of Metals in Proteins and Enzymes

Hans Thomann and Marcelino Bernardo

1. INTRODUCTION

EPR spectroscopy is a well-established method for characterizing the coordination structure of transition-metal ions and clusters of ions (Abragam and Bleaney, 1970; Pilbrow, 1991). In proteins, a metal site is often involved in an electron-transfer process. In enzymes, the metal sites are the active sites at which the substrate binds. EPR methods are a viable approach for probing the coordination structure of these active sites if the transition-metal sites are paramagnetic in one or more of the accessible oxidation states. An important advantage of EPR methods is that the signal emanates only from the specific site of interest in the macromolecule. Unlike many other spectroscopic methods, the EPR signals from the active site are not superimposed on a large background signal originating from other parts of the macromolecule.

EPR spectra for transition-metal sites in metalloproteins and metalloenzymes are usually inhomogeneously broadened (Mims, 1972a). Unresolved hyperfine splitting, both from the central metal nuclei and from ligand

Hans Thomann and Marcelino Bernardo • Exxon Corporate Research Laboratory, Annandale, New Jersey 08801. Hans Thomann is also a member of the Department of Chemistry, State University of New York at Stony Brook, Stony Brook, New York 11794.

Biological Magnetic Resonance, Volume 13: EMR of Paramagnetic Molecules, edited by Lawrence J. Berliner and Jacques Reuben. Plenum Press, New York, 1993.

nuclei in the coordination sphere, is one of the many sources of inhomogeneous broadening. With the exception of single-crystal studies, the line broadening is further exacerbated by the anisotropy of the *g* factor and of the hyperfine interaction. Multiple electron-spin resonance and pulsed EPR techniques have greatly extended the information which can be derived from such inhomogeneously broadened EPR spectra.

*E*lectron *n*uclear *do*uble *r*esonance (ENDOR) (Feher, 1956) and *e*lectron *s*pin *e*cho *e*nvelope *m*odulation (ESEEM) (Rowan *et al.*, 1965) spectroscopies are two of the more widely applied techniques in metalloprotein and metalloenzyme studies. In both ESEEM and ENDOR experiments, the NMR frequencies for paramagnetically coupled nuclei are detected indirectly through EPR transitions. NMR frequencies detected through EPR transitions are usually referred to as ENDOR frequencies or ESEEM frequencies, but there is an important distinction between ENDOR and ESEEM frequencies. The ENDOR signal arises from the combined direct irradiation of both the EPR and NMR transitions. The ENDOR transition therefore involves a change in the spin angular momentum quantum number for both the electron spin and the nuclear spin. In contrast, NMR transitions are not excited by direct rf irradiation in the ESEEM experiment. The NMR frequencies are instead observed by mixing the frequencies for the allowed and semiforbidden EPR transitions, which have been coherently excited using short, intense microwave excitation pulses. The NMR frequency observed as a peak in an ESEEM spectrum is therefore not correctly identified as an ESEEM transition but more correctly as simply an ESEEM frequency. This distinction between an ENDOR transition and an ESEEM frequency has practical consequences for the applications of these spectroscopic techniques.

The ESEEM experiment usually involves pulsed EPR, although ESEEM frequencies can also be observed using continuous-wave irradiation methods (Bowman and Yannoni, 1990). The two-pulse ESEEM experiment was first reported by Rowan, Hahn, and Mims in 1965 (Rowan *et al.*, 1965). Mims subsequently further developed the theory and technique (Mims, 1972). The application of ESEEM spectroscopy to biological studies was pioneered by Mims and Peisach who have written several reviews of the field (Mims and Peisach, 1981, 1989).

The ENDOR experiment was introduced by Feher (1956) for measuring the hyperfine coupling of the ^{13}P nucleus with the odd valence electron in phosphorous-doped silicon. Since then several texts have been written on ENDOR spectroscopy (Kevan and Kispert, 1979; Kurreck *et al.*, 1988). Reviews of ENDOR spectroscopy in studies of the coordination structure of organometallic complexes (Schweiger, 1982; Dorio and Freed, 1979), of metals in proteins and enzymes (Hoffman, 1992), and of organic radicals (Kurreck *et al.*, 1988) have been written.

In contrast to ESEEM experiments, continuous-wave irradiation (rather than pulsed excitation) has generally been used to excite both the NMR and EPR transitions in ENDOR experiments. Historically this is probably because of the greater instrumental complexity required to form the short, high-power microwave and rf pulses and deliver them to a sample. The commercial availability of ENDOR spectrometers (Bruker Instruments; Varian Instruments) employing CW microwave and rf technology has naturally provided greater accessibility to the CW-ENDOR experiment.

There are, however, several advantages to performing the ENDOR experiment using pulsed microwave and rf excitation for the EPR and NMR transitions. One of the most significant is that the amplitude of the ENDOR signal in the pulsed ENDOR experiment is not a function of the ratio of the electron- and nuclear-spin relaxation rates. This property of the amplitude in the CW-ENDOR experiment is an inherent consequence of using CW irradiation. A more detailed comparison between the CW and pulsed ENDOR experiments will be presented in Section 5. There are also several other advantages of the pulsed experiment, as we will demonstrate in this chapter.

Mims reported the first pulsed ENDOR experiment in 1965 (Mims, 1965). Relatively few pulsed ENDOR experiments followed this initial work (Brown *et al.*, 1969; Liao and Hartmann, 1973; Poel *et al.*, 1983; Mehring *et al.*, 1987). Until recently, most applications of pulsed ENDOR have been on samples with narrow ENDOR lines and long electron-spin relaxation times, such as single crystals or organic radicals in the solid state. The general methodology for pulsed ENDOR experiments has recently been reviewed (Gemperle and Schweiger, 1991). The first pulsed ENDOR experiment on a metalloprotein was reported in 1988 (Tindall *et al.*, 1988). In this chapter we focus specifically on those aspects of the pulsed excitation methodology pertinent for the study of metals in proteins and enzymes. The low symmetry of these metal sites leads to broad ENDOR lines and short spin-relaxation times compared to high-symmetry organometallic complexes. This complicates both the experiment and the analysis of the pulsed ENDOR experiment.

In this chapter, we will present spectroscopic results on the blue copper protein stellacyanin (Peisach *et al.*, 1967; Malmstrom *et al.*, 1970) and on the iron–sulfur enzyme hydrogenase (Erbes *et al.*, 1975; Chen *et al.*, 1976) to illustrate the pulsed ENDOR methodology and the comparison of these pulsed ENDOR methods to the ESEEM and CW-ENDOR experiments. Before continuing with the discussion of the spectroscopic methods, we present a brief introduction to this copper protein and iron–sulfur enzyme.

Blue copper proteins are identified by the characteristic spectroscopic and electronic properties originating from the coordination of the copper site (Malkin and Malmstrom, 1970; Fee, 1975; Gray and Solomon, 1981).

In the blue copper proteins plastocyanin (Coleman *et al.*, 1978) and azurin (Norris *et al.*, 1986), whose crystal structures have been reported, the chromophore is Cu[N(imid)]2S(cys)S(meth) in a rhombically distorted T_d coordination geometry. The intense adsorbance (molar extinction coefficient $\varepsilon \approx 20{,}000$) at 600 nm, which is a fingerprint for a blue copper protein, is attributed to a sulfur-to-copper charge-transfer transition of this chromophore. In EPR spectra, the blue copper proteins are identified by the smaller value of g_{max} and small Cu hyperfine splittings in the $g_{\parallel}$ region of the EPR spectrum compared to square-planar copper complexes. The copper chromophore in blue copper proteins exhibiting these spectral properties is known as a Type 1, or simply $T1$, copper site.

Among the blue copper proteins, stellacyanin, a protein isolated from the Japanese lacquer tree, *Rhus vernicifera*, is unusual in several respects (Peisach *et al.*, 1967; Malmstrom *et al.*, 1970). Stellacyanin has no protein methionine residues, has the lowest redox potential (+184 mv) of all blue copper proteins, exhibits a larger and more rhombic g anisotropy, and has its largest principal copper hyperfine coupling along g_{min} rather than g_{max}. Stellacyanin also exists in a reversible perturbed blue form between pH 8 and 11.5; this form has slightly different spectral properties but is still of predominantly $T1$ character. A third nitrogen ligand was identified in a recent pulsed ENDOR study of the high-pH form of stellacyanin (Thomann *et al.*, 1991b). A spectroscopic model for the copper site in the high-pH form of stellacyanin is shown in Figure 1.

Some of these spectral properties of stellacyanin are observed in the electron spin-echo detected EPR (ESE-EPR) spectra of stellacyanin at pH 7 and

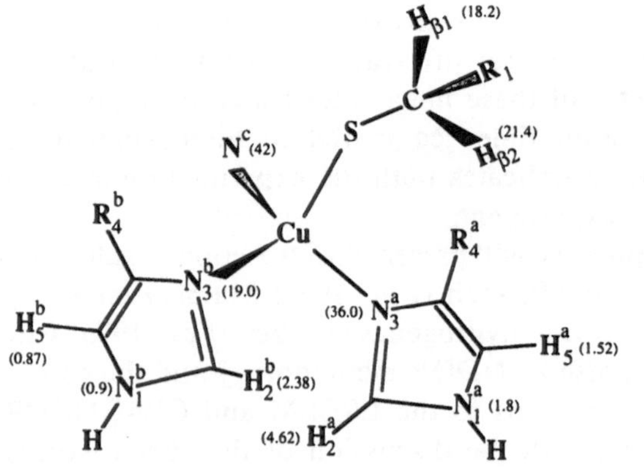

Figure 1. Spectroscopic model for the copper site in the high-pH form of stellacyanin. The numbers in parentheses are hyperfine coupling values A in MHz. Values for H_2^a, N_3^a, H_5^a, H_2^b, N_3^b, H_5^b, $H_{\beta1}$, $H_{\beta2}$, and N^c are from pulsed ENDOR and those for N_1^a and N_1^b are from ESEEM.

11 shown in Fig. 2. The ESE-EPR spectra, shown in Fig. 2a, are recorded by detecting the electron spin-echo intensity while incrementing the magnetic field stepwise on successive spin-echo-generating pulse sequence iterations. Since the echo intensity is proportional to the EPR susceptibility, this generates the EPR absorption spectrum directly rather than the derivative (actually, first harmonic of the Zeeman modulation frequency) display normally obtained in conventional EPR spectroscopy. The nuclear modulation effect, which arises from NMR active nuclei such as protons and nitrogen atoms in the coordination sphere of the cupric ion, is suppressed by using microwave pulses with weak magnetic field intensities. The modulation effect can superimpose amplitude oscillations on to the ESE-EPR

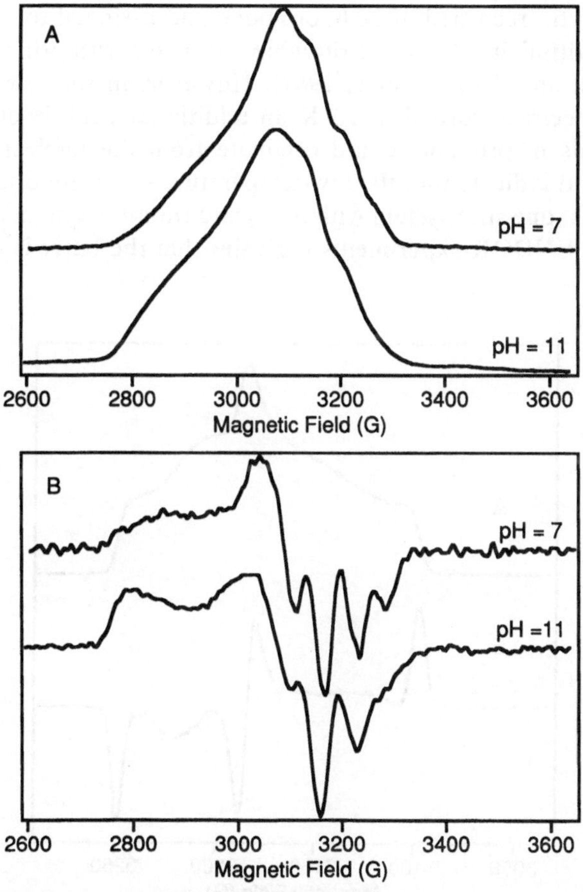

Figure 2. Electron spin-echo detected EPR (ESE-EPR) spectra of the native and high-pH forms of stellacyanin (A) and corresponding digital derivative spectra (B). Experimental conditions: $T = 1.6$ K; $\nu_{mw} = 8.884$ GHz; $t_p^{mw} = 0.02$ and 0.04 μs; $\tau = 0.57$ μs.

absorption spectrum. For ease of comparison to the standard EPR spectral display, the digital derivative of the ESE-EPR absorption envelopes are shown in Fig. 2B. These ESE-EPR spectra are very similar to the previously reported EPR spectra. This suggests that the nuclear modulation effect has been largely suppressed in the ESE-EPR spectra.

Hydrogenase catalyze the activation of molecular hydrogen according to the equation $H_2 = 2H^+ + 2e^-$. The proposed site for H_2 oxidation and production in the Fe-containing hydrogenases is a novel Fe–S center known as the hydrogenase or H cluster (Erbes *et al.*, 1975; Chen *et al.*, 1976; van Dijk *et al.*, 1980; Hagen *et al.*, 1986). The ESE-EPR absorption spectrum and digital derivative of the oxidized H cluster in hydrogenase I purified from the anaerobic N_2-fixing bacterium *Clostridium pasteurianum* are shown in Figs. 3A and B, respectively. The ESE-EPR spectrum closely resembles the EPR spectra recorded at 10 K or above and assigned to a spin system with spin multiplicity $S = \frac{1}{2}$ and rhombic symmetry with principal g values of 2.01, 2.04, and 2.10 (Adams, 1987). However, in the low-temperature ESE-EPR spectra recorded at 1.3 K an additional peak is observed near $g = 2.02$. This in principle could originate from the nuclear modulation effect or could indicate that the low-temperature spectrum does not simply arise from just one spin system with $S = \frac{1}{2}$ and rhombic symmetry. Evidence from pulsed ENDOR experiments indicates that the latter is correct.

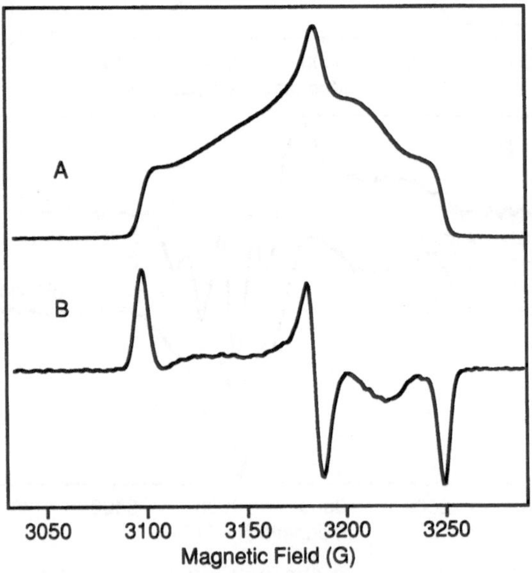

Figure 3. Electron spin-echo detected EPR (ESE-EPR) spectra (A) and corresponding digital derivative spectra (B) of the oxidized H cluster in hydrogenase I.

2. ENDOR ENERGY LEVELS AND TRANSITION FREQUENCIES

The pulsed ENDOR techniques described in this chapter will be demonstrated by considering the simple interaction of an electron with spin angular momentum of $S = \frac{1}{2}$ coupled to a nucleus with spin angular momentum of $I = \frac{1}{2}$. We further make the simplifying assumptions that the hyperfine coupling A between the electron and the nucleus is orientionally independent, so that only the isotropic coupling needs to be considered. The isotropic coupling, also known as the Fermi contact coupling, arises from the unpaired electron spin density at the nucleus. Finally we assume that the hyperfine coupling is small, so that second-order hyperfine interaction terms (of order $h^2A^2/g_e\beta_eB_0$) are negligible. With these assumptions, the system of electron and nucleus in a magnetic field can be represented by the Hamiltonian (Abragam and Bleaney, 1970; Wertz and Bolton, 1986)

$$\frac{\mathcal{H}_0}{h} = \nu_e S_z - \nu_n I_z + AS_zI_z \tag{1}$$

where $\nu_e = g_e\beta_eH_0/h$ and $\nu_n = g_n\beta_nH_0/h$ are the electron and nuclear Larmor frequencies, respectively. The negative sign for the nuclear Zeeman term accounts for the fact that the electron and proton are of opposite charge.

The Hamiltonian of Eq. (1) is applicable in the high-field limit where the applied magnetic field is sufficiently large that the electron S and nucleus I are separately quantized along the applied magnetic field direction. This is equivalent to stipulating that M_I and M_S are good quantum numbers. The spin eigenstates, in frequency units, are then given by

$$\frac{E}{h}(M_S, M_I) = \nu_e M_S - \nu_n M_I + AM_SM_I \tag{2}$$

The ordering of the spin energy levels depends on the relative magnitudes of hA and $h\nu_n$ and on the sign of A. $A < 0$ corresponds to a negative spin density on the nucleus. This is a consequence of electron spin polarization.

The spin energy-level diagram for $A < 0$, $|A|/ < 2\nu_n$, and $g_n > 0$ is shown in Fig. 4. For easy reference, the spin eigenstates are labeled in order of increasing energy as E1, E2, E3, and E4. The splitting of the electron Zeeman interaction by the nuclear Zeeman and hyperfine interactions produces the sublevels E1 and E2 in the lower electron spin manifold and the sublevels E3 and E4 in the upper manifold. Since $g_e\beta H_0 \gg g_n\beta H_0$, the dominant contribution to the electron spin polarization comes from the electron Zeeman interaction $\Delta E = h\nu = g_e\beta H_0$. (The polarization from the nuclear Zeeman and hyperfine interactions is negligible compared to the

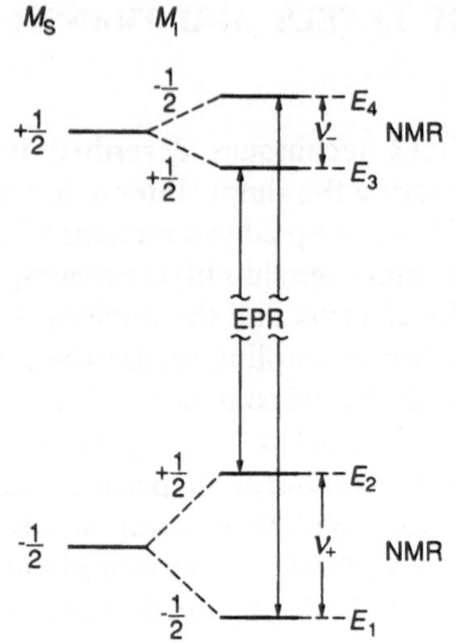

Figure 4. Energy-level diagram for the four-level system described in the text.

electron polarization and will be ignored.) According to the Boltzmann relation, the number of spins in the upper N_u and lower N_l electron Zeeman spin energy levels for a two-level system is given by

$$N_u = N_l \exp\left(\frac{-g\beta H_0}{kT}\right) \approx 1 - \frac{g\beta H_0}{kT} = 1 - \delta \tag{3}$$

where the total number of spins $N = N_l + N_u$. If the electron is coupled to nuclei, the polarization is distributed over the hyperfine sublevels in each electron spin manifold. In the high-temperature limit, $g_e\beta_e H_0/kT \ll 1$, and the equilibrium spin populations are given by $P1 \approx P2 \approx (1 + \delta/2)$ and $P3 \approx P4 \approx (1 - \delta/2)$. The excess spin population in the lower spin eigenstates at thermal equilibrium is represented by the boxes in Fig. 4. The sublevel population P_n represents the nuclear spin polarization in the energy sublevel E_n.

The allowed EPR transitions follow the selection rules $\Delta M_S = \pm 1$ and the allowed NMR transitions follow the selection rules $\Delta M_I = \pm 1$. These transitions are indicated in Fig. 4. The EPR transition frequencies are $\nu_\pm^{\text{EPR}} = \nu_e \pm A/2$, while the allowed NMR transition frequencies are $\nu_\pm^{\text{NMR}} = \nu_n \pm A/2$. In the present simple example, an inhomogeneous broadening of the EPR absorption would arise if the hyperfine splitting is smaller than the EPR linewidth. This is expressed by the inequality $AT_{2e} < 1$, where T_{2e} is the electron transverse relaxation time that characterizes the width of the homogeneously broadened line. If $AT_{2e} < 1$, the NMR transitions can potentially be observed indirectly in the ENDOR experiment.

ENDOR transitions are NMR transitions detected indirectly by an EPR transition. The observation of an ENDOR transition therefore requires both that $\Delta M_S = \pm 1$ and that $\Delta M_I = \pm 1$. This is achieved using microwave and rf magnetic fields to induce both EPR and NMR transitions. In the spin Hamiltonian formalism, the EPR and NMR transitions are represented by the terms $\omega_1 S_x$ and $\omega_2 I_x$, where $\omega_1 = g_e\beta_e H_1/\hbar$ and $\omega_2 = g_n\beta H_2/\hbar$, and H_1, H_2 are the magnitudes of the microwave and rf magnetic fields, respectively. In the present example with $\nu_n > A/2$, peaks in the ENDOR spectrum occur at $\nu_\pm^{\text{ENDOR}} = \nu_n \pm A/2$. When $\nu_n < A/2$, the nuclear Zeeman interaction is a perturbation on the hyperfine interactions and the ENDOR peaks are centered at $\nu_\pm^{\text{ENDOR}} = A/2 \pm \nu_n$.

The simple four-level system described by Eq. (1) and shown in Fig. 4 will be adequate to illustrate the basic concepts of polarization transfer pulsed ENDOR experiments. However, some additional interactions must be introduced in order to interpret the ENDOR spectra of metals in proteins and enzymes presented below. First, the hyperfine couplings are typically composed of both anisotropic and isotropic terms. The anisotropy can arise both from the classical dipolar interaction between the electron and the nucleus and from the pseudodipolar interaction arising from the distributed nature of the electron-spin wave function. The anisotropic coupling results in a distribution of ENDOR frequencies which to first order can be described by (Dalton and Kwiram, 1972)

$$\nu_\pm^0 = [(S_z A_{xx} - \nu_n)^2 \sin^2\theta \cos^2\phi + (S_z A_{yy} - \nu_n)^2 \sin^2\theta \sin^2\phi$$
$$+ (S_z A_{zz} - \nu_n)\cos^2\theta]^{1/2} \qquad (4)$$

where the superscript on $\nu_\pm^0$ is a reminder that second-order hyperfine interactions are assumed to be negligible. The angles θ and ϕ are the spherical polar coordinates which relate the principal components A_{xx}, A_{yy}, A_{zz} of the hyperfine matrix to the principal axis of the g matrix. For an isotropic g value, this is equivalent to relating the principal components of the hyperfine matrix to the external magnetic field. Each set (θ, ϕ) of spherical polar angles corresponding to a particular orientation of the hyperfine tensor results in a set of ENDOR lines. However, ENDOR lines arising from different sets of (θ, ϕ) pairs can be coupled if the rate of electron–nuclear cross-relaxation is fast compared to the electron spin-lattice relaxation time (Dalton and Kwiram, 1972). In this case a full powder-pattern spectrum corresponding to all orientations described by a range of angles (θ, ϕ) will be observed.

Many of the nuclei of interest in biological studies such as ^{14}N, ^{63,65}Cu, etc., have nuclear spin angular momentum $I > \frac{1}{2}$. In this case the additional level splitting from the nuclear quadrupole interaction must be taken into account. The first-order ENDOR frequencies are given by

$$\nu_{\text{ENDOR}} = \frac{|A_i|}{2} \pm \nu_n \pm \frac{3}{2}|P_i|(2m_I + 1) \qquad (|A| > 2\nu_n > 3|P_i|)$$
$$\nu_{\text{ENDOR}} = \frac{|A_i|}{2} \pm \frac{3}{2}|P_i|(2m_I + 1) \pm \nu_n \qquad (|A| > 3|P_i| > 2\nu_n) \tag{5}$$

where A_i and P_i denote the principal values of the **A** and **P** tensors along the axis i. The nuclear transitions in Eq. (5) are assumed to increase in the quantum number m_I, i.e., $m_I \rightarrow m_i + 1$.

For an $I = 1$ nucleus such as ^{14}N, four ENDOR transitions are expected. However, if the hyperfine splitting is resolved in the EPR spectrum, each m_I state can be selectively excited. In this case, less than the $4I$ ENDOR transitions expected may be observed, depending on the m_I levels involved in the EPR transition. A two-line ENDOR spectrum can be observed if the EPR transition connects the $m_I = 0$ to $m_I = 0$ states, while a four-line spectrum can be observed if the EPR transition connects either of the $m_I = \pm 1$ states. A smaller number of ENDOR lines than the $4I$ expected can also be observed for nuclei with $I > 1$ if the individual EPR transitions can be irradiated.

In the spin Hamiltonian formalism, the nuclear quadrupole interaction is described by a nuclear spin self-coupling $\mathbf{I} \cdot \mathbf{P} \cdot \mathbf{I}$. The anisotropy of the quadrupole interaction is described by the quadrupole tensor **P**, which is a traceless 3×3 matrix, so that it is fully characterized by five independent components. These are usually taken as the magnitude of the quadrupole coupling, e^2qQ/h; the quadrupole asymmetry η, and the three Euler angles α, β, γ that describe the orientation of the quadrupole tensor with respect to the principal axes of the g matrix. The quadrupole coupling constant and asymmetry parameter are related to the quadrupole tensor elements by the relations (Lucken, 1969)

$$e^2qQ = 2I(2I - 1)P_{zz}$$
$$\eta = \frac{P_{xx} - P_{yy}}{P_{zz}} \tag{6}$$

The elements P_{xx}, P_{yy}, and P_{zz} are the principal values of the quadrupole tensor **P** and are defined as $|P_{zz}| \geq |P_{yy}| \geq |P_{xx}|$.

In most ENDOR experiments, the microwave excitation excites only a small region of the EPR spectrum centered about the magnetic field selected (Mims, 1972a). This magnetic field position selects a subset of molecular orientations defined by the resonance condition: $h\nu = g\beta H_0$, where the orientation dependence of the anisotropic g factor is given by (Abragam and Bleaney, 1970; Atherton, 1973)

$$g(\theta, \phi) = \left[\sum_{i=1}^{3} (g_i h_i)^2 \right]^{1/2} \tag{7}$$

where the h_i are unit vectors for the directions of the magnetic field expressed in spherical polar coordinates:

$$(h_1, h_2, h_3) = (\cos\phi \sin\theta, \sin\phi \sin\theta, \cos\theta) \tag{8}$$

and where the angles (θ, ϕ) are the polar and azimuthal angles that relate the magnetic field vector to the coordinate system in which the g matrix is diagonal.

The number of molecular orientations that contribute to the ENDOR spectrum is minimum for values of (θ, ϕ) that correspond to the principal axes of the g matrix. This was first recognized by Rist and Hyde (1970) and is referred to as g-factor orientation selectivity. Higher spectral resolution is observed in ENDOR spectra recorded at magnetic field values corresponding to values of (θ, ϕ) that correspond to these principal axes. Orientation selectivity has been used to great advantage in ENDOR studies of transition metals in organometallic complexes and in studies of metals in enzymes and proteins (Hoffman *et al.*, 1984; Hurst *et al.*, 1985).

3. PULSE SCHEMES

Since the first pulsed ENDOR experiment was reported by Mims in 1965, a variety of pulse sequences have been proposed and demonstrated for pulsed ENDOR experiments (Brown *et al.*, 1969; Davies, 1974; Grupp and Mehring, 1990; Thomann and Bernardo, 1990a; Gemperle and Schweiger, 1991). It is instructive to categorize these pulse schemes according to whether the rf pulse(s) are applied in the presence of finite transverse electron spin magnetization $M_{x,y}(t)$, in contrast to schemes in which rf pulses are applied when $M_{x,y}(t) = 0$.

3.1. Coherence Transfer ENDOR

$M_{x,y}(t)$ magnetization represents electron spin coherence that decays with a phase memory time constant T_m. ENDOR experiments in which rf pulses are applied in this "T_m domain" affect the $M_{x,y}(t)$ coherence. Such experiments can therefore be referred to as coherence transfer ENDOR.

Coherence transfer ENDOR experiments were first demonstrated by Brown and co-workers in 1969 (Brown *et al.*, 1969) and more recently extended and developed by Mehring and coworkers (Mehring *et al.*, 1986, 1987). In coherence transfer ENDOR experiments a spin-echo sequence is used to create and detect electron spin coherence. One or more rf pulses are applied during the evolution time τ between the first and second microwave pulses and possibly during the evolution time τ between the second microwave pulse and the formation of the electron spin echo. The nuclear spin flips induced by these rf pulses interfere with the refocusing of the electron spin echo. This is manifest as a reduction in the electron spin-echo intensity $M_{x,y}(2\tau)$.

Several interesting applications of coherence transfer ENDOR have recently been reported. The spinor behavior for the spin wave function of a spin $\frac{1}{2}$ nucleus coupled to the paramagnetic electron has been observed (Mehring *et al.*, 1986). ENDOR spectra recorded by coherence transfer have also been observed for a phenoxy radical in solution at room temperature (Forrer *et al.*, 1990). In studies of transition-metal ions in proteins and enzymes, the applicability of coherence transfer ENDOR spectroscopy is severely constrained by the relatively short T_m typically observed for these samples. These short T_m result in broad ENDOR lines due to rf power broadening. In this chapter we therefore focus exclusively on polarization transfer pulsed ENDOR experiments.

3.2. Polarization Transfer ENDOR

Polarization transfer ENDOR experiments require the creation of a difference in the electron spin polarization at two positions in the EPR spectrum. The ENDOR signal arises from the transfer of electron spin polarization between these two positions within the EPR spectrum. The polarization is transferred by rf pulses which irradiate the NMR transitions of the hyperfine sublevels connecting the two positions in the EPR spectrum. In the absence of this polarization transfer, the decay of the electron spin polarization is governed by the electron spin–lattice relaxation time T_{1e}. At temperatures near that of liquid helium, T_{1e} values for transition-metal sites in proteins and enzymes are typically in the range of several hundred microseconds or longer. This is in contrast to the typical values of T_m, which are no longer than a few microseconds. ENDOR experiments in which rf pulses are applied in this "T_{1e} domain" are therefore known as polarization transfer ENDOR experiments. The long T_{1e} allows long rf pulses to be used, which do not power broaden the ENDOR lines.

It is convenient to partition the pulse sequences in pulsed ENDOR spectroscopy into preparation, sublevel mixing, evolution, and detection periods. This partitioning is similar to the approach used in pulsed NMR spectroscopy (Ernst *et al.*, 1987) and is applicable in coherence transfer ENDOR schemes as well as for polarization transfer ENDOR schemes. In Fig. 5 we show this partitioning for three experimental approaches for detecting sublevel polarization transfer. For pedagogical purposes, the conventional method of performing the ENDOR experiment using CW irradiation is also artificially partitioned into periods as shown in Fig. 5A. The sequence in Fig. 5B was proposed by Davies (1974) and has become known as Davies ENDOR. It is composed of a three-pulse inversion-recovery sequence

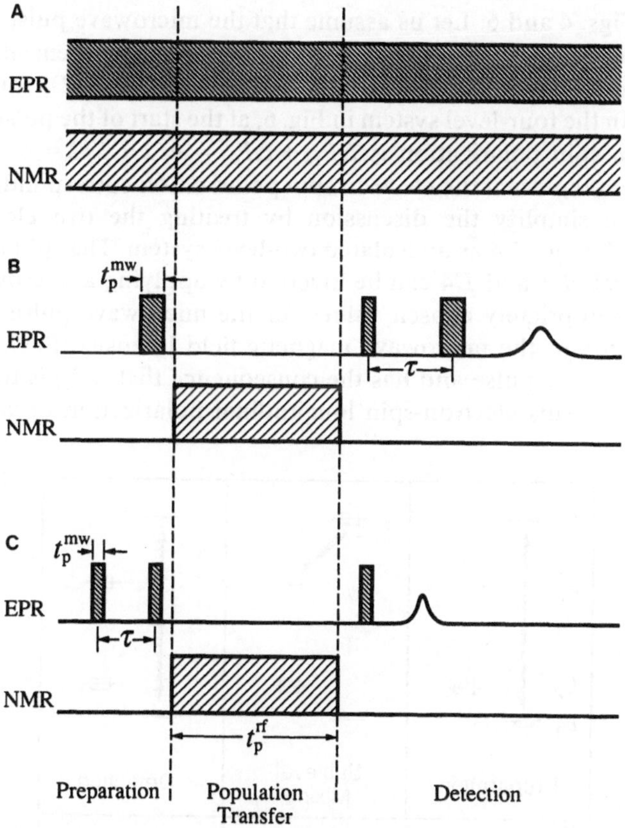

Figure 5. Sublevel polarization transfer ENDOR schemes: (A) CW ENDOR, (B) Davies ENDOR, and (C) Mims ENDOR partitioned into three periods, as described in text.

on the EPR channel and a single rf pulse on the NMR channel. The sequence in Fig. 5C was first demonstrated by Mims (1965) and has become known as Mims ENDOR. It is composed of a three-pulse stimulated echo sequence on the EPR channel and a single rf pulse on the NMR channel. We now describe each of these pulse periods in more detail.

3.2.1. Preparation Period

In the preparation period, sublevel nuclear spin polarization is created from electron spin polarization. This can be accomplished by selective excitation using CW ENDOR or irradiation using a single pulse, as shown in Figs. 5A and B, or by nonselective excitation using two pulses, as shown in Fig. 5C. The mechanism for creating the sublevel polarization can be understood by analyzing the effect of the selective pulse excitation on the population of the spin levels. Consider the transitions in the four-level system in Figs. 4 and 6. Let us assume that the microwave pulse in Fig. 5B selectively excites only the EPR transition between the eigenstates $E1$ and $E4$. The electron spin polarization for $E1$ is $P_1 \approx (1 + \delta/2)$; for $E4$, $P_4 \approx (1 - \delta/2)$. In the four-level system in Fig. 6, at the start of the pulse sequence the electron spin polarization is $\Delta P_{14} + \Delta P_{23} = \delta/2 + \delta/2 = \delta$. The boxes in Fig. 6 each represent the electron spin polarization corresponding to $\delta/2$.

We can simplify the discussion by treating the two electron spin eigenstates $E1$ and $E4$ as an isolated two-level system. The spin population between levels $E1$ and $E4$ can be inverted by applying a microwave pulse with the appropriately chosen values for the microwave pulse width and the magnitude of the microwave magnetic field intensity. Such a pulse is referred to as a π pulse and has the consequence that ΔP_{14} is transformed to $\Delta P_{14} = -\delta$. This electron-spin longitudinal polarization decays back to

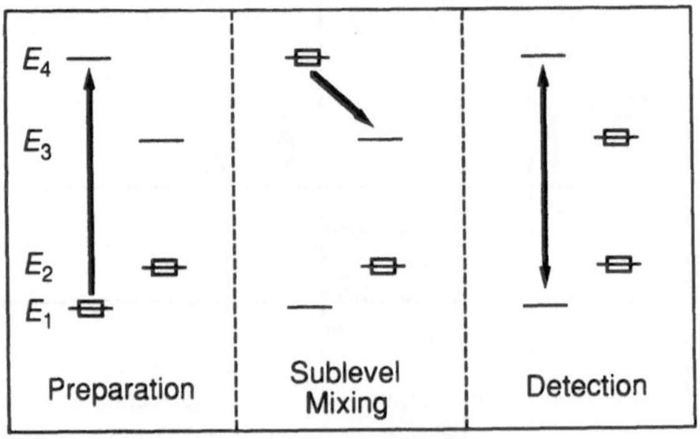

Figure 6. Evolution of the spin populations for the spin system in Fig. 4 under the Davies ENDOR pulse sequence in Fig. 5B.

its equilibrium value with a time constant T_{1e}, the electron spin-lattice or longitudinal relaxation time.

While no net polarization has been created by the preparation pulse, the selective microwave excitation has created a polarization among the hyperfine sublevels. Since we have ignored polarization arising from the nuclear Zeeman and hyperfine couplings, before the preparation pulse the nuclear polarization was $\Delta P_{12} = \Delta P_{34} = 0$. But after the preparation pulse, $\Delta P_{12} = \Delta P_{34} = -\delta$. In order to avoid possible confusion between net nuclear polarization and sublevel polarization, the latter is sometimes referred to as nuclear spin alignment. Sublevel polarization can also be created using two microwave pulses in the preparation period, as shown in the pulse sequence of Fig. 5B and first demonstrated by Mims (1965). One advantage of using a two-pulse excitation is that the microwave pulses need no longer correspond to selective EPR excitations. Sublevel polarization is created from the transverse components of the magnetization that evolves between the two microwave pulses. This implies that the two pulses should not be perfect π pulses if they are not selective excitations. In fact, the maximum sublevel polarization is created if the pulses are $\pi/2$ pulses. During the time τ following the first pulse, the local magnetic field arising from the hyperfine coupling causes the $M_{x,y}$ components of the magnetization corresponding to the two spin packets (of the two EPR transitions) to accumulate different phase factors. After a time τ, this accumulated phase is stored along the z axis by applying the second $\pi/2$ pulse. After this second pulse, the net electron spin polarization $[\Delta P_{14} + \Delta P_{23}] \approx \cos(A\tau/2)\cos(\Omega_s\tau)$, while the sublevel polarization $[\Delta P_{12} + \Delta P_{34}] \approx \sin(A\tau/2)\sin(\Omega_s\tau)$. Note that the creation of sublevel polarization also requires a resonance offset, denoted by Ω_s. Another constraint is that no sublevel polarization exists if the product $A\tau/2 = n\pi$, where n is an integer. These are known as blind spots in the Mims ENDOR experiment.

3.2.2. Mixing Period

The nuclear sublevel spin polarization created during the preparation period is transferred between hyperfine sublevels in the mixing period. This transfer is created by inducing a nuclear spin flip, i.e., an NMR transition. We can treat the magnetization from the two sublevels $E3$ and $E4$, and $E1$ and $E2$, as pairs of isolated two-level systems. The sublevel population differences ΔP_{34} and ΔP_{12} created in the preparation period then correspond to longitudinal sublevel magnetization $M_{z(34)}$ and $M_{z(12)}$ for these sublevels.

For the example in Fig. 6, we have selected the frequency of the rf mixing pulse to be on resonance with the sublevel transition $E3 \leftrightarrow E4$. The rf pulse transforms the longitudinal component $M_{z(34)}$ of this sublevel magnetization to $M_{z(34)} \cos\theta_R$. The nutation angle $\theta_R = \gamma_n \varepsilon H_2 t_p$, where γ_n

is the nuclear gyromagnetic ratio, H_2 is the rf magnetic field intensity, and t_p is the rf pulse length. Note that the sign of the nutation angle can be positive or negative since g_n can be positive or negative. The factor ε takes into account the enhancement of the nuclear transition rate due to the electronic magnetic field at the nucleus (Abragam and Bleaney, 1970). This electronic field arises from the hyperfine interaction and has a significant effect on the ENDOR amplitudes, as discussed in Section 4.3.

When the rf pulse is selected so that $\theta_R = \pi$, $M_{z(34)}$ is transformed from $-M_{z(34)}$ to $+M_{z(34)}$. This corresponds to the transformation of ΔP_{34} to $-\Delta P_{34}$, as indicated in the mixing period of Fig. 6. After the rf π pulse, the net electron spin polarization is $\Delta P_{14} + \Delta P_{23} = 0$, the net sublevel polarization is $\Delta P_{12} + \Delta P_{34} = 0$, but the net nuclear polarization now equals $-\delta$. In this simple four-level system, this nuclear spin flip corresponds to the transfer of electron polarization from one EPR line (the transition $E1 \leftrightarrow E4$) to the second EPR line (the transition $E2 \leftrightarrow E3$). An EPR absorption spectrum collected after the rf π pulse is applied would show no absorption from either allowed EPR transition.

3.2.3. Detection Period

The sublevel NMR transitions are typically in the rf frequency range. The sublevel polarization transfer is therefore transparent to the EPR spectrometer, which detects transitions at microwave frequencies. The NMR transitions could in principle be detected directly using a receiver tuned to the NMR transition frequency. However, direct detection suffers from the lower sensitivity of the NMR compared to the EPR transition. The hyperfine interaction also spreads the NMR transition frequencies over a much greater spectral bandwidth than can be irradiated by pulsed rf excitation. This necessitates sweeping the rf frequency, which in turn requires the use of an rf resonator with a wide bandwidth and concomitant loss of sensitivity. For these reasons the NMR transitions are usually detected indirectly by observing the change in electron spin polarization.

In principle the electron spin polarization can be detected using any method for observing an EPR transition. This includes sweeping the EPR absorption using CW irradiation as in conventional EPR spectroscopy, observing a free induction signal following excitation with a single microwave pulse (Wacker and Schweiger, 1992), and direct detection of the longitudinal electron spin polarization using the appropriate hardware configuration (Schweiger and Ernst, 1988). However, the most widely used method for observing the electron spin polarization is by detecting the amplitude of a primary electron spin echo using the Davies ENDOR pulse sequence (Fig. 5B) or of a stimulated spin echo when using the Mims ENDOR pulse sequence (Fig. 5C). The EPR excitations in the detection period can be selective or nonselective. If selective detection is used, then in order to generate the pulsed analog of the conventional ENDOR spectrum, the same EPR transitions must be irradiated in the preparation and detection periods. The selective detection of an EPR transition other than the EPR transition irradiated in the preparation period can generate a hyperfine-selective ENDOR spectrum, as discussed in Section 7.1.

The Davies and Mims ENDOR spectra are recorded by plotting the intensity of the primary or stimulated electron spin echo while incrementing the rf mixing pulse frequency stepwise on successive pulse sequence iterations. The rf and microwave pulse lengths and intensities and other timing parameters are held to fixed values. These parameters can affect the ENDOR amplitudes, as discussed in Section 4.

4. AMPLITUDES IN POLARIZATION TRANSFER PULSED ENDOR

The amplitudes in the ENDOR spectrum are determined by the transition probabilities based on the parameters of the spin Hamiltonian and also on the parameters in the pulsed experiment. The effect of these experimental parameters can be analyzed by separately discussing the contributions in the preparation, mixing, and detection periods. This analysis is facilitated by expanding the discussion in Section 3 to the more frequently encountered situation of excitations in an inhomogeneously broadened EPR spectrum.

For the simple four-level system discussed in Section 3, the selectivity of the excitation in the preparation period is readily defined since there are only two allowed EPR transitions. The excitation is selective if only one of the two allowed transitions is irradiated. In the case of an inhomogeneously broadened EPR line, an arbitrary position in the EPR absorption spectrum encompasses a large number of hyperfine couplings arising from nuclei with a widely varying range of values for the hyperfine interaction. In this case, a microwave pulse of width t_p irradiates a portion of the EPR spectrum determined by the Fourier components of the pulse. For a square pulse, this frequency is roughly given by $\Delta_p \approx 1/t_p$. The process of selectively irradiating a portion of the inhomogeneous linewidth is referred to as "burning a hole." A hole is burned into the EPR spectrum if Δ_p is less than the inhomogeneous linewidth.

The hole burned into the spectrum by a single microwave excitation pulse is schematically illustrated in Fig. 7A. In the example shown, it is assumed that the electron spin polarization for the EPR transitions corresponding to the center of the saturation hole have been inverted. In the four-level system discussed in Section 3, these are the EPR transitions that are on resonance and have evolved under a π pulse. The EPR transitions

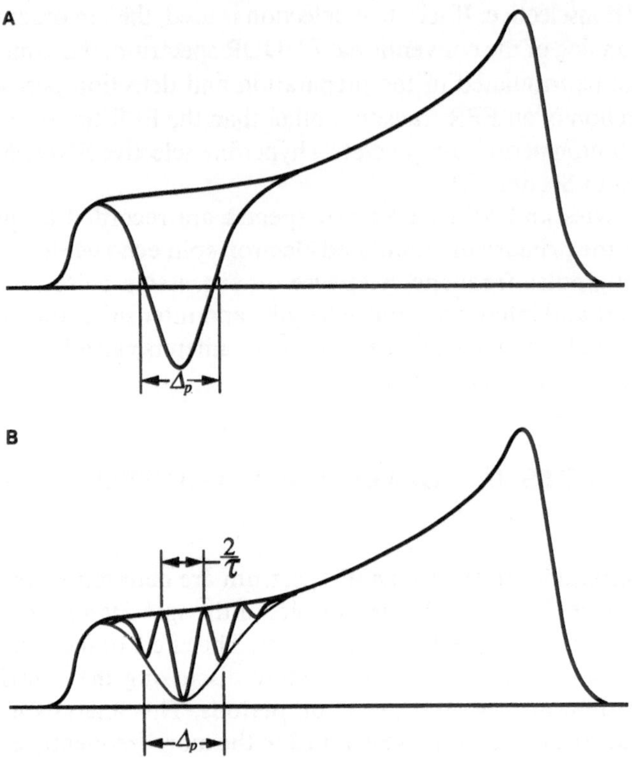

Figure 7. EPR spectra of an inhomogeneously broadened line after the preparation period in the (A) Davies ENDOR and (B) Mims ENDOR pulse sequences.

far away from the center of the hole are off resonance and not affected by the excitation pulse.

In principle, the excitation profile of the saturation hole should be directly determined by the microwave pulse waveform. In practice, the microwave transmitter pulses generally do not have ideal rectangular pulse shapes. The finite probe Q also tends to broaden the sharp transitions of a rectangular pulse. Furthermore, spin dynamics such as the spectral and instantaneous diffusion mechanisms tend to broaden the excitation profile burned into an inhomogeneously broadened EPR line. These spin-dynamics mechanisms distribute the electron spin polarization across the EPR spectrum. This spectral diffusion of spin energy may not be significant on the short time scale of the microwave excitation pulse in the preparation period but can become significant over the longer time scale during which the rf pulse is applied. As a result of these mechanisms, the electron spin polarization $P(\nu)$ created by the selective microwave preparation pulse can be described by a Lorentzian function (Mims *et al.*, 1961)

$$P(\nu_{\pm}, \Delta_p) = \frac{4(\nu_{\pm} - \nu_n)^2}{4(\nu_{\pm} - \nu_n)^2 + (2\Delta_p)^2} \tag{9}$$

where $P(\nu_{\pm}, \Delta_p)$ represents the sublevel spin polarization and Δ_p is the width of the Lorentzian hole. Under limiting conditions in which ideal $\pi/2$ and π pulses are assumed, the width can be related to the microwave magnetic field intensity: $\Delta_p = g_e\beta_e H_1/h$, where H_1 is the magnitude of the microwave magnetic field intensity.

The pattern of electron spin polarization imposed on an inhomogeneously broadened EPR line created by the pair of pulses is shown in Fig. 7B. For two-pulse excitation, the electron spin polarization that forms the saturation hole burned into the EPR line by the first microwave pulse is multiplied by $\cos(2\pi A_j \tau)$, where $A_j = (2n - 1)/2\tau$, n is a positive integer, and A_j is the hyperfine coupling of the jth nucleus. Mims (1972a) has referred to this sinusoidal modulation pattern shown in Fig. 7B as a serrated or sawtooth pattern of polarization.

4.1. Hyperfine Contrast Selectivity

In an inhomogeneously broadened EPR line, the transfer of polarization described in Section 3 of electron spin polarization between hyperfine sublevels in the mixing period corresponds to the transfer of electron spin polarization from within the saturation hole to another part of the EPR spectrum for which the electron polarization is at equilibrium. The sublevel transition or polarization transfer therefore corresponds to a displacement of the polarization within the EPR spectrum. If the displacement remains entirely within the hole, no change in EPR signal intensity can be observed. Of course, the polarization can be transferred only if the rf frequency matches a hyperfine sublevel transition. This hyperfine sublevel transition connects two EPR transitions separated by the hyperfine coupling. As was originally noted by Davies (1974), this establishes the important criterion that $A > \Delta_p/2$ in order to observe an ENDOR signal in the Davies ENDOR experiment. Nuclei with hyperfine couplings $A << \Delta_p/2$ do not contribute to the ENDOR spectrum.

If the excitation pulse were to define an infinitely narrow (i.e., a δ-function) saturation hole, the selectivity of the preparation pulse could be defined as in the four-level system discussed in Section 3. In general, the profile of the saturation hole is not known, but clearly it is not described accurately by a δ function. The width Δ_p of the saturation hole determines the minimum hyperfine coupling that can be observed. Since the slope of the saturation profile is a smoothly varying function, ENDOR lines corresponding to displacements of the polarization partially, but not completely, out of the hole have attenuated amplitudes in the ENDOR spectrum. The choice of the microwave pulse width in the preparation period therefore imposes a selectivity on the magnitude of the hyperfine couplings observed in the ENDOR spectrum. This *hyperfine contrast selectivity* can be used to suppress ENDOR amplitudes selectively for nuclei with small hyperfine couplings (Thomann and Bernardo, 1990). This suppression effect can be useful if ENDOR lines from nuclei with small hyperfine couplings overlap lines from nuclei with large hyperfine couplings (Thomann and Bernardo, 1990c; Doan *et al.*, 1991; Thomann and Mims, 1992).

Hyperfine contrast selectivity is illustrated in the Davies ENDOR spectra for stellacyanin shown in Fig. 8. The frequency sweep range between 1 and 31 MHz shown in Fig. 8 encompasses the spectral range expected for ENDOR transitions arising from ^1H and ^{14}N nuclei with both small and large hyperfine couplings (Rist *et al.*, 1970; Roberts *et al.*, 1980, 1984; Thomann *et al.*, 1991b). Except for the adjustments of the excitation bandwidth of the microwave pulses, the spectra were recorded under identical experimental conditions. Assuming for simplicity that the microwave pulses are ideally square, the excitation bandwidths in Figs. 8A–C are roughly $\Delta_p = 2.50$, 10.0, and 25.0 MHz, respectively. ENDOR transitions from protons with weak hyperfine couplings centered about the proton Larmor frequency of 12.23 MHz (indicated by the ν_H marker) are most intense in Fig. 8A.

Weaker, broader lines centered at 9, 18, and 22 MHz are also observed in Fig. 8A. The intensity of these lines can be significantly enhanced by increasing the excitation bandwidth of the microwave pulse in the preparation period, as shown in Fig. 8B. On further increasing the excitation bandwidth, the weakly coupled protons are almost completely suppressed, as illustrated in Fig. 8C.

Comparison of Figs. 8A–C reveals that the weakly coupled protons have partially obscured ENDOR lines from two broad lines denoted by N_a and N_b centered at roughly 9.5 and 17.5 MHz. These are nitrogen lines (see Section 7), which have been assigned to the imino nitrogens of two imidazole ligands on the cupric ion shown in Fig. 1. For these nitrogens, $|A|/2 \gg \nu_n$, so that the ENDOR transitions are centered at $|A|/2$. The splitting from the nitrogen Zeeman coupling shown in Fig. 8 identifies these ENDOR lines as arising from nitrogen nuclei. The ^{14}N ENDOR line for nitrogen N_a partly overlaps some residual part of the proton ENDOR spectrum, accounting for the increased amplitude at 10 MHz. The amplitude asymmetry of the ^{14}N ENDOR line for nitrogen N_b arises from electron–nuclear coherence effects as described in Section 4.4. The nitrogen quadrupole interactions are not resolved, presumably because the couplngs are small and lead only to line broadening.

The two ENDOR lines at 22.6 and 24.8 MHz, identified by $H_{\beta 1}$ and $H_{\beta 2}$ in Fig. 8C, are assigned to the ν_+ ENDOR transitions from the two meth-

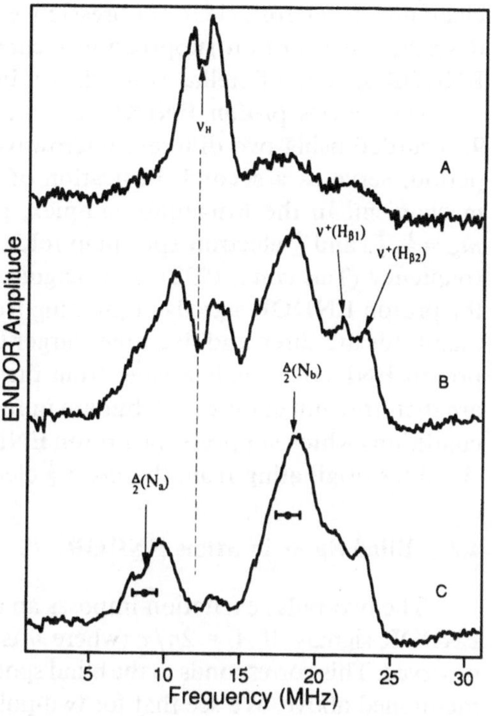

Figure 8. Hyperfine contrast selectivity illustrated in the Davies ENDOR spectrum of stellacyanin at $g = 2.20$ under different microwave pulse widths: t_p^{mw} (A) 0.40 μs, (B) 0.10 μs, and (C) 0.04 μs. Other experimental conditions: $T = 1.7$ K; $\nu_{mw} =$ 8.843 GHz; $t_p^{rf} = 5.50$ μs.

ylene protons on a thiolate ligand coordinated to the cupric ion (see Fig. 1). These protons have hyperfine couplings of roughly 18 and 21 MHz. The low-frequency ENDOR transitions ν_- for these methylene protons, which arise from the NMR transitions in the second electron-spin manifold, are expected at 1.7 and 4.0 MHz. These low-frequency ENDOR transitions are difficult to observe because of the small hyperfine enhancement factor at low rf frequencies.

Figure 8B represents an intermediate excitation condition, in which the weakly coupled protons are partially but not completely suppressed while the ENDOR lines arising from the strongly coupled nitrogen and proton nuclei are significantly enhanced. Under these conditions the spectral line assignments are complicated by the overlap of the weakly coupled proton lines with the strongly coupled nitrogen lines.

Note in Fig. 8 that the amplitude of the methylene proton ENDOR lines increases along with the increase in the nitrogen ENDOR amplitudes when a wider saturation hole is created in the preparation period. This follows since $A = 2(\nu_\pm - \nu_n)$, indicating that the hyperfine contrast selectivity mechanism is not dependent on the type of nucleus, i.e., on g_n. The mechanism for hyperfine contrast selectivity is therefore neither a homonuclear nor a heteronuclear suppression effect, and it is also not correctly described as a proton suppression technique, as is evident in the Davies ENDOR spectra of stellacyanin shown in Fig. 8.

The Davies proton ENDOR spectra of $Mn[(H_2O)_6]^{2+}$ shown in Fig. 9, recorded using two different microwave pulse widths in the preparation period, serve as a second illustration of the hyperfine contrast selectivity mechanism. In the Mn–aquo complex, proton ENDOR signals from the $m_S = \frac{1}{2}$, $\frac{3}{2}$, and $\frac{5}{2}$ electron-spin manifolds overlap near the proton Larmor frequency (Tan *et al.*, 1992). The magnitude of the hyperfine couplings for the proton ENDOR signals originating from the $m_S = \frac{3}{2}$ and $\frac{5}{2}$ electron-spin manifolds are three and five times larger than for the $m_S = \frac{1}{2}$ manifold. The proton ENDOR signals arising from these higher electron-spin manifolds are therefore not suppressed, but are in fact enhanced for preparation pulse conditions which suppress the proton ENDOR signals associated with small A values originating from the $m_S = \frac{1}{2}$ electron spin manifold.

4.2. Blind Spots in Mims ENDOR

The two-pulse excitation imposes an additional constraint for observing ENDOR signals. If $A = 2n/\tau$ (where n is an integer), no ENDOR signal is observed. This corresponds to the blind spots in the Mims ENDOR experiment mentioned above. We see that for two-pulse excitation, the condition that $A > \Delta_p/2$ is relaxed, but at the expense of introducing blind spots where no ENDOR is observed. The blind spots are only observed, however, when more selective microwave excitation pulses are used (Mims, 1965).

Although the blind spots are a complication in the Mims ENDOR experiment, they can also be an aid in the assignment of hyperfine coupling values. The blind spots can be easily identified by recording several spectra using different τ values. The upper bound on τ is determined by the electron spin phase memory time, which is usually on the order of 2–3 μsec in metalloproteins. The lower bound on the usable τ is determined by the spectrometer dead time.

The Mims ENDOR experiment offers excellent sensitivity for detecting ENDOR lines from nuclei with small hyperfine interactions. In the Mims ENDOR experiment, the interpulse delay time τ determines the relative amplitudes of the ENDOR transitions. The amplitudes of weakly coupled nuclei are suppressed least by choosing large values of τ, because such values create a sawtooth pattern in which the "teeth" of the pattern of polarization are more closely spaced. A direct comparison between the ENDOR amplitudes in the Davies and Mims experiments recorded under similar microwave excitation pulse conditions for the blue copper protein stellacyanin is shown in Fig. 10. The $n = 1$ blind spots are indicated by the arrows in Fig.

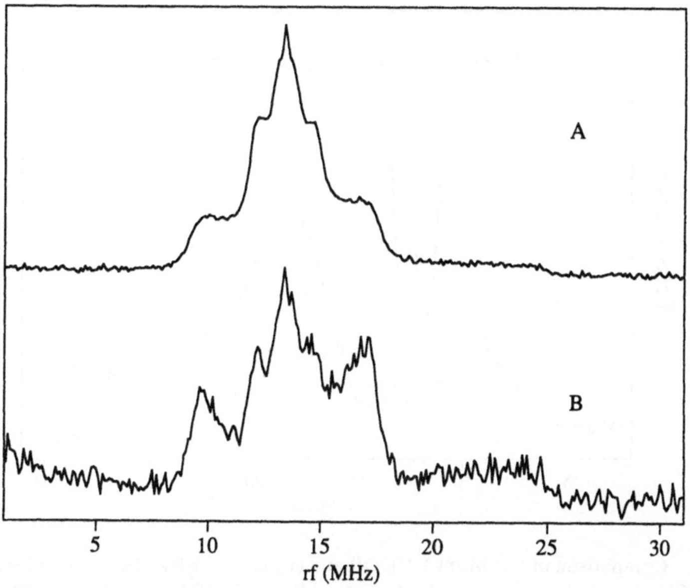

Figure 9. Davies ENDOR spectra of a frozen solution of $Mn(H_2O)_6^{2+}$ recorded with different microwave pulse widths: t_p^{mw} (A) 0.10, 0.10, 0.20 μs; (B) 0.03, 0.05, 0.03 μs. Other experimental conditions: $T < 1.3$ K; $\nu_{mw} = 8.835$ GHz; $t_p^{rf} = 7.50$ μs.

10A. The Mims ENDOR spectrum was recorded using microwave excitation pulses of 0.05 μsec and 0.30 μsec. This Mims spectrum may be contrasted with the Davies ENDOR spectrum in Fig. 10B, which was recorded with a similar microwave excitation pulse width. Note that the weakly coupled protons dominate the Mims ENDOR spectrum, while the ENDOR signals from these protons are almost completely suppressed in the Davies ENDOR spectrum.

The Mims ENDOR experiment is also more effective in detecting ENDOR signals from nuclei which have both small g_n and small hyperfine coupling values. For example, a ^{14}N nucleus with a very small hyperfine coupling such that $A \ll 2\Delta_p$ can be more readily detected by Mims ENDOR because the Davies experiment would require a very long preparation pulse. In the Mims experiment short pulses can be used by choosing a long interpulse delay time τ. ^{14}N ENDOR for ^{14}N nuclei with very small hyperfine couplings has been observed in Mims ENDOR studies of copper model complexes (Poel *et al.*, 1983; Reijerse *et al.*, 1986) and for nitrogen ligands coupled to an iron-sulfur center in the enzyme hydrogenase (Thomann *et al.*, 1991a).

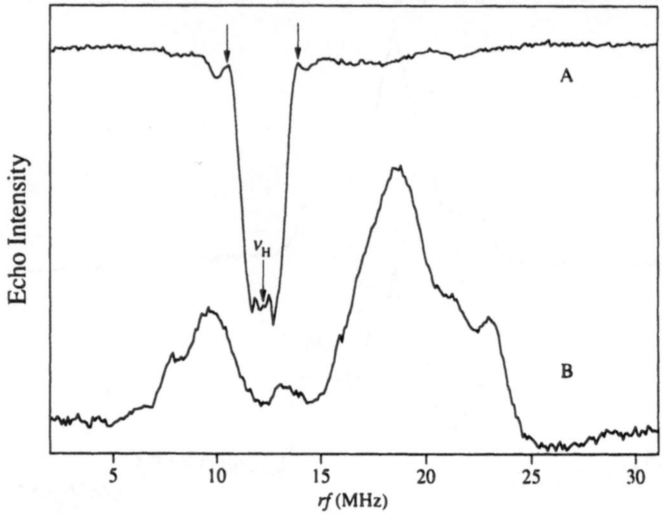

Figure 10. Comparison of (A) Mims ENDOR and (B) Davies ENDOR spectra of stellacyanin at $T = 1.6$ K, $g = 2.20$, and $\nu_{mw} = 8.843$ GHz. Mims ENDOR conditions: $t_p^{mw} = 0.05\ \mu s$; $t_p^{rf} = 5.50\ \mu s$; $\tau = 0.30\ \mu s$. Davies ENDOR conditions: $t_p^{mw} = 0.04, 0.02, 0.04\ \mu s$; $t_p^{rf} = 5.50\ \mu s$; $\tau = 0.50\ \mu s$.

4.3. Hyperfine Enhancement Factor

The polarization transfer in the mixing period is proportional to $\cos(\theta_{R\pm})$, where $\theta_{R\pm} = \gamma_{eff} H_2 t_{rf}$ is the Rabi nutation angle (Rabi, 1957), γ_{eff} is the effective nuclear gyromagnetic ratio (in which the hyperfine enhancement factor is taken into account), H_2 is the rf magnetic field intensity, and t_{rf} is the rf pulse length. The $\pm$ subscript in $\theta_{r\pm}$ refers to the two electron-spin manifolds corresponding to the eigenvalues m_S. For simplicity, we drop this subscript in subsequent discussions below. The effective nuclear gyromagnetic ratio is related to the nuclear gyromagnetic ratio by the hyperfine enhancement factor ε. For an isotropic hyperfine interaction A, the enhancement factor is given by (Abragam and Bleaney, 1979)

$$\varepsilon = \left| 1 + \frac{m_S A}{\nu_n} \right| \tag{10}$$

where ν_n is the nuclear Larmor frequency. For proton A values typically encountered in ENDOR studies of metals in proteins and enzymes, the enhancement factor is in the range $0 < \varepsilon < 2$. Nitrogen nuclei directly bound to a metal ion can have large hyperfine interactions, so that the enhancement is typically in the range $2 < \varepsilon < 20$. Significant effects on the ENDOR amplitudes are observed if $\varepsilon \neq 1$.

The periodicity of the polarization transfer in the mixing period can be observed as transient NMR nutations. These transient nutations are known as Rabi oscillations. Of course, in the ENDOR experiment these nutations are detected indirectly through EPR transitions in the detection period. The Rabi oscillations can be observed either by incrementing the rf magnetic field intensity stepwise or by incrementing the rf pulse length on successive iterations of the Davies ENDOR pulse sequence. If the pulse length is incremented, the Rabi oscillations are superimposed on the overall net loss of the electron polarization due to spectral diffusion and spin-lattice relaxation processes. The nutation pattern obtained by incrementing the rf pulse length observed for the ENDOR line at 21.94 MHz in the ENDOR spectrum of stellacyanin is shown in Fig. 11.

The effect of the electron polarization decay can be removed by digital filtering or by subtracting the signal obtained by repeating the Davies ENDOR pulse sequence with no rf power applied. The Rabi oscillations for several rf frequencies in the ENDOR spectrum of stellacyanin are shown in Fig. 12. The sudden decrease in the ENDOR signal intensity at the end of the waveform is a measure of the signal baseline. The advantage of measuring the transient nutation by subtracting the waveform with no rf power applied from the signal with rf power on is that the absolute ENDOR enhancement, defined by the change in EPR susceptibility according to

$$\chi_{\text{ENDOR}} = \Delta\chi_{\text{EPR}} = \frac{\chi_{(\text{rf on})} - \chi_{(\text{rf off})}}{\chi_{(\text{rf off})}} \tag{11}$$

can be measured accurately.

The first maximum in each waveform shown in Fig. 12 corresponds to the maximum polarization transfer for that hyperfine sublevel. The rf pulse length at this maximum corresponds to a π pulse. Since the hyperfine enhancement factor depends on the electron spin state m_s as well as the hyperfine coupling, it can be expected that the rf mixing pulse length at fixed power will not be constant across the ENDOR spectrum. The Rabi oscillations shown in Fig. 12 provide a direct method for assessing the effect of the NMR nutation on the ENDOR amplitudes as a function of the ENDOR frequency.

From Fig. 12 it appears that the effect of a slight missetting from the optimum π pulse condition has much less effect on the ENDOR amplitude than would be expected on the basis of the analysis of the two-level system discussed in Section 3. This is a consequence of the inhomogeneous broaden-

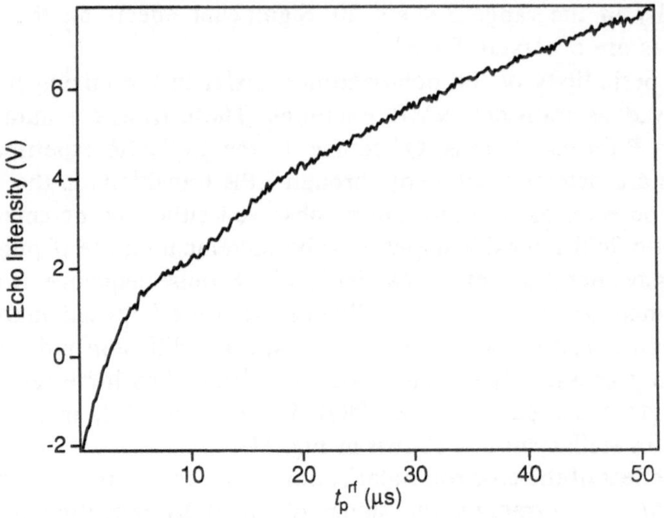

Figure 11. Rabi oscillation superimposed on the electron polarization recovery observed for the ENDOR line of stellacyanin at 21.94 MHz. Experimental conditions are identical to those of Fig. 10B.

ing of the ENDOR lines. However, a more dramatic effect on the ENDOR amplitudes is observed when wider frequency ranges are considered. This becomes evident if one compares the nutation angles θ_2 for protons and nitrogen nuclei. Suppose that the proton and nitrogen nuclear Larmor frequencies are 13 and 1 MHz, respectively. For a proton hyperfine coupling of 10 MHz, $\theta_2 = 0.62\pi$ and 1.38π in the two electron-spin manifolds, where $\theta_2 = \pi$ at 13 MHz. Assuming the same rf pulse length and magnetic field intensity for the nitrogen nuclei as for protons, the flip angles for nitrogen are multiplied by $\gamma_N/\gamma_H \approx 0.07$. For a nitrogen hyperfine coupling of 36 MHz, $\theta_2 = 1.23\pi$ and 1.37π, while for a nitrogen coupling of 18 MHz, $\theta_2 = 0.58\pi$ and 0.72π. This is in good agreement with the approximately 2:1 intensity ratio observed for the two nitrogen ENDOR lines in Fig. 8. The intensities of the proton ENDOR lines for the thiolate methylene protons are lower than expected because of hyperfine anisotropy. The nitrogen and proton ENDOR lines are approximately equal in intensity for Davies ENDOR spectra recorded at g_y.

Davies and Mims ENDOR spectra are typically recorded using a fixed rf pulse length and power. The frequency-dependent nutation angle for the sublevel magnetization is directly manifest in the ENDOR amplitudes observed. At lower rf frequencies, longer rf pulses are required to compensate for the smaller hyperfine enhancement factor. Unfortunately, this

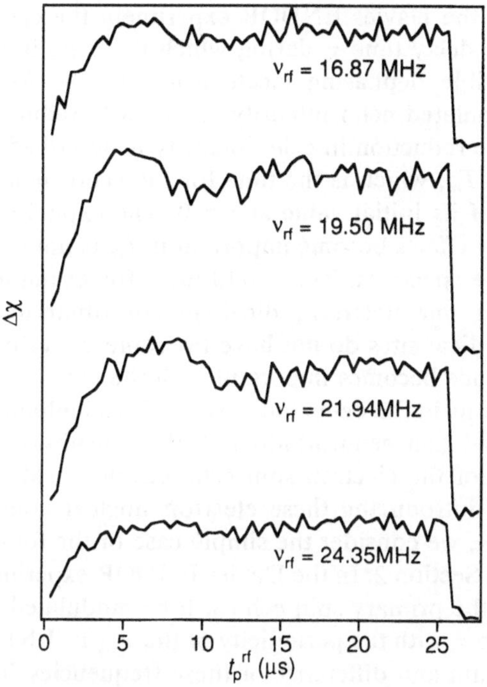

Figure 12. Rabi oscillations observed at several rf frequencies (16.87, 19.50, 21.94, and 24.35 MHz) observed by subtracting the electron polarization recovery as described in the text. Experimental conditions are identical to those of Fig. 10B.

results in a more significant loss of electron polarization, as is evident from the rapid loss in electron polarization in Fig. 11. Note also from Figs. 11 and 12 that the Rabi oscillations tend to dampen rapidly with increasing pulse lengths. This damping can arise from nuclear spin relaxation, from rf field inhomogeneity, or from the dephasing caused by the destructive interference arising from a distribution of ENDOR frequencies. The observation of multiple oscillations in the transient nutation patterns in studies of single crystals (Gemperle *et al.*, 1988; Hofer, 1988; Thomann and Bernardo, 1990a) suggests that the damping is in general not caused by rf field inhomogeneity. It is likely that both spin relaxation and the inhomogeneity of the ENDOR line contribute to the damping.

4.4. Electron and Nuclear Coherence Effects

The relative ENDOR amplitudes can be affected by electron or electron–nuclear coherence effects in the detection period (Thomann and

Bernardo, 1992). In the Davies ENDOR experiment the echo is detected at a finite interpulse delay time τ, during which the echo intensity can be reduced by irreversible dephasing mechanisms. In the Mims ENDOR experiment, the stimulated echo intensity is similarly reduced during the evolution time τ. The reduction in echo intensity is described by the phase memory decay time T_m, which is the time for the echo signal to decay to roughly 37% (e^{-1}) of its initial value at $\tau = 0$. The explicit consideration of the phase memory effects become important if T_m is not constant for all ENDOR lines in the spectrum. This could arise for example if ENDOR lines from more than one electron radical site contribute to the ENDOR spectrum and the radical sites do not have the same T_m values.

Nuclear coherence becomes important with the excitation of semiforbidden EPR transitions in the detection period. The simultaneous coherent excitation of allowed and semiforbidden EPR transitions generates the nuclear modulation of the electron spin echo envelope (Mims, 1972a, b). For the purpose of introducing these electron–nuclear coherence effects into pulsed ENDOR, we consider the simple case of the four-level system (Fig. 4) discussed in Section 2. In the Davies ENDOR experiment, the echo envelope traced by the primary spin echo will be modulated as a function of the evolution time τ, with the periodicity of the ν_+, ν_- ENDOR frequencies as well as the sum and difference of these frequencies. This is known as electron spin-echo envelope modulation, or simply ESEEM. The ESEEM amplitudes have been calculated by Mims (Rowan *et al.*, 1965; Mims, 1972a, b):

$$E_{\text{mod}}(\tau) = 1 - \frac{k}{2} - \frac{k}{2}\left\{\cos(2\pi\nu_+\tau) + \cos(2\pi\nu_-\tau) - \frac{1}{2}\cos[(2\pi\nu_+ + 2\pi\nu_-)\tau]\right.$$

$$\left. - \frac{1}{2}\cos[(2\pi\nu_+ - 2\pi\nu_-)\tau]\right\} \tag{12}$$

where ν_+ and ν_- are the hyperfine frequencies in the two electron-spin manifolds. The modulation depth factor k is given by

$$k = \left[\frac{2\pi\nu_n B}{(2\pi\nu_+)(2\pi\nu_-)}\right]^{1/2} \tag{13}$$

where $B = (1/h)(g_e g_n \beta_e \beta_n)(3\cos\theta \sin\theta)/r^3$. This B term arises from hyperfine anisotropy and must be nonzero in order for envelope modulation to be observed. A quantitative description of the nuclear modulation of the electron spin-echo envelope also requires consideration of additional interactions such as the nuclear quadrupole interaction for nuclei with $I > \frac{1}{2}$ and must include orientational averaging and orientation selectivity effects (Mims, 1972b; McCracken *et al.*, 1988; Dikanov, 1992).

4.4.1. Orientation Selectivity by ESEEM

There are several circumstances in which the nuclear modulation effect can be important. One occurs when the ESEEM arises from one nucleus and the ENDOR spectrum arises from a second nucleus. Even though both nuclei may be coupled to the same electron, it is likely that the orientations and magnitudes of the hyperfine and possibly the quadrupole tensors will not be the same for the two nuclei. There is a correspondence between a given τ value in the spin-echo readout period and an orientation or range of orientations in the ESEEM spectrum. This gives rise to an ESEEM orientation selectivity. This type of orientation selectivity is in addition to the g-factor orientation selectivity that is a consequence of the magnetic field position at which the ENDOR spectrum is recorded.

4.4.2. Partial Excitation Effects

Another mechanism in which the ESEEM effect can determine the ENDOR amplitudes is by the partial excitation of lines in an ESEEM spectrum. This mechanism can be operative even if only one nucleus is responsible for the ESEEM and ENDOR spectra. This situation can be encountered for protons where the hyperfine interaction is large and of the order of $A \approx 2\nu_n$. It is then likely that the ν_- ENDOR frequency gives rise to ESEEM while the ν_+ ENDOR frequency is observed only as an ENDOR transition. The amplitude of the ν_+ ENDOR line in Davies ENDOR spectra recorded at several values of τ is then amplitude modulated with periodicity $\cos[2\pi(\nu_-)\tau]$. If the modulation arises from more than one nucleus, then the relative amplitudes in the ENDOR spectrum can depend on the periodicity of the ESEEM frequencies for the individual nuclei as well as on the combination frequencies that arise from the product function E_{mod} of the modulation. If the nuclei are not coupled to each other, the modulation for more than one nucleus is given simply by the product of E_{mod} for the individual nuclei (Mims, 1972b):

$$E_{\text{mod}} = \Pi_j[E_{\text{mod}}(I_j)] \tag{14}$$

which contains the modulation frequencies for the individual nuclei as well as combination frequencies from the product.

The effects of partial excitation in one electron-spin manifold and of ESEEM orientation selectivity are illustrated in Fig. 13 for two Davies ENDOR spectra of stellacyanin. These spectra were recorded using identical experimental conditions except for the spin-echo delay time in the detection period. The spin-echo envelope modulation waveform recorded using the same microwave pulse conditions used in the detection period of the Davies ENDOR experiment is shown as an inset in Fig. 13.

One pronounced difference between these two spectra is that the relative ENDOR amplitudes for the two methylene proton lines at 21.4 and 23.1 MHz are observed to depend on τ. The low-frequency partners for these proton ENDOR lines are expected at 3.0 and 1.3 MHz, respectively. For this small splitting, the microwave pulses used in the detection period have sufficient bandwidth to excite the allowed and semiforbidden EPR transitions coherently. If the nuclear modulation arises from the 1.3-MHz line, the minima and maxima for the intensities of the 23.1-MHz line would be separated by roughly 0.38 μsec, as is qualitatively consistent with the observed results. Another possibility is that the τ-dependent amplitudes shown in Fig. 13 arise from the excitation of semiforbidden EPR transitions

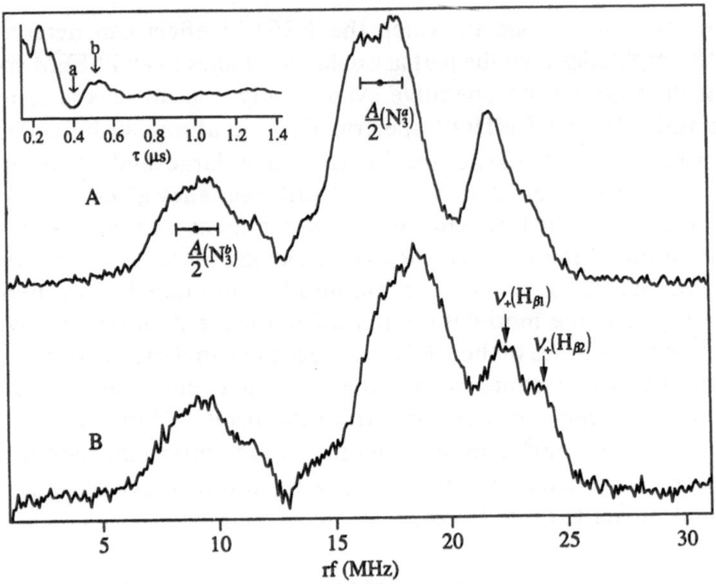

Figure 13. Davies ENDOR spectra of stellacyanin at $g = 2.20$ recorded (A) at the two-pulse modulation minima at $\tau = 0.40$ μs and (B) at the maxima at $\tau = 0.51$ μs, as indicated in the inset. Other experimental conditions: $T < 1.3$ K; $\nu_{mw} = 9.236$ GHz; $t_p^{mw} = 0.05, 0.03, 0.05$ μs; $t_p^{rf} = 2.00$ μs.

associated with the ^{14}N nucleus from the remote amino nitrogen of the imidazole ligand.

The shape of the ^{14}N ENDOR line centered at 17.5 MHz is also a function of the echo delay time τ in the detection period. This serves as an example of the ESEEM orientational selectivity effect discussed above. The ^{14}N ENDOR line has been assigned to the histidine imidazole imino nitrogen N_{3a}, as shown in Fig. 1 (Rist *et al.*, 1970; Roberts *et al.*, 1980). The dominant contribution to the modulation of the echo waveform comes from the amino nitrogen, N_{1a} in Fig. 1, on the same histidine imidazole ligand (Mims and Peisach, 1978). The ESEEM orientation selectivity improves the resolution of the ENDOR line because a smaller number of nuclei corresponding to a subset of the hyperfine (and quadrupolar) orientations contribute to a given spectral range in the ENDOR spectrum.

5. PULSED VERSUS CW ENDOR

In principle, the Davies ENDOR experiment is the pulsed analog of the CW ENDOR experiment. In both experiments the NMR transitions for paramagnetically coupled nuclei are detected indirectly by electron-spin polarization transfer between hyperfine sublevels. In practice the use of pulsed excitation for both the EPR and NMR channels often offers distinct advantages over CW excitation methods. One of the most significant is that the ENDOR amplitudes in the pulsed experiment do not depend on the detailed balance between the electron and nuclear spin relaxation rates. Another important advantage is that the baseline in the pulsed ENDOR spectrum is generally flat because the rf power is not applied during the detection period. Rolling baselines are a difficult and common problem in CW ENDOR studies of transition-metal complexes. As we demonstrate in this section, these advantages make it possible to record extremely wide frequency sweeps that encompass the ^{1}H, ^{14}N, and central metal $^{63,65}Cu$ ENDOR for a blue copper protein. However, the pulsed experiment is also experimentally more complex, requires that the sample have a sufficiently long electron-spin coherence lifetime so that one can observe a free-induction or spin-echo signal, and (as of this writing) requires the construction of a spectrometer.

The artificial partitioning of the CW ENDOR experiment into preparation, mixing, and detection periods, as shown in Fig. 5A, serves as a useful pedagogical approach, especially when comparing the CW and pulsed ENDOR experiments. The CW ENDOR experiment can be described as a polarization transfer pulsed ENDOR experiment in which the three separate preparation, mixing, and detection periods are all combined into a single period in which all three processes occur simultaneously. For the preparation period we focus on the effects of the microwave irradiation, ignoring for the moment the rf irradiation field. The microwave irradiation field is essentialy monochromatic (i.e., single frequency), so that it corresponds to a selective excitation. As an example, in Fig. 4, assume that transition $E1 \leftrightarrow E4$ is irradiated. The microwave field partially saturates the EPR transition, creating sublevel polarization. This is analogous to the effect of a microwave pulse in the preparation period of a pulsed ENDOR experiment such as the Davies ENDOR sequence shown in Fig. 5B. If the EPR transition were completely saturated, the populations of $E1$ and $E4$ would become equal and the EPR signal would vanish. Intense rf irradiation applied to one of the two NMR transitions, either $E1 \leftrightarrow E2$ or $E3 \leftrightarrow E4$, can partially restore the electron-spin polarization between levels $E1$ and $E4$. This irradiation is the equivalent of the polarization transfer in the mixing period in a pulsed ENDOR experiment. The change in electron polarization is detected by the change in EPR signal intensity on the transition $E1 \leftrightarrow E4$.

In the CW ENDOR experiment, the magnitude of the ENDOR enhancement depends on the saturation of the EPR and NMR transitions (Kevan and Kispert, 1979). From the Bloch equations, it is readily shown that the condition for partially saturating the EPR transition is $\omega_1^2 T_{1e} T_{2e} \approx 1$, while the condition for strongly saturating the NMR transition is $\omega_2^2 T_{1n} T_{2n} \gg 1$. Since $T_{1e}^{-1} \gg T_{1n}^{-1}$, achieving strong ENDOR signals requires the use of strong rf fields to drive the NMR transitions. This criterion naturally suggests the use of pulsed excitation, since $\omega_1^{-2} \ll T_{1e} T_{2e}$ and $\omega_2^{-2} \ll T_{1n} T_{2n}$ for coherent pulsed excitation of the EPR and NMR transitions, respectively.

The ENDOR enhancement in the CW ENDOR experiment also depends on the detailed balance between the electron and nuclear relaxation rates (Allendorfer and Maki, 1970). The EPR transition must be at least partially saturated, while the NMR transition (i.e., transitions between the nuclear hyperfine sublevels) must be strongly saturated. At the same time, some faster spin-relaxation pathways such as electron–nuclear cross relaxation must exist so that the equilibrium spin polarization can be reestablished. Otherwise, the ENDOR signal is rapidly quenched. However these relaxation pathways also compete with the sublevel polarization transfer. This means that the CW ENDOR signal depends on the detailed balance between the effective electron and nuclear relaxation rates. Here "effective" refers to all relaxation pathways that tend to restore the equilibrium spin polarization. Since the electron and nuclear relaxation rates usually have very different temperature dependences, this sensitivity to the detailed balance between relaxation rates often restricts the observation of the ENDOR signal to a narrow temperature range.

In the pulsed experiments, the polarization transfer occurs on a time scale shorter than T_{1e}. The ENDOR amplitudes are reduced if the cross-relaxation is appreciable during the mixing period, but the amplitudes do not depend on the ratio of electron and nuclear relaxation rates. The temporal separation of the processes of creating polarization, polarization transfer, and detection also provides the advantage of optimizing the efficiency of the experiment. In the CW experiment, the microwave irradiation should be strongly saturating the EPR transition when creating polarization but nonsaturating for the detection of the polarization transfer. This idealized situation is more closely approached in the pulsed experiment.

In the pulsed experiment it is usually not possible to observe an EPR free induction for a metal ion in a protein or enzyme. In some cases it may also not be possible to observe an electron spin echo. This occurs if the electron phase memory relaxation time is shorter than the receiver recovery time (known as the dead time of the spectrometer). In this case, the CW ENDOR experiment may be possible and in fact preferred, while the pulsed ENDOR experiments based on polarization transfer may fail. This situation is most likely to be encountered in ENDOR studies of the liquid state. Coherence transfer ENDOR spectra of phenoxy radicals in the liquid state have been reported (Forrer *et al.*, 1990), but the ENDOR lines in these spectra suffer from power broadening compared to the CW ENDOR experiment.

5.1. Cu ENDOR

The experimental advantages of the pulsed excitation methods in ENDOR spectroscopy discussed immediately above are illustrated by the Davies ENDOR spectrum, shown in Fig. 14, of the blue copper protein stellacyanin. This spectrum was recorded over the frequency range from 1 to 200 MHz near g_x in the EPR spectrum using similar pulse conditions to those of the spectrum shown in Fig. 8C. A distinct advantage of pulsed ENDOR methods is that large microwave excitation bandwidths can be used without the line broadening arising from coherence effects. As discussed in Section 4.1, these large microwave intensities significantly enhance the ENDOR amplitudes for nuclei with very large hyperfine couplings, such as for the ^{63,65}Cu ENDOR transitions. In the low-frequency range of the spectrum in Fig. 14, the two strongly coupled nitrogen ENDOR lines are observed at 9.5 and 17.5 MHz and the two methylene proton lines are observed near 23 and 25 MHz. The amplitudes from the weakly coupled protons are strongly suppressed using this microwave preparation pulse condition.

The broad lines at frequencies above 30 MHz in Fig. 14 are assigned to the ENDOR transitions from the Cu nucleus. The lines are rather broad,

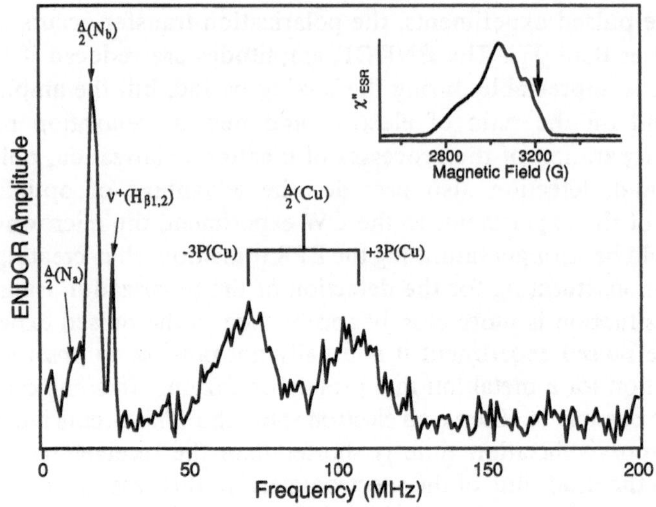

Figure 14. Davies ENDOR spectrum of stellacyanin emphasizing the copper ENDOR signals. Inset shows the position in the EPR spectra where the ENDOR spectrum was recorded. Experimental conditions: ν_{mw} = 8.884 GHz; t_p^{mw} = 0.04, 0.02, 0.04 μs; t_p^{rf} = 2.00 μs; T = 1.65 K; rf step size, 1.00 MHz/point; pulse repetition time, 30 ms; total acquisition time, 80 min.

so that the ENDOR lines from the ^{63}Cu and ^{65}Cu nuclei could be not be separately resolved. Recording of the ^{1}H, ^{14}N, and Cu ENDOR transitions in one wide frequency spectrum has the advantage of making possible a more direct comparison of the relative ENDOR intensities for these nuclei. The advantage of having a good baseline becomes especially important when comparing relative ENDOR amplitudes or when the ENDOR lines are broad and cover a wide frequency range, as in Fig. 14. The Cu ENDOR transitions are very intense, actually comparable to the amplitudes of the strongly coupled nitrogen and methylene proton ENDOR transitions. Note also that the ^{63,65}Cu, ^{1}H, and ^{14}N ENDOR lines all have the same phase. This is in contrast to the CW ENDOR experiment on stellacyanin, where the ^{63,65}Cu ENDOR lines have opposite phases to those of the ^{1}H and ^{14}N ENDOR lines (Roberts *et al.*, 1980).

The assignment of the ENDOR lines to the ^{63,65}Cu, ^{1}H, and ^{14}N ENDOR transitions is indicated in Fig. 14. For the cupric ion, the ^{63}Cu and ^{65}Cu nuclei, both with $I = \frac{3}{2}$, are expected to give rise to three ENDOR transitions in each of the two electron-spin manifolds, so that up to six lines could in principle be observed in the ENDOR spectrum. At magnetic field strengths typical in X-band ENDOR studies, the Cu hyperfine interaction is significantly larger than the nuclear Zeeman and quadrupole interactions. The Cu ENDOR transitions are therefore expected to be centered at $|A|/2$. Additional line splittings are expected from the nuclear Zeeman and quadrupole couplings, as described in Section 2. However, only two Cu ENDOR lines are resolved in the ENDOR spectra recorded at $g = 1.967$ ($g_x = 2.03$). The value $A_x(\text{Cu}) = 176$ MHz is obtained by assigning the center of the doublet to $|A|/2$, as shown in Fig. 14. This is consistent with the hyperfine splitting observed in the EPR spectrum at g_x. The splitting of the doublet is much larger than $2\nu_{\text{Cu}}$ (which is approximately 7.30 MHz for ^{63}Cu and 7.81 MHz for ^{65}Cu at this magnetic field). This splitting is therefore assigned to the quadrupole interaction, which gives $P_x(\text{Cu}) = 5.8$ MHz. The unresolved Cu Larmor-frequency splitting results in the broadening of the two ENDOR lines centered at $|A|/2 \pm 3|P|/2$.

The hyperfine coupling for the Cu nucleus is much larger than the excitation bandwidth arising from the microwave pulse in the preparation period. Thus the four possible allowed EPR transitions are not excited with equal probability by the microwave pulse. Consequently, for spectra recorded near g_x, the intensity of the ENDOR transitions between $m_I = \frac{1}{2}$ and $-\frac{1}{2}$ are expected to be suppressed compared to the transitions between $m_I = \frac{3}{2}$ and $\frac{1}{2}$ and between $m_I = -\frac{3}{2}$ and $-\frac{1}{2}$.

5.2. Amplitudes in CW and Pulsed ENDOR

In conventional CW ENDOR spectroscopy, the relative ENDOR amplitudes generally are not simply related to the number of nuclei contributing to the ENDOR transition (Kevan and Kispert, 1979). This arises for several reasons. First, the observation of an ENDOR enhancement in the CW experiment requires that both the EPR and NMR transitions are strongly irradiated. Consequently, the transition intensities are no longer linearly proportional to the applied power of the driving radiation fields. Second, as has been discussed above, the amplitudes in CW ENDOR depend on the sensitive balance between the effective electron and nuclear spin relaxation rates. As a result, the ENDOR amplitudes do not simply correlate to the number of nuclei that contribute to the ENDOR line. In samples with strongly inhomogeneously broadened EPR spectra, the spin dynamics can be further complicated by spectral diffusion. Spectral diffusion disperses the energy of the applied microwave power across the EPR spectrum and consequently diminishes the ENDOR amplitude. The complex spin dynamics in CW ENDOR spectra often result in ENDOR lines which are distorted by rapid-passage effects (Weger, 1960). Rapid-passage conditions result in lines that are frequency-shifted in the direction of the rf sweep and a frequency-dependent phase shift.

In the limiting case of negligible relaxation effects, the amplitudes of the CW ENDOR lines have been shown to increase with the square of the ENDOR frequency (Dalton and Kwiram, 1972). This result can be easily derived from the hyperfine enhancement factor in the limit of small sublevel

Rabi nutation angles. The use of a low rf power level corresponds to a small Rabi nutation angle. The ENDOR amplitude EA is proportional to $\cos(\theta_R)$. For small $\theta_{R\pm}$, $\cos(\theta_{R\pm})$ can be expanded in the power series:

$$\begin{aligned} EA(\nu_{rf}) &= \frac{1}{2}[1 - \cos(\theta_{R\pm})] \\ &\approx \frac{\theta_{R\pm}^2}{4} \\ &\approx \frac{(m_S A)^2}{(2\nu_n)^2} \\ &= \left\{\frac{m_S^2}{\nu_n^2}\right\}(\nu_\pm - \nu_n)^2. \end{aligned} \tag{15}$$

The last equality explicitly indicates that the amplitude of the two ENDOR lines at ν_+ and ν_- for a given $I = \frac{1}{2}$ nucleus are related by the square of the rf frequency.

In principle, the amplitudes in pulsed ENDOR spectra are directly related to the number of nuclei that contribute to the transition. The ENDOR lines are not affected by rapid-passage conditions and are not a function of the ratio of electron and nuclear spin relaxation rates. The amplitudes in pulsed experiments are a function of the hyperfine contrast selectivity, hyperfine enhancement factor, relaxation processes that result in the loss of electron polarization during the mixing period, and finally of electron and nuclear coherence effects in the detection period. In practice, the amplitudes in the pulsed ENDOR spectrum can be quantitatively analyzed within the accuracy that these experimental conditions can be measured.

The correspondence between signal amplitudes and number of nuclei is significantly improved in pulsed ENDOR experiments compared to the CW ENDOR experiment. However some experimental parameters are not readily amenable to quantitation. For example, spectral diffusion affects the efficiency with which weakly coupled nuclei are suppressed by the hyperfine contrast mechanism. The ENDOR amplitudes of weakly coupled nuclei will not be as effectively suppressed if spectral diffusion fills in the polarization hole burned into the EPR line by the microwave preparation pulse. Thus spectral diffusion affects the amplitudes of weakly coupled nuclei to a greater extent than the more strongly coupled nuclei.

5.3. Linewidths

The rf pulse length determines the ultimate resolution that can be obtained in the pulsed ENDOR spectrum. The maximum resolution that can be obtained for an rf pulse length t_p is roughly given by $2/t_p$. The increased spectral resolution that can be obtained using "softer" (longer with lower-power) rf pulses can readily be observed in single-crystal studies. An ENDOR linewidth completely free of the pulse-induced broadening effects can be obtained by the indirect detection of the sublevel free-induction signal (Hofer *et al.*, 1986). However, in both cases this increased spectral resolution is likely to be obtained at the expense of lower sensitivity. This has been demonstrated in a comparative study of a single crystal (Thomann and Bernardo, 1990a). The lower sensitivity occurs because electron polarization will be lost during the time of the longer rf pulse length or during the time of free precession. This is particularly true in ENDOR studies of metal sites in metal proteins and enzymes, where the ENDOR lines are likely to be inhomogeneously broadened by the anisotropy of the hyperfine and quadrupole interactions.

6. ESEEM VERSUS ENDOR SPECTROSCOPY

The combination of ENDOR and ESEEM spectroscopies can be particularly powerful as complementary approaches for studying the coordination of ligand nuclei to transition-metal ions in proteins and enzymes (Reijerse *et al.*, 1986; Thomann and Mims, 1993). This is demonstrated in a combined ESEEM and Mims ENDOR study of the iron–sulfur active site in the hydrogen-activating enzyme hydrogenase (Thomann *et al.*, 1991a). ESEEM spectra obtained by cosine Fourier transformation of the three pulse-stimulated echo waveforms and recorded at two microwave excitation frequencies, 7.930 and 9.087 GHz, are shown in Fig. 15A and C, respectively. The ESEEM spectra were recorded near the extrema ($g_{max} = 2.10$) of the EPR spectrum to take advantage of the enhanced resolution afforded by g-factor orientation selectivity (Jin, 1989; Cornelius *et al.*, 1990).

Pure nuclear quadrupole frequencies can be observed in ESEEM spectroscopy when the local magnetic field vanishes in one of the electron-spin manifolds (Mims and Peisach, 1978; Flanagan and Singel, 1987). This occurs when the nuclear Zeeman field is cancelled by the hyperfine field. The three ESEEM frequencies in this manifold are pure ^{14}N nuclear quadrupole frequencies ν_0, ν_-, ν_+, from which the quadrupole coupling parameters can be directly calculated. The ^{14}N quadrupole coupling constants are related to the ESEEM frequencies by (Lucken, 1969) $\nu_+ = K(3 + \eta)$, $\nu_- = K(3 - \eta)$, $\nu_0 = 2K\eta$, where $e2qQ/h = 4K$. At exact cancellation, the ν_+ frequency reaches its minimum value (Mims and Peisach, 1978; Flanagan and Singel, 1987; Reijerse and Keijzers, 1987). Based on these considerations, the three ESEEM lines at 0.50, 3.39, and 3.89 MHz recorded at 2706 G (Fig. 15A), which shift to

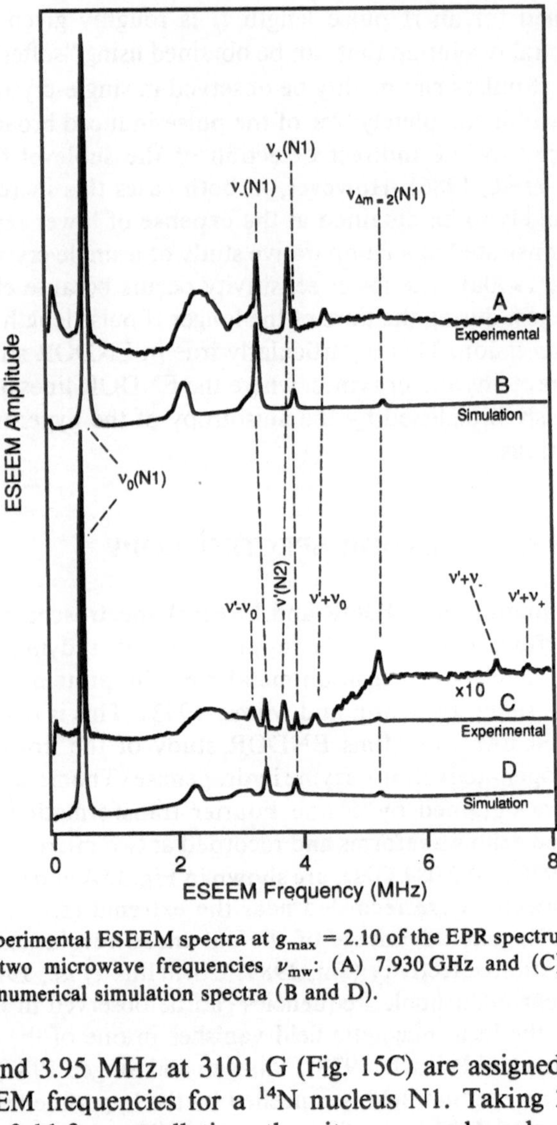

Figure 15. Experimental ESEEM spectra at $g_{max} = 2.10$ of the EPR spectrum of hydrogenase I recorded at two microwave frequencies ν_{mw}: (A) 7.930 GHz and (C) 9.087 GHz, and corresponding numerical simulation spectra (B and D).

0.58, 3.37, and 3.95 MHz at 3101 G (Fig. 15C) are assigned to the ν_0, ν_-, and ν_+ ESEEM frequencies for a ^{14}N nucleus N1. Taking 2706 G as the approximate field for cancellation, the nitrogen quadrupole parameters are $K = 1.21$ MHz, $n = 0.21$.

The ESEEM lines at 2.4 and 5.4 MHz (Fig. 15A) arise from the second electron-spin manifold in which the hyperfine field adds to the Zeeman field. The broader 2.4-MHz peak is assigned to the overlapping (ν_0, ν_-) lines. These ESEEM lines are usually broadened beyond detectability. In the present case, however, they are most likely observed because of g-factor orientation selectivity. The 5.23-MHz peak, which shifts to 5.36 MHz at 3101 G, is assigned to the ν_+ ESEEM frequency from this second electron spin manifold.

The ESEEM line assignments for N1 were further confirmed by numerical simulations in which g-factor orientation selectivity is explicitly incorporated (Thomann *et al.*, 1991a). Using the quadrupole parameters obtained directly from the experimental spectra, the simulated spectra, Fig. 15B and D, quantitatively reproduce both the ESEEM frequencies and intensities and the correct shift of the ESEEM frequencies with a shift in microwave excitation frequency. No adjustment of simulation parameters was made between the simulations at the two excitation frequencies.

The line at 3.68 MHz, $\nu'(N2)$ in Fig. 15A, shifts to 3.87 MHz in Fig. 15C. This is a shift of 0.44 kHz/G, which is close to the gyromagnetic value for the nitrogen nucleus (0.31 kHz/G). The peak $\nu'(N2)$ is therefore assigned to a second nitrogen, N2, which is far from the cancellation condition, so that pure nuclear quadrupole frequencies can not be observed. The additional lines in Fig. 15A, at 3.18 and 4.18 MHz, are sum and difference ESEEM combination peaks $\nu_0(N1) \pm \nu'(N2)$ between N1 and N2. Sum peaks between $\nu'(N2)$ and the ν_- and ν_+ lines from N1 can also be observed at 7.1 and 7.6 MHz, respectively. These are shown on the expanded vertical scale of the spectrum above Fig. 15C. At 3101 G, the $\nu'(N2)$ line shifts to 3.87 MHz so that, as expected, the $\nu_0(N1) \pm \nu'(N2)$ peaks shift to 4.45 MHz and 3.29 MHz, respectively.

A direct comparison of the ESEEM and Mims ENDOR spectra is shown in Fig. 16. Nitrogen ENDOR lines in hydrogenase I are not observed by conventional ENDOR spectroscopy (Telser *et al.*, 1987). This is likely due to the small magnitude of the ^{14}N hyperfine coupling interaction, which is close to the ^{14}N nuclear Larmor frequency at this magnetic field. Nitrogen ENDOR lines from both nitrogens, N1 and N2, are however observed in the Mims ENDOR spectrum, as shown in Fig. 16A.

The combined ESEEM and pulsed ENDOR spectroscopies illustrate the complementary nature of the two techniques. ENDOR amplitudes for ^{14}N nuclei with small hyperfine couplings are strongly attenuated, because of the low hyperfine enhancement factor. In contrast, the ESEEM intensity is largest when the hyperfine coupling cancels the nuclear Zeeman coupling so that the nucleus is effectively in zero magnetic field. This has the consequence (given other conditions equal) that ESEEM lines are more intense at lower frequencies, while ENDOR lines are more intense at higher frequencies. Thus the peak at 5.4 MHz is much weaker in the ESEEM spectrum than in the ENDOR spectrum. The intensity of these peaks is also determined by the hyperfine anisotropy and the orientation of the quadrupole tensor's principal axes with respect to the g tensor. Thus, the relative intensity of this line in the ESEEM and ENDOR spectra will depend on the g value at which the spectra are recorded.

The combination of the ESEEM and pulsed ENDOR spectroscopies provides a good method for corroborating and cross-checking spectral line

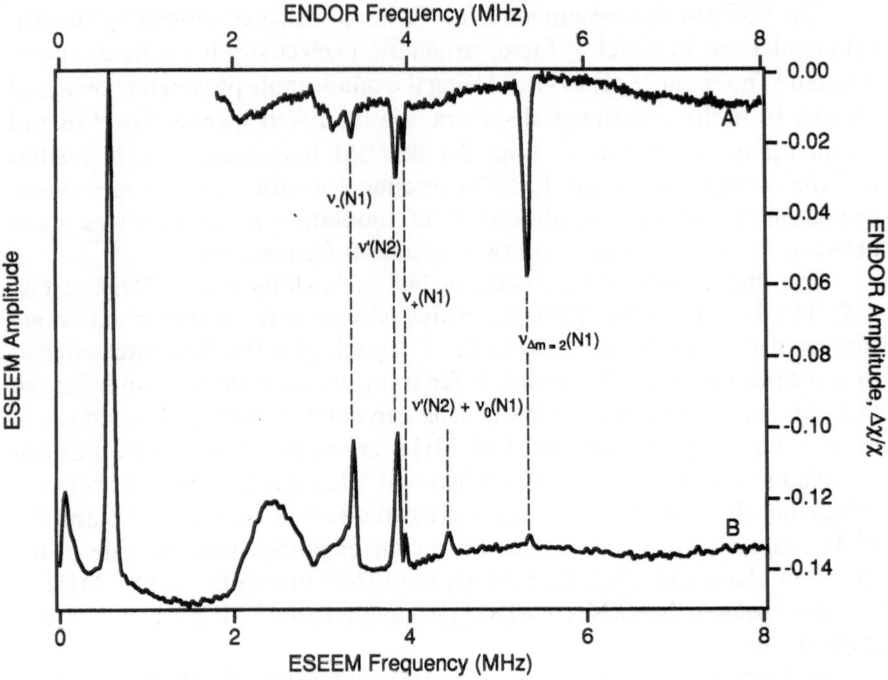

Figure 16. Comparison of (A) Mims ENDOR spectrum and (B) three-pulse ESEEM spectrum of hydrogenase I at the $g = g_{max} = 2.10$ position in the EPR spectrum.

assignments. Observation of ESEEM lines requires the coherent excitation of allowed and semiforbidden EPR transitions. This has the consequence that combination frequency lines between nitrogens N1 and N2 can be observed in the ESEEM spectrum. On the other hand, ENDOR transitions are observed by the excitation of allowed EPR and NMR transitions. As a result, ESEEM peaks arising from combination frequencies between nitrogens N1 and N2 observed in the ESEEM spectrum are not observed in an ENDOR spectrum. The absence in the Mims ENDOR spectrum of the peak observed at 4.45 MHz in the ESEEM spectrum is consistent with its assignment as a combination frequency between the two nitrogens N1 and N2.

7. ELECTRON NUCLEAR ELECTRON TRIPLE RESONANCE: 2D ENDOR

A characteristic feature of ENDOR spectra of transition-metal ions in proteins and enzymes is that the peaks are broad with significant overlap (Rist *et al.*, 1970; Stevens *et al.*, 1982; Roberts *et al.*, 1984; Hoffman, 1992).

The ENDOR spectra of the copper protein stellacyanin presented throughout this chapter serves as a typical example. ENDOR lines from protons with small as well as large hyperfine couplings may overlap the nitrogen ENDOR lines. As we demonstrated in Section 4.1, the overlap between ENDOR lines from nuclei with large and small hyperfine couplings may be reduced by the hyperfine contrast selectivity mechanism. Another approach is to record the ENDOR spectra at higher microwave excitation frequencies (Box, 1977; Burghaus, 1988). This is a very effective method for eliminating the overlap of nitrogen and proton ENDOR lines, as has recently been demonstrated for copper proteins by Werst *et al.* (1991). ENDOR lines from protons with small hyperfine couplings shift to the higher ENDOR frequency centered about the proton Larmor frequency, while those of the strongly coupled nitrogen nuclei will remain centered at $A/2$. However, the overlap of ENDOR lines from two nuclei with large hyperfine couplings whose ENDOR frequencies are centered at $A/2$ can not be eliminated either by hyperfine contrast selectivity or by recording the ENDOR spectrum at higher microwave frequencies.

Overlapping ENDOR peaks from strongly coupled nuclei centered at their hyperfine frequencies $A/2$ can be resolved using a pulsed electron nuclear electron triple resonance method (Buhlmann *et al.*, 1989; Thomann and Bernardo, 1990b). The pulse sequence is shown in Fig. 17. It is based on a modification of the Davies ENDOR pulse sequence shown in Fig. 5B, in which separate EPR transitions are irradiated in the preparation and detection periods. The hyperfine sublevel irradiated in the mixing period connects these two EPR transitions. With reference to the energy-level diagram in Fig. 4, suppose the difference, denoted as Δ, between the

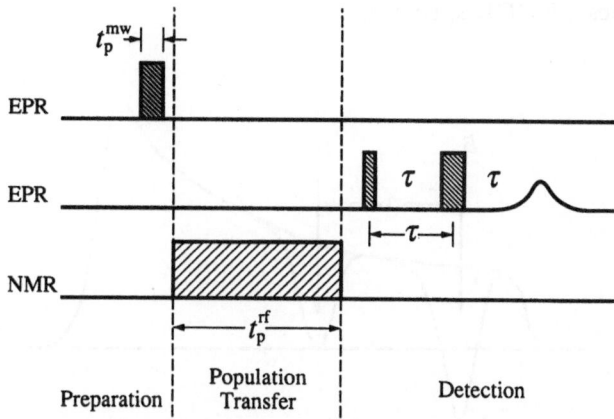

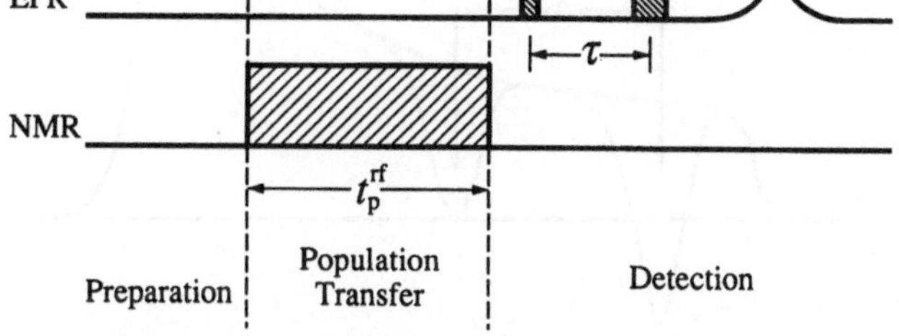

Figure 17. Electron nuclear electron triple resonance pulse sequence partitioned into three periods, as in Fig. 5.

frequencies used to irradiate the EPR transitions in the preparation and detection periods is chosen to be equal to the hyperfine splitting A that separates these EPR transitions. The saturation holes burned into the inhomogeneously broadened EPR spectrum are shown in Fig. 18. The hole from the preparation period is indicated at B_p, and the hole from the detection period is indicated by B_d. These positions in the EPR spectrum are separated by Δ.

The combination of exciting two separate EPR transitions and one nuclear hyperfine transition constitutes a triple resonance experiment or, effectively, a combined ELDOR (electron electron double resonance) and ENDOR experiment, i.e., an ELDOR–ENDOR experiment. This is a 2D ENDOR experiment which can be performed in one of two modes. One mode generates a hyperfine-selective (HS) ENDOR spectrum. The other mode generates an EPR subspectrum.

7.1. Hyperfine Selective ENDOR Spectroscopy

The HS ENDOR spectrum is recorded in a manner analogous to the Davies ENDOR experiment. The HS ENDOR experiment proceeds by incrementing the rf frequency of the sublevel mixing pulse stepwise on successive pulse-sequence iterations. A sublevel polarization transfer is detected only for those nuclei for which $A = \Delta$. The polarization transfer is therefore observed as a decrease in the EPR susceptibility of the transition irradiated in the detection period. This is in contrast to the effect of the polarization transfer in the Davies ENDOR experiment, where an increase in the EPR susceptibility of the transition irradiated in the detection period is observed. Transitions in the HS ENDOR spectrum therefore have the appearance of an emission line rather than an absorption line as observed in the Davies ENDOR spectrum.

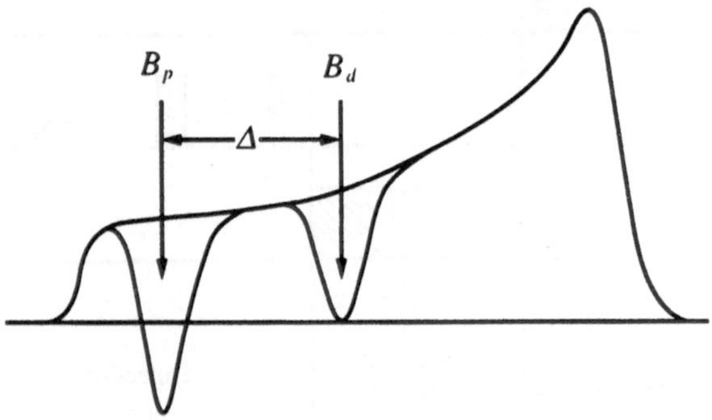

Figure 18. The EPR spectrum of an inhomogeneously broadened line, showing the inversion hole at B_p after the preparation period and the position B_d of the detection pulses.

The HS ENDOR spectrum obtained with the triple resonance pulse sequence in Fig. 17 preselects ENDOR transitions from those nuclei with $A = \Delta$. When more than one hyperfine transition is possible in each electron-spin manifold, each HS ENDOR spectrum is composed of sublevel (hyperfine) transitions from those nuclei with a preselected hyperfine coupling determined by $\Delta = A_i$, where A_i is the hyperfine coupling for the ith nucleus. Each HS ENDOR spectrum is therefore a slice of a two-dimensional ENDOR spectrum where one axis corresponds to the sublevel mixing frequency, as in the standard ENDOR experiment, while the other axis corresponds to the difference frequency Δ. The spectral resolution along the Δ axis is determined by the bandwidths selected for both the microwave inversion and spin-echo detection microwave pulses. In this 2D ENDOR plot, the conventional ENDOR spectrum corresponds to the spectrum along the $\Delta = 0$ axis. It is the inverted image of the projection onto the $\Delta = 0$ axis of all HS ENDOR spectra.

Hyperfine-selective ENDOR spectra can be recorded by switching between two microwave frequencies (Thomann and Bernardo, 1990b), as indicated in Fig. 17, or by jumping the magnetic field during the mixing period (Buhlmann *et al.*, 1989). The change in the magnetic field preselects the magnitude of the hyperfine coupling for the nucleus observed in the HS ENDOR experiment. The triple resonance technique eliminates the possible signal degradation associated with rapid magnetic field switching. On the other hand, the microwave bandwidth of the ENDOR probe must be large enough to accommodate the frequency difference for the two EPR transitions. For large bandwidths, the lower Q required reduces the sensitivity.

Several HS ENDOR spectra are shown in Fig. 19, along with the Davies ENDOR spectrum ($\Delta = 0$), Fig. 20A, for the high-pH form of stellacyanin (Thomann *et al.*, 1991b). When $\Delta = 19$ MHz, Fig. 19B, two lines are observed in the HS ENDOR spectrum. The line centered at 9.5 MHz when $\Delta = 19$ MHz is readily identified as arising from a nitrogen with $A = 19$ MHz. The second line, at 22.4 MHz, must arise from a proton whose low-frequency partner is not observed because of the weak hyperfine enhancement factor at low rf frequency. The intensity of the lower-frequency nitrogen line centered at 9.5 MHz is weak in Fig. 19B in part because of the lower hyperfine enhancement factor and in part because of hyperfine anisotropy. The HS ENDOR spectrum observed for $\Delta = 36$ MHz is shown in Fig. 19C. A broad line centered at 17.5 MHz and split by twice the nitrogen Larmor frequency is observed. This splitting clearly identifies this as a nitrogen ENDOR line with $A = 35$ MHz.

The Davies ENDOR spectra for the native (pH = 7) and high-pH forms of stellacyanin are identical except for a broad peak at 21 MHz observed in the high-pH form. This ENDOR line overlaps both the ^{14}N ENDOR

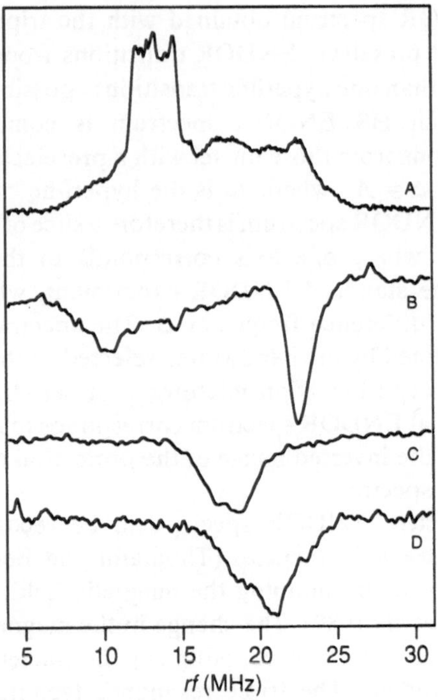

Figure 19. HS ENDOR spectra of the high-pH form of stellacyanin obtained with the pulse sequence in Fig. 17 under several values of the microwave frequency difference Δ: (A) 0 MHz, (B) 19 MHz, (C) 36 MHz, and (D) 42 MHz.

line at 17.5 MHz and the ^{1}H ENDOR lines near 22 MHz. If this new line arises from a new proton coupling it would correspond to $A = 20$ MHz. If it arises from a new nitrogen coupling it would correspond to $A = 42$ MHz. As shown in Fig. 19B, this ENDOR line is not observed in the HS ENDOR spectrum with $\Delta = 19$ MHz. On the other hand, it is observed in the HS ENDOR spectrum recorded with $\Delta = 42$ MHz. The line splitting identifies this line as a nitrogen with a hyperfine coupling of roughly $A = 42$ MHz. This ENDOR line is assigned to an amide nitrogen, based on the similarity of the magnitude and anisotropy of the hyperfine coupling for this third nitrogen with those reported for the nitogen hyperfine couplings in Cu(II)-amide complexes (Calvo *et al.*, 1980). The amide coordination is also consistent with resonance Raman (Penfield, 1984) and molecular modeling data (Fields *et al.*, 1992).

7.2. EPR Subspectra

A variation of the 2D ENDOR experiment is convenient for identifying the hyperfine coupling for selected ENDOR lines without the need to first

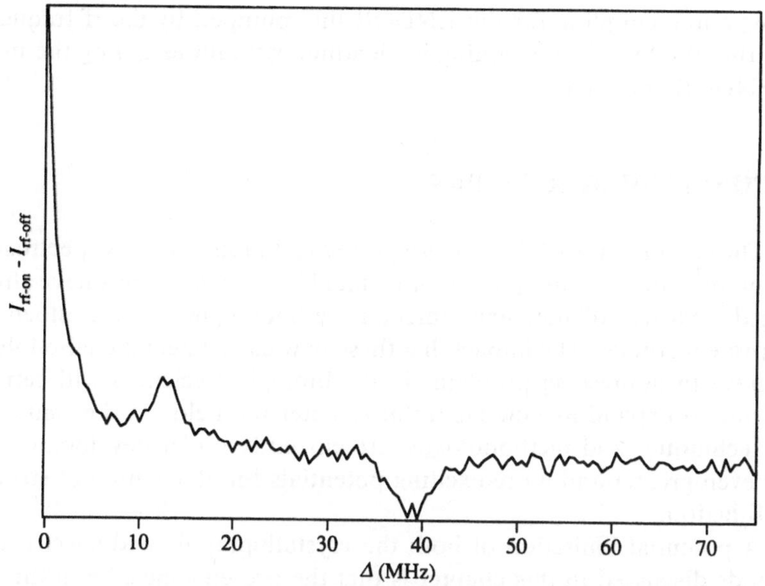

Figure 20. EPR subspectrum for the 20-MHz ENDOR signal of stellacyanin, recorded using the pulse sequence in Fig. 17. Experimental conditions: $T = 1.7$ K; $g = 2.07$; $\nu_{mw} = 9.080$ GHz (detection); $t_p^{mw} = 0.80, 0.30, 0.60$ μs: $t_p^{rf} = 8.00$ μs; $\tau = 1.15$ μs.

assign the ENDOR transitions. This is accomplished by setting the rf frequency to the ENDOR line whose hyperfine coupling is desired. The experiment proceeds by repeating the pulse sequence shown in Fig. 17 and incrementing the difference Δ between the preparation and detection microwave frequencies stepwise on successive pulse sequence iterations. A polarization transfer is observed only if $\Delta = A$. Since Δ is known, A is immediately known.

The spectrum obtained by sweeping Δ generates an EPR subspectrum. In practice, it is convenient to isolate the polarization transfer that occurs in the mixing period by taking the difference between signals recorded with the rf power on and with the rf power off. The EPR subspectra then correspond to a polarization difference, and the polarization transfer is observed as a negative excursion from the baseline. An EPR subspectrum recorded for the blue copper protein stellacyanin is shown in Fig. 20. This EPR subspectrum is observed with the rf frequency set to 20 MHz in the ENDOR spectrum. The large positive peak centered at $\Delta = 0$ is one-half of the width of the hole burned into the EPR line by the preparation pulse. The positive peak at roughly 13 MHz is at the proton nuclear Larmor frequency. This peak arises from proton nuclear spin flips due to semiforbidden EPR transitions. The side hole at $\Delta = 39$ MHz corresponds directly to the hyperfine coupling for the ENDOR line pumped by the rf frequency. Note that the hyperfine coupling is identified without assigning the line in the ENDOR spectrum.

8. CONCLUDING REMARKS

The combination of the modern pulsed techniques in EPR spectroscopy with modern pulse techniques in solid-state NMR spectroscopy have already created a variety of new experimental methodologies for metalloenzyme and protein studies. The impact that these new experimental methodologies will have in addressing problems in the biological sciences will certainly continue to expand as new laboratories enter the field. At the same time, new techniques and methodologies are currently under development that offer even greater and more exciting potentials for elucidation of structure and function.

A potential limitation of both the crystallographic and spectroscopic methods discussed in this chapter is that the proteins must be in the solid state. For crystallographic studies, the protein must also, obviously, be crystallized. For the spectroscopic studies, single crystals are not necessary, but most experiments must be performed at low temperature. Structural data deduced from physical methods must be interpreted with the caveat that the data on the solid state may not be entirely relevant to the physiologic state. While it is unlikely that the protein coordination structure near the metal sites is perturbed in the frozen solution state, it is nevertheless always a possibility.

REFERENCES

Abragam, A., and Bleaney, B., 1970, *Electron Paramagnetic Resonance of Transition Metal Ions*, Clarendon Press, Oxford.

Adams, M. W. W., 1987, *J. Biol. Chem.* **262**:15,054–15,061.

Allendorfer, R. D., and Maki, A. H., 1970, *J. Magn. Reson.* **3**:396.

Atheron, N. M., 1973, *Electron Spin Resonance: Theory and Applications*, Halsted Press (Wiley), New York.

Bowman, M. K., Massuth, R. J., and Yannoni, C. S., 1993, in *Pulsed Magnetic Resonance: NMR, ESR, Optics* (D. M. S. Bagguley, ed.), Oxford Science Publishers, New York, pp. 423–445.

Box, H. C., 1977, *Radiation Effects: ESR and ENDOR Analysis*, Academic Press, New York.

Brown, I. M., Sloop, D. J., and Ames, D. P., 1969, *Phys. Rev. Lett.* **22**:324–326.

Buhlmann, C., Schweiger, A., and Ernst, R. R., 1989, *Chem. Phys. Lett.* **154**:285-291.

Burghaus, O., Toth-Kischkat, A., Klette, R., and Möbius, K., 1988, *J. Magn. Reson.* **80**:383.

Calvo, R., Oseroff, S. B., and Abacke, H. C., 1980, *J. Chem. Phys.* **72**:760–767.

Chen, J.-S., Mortenson, L. E., and Palmer, G., 1976, *Iron and Copper Proteins*, Plenum Press, New York, pp. 68–86.

Coleman, P. M., Freeman, H. C., Guss, J. M., Murata, M., Morris, V. A., Ramshaw, J. A. M., and Vekatappa, M. P., 1978, *Nature* (*London*) **272**:319–324.

Cornelius, J. B., McCracken, J., Clarkson, R. B., Belford, R. L., and Peisach, J., 1990, *J. Phys. Chem.* **94**:6977–6982.

Dalton, L. R., and Kwiram, A. L., 1972, *J. Chem. Phys.* **57**:1132–1145.

Davies, E. R., 1974, *Phys. Lett. A* **47**:1–4.

Dikanov, S., 1992, *Electron Spin Echo Spectroscopy*, CRC Press, Boca Raton, Florida.

Doan, P. E., Fan, C., Daroust, C., and Hoffman, B., 1991, *J. Magn. Reson.* **95**:196.

Dorio, M. M., and Freed, J. H. eds., 1979, *Multiple Electron Resonance Spectroscopy*, Plenum Press, New York.

Erbes, D. L., Burris, R. H., and Orme-Johnson, W. H., 1975, *Proc. Natl. Acad. Sci. USA* **72**:4795–4799.

Ernst, R. R., Bodenhausen, G., and Wokaun, A., 1987, *Principles of NMR in One and Two Dimensions*, Clarendon Press, Oxford.

Fee, J. A., 1975, *Structural Bonding* (*Berlin*) **23**:1–60.

Feher, G., 1956, *Phys. Rev.* **103**:834.

Fields, B. A., and Guss, J. M., 1992, *J. Mol. Biol.* (preprint).

Flanagan, H. L., and Singel, D. J., 1987, *J. Chem. Phys.* **87**:5606–5611.

Forrer, J., Pfenninger, S., Eisenegger, J., and Schweiger, A., 1990, *Rev. Sci. Instrum.* **61**:3360.

Gemperle, C., and Schweiger, A., 1991, *Chem. Rev.* **91**:1481.

Gemperle, C., Schweiger, A., and Ernst, R. R., 1988, *Chem. Phys. Lett.* **145**:1.

Gray, H. B., and Solomon, E. I., 1981, *Copper Proteins*, Wiley-Interscience, New York.

Grupp, A., and Mehring, M., 1990, *Modern Pulse and Continuous Wave Electron Spin Resonance Spectroscopy*, Wiley, New York.

Hagen, W. R., van Berkel-Arts, A., Kruse-Wolters, K. M., Dunham, W. R., and Veeger, C., 1986, *FEBS Lett.* **201**:158–162.

Höfer, P., 1988, *Development of Pulsed ENDOR and Applications to Polyacetylene* (Ph.D. thesis), University of Stuttgart, Stuttgart, Germany.

Höfer, P., Grupp, A., and Mehring, M., 1986, *Phys. Rev. A* **33**:3519–3522.

Hoffman, B. M., 1992, *Acts. Chem. Res.* **24**:164.

Hoffman, B. M., Martinsen, J., and Veuters, R. A., 1984, *J. Magn. Reson.* **59**:110–123.

Hurst, G. C., Henderson, T. A., and Kreilick, R. W., 1985, *J. Am. Chem. Soc.* **107**:7294–7299.

Jin, H.-Y., 1989, *Electron Spin Echo Studies of Metalloproteins* (Ph.D. thesis), City University of New York, New York.

Kevan, L., and Kispert, L. D., eds., 1979, *Electron Spin Double Resonance Spectroscopy*, Wiley-Interscience, New York.

Kurreck, H., Kirste, B., and Lubitz, W., 1988, *ENDOR Spectroscopy of Radicals in Solution*, VCH Publishers, New York.

Liao, P. F., and Hartmann, S. R., 1973, *Phys. Rev. B* **8**:69–80.

Lucken, E. A. C., 1969, *Nuclear Quadrupole Coupling Constants*, Academic Press, London.

Malkin, R., and Malmstrom, B. G., 1970, *Adv. Enzymol.* **33**:177–244.

Malmstrom, B. G., Reinhammer, B., and Vanngard, T., 1970, *Biochim. Biophys. Acta* **205**:48–57.

McCracken, J., Pember, S., Benkovic, S. J., Villafranca, J. J., Miller, R. J., and Peisach, J., 1988, *J. Am. Chem. Soc.* **110**:1069–1074.

Mehring, M., Hofer, P., and Grupp, A., 1986, *Phys. Rev. A* **33**:3523.

Mehring, M., and Hofer, P., 1987, *Ber. Bun. Phys. Chem.* **91**:1132.

Mims, W. B., 1965, *Proc. Royal Soc. London* **283**:452–457.

Mims, W. B., 1972a, *Electron Paramagnetic Resonance*, Plenum Press, New York, pp. 263–285.

Mims, W. B., 1972b, *Phys. Rev. B* **5**(3):2409–2419.

Mims, W. B., and Peisach, J., 1978, *J. Chem. Phys.* **69**(11):4921–30.

Mims, W. B., and Peisach, J., 1981, *Biological Magnetic Resonance*, Vol. 3 (L. J. Berliner and J. Reuben, eds.), Plenum Press, New York.

Mims, W. B., and Peisach, J., 1989, *Advanced EPR Techniques: Applications in Biology and Biochemistry*, Elsevier, Amsterdam.
Mims, W. B., Nassau, K., and McGee, J. P., 1961, *Phys. Rev.* **123**:2059.
Norris, G. E., Anderson, B. F., and Baker, E. N., 1986, *J. Am. Chem. Soc.* **108**:2784–2785.
Peisach, J., Levine, W. G., and Blumberg, W. E., 1967, *J. Biol. Chem.* **242**:2847–2858.
Penfield, K. W., 1984, Ph.D. thesis, Massachusetts Institute of Technology, Cambridge, Massachusetts.
Pilbrow, J. R., 1991, *EPR of Transition Metal Ions,* Clarendon Press, Oxford.
Poel, W., Singel, D. J., Schmidt, J., and van der Weals, J. H., 1983, *Mol. Phys.* **49**:1017–1028.
Rabi, I. I., 1957, *Phys. Rev.* **51**:652.
Reijerse, E. J., and Keijzers, C. P., 1987, *J. Magn. Reson.* **71**:83–96.
Reijerse, E. J., van Earle, N. A. J. M., and Keijzers, C. P., 1986, *J. Magn. Reson.* **67**:114–124.
Rist, G.H., and Hyde, J. S., 1970, *J. Chem. Phys.* **52**(9):4633–4643.
Rist, G. H., Hyde, J. S., and Vanngard, T., 1970, *Proc. Natl. Acad. Sci. USA* **67**:79–86.
Roberts, J. E., Brown, T. G., Hoffman, B. M., and Peisach, J., 1980, *J. Am. Chem. Soc.* **102**:825–829.
Roberts, J. E., Cline, J. F., Lum, V., Freeman, H., Gray, H. B., Peisach, J., Reinhammer, B., and Hoffman, B. M., 1984, *J. Am. Chem. Soc.* **106**:5324–5330.
Rowan, L. G., Hahn, E. L., and Mims, W. B., 1965, *Phys. Rev. A* **137**:61–74.
Schweiger, A., 1982, *Structure and Bonding* **51**:1.
Schweiger, A., and Ernst, R. R., 1988, *J. Magn. Reson.* **77**:512.
Stevens, T. H., Martin, C. T., Wang, H., Brudvig, G. W., Scholes, C. P., and Chan, S. I., 1982, *J. Biol. Chem.* **257**:12,106–12,113.
Tan, X., Bernardo, M., Thomann, H., and Scholes, C. P., 1993, *J. Chem. Phys.* **98** (in press).
Telser, J., Benecky, M. J., Adams, M. W. W., Mortensen, L. E., and Hoffman, B. M., 1987, *J. Biol. Chem.* **262**:6589–6594.
Thomann, H., and Bernardo, M., 1990a, *Spectrosc. Int. J.* **8**:119–48.
Thomann, H., and Bernardo, M., 1990, *Chem. Phys. Lett.* **169**(1,2):5–11.
Thomann, H., and Bernardo, M., 1990, *Pulsed ENDOR Methods for Metalloproteins and Enzymes*, Annual Rocky Mountain Conference on Applied Spectroscopy, Denver.
Thomann, H., and Bernardo, M., 1992, in preparation.
Thomann, H., and Mims, W. B., 1993, *Pulsed Magnetic Resonance: NMR, ESR, and Optics,* Clarendon Press, Oxford, pp. 362–389.
Thomann, H., Bernardo, M., and Adams, M. W. W., 1991a, *J. Am. Chem. Soc.* **113**(18):7044–7046.
Thomann, H., Bernardo, M., Baldwin, M. J., Lowery, M. D., and Solomon, E. I., 1991b, *J. Am. Chem. Soc.* **113**(15):5911–5913.
Tindall, P. J., and Bernardo, M., 1988, *Time Domain ENDOR Studies of Disordered Solids,* 29th Experimental NMR Conference, Rochester, New York.
Van Dijk, C., Grande, H. J., Mayhew, S. G., and Veeger, C., 1980, *Eur. J. Biochem.* **107**:251–261.
Wacker, T., and Schweiger, A., 1992, *Chem. Phys. Lett.* (in press).
Weger, M., 1960, *Bell System Tech. J.* **39**:1013.
Werst, M. M., and Davoust, C. E., 1991, *J. Am. Chem. Soc.* **113**:1533–1538.
Wertz, J. E., and Bolton, J. R., 1986, *Electron Spin Resonance,* Chapman and Hall, London.

8

Transient EPR of Spin-Labeled Proteins

David D. Thomas, E. Michael Ostap, Christopher L. Berger, Scott M. Lewis, Piotr G. Fajer, and James E. Mahaney

1. INTRODUCTION

A central goal in modern molecular biophysics is to understand the physical mechanisms of protein function. This problem is often stated as an issue of structure and function, but static structural information, from electron microscopy or X-ray diffraction, has consistently proven to be insufficient for the understanding of enzyme or receptor mechanisms. This should not be surprising, since each of these mechanisms involves a dynamic sequence of structural changes. In order to elucidate these essential molecular motions, direct measurements of chemical and structural dynamics are required.

The two principal sources of information about dynamics are spectroscopy and biochemical kinetics. The most effective use of these techniques in elucidating biochemical mechanisms occurs when they are applied in parallel (preferably simultaneous) experiments that are time-resolved, usually on the millisecond scale. Most work of this kind has involved the detection of endogenous absorption of fluorescence as a function of time after rapid mixing (usually stopped-flow) of protein and substrates.

David D. Thomas, E. Michael Ostap, Christopher L. Berger, Scott M. Lewis, Piotr G. Fajer, and James E. Mahaney • Department of Biochemistry, University of Minnesota Medical School, Minneapolis, Minnesota 55455.

Biological Magnetic Resonance, Volume 13: EMR of Paramagnetic Molecules, edited by Lawrence J. Berliner and Jacques Reuben. Plenum Press, New York, 1993.

Electron paramagnetic resonance spectroscopy, using nitroxide spin labels, offers three potential advantages over these intrinsic optical signals. First, the use of extrinsic probes permits the labeling of a specific site on a specific protein in a complex multicomponent system. This labeling requires the synthesis of a nitroxide spin label with an appropriate reactive or affinity group, and it sometimes also involves the molecular engineering of a labeling site on the protein (Hubbell *et al.*, 1989). Thus a sufficiently clever chemist can obtain spectroscopic signals from a chosen site in a complex biological system. Second, the EPR spectra of nitroxide spin labels offer high sensitivity to orientation and to rotational motion over a very wide dynamic range (picosecond to millisecond). These are essential physical parameters for defining protein structure and conformation. Third, these EPR spectra offer sufficiently high resolution to study several labeled sites or conformations simultaneously. This is important for the analysis of kinetic mechanisms, involving transitions among several conformational states.

A potential disadvantage of EPR, compared with optical techniques, is its lower absolute sensitivity, which could limit its ability to obtain sufficient signal-to-noise ratios to acquire millisecond time-resolved signals, as are required to monitor many biochemical reactions. However, previous transient EPR studies have shown that acceptable signal-to-noise ratios can be obtained with time resolution less than 10 ms on submillimolar spin concentrations (Hiller and Carmeli, 1990; Hartsel *et al.*, 1987; Hubbell *et al.*, 1987). There have been two principal means of achieving millisecond time resolution in EPR, rapid mixing and light-pulse-induced photochemistry. Rapid-mixing experiments, employing either stopped flow or continuous flow, have been used to detect the binding of nitroxide spin-labeled phosphonium ions to acetylcholine receptors (Hartsel *et al.*, 1987); the reduction of nitroxides by ascorbic acid (Hubbell *et al.*, 1987); the binding of manganese ions to chloroplast H^+ATPase (Hiller and Carmeli, 1990); the substrate-induced reduction of enzyme-bound Cu^{2+} in dopamine β-monooxygenase (Brenner *et al.*, 1989); and the reduction of nitrite by nitrite reductase (Silvestrini *et al.*, 1990). Laser-light pulses have been used with time-resolved EPR to detect transient free radicals in solution (Atkins *et al.*, 1970; reviewed by Poole, 1983) and to detect transient protein-bound paramagnetic intermediates in photosynthesis (reviewed by Miller and Brudvig, 1991; and by Budil and Thurnauer, 1991). These flash-induced transients usually offer higher time resolution than rapid-mixing methods, especially in viscous solutions or intact cells, but they have usually been limited to the study of systems that undergo intrinsic photochemical reactions.

The development of "caged" compounds, i.e., photochemically labile precursors of biologically active compounds (reviewed by McCray and

Trentham, 1989), offers an opportunity to combine key advantages of the rapid-mixing and flash-photolysis techniques. A biologically inert but photoreactive caged compound is incubated with the system in the dark, and a laser flash photochemically removes the cage and produces a rapid (usually submillisecond) and spatially uniform increase in the concentration of the biologically active ligand. Since the caged compound has plenty of time to diffuse throughout the sample in the dark, this technique is applicable to fragile or viscous or cellular systems to which the rapid-mixing technique is not applicable. The dead time at the beginning of the transient is typically much less when the transient is initiated by a light pulse than by rapid mixing. The availability (commercial and otherwise) of a wide range of caged substrates and other ligands opens up the flash technique to a wide range of systems. This technology has been used in the past decade to study biochemical and physiological transients (reviewed by McCray and Trentham, 1989). These physiologically important measurements have made important contributions to the understanding of biological mechanisms, but in order to make further progress in understanding the *molecular* bases of these processes, it is essential to obtain quantitative information about protein structure and dynamics during the transient phase. Spectroscopic techniques are ideal for this type of data acquisition, and nitroxide EPR has proven to be particularly effective, as discussed below.

The present review describes recent work in this laboratory that combines the use of caged compounds with nitroxide spin-label EPR. We have used caged nucleotides to study the orientation and rotational motion of spin labels attached to specific protein components in muscle fibers and membranes. These are complex assembled systems that do not lend themselves to rapid-mixing techniques, so the use of caged nucleotides allows us to correlate the biochemical kinetics of energy-transducing ATPase systems with the underlying protein motions.

2. METHODS

2.1. Caged Compounds

In recent years, experiments with photoreactive caged compounds—biologically inactive photolabile precursors of biologically active substrates—have made it possible to perform transient kinetic experiments in systems where conventional mixing techniques are not rapid enough to match the key biochemical rate constants (reviewed by McCray and Trentham, 1989). These have included caged precursors of ion-channel inhibitors (Lester *et al.*, 1979), metal ions and their chelators (reviewed by Kaplan, 1990), and nucleotides (Kaplan *et al.*, 1978; McCray *et al.*, 1980; reviewed

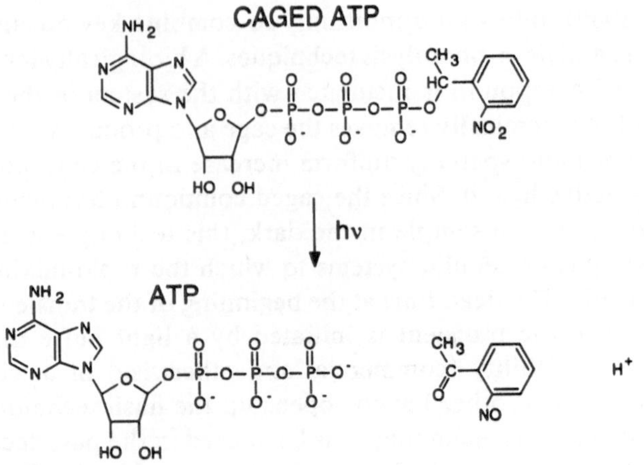

Figure 1. The conversion of caged ATP to ATP by flash photolysis.

by McCray and Trentham, 1989). For example, caged ATP (P^3-1-[2-nitro]phenylethyl-ATP, shown in Fig. 1) is essentially inert in the presence of most ATP-binding proteins. However, when this compound is excited by a pulse of UV light (optimum wavelength around 350 nm), the "cage" is photolytically cleaved, leaving ATP plus a relatively innocuous 2-nitrosoacetophenone group. Under typical physiological (intracellular) solution conditions, this reaction is complete within 1 ms or less, and a single pulse (from a flash lamp or laser) can produce several millimoles of ATP, more than enough to saturate most ATP-binding sites in proteins (McCray and Trentham, 1989). The chemistry used to prepare caged ATP can be used to cage virtually any compound with a terminal phosphate group (Walker *et al.*, 1989).

2.2. Laser Photolysis

Figure 2 illustrates the apparatus used to obtain transient EPR signals by photolysis of caged compounds. The EPR spectrometer can be tuned to record either conventional EPR, which is sensitive to nitroxide orientation and ns rotational motion, or saturation-transfer EPR (ST-EPR), which is sensitive to μs–ms rotational motion (Thomas, 1986; Squier and Thomas, 1986). The sample is mixed with the caged compound in the dark and placed in the appropriate EPR cell and cavity/resonator (allowing optical access). The spectrometer is tuned to an EPR spectral position that offers optimal sensitivity to the expected orientational or motional change. Spectral acquisition is initiated and recorded as a function of time. After a brief initial period to establish the prephotolysis signal level, the caged compound

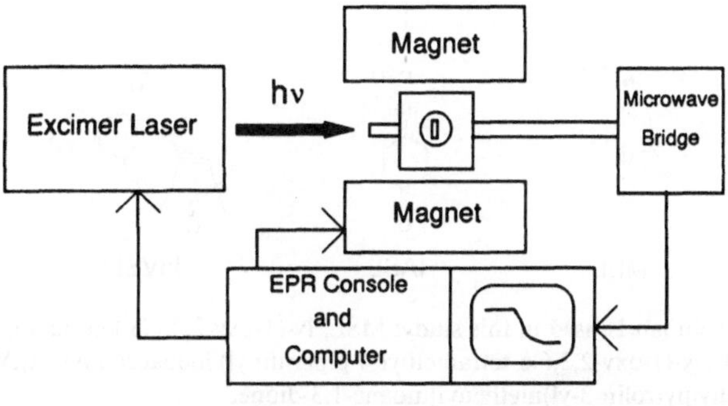

Figure 2. Schematic of the instrumentation used in these studies. Steady-state and transient EPR spectra were obtained with a Bruker ESP 300 EPR spectrometer. The samples were contained in a 20-μl well on a TPX cover plate attached to a quartz flat cell. Sample temperature was maintained to within 0.5 °C using a Bruker temperature controller. Caged ATP was photolyzed using a 10-ns pulse of 351-nm light from a Lambda Physik LPX200i XeF excimer laser, triggered at a preset time by the EPR spectrometer. The laser pulse entered the EPR cavity through an optical port affixed to the front of the cavity.

is photolyzed with a single pulse from an excimer laser. The energy density (mJ/cm^2) is attenuated with glass plates to achieve the desired photochemical conversion, which is measured after each experiment (Berger *et al.*, 1989; Fajer *et al.*, 1990b). For example, with 5-mM caged ATP, 150 mJ/cm^2 results in the production of 600-μM ATP. The resulting transient is then digitized with 5–10 ms time resolution. The entire experimental process is controlled by a microcomputer, which triggers the laser pulse at a specified time during data acquisition. This increases the accuracy of the time base in the recordings, making it possible to perform signal averaging of several repeated transients, thus also increasing the signal-to-noise ratio and making experiments possible under less favorable conditions.

2.3. EPR Instrumentation

Muscle fibers and membranes are macromolecular assemblies offering two challenges that differentiate them from simpler systems capable of being studied with standard solution magnetic resonance methods. First, their complex composition requires site-specific techniques that allow the selective study of a single component under physiological conditions, and second, the molecular interactions in these assemblies often result in motions much slower than the ns motions typical of individual proteins in solution. Introducing site-specific spectroscopic probe molecules, such as nitroxide spin labels, has proven effective for obtaining information about individual protein dynamics in such systems. Examples of these probes are shown in Fig. 3. Further, the development of new biophysical EPR spectroscopic techniques to utilize the information provided by these probes has resulted

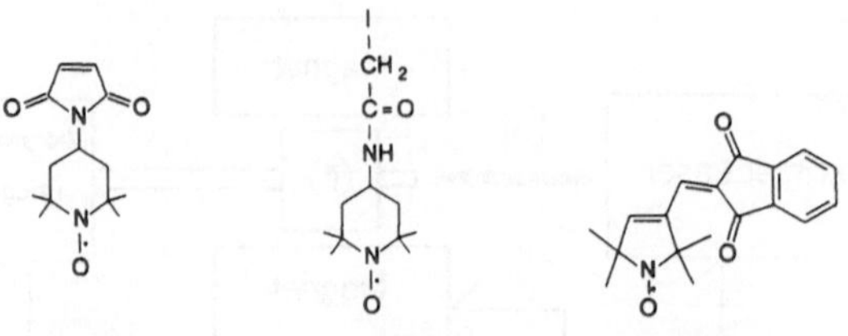

Figure 3. The spin labels used in this study: MSL, N-(1-oxy-2,2,6,6-tetramethyl-4-piperidinyl) maleimide; IASL, N-(1-oxy-2,2,6,6-tetramethyl-4-piperidinyl) iodoacetamide; InVSL, 2-[(1-oxyl-2,2,5,5-tetramethylpyrrolin-3-yl)methenyl]indane-1,3-dione.

in progress toward understanding molecular dynamics in muscle proteins and membranes (Thomas, 1986, 1987). The conventional EPR spectrum of a nitroxide spin label is extremely sensitive to orientation and to ns rotational motion. Nitroxide lipid analogs have provided unique information about the ns motions of lipid hydrocarbon chains, and spin labels attached to protein side chains can be sensitive monitors of protein conformation. However, many protein motions in muscle fibers and membranes are too slow to be measured by conventional EPR. We have shown that ST-EPR (Thomas *et al.*, 1976; Squier and Thomas, 1986) provides the needed sensitivity to slower (μs) rotational motion, and it has proven to be a powerful tool in the study of muscle proteins and membranes (Thomas, 1986, 1987). Recent advances in ST-EPR methodology have improved the precision, sensitivity, and reliability of the method (Thomas and Squier, 1986; Fajer *et al.*, 1986).

Two major problems, however, must be overcome in the use of caged compound technology with spin-label EPR: free radical production and photochemical damage. An intermediate species of the photolysis of caged compounds is a reactive free radical precursor of the 2-nitrosophenone by-product (Fig. 1), which is capable of modifying —SH residues in muscle proteins. To circumvent this problem, the free radical scavenger glutathione is usually added to samples before photolysis of the caged compounds (McCray and Trentham, 1989). Unfortunately, glutathione also reduces spin labels, making it useless for EPR experimental applications. A separate but related problem is that the intense laser flash could be absorbed by the aromatic residues of the proteins, compromising biological function. Therefore, light intensity and time exposure must be carefully adjusted and the resulting biological activity monitored to ensure that the photolysis of caged compounds does not impair function. We have consistently found that substantial concentrations of substrate (e.g., enough ATP to saturate ATPase active sites) can be produced without significant inhibition of biological function (e.g., ATPase activity).

In order to ensure that the flash produces a uniform spatial distribution of the substrate, it is essential to maintain uniform power density over the entire volume of the sample inside the EPR resonator or cavity. We have found that an EPR cavity with an unobstructed optical port provides the best means of delivering a uniform laser pulse to the sample (Fajer *et al.*, 1990b), although the optical slits in the front of standard EPR cavities (accessible upon removal of the front plate) can be used if steps are taken to ensure uniform scattering of the light over the sample (Berger *et al.*, 1989; Lewis and Thomas, 1991). The EPR sample cell is positioned so that the entire face of the sample is illuminated by the laser. Sample turbidity and optical density are also crucial, since these factors attenuate the laser light as it passes through the sample. Therefore, we use a very thin quartz flat cell with a solution path length of 0.5 mm. Since the extinction coefficients of most proteins and lipids are negligible at 351 nm, the limiting factor is usually the concentration of the caged compound. For caged ATP the extinction coefficient is 600 M^{-1} cm^{-1} at 351 nm, so a concentration of 5 mM produces only an absorbance of 0.3, indicating that the light intensity at the back of the cell is reduced by only a factor of about two relative to the front of the cell. Since the mean concentration of ATP produced by a single flash under these conditions is typically about 1 mM, and the K_M(ATP) for our ATPase enzymes is less than 0.05 mM, the enzyme active sites are saturated throughout the sample volume. In samples containing membrane vesicles or myofibrils, the sample concentration must be kept low enough that turbidity does not significantly attenuate the laser beam. In principle, heating of the sample due to intense light flashes could also be a consideration, but we have found that conventional EPR temperature control is sufficient to prevent significant temperature changes.

3. TIME-RESOLVED EPR OF CONTRACTILE PROTEINS

3.1. Background

Force is generated during muscle contraction by the interaction of two proteins, actin and myosin, coupled to a cycle of ATP hydrolysis. The ATPase sites of myosin reside in two globular heads (S1) that project radially from the thick filament. These heads interact with the actin of the thin filament, forming cross-bridges between the two filaments. Molecular reorientation of the myosin cross-bridge in active muscle is an essential feature of many current models for the molecular mechanism of muscle contraction (Huxley, 1969; Huxley and Simmons, 1971). A simplified illustration of the rotating cross-bridge model, which is based on structural, mechanical, and ATPase kinetics data, is shown in Fig. 4 (Taylor, 1979;

Eisenberg and Hill, 1985; Hibberd and Trentham, 1986). The general hypothesis, which is almost certainly correct, is that the changes in the occupancy of the active site induce changes in the protein conformation and motion, which in turn change the interaction between myosin and actin in a way that results in directional sliding of the actin filament past the myosin filament. The specific hypothesis of the model pictured is that the active-site ligand determines the orientation of the myosin head. The head

1. is dissociated from actin at a 45° angle by ATP,
2. cleaves ATP,
3. binds to actin at a 90° angle, and
4. releases products accompanied by rotation to a 45° angle, producing strain, which is then relieved by the sliding motion of the myosin and actin filaments.

A wide range of physical techniques (e.g., electron microscopy, X-ray diffraction, fiber mechanics, light scattering, fluorescence, and NMR) have been applied to test this prediction and to refine the models of muscle contraction (reviewed by Cooke, 1986). Because molecular dynamics plays a central role in muscle contraction, techniques must be used that allow for the high-resolution detection of complex changes in orientational and rotational dynamics. EPR, to measure orientation and ns rotational motion, and ST-EPR, to measure μs rotational motion, have proven to be particularly useful in studies of the molecular dynamics of both myosin and actin in solution and in fibers. EPR and ST-EPR allow for the detection of site-specific probes (spin labels) that provide the kind of selectivity not possible with mechanical or X-ray diffraction studies (reviewed by Thomas, 1987).

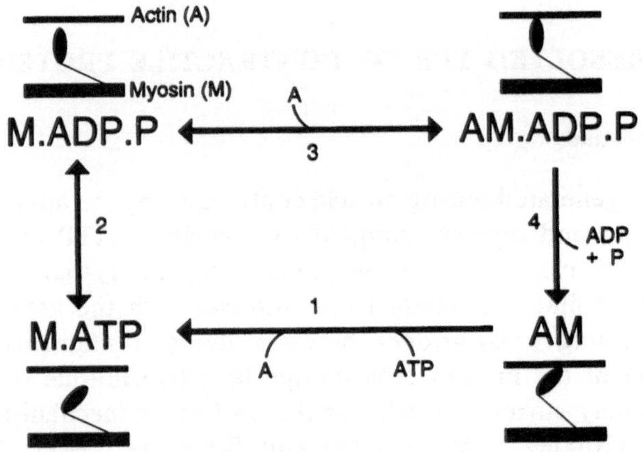

Figure 4. Model describing force generation during muscle contraction.

Spectra of nitroxide spin labels bound to the fast-reacting thiol SH1 (cys-707) on myosin heads report conformational states in purified myosin due to step 2 of the ATPase cycle (Barnett and Thomas, 1987) and verify the high orientational order of heads in muscle fibers in the absence of ATP (Thomas and Cooke, 1980). However, the myosin heads that are detached from actin are highly disordered (Thomas and Cooke, 1980; Barnett and Thomas, 1984; Fajer *et al.*, 1988) and have μs flexibility (Thomas *et al.*, 1980; Fajer *et al.*, 1988), regardless of the ligand bound to the active site (Barnett and Thomas, 1984). In the steady state of contraction, EPR shows that most heads are dynamically disordered, with no preferred angle other than the rigor (AM) angle (Thomas and Cooke, 1980; Barnett and Thomas, 1989). The addition of various nucleotide analogs designed to mimic intermediates in the ATPase cycle also shows no evidence for a well-defined orientation other than that of rigor (Thomas *et al.*, 1988; Fajer *et al.*, 1988). The addition of nucleotide analogs also shows that, in solution, the weakly bound actin-attached states undergo μs rotational motions, while the strongly bound intermediates do not (Berger and Thomas, 1991). The dynamic disorder in steady-state contraction has been further quantitated by time-resolved phosphorescence anisotropy (Stein *et al.*, 1990). The current picture of the cross-bridge cycle is thus much more dynamic than pictured in Fig. 4. The orientation of the head is not rigidly coupled to the identity of the active-site ligand, and the μs rotational transitions appear to be occurring in both the detached (left side of Fig. 4) and attached (right side of Fig. 4) phases of the contraction cycle. The rapid rotations of attached cross-bridges are consistent with submillisecond mechanical transients (Huxley and Simmons, 1971; Brenner, 1987), and with the conclusion from the *in vitro* motility assay that a single ATPase cycle can drive many actomyosin interactions (Harada *et al.*, 1990).

These EPR studies have probed the molecular dynamics of muscle proteins in the absence of nucleotide (rigor), in the presence of nucleotide analogs, and during steady-state ATP hydrolysis. They have yielded a great deal of information about the molecular dynamics of the cycling energy-transduction process. However, measurements during the transient phase and measurements of steady-state phases in preparations that exist too briefly to detect by conventional techniques, are needed to obtain detailed information about the mechanisms of force generation. The use of photoreactive compounds to study transient changes in muscle proteins has been extensive (for reviews, see McCray and Trentham, 1989; Homsher and Millar, 1990). This technology has been used to study the transient kinetics of ATP hydrolysis in skinned muscle fibers (Ferenczi *et al.*, 1984), mechanical transients during different activation states of the muscle fiber (e.g., Goldman *et al.*, 1984; Dantzig *et al.*, 1988, 1991), X-ray diffraction

(Poole *et al.*, 1991), birefringence (Irving *et al.*, 1988), and time-resolved cryoelectron microscopy (Menetret *et al.*, 1991).

Although these investigations have provided detailed information about the transient phases of muscle contraction, they do not directly probe the molecular dynamics of the cross-bridge. A closer correlation between the functional and molecular events is needed for a description of the mechanisms of muscle contraction. This correlation can be achieved by performing EPR measurements on site-directed spin labels during the transient phase of the ATPase energy transduction cycle. Transient EPR allows for both ms time resolution and the orientational and motional resolution inherent in the EPR spectrum.

3.2. Time-Resolved EPR of Myosin and Actin in Solution

EPR experiments on spin-labeled myosin heads in skinned muscle fibers have shown that most myosin heads are undergoing large-amplitude submillisecond rotations during isometric contraction (Cooke *et al.*, 1982; Barnett and Thomas, 1989). Stiffness measurements suggest that most myosin heads (60–80%) are attached to actin in contraction during the spectroscopic studies (Fajer *et al.*, 1990b; Stein *et al.*, 1990). However, lacking a direct measurement of myosin head binding in fibers, these EPR measurements of fibers cannot directly correlate rotationally mobile heads with actin-attached heads. Also, in muscle fibers, covalent attachment of spin labels has been limited to myosin heads. Because it has been suggested that all or some of the ATP-induced microsecond motions of the myosin head originate with actin itself, it is necessary to attach spin labels specifically to actin to observe actin rotational dynamics during ATP hydrolysis.

There are three principal advantages to working with solutions of purified actin and myosin heads (S1):

1. Centrifugation experiments can be performed under spectroscopy conditions to provide a direct and quantitative measure of the fraction of myosin heads bound to actin.
2. Spin labels can be specifically bound to either actin or myosin heads, minimizing signals due to nonspecifically bound probes.
3. The kinetic rate constants for the acto-myosin ATPase cycle have been determined for these proteins in solution.

3.2.1. Global Rotations of Actin and Myosin in Solution during ATP Hydrolysis

ST-EPR was performed using solutions of actin and of myosin heads (S1) spin-labeled with MSL; this experiment measured the ATP-induced global rotational dynamics, as opposed to internal dynamics, of actin-attached myosin heads during steady-state ATP hydrolysis (Berger *et al.*, 1989). To ensure that a significant number of myosin heads were attached to actin in the presence of ATP, low ionic strength solutions ($\mu = 36$ mM) and high actin concentrations (200 μM) were used to increase the affinity of myosin heads for actin. Direct binding measurements reported that $52 \pm 2\%$ of the myosin heads bound to actin during steady-state ATP hydrolysis under spectroscopy conditions ($K_b = 7.42 \pm 0.41 \times 10^3\ M^{-1}$; Berger *et al.*, 1989). This binding measurement allows for the unambiguous determination of the rotational correlation time of the actin-attached myosin heads during steady-state ATP hydrolysis.

Because of the rapid hydrolysis of ATP by myosin, there are only a few seconds for EPR data acquisition after the addition of ATP. This is not enough time for mixing the sample, putting it into the EPR cell, and tuning the EPR spectrometer. Moreover, an ATP-regenerating system cannot be used, due to its high ionic strength, and rapid mixing (stopped flow) is not feasible with such a viscous solution. Therefore, the photolysis of caged ATP was used to rapidly produce millimolar concentrations of ATP during ST-EPR data acquisition. Because of the time limitation, the entire ST-EPR spectrum could not be acquired, so the intensity at a single field position in the spectrum was transiently monitored. The effective rotational correlation time of the MSL-S1 was obtained from this field position by comparing its intensity with a calibration curve obtained from a series of spin-labeled hemoglobin samples with known correlation times (Squier and Thomas, 1986; Berger *et al.*, 1989).

The ST-EPR spectra of 200 μM actin and 100 μM MSL-S1 (Fig. 5, left, solid line) or 100 μM MSL-S1 (Fig. 5, dotted line, left) are unaffected by the addition of caged ATP, as expected, since unphotolyzed caged ATP is not a substrate for myosin. Before photolysis, the ST-EPR spectrum of actin and MSL-S1 is very intense, indicating that the MSL-S1 is rigidly bound to actin ($\tau_r \geq 100\ \mu$s). After the laser pulse, which produced 1–2 mM ATP, the ST-EPR signal dropped rapidly to a point intermediate between that of the rigor complex (no ATP) and free MSL-S1 (Fig. 5, right). This rapid decrease in intensity indicates increased μs rotational motion of the S1. The ST-EPR spectral intensity in the presence of saturating ATP is a linear combination of the spectral intensity of the bound heads (52%) and free heads (48%). The effective rotational correlation time of the bound heads can be determined by correcting the intensity for the contribution of the free MSL-S1. It can be concluded that actin-bound MSL-S1 undergoes rotational motion with an effective rotational correlation time of 17 μs.

The most straightforward interpretation of this result is that the S1, or at least the labeled part of the S1 head, rotates relative to the actin filament. An alternative interpretation is that some or all of the ATP-induced μs

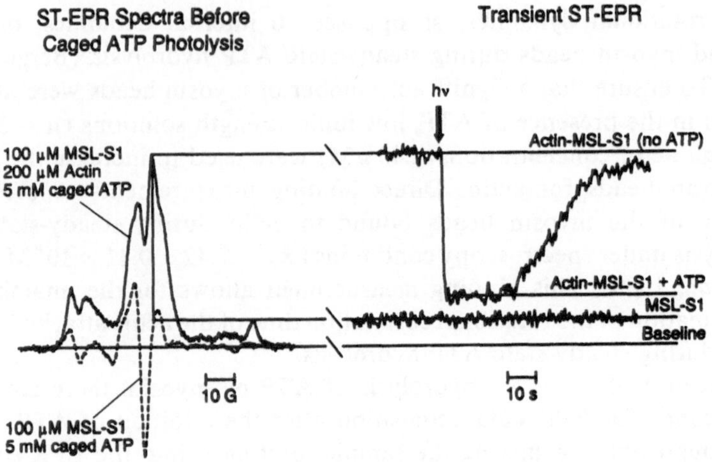

Figure 5. Saturation Transfer EPR spectra of actin-MSL-S1 and MSL-S1 at low ionic strength (μ = 36 mM) under steady-state conditions (left) and as a function of caged ATP photolysis (right) at pH 7.0, 25 °C. Samples contained either 100-μM MSL-S1 with 200-μM actin (solid spectrum, left) or 100-μM MSL-S1 alone (dashed spectrum, left), and 5-mM caged ATP. The ST-EPR transient signals (right) show the change in spectral intensity at a fixed field position (defined, left) as a function of time with 0.5-s time resolution. ATP (1–2 mM) was released by a 1-s burst (100 Hz) of 10-ns light pulses from an excimer laser at time 0 (arrow). Photolysis of caged ATP had no effect on the intensity of MSL-S1 alone (bottom trace, right), and identical experiments obtained in the absence of caged ATP indicated there were no significant effects of the laser flash itself on the ST-EPR intensity.

motions of the myosin head originate with the motions of actin itself. Experiments using the same solution conditions and protein concentrations as above, except with the spin label attached to actin instead of S1, have been done to clarify the interpretation (Ostap and Thomas, 1991; Ostap *et al.*, 1992). At low ionic strength, when about half of the S1 heads remain bound during ATP hydrolysis, no change in actin mobility was detected, despite much faster motions of labeled S1 bound to actin. Therefore, we concluded that the active interaction of S1, actin, and ATP induces rotation of myosin heads relative to actin, but does not affect the μs rotational motions of actin itself.

3.2.2. Internal Rotations of Myosin in Solution during ATP Hydrolysis

Because force production is almost certainly the result of dynamic changes in the myosin head, it is necessary to investigate the internal dynamics of the myosin head as a function of nucleotide occupancy of the active site. The spin label IASL [iodoacetamide spin label, 2,2,6,6-tetramethyl 4-amino (N-iodoacetamide); see Fig. 3], which is specific for cys-707 (SH1) on the myosin head, has been used to study myosin's internal rotational motions (Barnett and Thomas, 1987; Ostap and Thomas, 1992). It has been demonstrated that this spin label reports nucleotide-dependent changes within the myosin head. In the absence of nucleotides, IASL is rigidly attached ($\tau_r > 10^{-7}$ s) to the myosin head, so the motion of the probe reports the overall motion of the myosin head. However, upon the addition of ADP, the probe mobility increases slightly ($\tau_r \sim 80$ ns). During steady-state ATP hydrolysis, the EPR spectrum has two resolved spectral components, which represent two motional classes separated by approximately an order of magnitude in effective rotational correlation time (Fig. 6, top left). Component 1 is indistinguishable from the spectrum of IASL-myosin in the presence of ADP, and component 2 is representative of probes undergoing large-amplitude ns rotational motion. By quantifying the relative contributions of the two spectral components and comparing the contributions with the populations of kinetic states predicted by kinetic simulations, the spectral components were tentatively assigned to specific kinetic intermediates. Component 1 has been assigned to the M*ATP state, and component 2 has been assigned to the M**ADP·P$_i$ state (Barnett and Thomas, 1987):

$$M + ATP \rightleftharpoons M^{*}ATP \rightleftharpoons M^{**}ADP \cdot P_i \rightleftharpoons M^{*}ADP \cdot P_i \rightleftharpoons M^{*}ADP \rightleftharpoons M + ADP$$

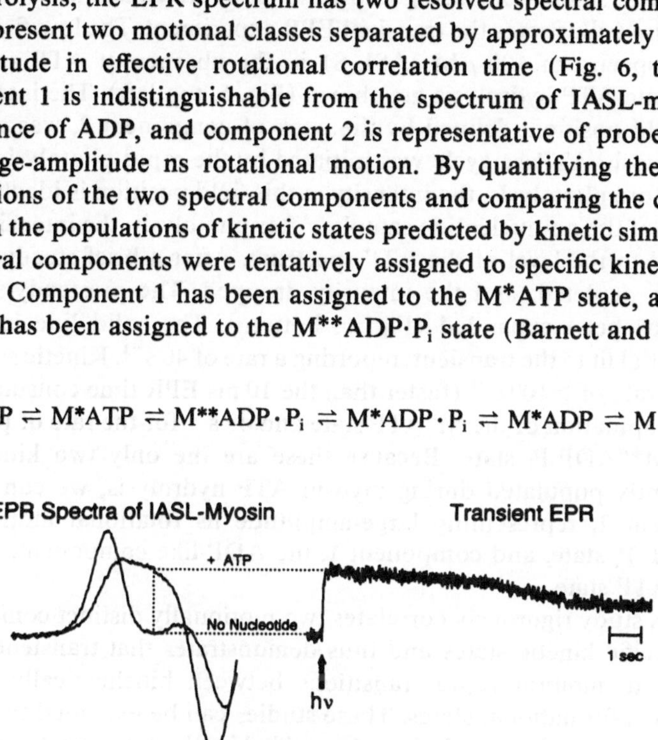

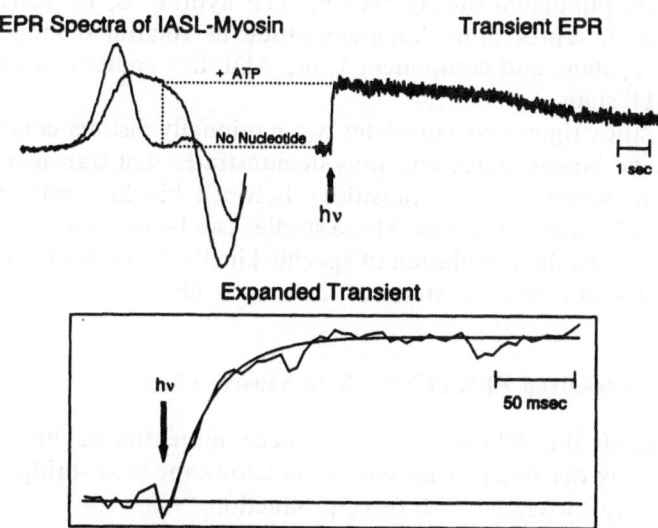

Figure 6. Conventional steady-state (left) and transient EPR spectra (right and bottom) of 100-μM IASL-S1 at 22 °C, pH 8.0. Left: Overlays of the low-field portions of conventional EPR spectra in the presence and absence of 5-mM ATP. The dotted line identifies the field position monitored during the transient EPR experiment. Right: Transient EPR spectrum monitoring a single field position with a time resolution of 10 ms. ATP (1 mM) was released by a single 10-ns pulse from an excimer laser at time 0 (arrow). Bottom: Expanded EPR transient showing the exponential fit.

It is necessary to perform transient EPR studies to rigorously correlate these motionally distinct states with kinetic states on the myosin ATPase pathway. Component 2, tentatively assigned to the $M^{**}ADP{\cdot}P_i$ state, must be monitored during the presteady-state phase of ATP hydrolysis. The rate of appearance and the extent of population of component 2 can be correlated with the rate and extent of population of a kinetic state (or states) obtained from kinetic simulations using known kinetic rate constants.

Figure 6 illustrates the transient EPR experiment. The low-field regions of EPR spectra of 100-μM IASL-S1 in the absence of ATP and during steady-state ATP hydrolysis are shown (Fig. 6, top left). The intensity at a single field position, defined by the peak of component 2, was monitored as the myosin ATPase cycle was initiated by the rapid photolysis of caged ATP. Before photolysis, the intensity at this field position is identical to the intensity of the no-nucleotide sample. After photolysis, the intensity rapidly increased to the level of the ATP spectrum. After a brief steady state, the ATP was depleted, and the intensity dropped. The presteady-state EPR signal has been expanded (Fig. 6, bottom). The solid line is a single-exponential fit to the transient, reporting a rate of 40 s^{-1}. Kinetic simulations report a rate of $>100\ s^{-1}$ (faster than the 10 ms EPR time constant) for the rate of population of the M*ATP state and 43 s^{-1} for the rate of population of the $M^{**}ADP{\cdot}P_i$ state. Because these are the only two kinetic states significantly populated during myosin ATP hydrolysis, we conclude that component 2, representing large-amplitude ns rotational motion, is the $M^{**}ADP{\cdot}P_i$ state, and component 1, the ADP-like component, represents the M*ATP state.

This study rigorously correlates two motionally distinct conformations with specific kinetic states and thus demonstrates that transient EPR can be used to monitor rapid transitions between biochemically important protein-conformational states. These studies can be extended to myosin in muscle fibers, so the population of specific kinetic states can be determined during different activation states of the muscle fiber.

3.3. Time-Resolved EPR of Myosin in Muscle Fibers

Although the ATPase cycle has been more thoroughly studied in solution, only the study of muscle fibers allows the cross-bridge dynamics to be directly correlated with force production.

3.3.1. Global Orientation of Myosin Heads in Muscle Fibers

The highly organized and oriented filament lattice in the muscle fiber adds a new dimension—orientation—to the EPR experiments. Since MSL

(Fig. 3) remains rigidly fixed to the myosin head throughout the ATPase cycle and since conventional EPR spectra of strongly immobilized spin labels depend directly on the orientational distribution of the spin label's principal axes relative to the fiber axis, the EPR spectra of MSL are highly sensitive to the orientation of heads relative to the fiber axis (Thomas and Cooke, 1980; Fajer *et al.*, 1990b).

Figure 7 illustrates transient EPR experiments on MSL-labeled myosin in oriented muscle fibers in the absence (top) and presence (bottom) of calcium. At the left, the low-field regions of the EPR spectra are shown before the photoproduction of ATP. Under these conditions ("rigor"), the intense line at P2 indicates that most of the spin labels, and thus the heads, are highly oriented, due to their stereospecific attachment to actin. For the transient experiments (Fig. 7, right), the magnetic field is fixed at P2, permitting the orientation to be followed as a function of time. In the absence of calcium (Fig. 7, top, relaxation), the photoproduction of ATP drops the intensity rapidly to a low steady-state level that indicates highly

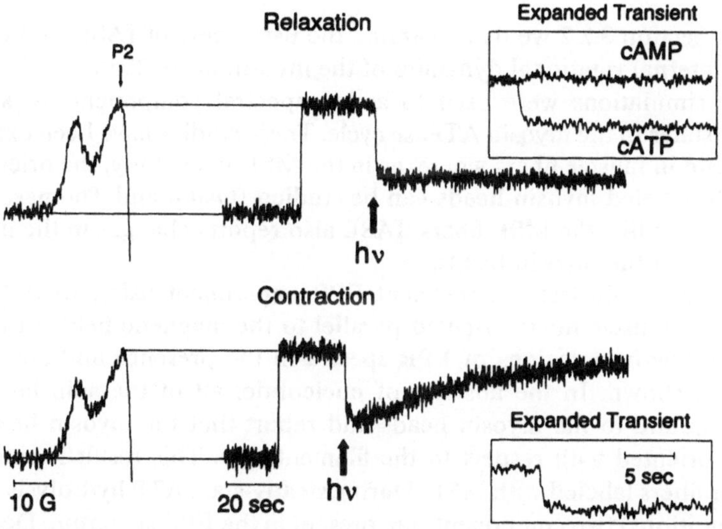

Figure 7. Conventional EPR spectra of MSL-labeled muscle fibers oriented parallel to the magnetic field under steady-state conditions (left) and as a function of caged ATP photolysis (right), either in the absence of Ca^{2+} (upper) or in the presence (lower) of Ca^{2+}, at 5 °C. (Left): The low-field region of the EPR spectrum prior to caged ATP photolysis. (Right): The EPR signal transients, monitored at position P_2 (left), were obtained wtih 24-ms time resolution. ATP (2–3 mM) was released by a 2-s burst (20 Hz) of 10-ns light pulses from an excimer laser at time 0 (arrow). The insets show an expanded time base for the transients. Identical experiments obtained in the absence of caged ATP indicated there were no significant effects of the laser flash itself on the ST-EPR intensity.

disordered myosin heads, presumably due to the dissociation of myosin from actin in these relaxed muscle fibers. In the presence of calcium (Fig. 7, bottom, contraction), the signal also drops rapidly, but the steady-state level is not quite as low, indicating slightly less disorder than in relaxation. The signal in contraction recovers much more rapidly than in relaxation, because the calcium-activated (isometrically contracting) fibers exhaust their ATP rapidly and return to the highly oriented rigor state. The high degree of disorder observed during contraction is consistent with the high degree of microsecond rotational mobility observed by ST-EPR in solution (Fig. 5). This provides strong evidence for the general notion of ATP-driven myosin head rotations in contracting muscle, but both the relaxed and contracting states are much more dynamic and disordered than was depicted simplistically in Fig. 4. The rotations are not tightly coupled to force generation, supporting the proposal that thermally driven motions play an important role in contraction (Harada *et al.*, 1990).

3.3.2. Internal Rotations of Myosin Heads in Muscle Fibers

In Section 3.2.2 we demonstrated the usefulness of IASL in the study of the internal rotational dynamics of the myosin head. Transient EPR and kinetic simulations were used to assign spectral components to specific kinetic states in the myosin ATPase cycle. These studies have been extended to myosin in muscle fibers where, as in the MSL-fiber study, the orientation of IASL-labeled myosin heads can be studied (Ostap and Thomas, 1992). However, unlike the MSL fibers, IASL also reports changes in the internal dynamics of the myosin heads.

Figure 8 illustrates a transient EPR experiment using IASL-labeled myosin in muscle fibers oriented parallel to the magnetic field. At the top left, the low-field regions of EPR spectra in the presence and absence of ATP are shown. In the absence of nucleotide, all of the spin labels are rigidly bound to the myosin heads and report that the myosin heads are highly oriented with respect to the filament axis. This result is similar to that for fibers labeled with MSL. During steady-state ATP hydrolysis (relaxing conditions) two components are present in the EPR spectrum. Deconvolution of the spectrum indicates the presence of a disordered component (similar to relaxed MSL fibers) and a component that is rotationally dynamic on the ns time scale.

In the transient experiments (Fig. 8, right and bottom), the intensity at a single field position, defined by the oriented peak, was monitored as the ATPase cycle was initiated by the photolysis of caged ATP. The pre-steady-state EPR signal can be described as a transient with three phases (Fig. 8, bottom). During phase A, the intensity of the peak rapidly drops

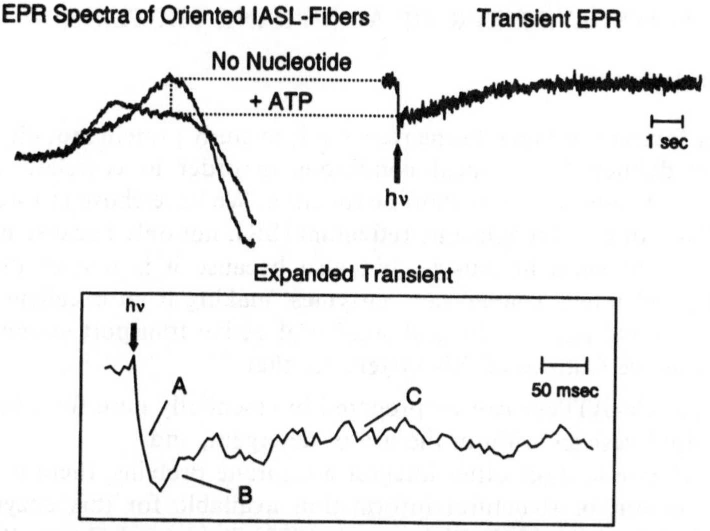

Figure 8. Conventional steady-state (left) and transient EPR spectra (right and bottom) of IASL fibers oriented parallel to the magnetic field in the absence of Ca^{2+} at 22 °C, pH 7.0. (Left): Overlays of the low-field portions of conventional EPR spectra in the presence and absence of ATP. The dotted line identifies the field position monitored during the transient EPR experiment. (Right): Transient EPR spectrum monitoring a single field position with a time resolution of 10 ms. ATP (1 mM) was released by a single 10-ns pulse from an excimer laser at time 0 (arrow). (Bottom): Expanded EPR transient showing the three phases of the transient. (A) Disordering of the myosin heads. (B) Population of the $M^{**}ADP{\cdot}P_i$ state. (C) Steady-state ATPase level.

to a point below the steady-state ATPase level. This rapid decrease occurs within the EPR time constant and represents the disordering of the myosin heads upon nucleotide binding. During phase B, the intensity transiently increases to the steady-state ATPase level. This increase represents the transient population of the ns mobile $M^{**}ADP{\cdot}P_i$ state at a rate similar to that determined for myosin in solution (see Section 3.2.2). The intensity of phase C represents the steady-state ATPase level. Upon ATP depletion, the intensity returns to the pre-photolysis level.

The transient data clearly indicate that

1. the myosin heads become disordered upon nucleotide binding,
2. disorder of the myosin heads precedes ATP hydrolysis, and
3. ATP is hydrolyzed by myosin in relaxed muscle fibers at the same rate as myosin heads in solution (Ostap and Thomas, 1992).

These studies will be extended to muscle fibers contracting isometrically and isotonically.

4. TIME-RESOLVED EPR OF MEMBRANE PROTEINS

4.1. Background

The focus of our membrane research is to study protein mobility under precisely defined biochemical conditions in order to correlate specific molecular dynamics with membrane function. We have chosen to study the Ca–ATPase of the sarcoplasmic reticulum (SR), not only because it is one of the key enzymes in muscle, but also because it is one of the best-characterized integral membrane enzymes, making it an excellent model for biochemical and biophysical studies of active transport systems. The most attractive features of this system are that

1. the Ca–ATPase can be prepared in essentially pure form in native lipid vesicles without the use of detergent, and
2. relative to most other integral membrane proteins, there is a large amount of structural information available for this enzyme (*cf.* MacLennan, 1990; Blaise *et al.*, 1990; Stokes and Green, 1990).

The biochemical kinetics of this enzyme have been studied thoroughly (Inesi, 1985; Inesi and de Meis, 1985; Froehlich and Heller, 1985; Jencks, 1989), and detailed models have been proposed for the mechanism of active Ca^{2+} transport (Tanford, 1984; Jencks, 1989; Inesi *et al.*, 1990). The most widely discussed model is illustrated in Fig. 9:

1. Ca^{2+} and ATP bind quickly and tightly to the outside surface,
2. ATP is cleaved, producing
3. a "high-energy" covalent phosphoenzyme E ~ P that is "ADP-sensitive" (Froehlich and Heller, 1985),

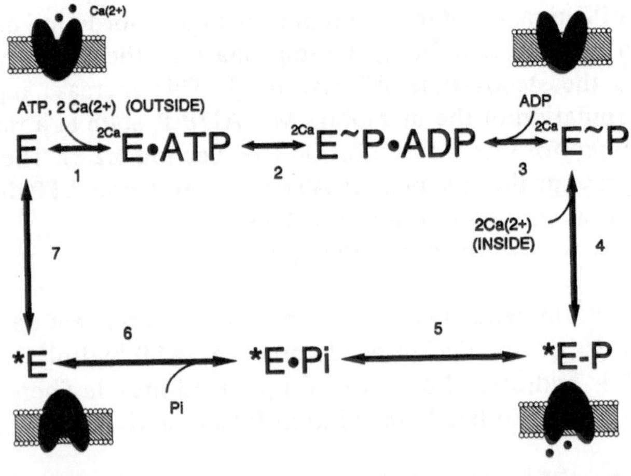

Figure 9. Model describing calcium translocation by the Ca-ATPase of sarcoplasmic reticulum membranes.

4. a presumed (but poorly defined) conformational change occurs (E → *E, rate-limiting under some conditions), in which the phosphate bond strengthens (~ P → −P) while Ca^{2+} affinity decreases and the Ca^{2+} binding sites become accessible to the inside, so that Ca^{2+} is released to the inside,
5. the phosphoenzyme is hydrolyzed (rate-limiting under some conditions),
6. phosphate is released, and
7. the conformational change is reversed.

Many key details of this model remain controversial, such as the points in the cycle where key conformational changes occur (Jencks, 1989) and the structural bases of these changes (Blasie *et al.*, 1990). There is evidence suggesting that enzyme oligomers are involved and that the oligomeric state may change during the cycle (Vanderkooi *et al.*, 1977; Taylor *et al.*, 1984; Froehlich and Heller, 1985). Lipid composition and properties have profound effects on the enzyme's activity (Hidalgo, 1985, 1987; Lee, 1988). The resulting questions about molecular structures, motions, and interactions require direct measurements of these physical parameters at the molecular level, correlated with enzymatic activity. Site-specific spectroscopic probe techniques are essential tools in obtaining this crucial information, since the time scales of lipid and protein motions are within the range of optimum sensitivity of these techniques.

Conventional EPR has been used to probe ns motions of both lipid (Hidalgo, 1985, 1987; Bigelow and Thomas, 1987) and protein (Coan and Inesi, 1977; Coan *et al.*, 1979, 1986; Champeil *et al.*, 1981; Coan and Keating, 1982; Coan, 1983; Chen *et al.*, 1991; Lewis and Thomas, 1992) in native SR, and the temperature dependences of both types of probes correlate roughly with that of Ca–ATPase activity. Lipid spin labels have been used to show that an annulus of 20–30 restricted (boundary) lipids surrounds the enzyme (Thomas *et al.*, 1982; Hidalgo, 1985, 1987). We used ST-EPR to show that the spin-labeled Ca–ATPase rotates globally in the μs time range (Thomas and Hidalgo, 1978), that these motions are quite sensitive to protein–protein and protein–lipid interactions (Squier and Thomas, 1988; Squier *et al.*, 1988a, b), and that inhibition of these motions inhibits the cycle at the level of phosphoenzyme decomposition (Squier and Thomas, 1988). We have proposed that the requirement for lipid fluidity is based on a primary requirement for protein mobility and dynamic protein–protein interactions (Squier *et al.*, 1988b).

Blasie and co-workers have used X-ray and neutron diffraction to study the time-averaged electron density profiles of Ca–ATPase and the surrounding phospholipid bilayer during enzyme cycling, initiated by the flash photolysis of caged ATP (Blasie *et al.*, 1986). Their results show that the

ATPase undergoes significant structural rearrangement during cycling, consistent with conformational changes associated with specific enzymatic steps. Prior to the structural studies, the behavior of caged ATP photolysis-induced calcium uptake by SR samples under their specific experimental conditions was studied (Pierce *et al.,* 1983a, b). By combining the results obtained from the kinetic and structural studies, Blasie *et al.* (1986) assigned the observed structural changes to the formation of the first phosphoenzyme intermediate, $E \sim P \circ Ca_2$. Blasie and co-workers have described similar kinetic measurements utilizing caged calcium (DeLong *et al.,* 1990; DeLong and Blaise, 1993).

Mäntele and co-workers have used Fourier-transform infrared (FTIR) spectroscopy to monitor ligand binding and conformational changes in Ca-ATPase during enzyme cycling, initiated by the flash photolysis of caged ATP (Barth *et al.*, 1990, 1991). Their data reported molecular changes in the ATPase during cycling relative to the resting enzyme prior to ATP liberation, and they assigned these changes to the kinetic step of phosphoenzyme formation. The observed changes were small, and Mäntele and co-workers concluded that the Ca-ATPase undergoes no major conformational rearrangement during enzyme cycling.

4.2. Global Rotation of Ca–ATPase during ATP Hydrolysis

MSL binds rigidly to the Ca-ATPase and is sensitive to the enzyme's μs global rotation in the membrane (Lewis and Thomas, 1991; Squier and Thomas, 1988). The global rotations, probed by MSL, are sensitive to the lipid fluidity (Squire *et al.*, 1988b; Squier and Thomas, 1989) and protein-protein interactions (Lewis and Thomas, 1986; Squier and Thomas, 1988; Birmachu and Thomas, 1990; Mahaney and Thomas, 1991; Voss *et al.*, 1991). Since there is evidence that protein-protein interactions may change during the Ca-ATPase cycle, we measured the rotational dynamics of MSL Ca-ATPase after caged ATP photolysis, with a time resolution of 2 s (Lewis and Thomas, 1991). When 1-mM ATP was produced by photolysis of caged ATP in the presence of calcium, no significant change in the ST-EPR signal was observed (Fig. 10), suggesting that the Ca-ATPase does not change its oligomeric state, shape, or protein-lipid interactions during the ATPase cycle. New studies, which build on this foundation, are now in progress. First, similar measurements are being made using Ca-ATPase labeled with a novel vinyl ketone spin label, InVSL 2-[(1-oxyl-2,2,5,5-tetramethylpyrrolin-3-yl) methenyl] indane-1,3-dione (Fig. 3), as described by Hankovszky *et al.* (1989). The advantage of InVSL is that this label is more rigidly immobilized when bound to the enzyme than is MSL (Horváth *et al.*, 1990), suggesting that it may better report the global rotation of the Ca-ATPase and thus be more sensitive to changes in this rotation as a function of

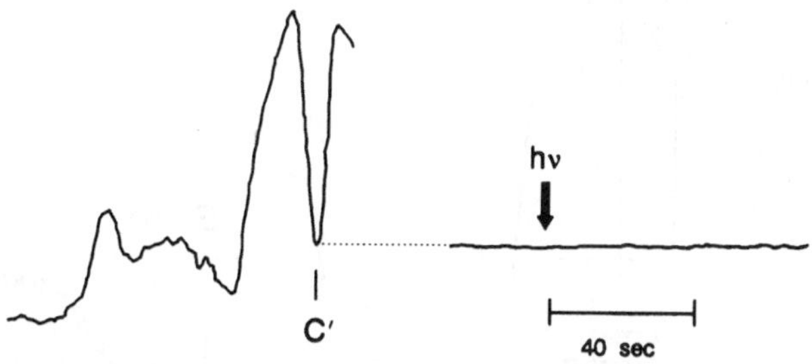

Figure 10. Saturation Transfer EPR spectra of MSL Ca-ATPase under steady-state conditions and during active Ca-ATPase cycling induced by photolysis of caged ATP at 4 °C. The ST-EPR spectrum (first 50 G) was recorded using SR membranes (50 mg/ml) in SRB, pH 7.4, containing 1 μg A23187/0.05 mg SR protein, and 5-mM caged ATP. The ST-EPR transient signal, showing C′ versus time, was recorded with a 2.0-s time resolution. ATP (1.0 mM) was released by a 1-s burst (50 Hz) from an excimer laser at time 0 (arrow). Identical experiments obtained in the absence of caged ATP indicated that there were no significant effects of the laser flash itself on the EPR intensity.

cycling. Second, based on substantial improvements in the transient ST-EPR experimental methodology (Ostap and Thomas, 1991), we are repeating the experiments of Lewis and Thomas (1991) with a time resolution of 300 ms per data point in order to more completely capture the transient phase of ATP hydrolysis and phosphoenzyme formation. These experiments should provide new insight into whether there are any changes in global dynamics in the cycle.

4.3. Internal Rotations of Ca–ATPase during ATP Hydrolysis

Internal rotational dynamics of Ca–ATPase during enzymatic activity are measured using the IASL bound to a specific SH group on the Ca–ATPase (Lewis and Thomas, 1992). IASL is more loosely bound to the protein and is sensitive to the protein's internal (ns) rotational dynamics (Coan and Inesi, 1977; Coan *et al.*, 1979; Coan and Keating, 1982; Lewis and Thomas, 1992). The spectrum of IASL-SR is a composite of two predominant label mobilities: a mobile component, characterized by an outer splitting of 52 G, and a restricted component, characterized by an outer splitting of 63 G. For improved resolution of these spectral components (Fig. 11), we use a perdeuterated derivative of IASL (referred to as *d*-IASL). Some of the ligands involved in kinetic steps in the Ca–ATPase cycle (e.g., ATP, ATP analogues, Ca^{2+}, P_i) induce changes in these rotations, as determined by an increased immobilization of the IASL on the enzyme (Coan and Inesi, 1977; Coan *et al.*, 1979,

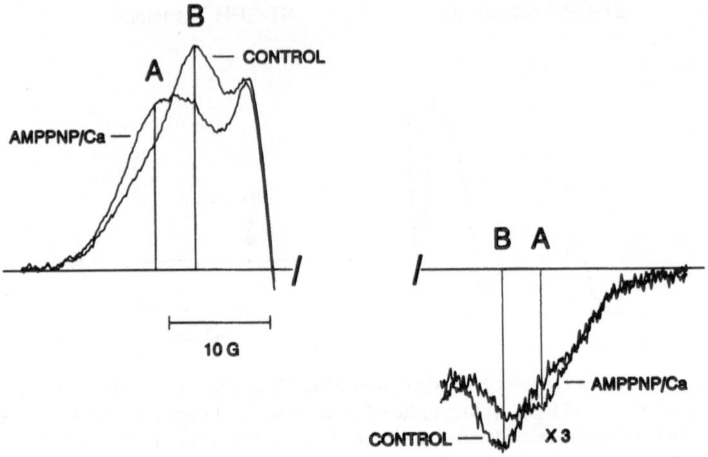

Figure 11. Conventional EPR spectra of deuterated IASL-Ca-ATPase demonstrating the effects of 5-mM AMPPNP [adenosine 5′-(β,γ-iminotriphosphate)], which model the effects of ATP. Spectra were recorded at 4 °C using 10-mg/ml *d*-IASL-SR protein in SRB, pH 7.0. For clarity, only the low-field and high-field peaks are shown, and the high-field region is magnified three-fold. The line-height parameters A and B show the spectral splittings of the less mobile (A) and more mobile (B) probes.

1986; Coan and Keating, 1982; Coan, 1983; Lewis and Thomas, 1992), and thus shift the distribution between the two populations. Using high-resolution EPR and quantitative spectral subtraction techniques (Bigelow and Thomas, 1987; Mahaney *et al.*, 1992; Lewis and Thomas, 1992), we have quantified the effects of ATPase cycle ligands on the distribution between the two populations. Since the EPR spectrum appears always to consist of a linear combination of these two spectral components, we are probably resolving the two principal conformational states of the enzyme. Ligands that combine with the enzyme to promote formation of the first phosphoenzyme intermediate ($E \sim P \circ Ca_2$, Fig. 9) induce significant changes in the distribution of the two label populations, whereas ligands that promote the formation of the second phosphoenzyme intermediate (*E-P, Fig. 9) through cycle reversal (Barrabin *et al.*, 1984) induce only very small changes in the distribution. This suggests that the change in probe mobility most likely occurs in conjunction with the formation of the first phosphoenzyme intermediate and is reversed as the enzyme undergoes the conformational changes associated with phosphoenzyme isomerization and enzyme turnover.

In order to resolve these conformational changes in the transient phase of the ATPase cycle (Lewis and Thomas, 1992; Mahaney and Thomas, 1990), we detected the EPR signal in the transient phase after caged ATP photolysis, with a time resolution of 100 ms per data point (Fig. 12). We

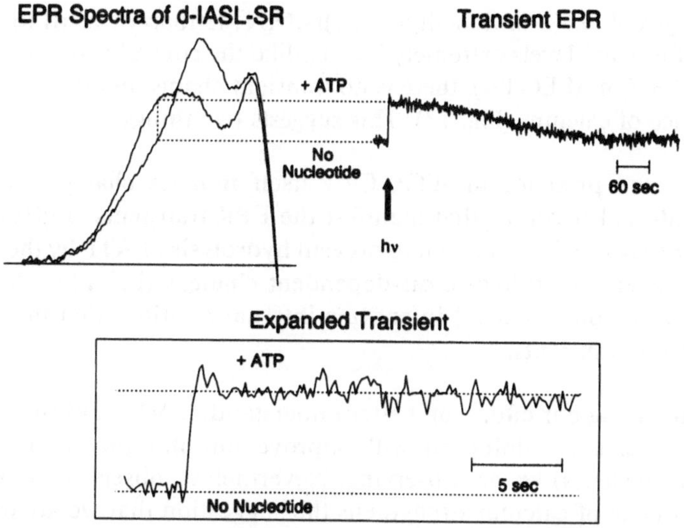

Figure 12. Conventional EPR spectra of *d*-IASL-Ca-ATPase under steady-state conditions and during active Ca-ATPase cycling induced by photolysis of caged ATP at 4 °C. The EPR spectrum denoted no nucleotide was recorded using SR membranes (10 mg/ml) in SRB, pH 7.0, containing 1 μg A23187/0.05 mg SR protein, and 5-mM caged ATP. The EPR spectrum denoted +ATP was recorded under the same conditions, but after photolysis of caged ATP. For clarity, only the low-field regions are shown. The EPR transient signal, monitored at position A (see Fig. 11), was recorded with a 40-ms time resolution. ATP (0.6 mM) was released by a single 10-ns pulse from an excimer laser at time 0 (arrow). Identical experiments performed in the absence of caged ATP indicated that there were no significant effects of the laser flash itself on the EPR intensity.

monitored the EPR signal in the low-field region of the spectrum, which is the region of maximum sensitivity to this change (Lewis and Thomas, 1992). Experiments were carried out at 4 °C to slow the kinetics and to maximize the formation of the ADP-sensitive phosphoenzyme ($E \sim P \circ Ca_2$ in Fig. 9). The photogeneration of 1-mM ATP, in the presence of Ca^{2+}, produces a rapid rise in the signal, indicating an increased population of the less mobile state (Fig. 9). The time course of this change is remarkably like that observed for the ADP-sensitive phosphoenzyme in rapid-quench biochemical measurements (Froehlich and Heller, 1985), supporting the proposal that this spectroscopic transition corresponds to phosphoenzyme formation (step 4 in Fig. 9) and isomerization in the Ca–ATPase cycle.

A firm interpretation of which kinetic steps or intermediates we are observing during our EPR transients requires further study, but we describe here some of our results. Since Ca^{2+} is required for enzyme cycling to occur (to promote formation of phosphoenzyme intermediates by ATP hydrolysis), we omitted Ca^{2+} in one series of transient experiments, and used EGTA [ethyleneglycol-*bis*-(β-aminoethyl ether)N,N,N′,-tetraacetic acid] to keep the residual free Ca^{2+} levels extremely low. Unlike the control experiment (0.45-mM Ca^{2+} + 5-mM EGTA), there is no transient change in the EPR signal in the absence of calcium (Fig. 13). This suggests two things:

1. In the presence of ATP, Ca^{2+} itself induces changes in enzyme internal rotations that manifest the EPR transients, and/or
2. the lack of Ca^{2+}, known to prevent hydrolysis of ATP by the enzyme, prevents ATP hydrolysis-dependent changes (i.e., phosphoenzyme formation or decay) in enzyme internal rotations that manifest the EPR transients.

We are focusing our efforts on better understanding which of these suggestions is true, since doing so will improve our interpretation of which mechanistic step(s) we are observing. Nevertheless, observing no transient in the absence of calcium strengthens the suggestion that we are observing Ca-ATPase structural changes associated with phosphoenzyme forma-

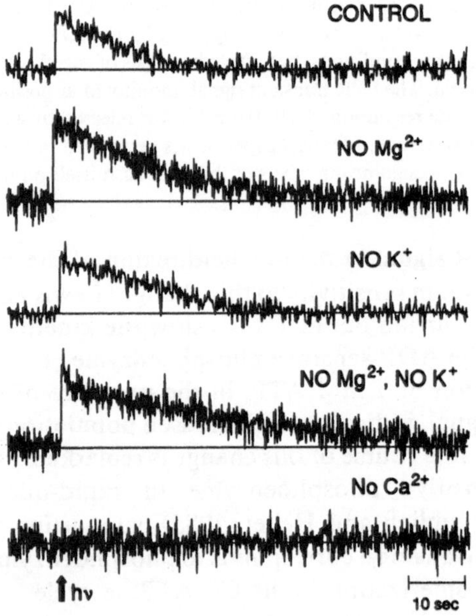

Figure 13. EPR transients obtained from *d*-IASL-Ca-ATPase at 4 °C, as described in Fig. 12. The no-calcium transient was obtained using SRB containing 5-mM EGTA and no added $CaCl_2$. ATP (0.4 mM) was released by a single 10-ns laser pulse at time 0 (arrow). Identical experiments obtained in the absence of caged ATP indicated that there were no significant effects of the laser flash itself on the EPR intensity.

tion/degradation. In another set of experiments, we omitted either Mg^{2+} (the essential catalytic divalent metal ligand) or K^+ (which activates the ATPase cycle) from the sample buffer. Identical to the control (1-mM free Mg^{2+}, 60-mM KCL) experiment, an intensity change occurs, indicating that Ca^{2+}/ATP-dependent changes in enzyme internal rotations were not inhibited (Fig. 13). However, unlike the control, there is a transient "overshoot" in the absence of Mg^{2+} or K^+, and the decay of the signal to the original intensity in both samples is much slower. These effects are accentuated in the transient obtained from samples containing neither Mg^{2+} nor K^+. Control experiments using caged ATP alone, caged AMP, and IASL-Ca-ATPase without caged ATP indicate that the observed overshoot is a real, nonartifactual component of the EPR signal transient.

Previous kinetic studies of the Ca-ATPase mechanic cycle at lower temperatures (0–4 °C) have shown that in the absence of Mg^{2+} or K^+, the rate of ATPase turnover is diminished, and the fraction of the first phosphoenzyme intermediate ($E \sim P \circ 2Ca_2$) is enhanced at the expense of the second phosphoenzyme intermediate prior to P_i hydrolysis (*E-P) (Shigekawa and Dougherty, 1978a, b; Shigekawa *et al.*, 1978; Wakabayashi *et al.*, 1986). The transients in Fig. 13 agree with these kinetic studies. First, in the absence of Mg^{2+} or K^+, the steady-state phase of the transient is longer, and the decay of the transient is slower, consistent with the reduced rate of ATP hydrolysis in the absence of these ions. Second, the change in intensity from the initial value (prior to ATP liberation) to the steady-state value (during enzyme cycling) is greater in the absence of Mg^{2+} or K^+ than in the control, consistent with an increased fraction of the first phosphoenzyme intermediate ($E \sim P \circ Ca_2$). While unequivocal assignment of which phosphoenzyme state the EPR transients are reporting is not yet possible, the data suggest that we are observing changes in enzyme internal rotations corresponding to (distinct) step(s) in the Ca-ATPase kinetic cycle related directly to phosphoenzyme formation. This assignment is strengthened by the conventional EPR experiments (above) that demonstrated that the V_1 spectrum of the enzyme in an E_2-P state is nearly identical to the corresponding spectra of the enzyme in the E_2 and $E_1/E_1 \circ Ca_2$ states.

As our research continues, we will continue to vary the conditions known to affect the rates of formation and decomposition of the proposed phosphoenzyme conformational states. Transient EPR data obtained under these conditions, in conjunction with quantitative transient chemical measurements and kinetic simulations, should allow us to assign our signal changes to specific biochemical events. It is our hope to define clearly the rotational dynamics associated with conformational intermediates, thus providing a direct means of testing and refining proposed mechanistic models.

5. CONCLUSIONS

The use of caged compounds in parallel with time-resolved EPR spectroscopy is a powerful method for obtaining insight into the dynamic sequence of structural changes during enzymatic cycling in muscle protein systems, where conventional rapid-mixing techniques are not ideal. Our progress in using caged compounds to detect spectroscopic transients in protein internal and global rotational dynamics during enzymatic cycling indicates that this methodology provides a valid means for physically detecting and quantifying key muscle ATPase conformational transitions. In conjunction with kinetic perturbations and both kinetic and EPR transient simulation, these methods are providing critical tests for alternative models of energy transduction in muscle. Given the wide variety and availability of caged compounds, and the diversity of EPR spectroscopic techniques for studying specific dynamic events, this methodology should continue to provide excellent opportunities for similar kinetic–dynamic studies in many other systems as well.

ACKNOWLEDGMENTS. We thank Yale Goldman for providing caged compounds and advice about their use, Al Beth for deuterated spin labels, Robert L. H. Bennett for setting up and maintaining the instrumentation, and Franz L. Nisswandt for computer programming and network support.

This work was supported by a grant to the first author from the National Institutes of Health (AR39754). The second author was supported by a predoctoral training grant from NIH. The third author was supported by a predoctoral training grant from NIH and a doctoral dissertation fellowship from the University of Minnesota. The fourth author was supported by a scholarship from the Life and Health Insurance Medical Research Fund. The fifth author was supported by fellowships from the Muscular Dystrophy Association and the Minnesota Supercomputer Institute. The sixth author was supported by a postdoctoral fellowship from the American Heart Association, Minnesota Affiliate.

REFERENCES

Atkins, P. W., McLauchlan, K. A., and Simpson, A. F., 1970, *J. Phys. E.* **3**:547.
Barnett, V. A., and Thomas, D. D., 1984, *J. Mol. Biol.* **179**:83.
Barnett, V. A., and Thomas, D. D., 1987, *Biochemistry* **26**:314.
Barnett, V. A., and Thomas, D. D., 1989, *Biophys. J.* **56**:517.
Barrabin, H., Scofano, H. M., and Inesi, G., 1984, *Biochemistry* **23**:1542.
Barth, A., Kreutz, W., and Mäntele, W., 1990, *FEBS Lett.* **277**:147.
Barth, A., Mäntele, W., and Kreutz, W., 1991, *Biochim. Biophys. Acta* **1057**:115.
Berger, C. L., and Thomas, D. D., 1991, *Biochemistry* **30**:11036.

Berger, C. L., Svensson, E. C., and Thomas, D. D., 1989, *Proc. Natl. Acad. Sci. USA* **86**:8753.
Bigelow, D. J., and Thomas, D. D., 1987, *J. Biol. Chem.* **262**:13,449.
Birmachu, W., and Thomas, D. D., 1990, *Biochemistry* **29**:3904.
Blasie, J. K., Pascolini, D., Herbette, G., Pierce, D., Itshak, F., Skita, V., and Scarpa, A., 1986, *Biophys. J.* **49**:110.
Blasie, J. K., Pascolini, D., Asturias, F., Herbette, L. G., Pierce, D., and Scarpa, A., 1990, *Biophys. J.* **58**:687.
Brenner, B., 1987, *Ann. Rev. Physiol.* **49**:655.
Brenner, M. C., Murray, C. J., and Klinman, J. P., 1989, *Biochem.* **28**:4656.
Budil, D. E., and Thurnauer, M. C., 1991, *Biochim. Biophys. Acta* **1057**:1.
Champeil, P., Büshlen, S., and Guillian, F., 1981, *Biochemistry* **20**:1520.
Chen, Z., Coan, C., Fielding, L., and Cassafer, G., 1991, *J. Biol. Chem.* **266**:12,386.
Coan, C., 1983, *Biochemistry* **22**:5826.
Coan, C., and Inesi, G., 1977, *J. Biol. Chem.* **252**:3044.
Coan, C., and Keating, S., 1982, *Biochemistry* **21**:3214.
Coan, C., Verjovski-Almeida, S., and Inesi, G., 1979, *J. Biol. Chem.* **254**:2968.
Coan, C., Scales, D. J., and Murphy, A. J., 1986, *J. Biol. Chem.* **261**:10,394.
Cooke, R., 1986, *CRC Crit. Rev. Biochem.* **21**:53.
Cooke, R., Crowder, M. S., and Thomas, D. D., 1982, *Nature* **300**:776.
Dantzig, J. A., Walker, J. W., Trentham, D. R., and Goldman, Y. E., 1988, *Proc. Natl. Acad. Sci. USA.* **85**:6716.
Dantzig, J. A., Hibberd, M. G., Trentham, D. R., and Goldman, Y. E., 1991, *J. Physiol.* **432**:639.
DeLong, L. J., and Blasie, J. K., 1993, *Biophys. J.*, in press.
DeLong, L. J., Philips, C. M., Kaplan, J. H., Scarpa, A., and Blasie, J. K., 1990, *J. Biochem. Biophys. Meth.* **21**:333.
Eisenberg, E., and Hill, T. L., 1985, *Science* **227**:999.
Fajer, P. G., Thomas, D. D., Feix, J. B., and Hyde, J. S., 1986, *Biophys. J.* **50**:1202.
Fajer, P. G., Fajer, E. A., Brunsvold, N. J., and Thomas, D. D., 1988, *Biophys. J.* **53**:513.
Fajer, P. G., Fajer, E. A., Matta, J. M., and Thomas, D. D., 1990a, *Biochemistry* **29**:5865.
Fajer, P. G., Fajer, E. A., and Thomas, D. D., 1990b, *Proc. Natl. Acad. Sci. USA* **87**:5538.
Ferenczi, M. A., Homsher, E., and Trentham, D. R., 1984, *J. Physiol.* **352**:575.
Froehlich, J. P., and Heller, P. F., 1985, *Biochemistry* **24**:126.
Goldman, Y. E., Hibberd, M. G., and Trentham, D. R., 1984, *J. Physiol.* **354**:605.
Hankovszky, O. H., Hideg, K., and Jerkovich, G., 1989, *Synthesis* **7**:526.
Harada, Y., Sakurada, K., Aokik, T., Thomas, D. D., and Yanagida, Y., 1990, *J. Mol. Biol.* **216**:49.
Hartsel, S. C., Moore, C. R., Raines, D. E., and Cafiso, D. S., 1987, *Biochemistry* **26**:3253.
Hibberd, M. G., and Trentham, D. R., 1986, *Ann. Rev. Biophys. Chem.* **15**:119.
Hidalgo, C., 1985, Membrane fluidity and the function of the Ca^{2+}-ATPase of sarcoplasmic reticulum, in *Membrane Fluidity in Biology*, Vol. 4 (J. M. Boggs, and R. C. Aloia, eds.), Academic Press, New York, pp. 51–96.
Hidalgo, C., 1987, *CRC Crit. Rev. Biochem.* **21**:319.
Hiller, R., and Carmeli, C., 1990, *Biochem.* **29**:6189.
Homsher, E., and Millar, N. C., 1990, *Ann. Rev. Physiol.* **52**:875.
Horváth, L. I., Dux, L., Hankovszky, O. H., Hideg, K., and Marsh, D., 1990, *Biophys. J.* **58**:231.
Hubbell, W. L., Froncisz, W., and Hyde, J. S., 1987, *Rev. Sci. Instrum.* **58**:1879.
Hubbell, W. L., Khorana, G. H., Flitsch, S. L., and Altenbach, C., 1989, *Biochemistry* **28**:7806.
Huxley, H. E., 1969, *Science* **164**:1356.
Huxley, A. F., and Simmons R. M., 1971, *Nature* **233**:533.
Inesi, G., *Ann. Rev. Physiol.* **47**:574.
Inesi, G., and de Meis, L., 1985 in *Enzymes of Biological Membranes*, Vol. 3 (A. Martonosi, ed.), Plenum Press, New York, pp. 157–191.

Inesi, G., Sumbilla, C., and Kirtley, M. E., 1990, *Physiol. Rev.* **70**:749.
Irving, M., Peckham, M., and Ferenczi, M. A., 1988, *Adv. Exp. Med. Biol.* **226**:299.
Jencks, W. P., 1989, *J. Biol. Chem.* **264**:18,855.
Kaplan, J. H., 1990, *Ann. Rev. Physiol.* **52**:897.
Kaplan, J. H., Forbush, B., and Hoffman, J. H., 1978, *Biochemistry* **17**:1929.
Lee, A. G., 1988, Annular lipids and the activity of the calcium-dependent ATPase, in *Advances in Membrane Fluidity*, Vol. 2 (R. C. Aloia, C. C. Curtain, and L. M. Gordon, eds.), Liss, New York, pp. 111–139.
Lester, H. A., Krouse, M. E., Nass, M. M., Wassermann, N. H., and Erlanger, B. F., 1979, *Nature* **280**:509.
Lewis, S. M., and Thomas, D. D., 1986, *Biochemistry* **25**:4615.
Lewis, S. M., and Thomas, D. D., 1991, *Biochemistry* **30**:8331.
Lewis, S. M., and Thomas, D. D., 1992, *Biochemistry*, **31**:7381.
MacLennan, D. H., 1990, *Biophys. J.* **58**:1355.
Mahaney, J. E., and Thomas, D. D., 1990, *Biophys. J.*, **57**:499a.
Mahaney, J. E., and Thomas, D. D., 1991, *Biochemistry* **30**:7171.
Mahaney, J. E., Kleinschmidt, J., Marsh, D., and Thomas, D. D., 1992, *Biophys. J.* **63**:1513.
McCray, J. A., and Trentham, D. R., 1989, *Ann. Rev. Biophys. Biophys. Chem.* **18**:239.
McCray, J. A., Herbette, L., Kihara, T., and Trentham, D. R., 1980, *Proc. Natl. Acad. Sci. USA.* **77**:7237.
Menetret, J. F., Hofman, W., Schroder, R. R., Rapp, G., and Goody, R. S., 1991, *J. Mol. Biol.* **219**:139.
Miller, A., and Brudvig, G. W., 1991, *Biochim. Biophys. Acta* **1056**:1.
Ostap, E. M., and Thomas, D. D., 1991, *Biophys. J.* **59**:1235.
Ostap, E. M., and Thomas, D. D., 1992, *Biophys, J.*, **61**:303a.
Ostap, E. M., Yanagida, T., and Thomas, D. D., 1992, *Biophys. J.* **63**:966.
Pierce, D. H., Scarpa, A., Topp, M. R., and Blasie, J. K., 1983a, *Biochemistry* **22**:5254.
Pierce, D. H., Scarpa, A., Trentham, D. R., Topp, M. R., and Blasie, J. K., 1983b, *Biophys. J.* **44**:365.
Poole, C. P., 1983, *Electron Spin Resonance*, Wiley, New York.
Poole, K. J. V., Maeda, Y., Rapp, G., and Goody, R. S., 1991, *Adv. Biophys.* **27**:63.
Shigekawa, M., and Dougherty, J. P., 1978a, *J. Biol. Chem.* **253**:1451.
Shigekawa, M., and Dougherty, J. P., 1978b, *J. Biol. Chem.* **253**:1458.
Shigekawa, M., Dougherty, J. P., and Katz, A. M., 1978, *J. Biol. Chem.* **253**:1442.
Silvestrini, M. C., Tordi, M. G., Musci, G., and Brunor, M., 1990, *J. Biol. Chem.* **265**:11,783.
Squier, T. C., and Thomas, D. D., 1986, *Biophys. J.* **49**:921.
Squier, T. C., and Thomas, D. D., 1988, *J. Biol. Chem.* **263**:9171.
Squier, T. C., and Thomas, D. D., 1989, *Biophys. J.* **56**:735.
Squier, T. C., Hughes, S. E., and Thomas, D. D., 1988a, *J. Biol. Chem.* **263**:9162.
Squier, T. C., Bigelow, D. J., and Thomas, D. D., 1988b, *J. Biol. Chem.* **263**:9178.
Stein, R. A., Ludescher, R. D., Dahlberg, P. S., Fajer, P. G., Bennett, R. L. H., and Thomas, D. D., 1990, *Biochemistry* **29**:10,023.
Stokes, D. L., and Green, N. M., 1990, *J. Mol. Biol.* **213**:529.
Tanford, C., 1984, *CRC Crit. Rev. Biochem.* **17**:123.
Taylor, E. W., 1979, *CRC Crit. Rev. Biochem.* **6**:103.
Taylor, K., Dux, L., and Martonosi, A., 1984, *J. Mol. Biol.* **174**:193.
Thomas, D. D., 1986, in *Techniques for the Analysis of Membrane Proteins* (C. I. Ragan and R. J. Cherry, eds), Chapman and Hall, London, pp. 377–431.
Thomas, D. D., 1987, *Ann. Rev. Physiol.* **49**:691.
Thomas, D. D., and Cooke, R., 1980, *Biophys. J.* **32**:891.
Thomas, D. D., and Hidalgo, C, 1978, *Proc. Natl. Acad. Sci. USA.* **75**:5488.
Thomas, D. D., Dalton, L. R., and Hyde, J. S., 1976, *J. Chem. Phys.* **65**:3006.

Thomas, D. D., Ishiwata, S., Seidel, I., and Gergely, J., 1980, *Biophys. J.* **32**:873.
Thomas, D. D., Bigelow, D. J., Squier, T. C., and Hidalgo, C., 1982, *Biophys. J.* **37**:217.
Thomas, D. D., Fajer, P. G., Fajer, E. A., Brunsvold, N. J., and Svensson, E. C., 1988, in *Molecular Mechanisms of Muscle Contraction* (H. Sugi and G. H. Pollack, eds.), Plenum, New York, pp. 241-253.
Vanderkooi, J. M., Nakamura, H., Ierokomas, A., and Martonosi, A., 1977, *Biochemistry* **16**:1262.
Voss, J., Birmachu, W., Hussey, D., and Thomas, D. D., 1991, *Biochemistry* **30**:7498.
Wakabayashi, S., Ogarusu, T., and Shigekawa, M., 1986, *J. Biol. Chem.* **261**:9762.
Walker, J. W., Reid, G. P., and Trentham, D. R., 1989, *Methods in Enzymol.* **172**:288.

9

ESR Spin-Trapping Artifacts in Biological Model Systems

Aldo Tomasi and Anna Iannone

1. INTRODUCTION

The direct observation of free radicals in complex biological systems is often hampered because of the low concentration and high reactivity of the primary free radical species.* Spin trapping is a powerful tool that facilitates the visualization of free radicals, including those formed in such complex biosystems. The spin trap is a diamagnetic compound that reacts with a reactive free radical to form a more stable radical adduct. Although detection through electron-spin resonance spectroscopy offers several distinct advantages in its high sensitivity, and in some cases its specificity toward various radical species, there are also several drawbacks to using this technique.

The technique was developed in the late 1960s by several laboratories (reviewed by Janzen, 1980). Two categories of compounds are commonly utilized as spin-trapping agents: nitroso and nitrone compounds. The nitrogen atom of the nitroso spin trap reacts directly with the free radical species, giving distinctive spectral features. Two nitroso compounds are

* Dedicated to the memory of Prof. Trevor F. Slater.

Aldo Tomasi and Anna Iannone • Istituto di Patologia generale, Università degli Studi di Modena, 41100 Modena, Italy.

Biological Magnetic Resonance, Volume 13: EMR of Paramagnetic Molecules, edited by Lawrence J. Berliner and Jacques Reuben. Plenum Press, New York, 1993.

currently used in biological investigation: 2-methyl-2-nitroso propane (MNP) and 3,5-dibromo-4-nitrosobenzenesulfonate (DBNBS).

The nitrone spin traps each have a radical added to the carbon; the radical will be β with respect to the nitroxide radical center. The lack of spectral information about the trapped radical is the major drawback of this class of spin trap. Three commonly used spin traps will be discussed: phenyl-*t*-butyl nitrone (PBN), α(4-pyridyl-1-oxide)-N-*t*-butyl nitrone (POBN), and 5,5-dimethyl-1-pyrroline N-oxide (DMPO).

The technique is still quite young and has been associated with various cases of incorrect interpretations; these generally can be attributed to

1. changes in the spin trap (nonradical, chemical, photochemical, enzymatic reactions),
2. perturbation of the biological system by the probe, or
3. artifactual reporting associated with intrinsic properties of the probe.

The word *artifact* literally means a product of human endeavor, as distinguished from a natural object; from the Latin *arte* (by art, by skill) and *factum* (something done). The word has often assumed a negative nuance, meaning an inconstant product resulting from "artificial" manipulation. Spin trapping is a good example of the complex interaction between the model system and the artificial addition of a probe, with a high possibility of recording "artifactual" results.

This review will discuss both advantages and limitations of individual spin traps; the pharmacologic and toxicologic aspects will be also covered. Many published reviews have been centered on the trapped radical species. Thinking of the central role played by the spin trap, we thought that it would be useful to observe the matter from the point of view of the trapping agent; thus, each spin trap will be individually examined and discussed.

Reading this chapter requires a certain familiarity with the spin-trapping technique and its vocabulary (Janzen, 1990). An ample literature concerning spin trapping, with a particular eye on its application to biological systems, is available. Starting with the first to propose the term *spin trapping* (Janzen, 1971; Janzen, 1980) and its application to biological systems; excellent coverage of the chemical applications can be found in the reviews by Evans (1979) and Perkins (1980). Spin trapping in biology is covered by various reviews (McCay *et al.*, 1980; Mason, 1984; Rosen, 1985; and Mottley and Mason, 1989). An extensive literature survey has been carried out by Dodd (1990). Specifically devoted to examining the problems associated with the spin trapping of oxygen-centered free radicals are the reviews of Finkelstein *et al.* (1980), Rosen and Rauckman (1984), Rosen (1990), and Pou and Rosen (1990).

Invaluable help in disentangling the number of spectra and attributions is given in the database for spin-trapping by Li *et al.* (1989), which has been made freely available to all those interested in the field. Further development of an ESR spin-trapping database, to include mass spectral data, is in preparation (DuBose *et al.*, 1991).

2. NITROSO SPIN TRAPS

Nitroso compounds have an inherent advantage over nitrones for radical identification in that the added group lies immediately adjacent to the nitroxide center and therefore can easily give rise to additional hyperfine splitting, helping to identify the radical trapped.

One of the main drawbacks to using nitroso compounds is that they are thermally and photochemically unstable. Additionally, they are slightly soluble in water and tend to dimerize. In solution at room temperature only a small amount of the monomer, which is the actively trapping species, is present. Since the radical is bonded directly to the nitroxide function, the possibility of a reverse or cleavage reaction of the nitroxide is also favorable.

2.1. 2-methyl-2-nitroso Propane (MNP)

2.1.1. General Considerations

MNP was probably the first nitroso compound to be studied extensively and the first to be proposed for use in biology (Mackor *et al.*, 1966, 1967; Wajer *et al.*, 1967; Leaver *et al.* 1969); the difficulties arising when using this compound were immediately made clear. MNP, a solid dimer, gives rise in solution to a monomer–dimer equilibrium (Sergeev and Leenson, 1978). Unfortunately the dissociation of dimeric MNP to its active monomeric form is slow and is accompanied by a slow and complicated decomposition even in the dark (Kirino *et al.*, 1981). Isobutene, *t*-butylnitrosohydroxylamine, and *t*-butyl alcohol have been found in relevant amount in aqueous solution. Artifactual radical adducts are formed when the compound is exposed to visible light, giving the monomeric form an optical absorption band at 680 nm (Mackor *et al.*, 1966). The stable radical species observed is due to di-*t*-butyl nitroxide (DTBN), and the ESR triplet often overlaps the absorption of radical adducts of interest.

MNP is very volatile, to such an extent that it is readily lost when the solution is deoxygenated by nitrogen gas bubbling.

Makino *et al.* (1981), in a fundamental study on MNP, observed that the decomposition of MNP dimer can not be suppressed even at low temperature. In addition, a low yield of the monomer is obtained. It is hence advisable to obtain a solution in which the concentration of the monomer is as high and that of the decomposition products is as low as possible. The best compromise has been obtained by preparing MNP in aqueous solution by stirring in the dark for less than 2 hr at 45 °C.

In case of a large DTBN signal preventing the identification of an underlying spectrum, it may be possible to remove DTBN signal by decreasing the pH to about 4, where DTBN generally decays faster than other species (Janzen *et al.,* 1973). Extraction in petroleum ether succeeded in removing DTBN. Naturally such an approach, routinely done in a chemical model system, is not always applicable to a biological one.

2.1.2. Artifacts in Biological Applications

Since the early chemical studies, MNP has given rise to a somewhat unexpected and often challenging chemistry. Hartgerink *et al.* (1971), for example, reported that the trapping of trichloromethyl radicals by monomeric MNP would eventually give rise to a complex bimolecular oxidation–reduction reaction in solution, in parallel to other decomposition reactions, giving rise to a host of intermediates, ultimately leading to a more stable species identified as an acyl alkyl nitroxide. Notwithstanding the delineated difficulties, MNP has been used in various biological model systems. In what is probably the first attempted assignment of an MNP-derived signal, Ingall *et al.* (1978) recorded a triplet characterized by $a_N = 15$ G and broad lines during the incubation of isolated microsomes in the presence of NADPH, CCl_4, and MNP under anaerobic conditions. It was proposed that a secondary lipid peroxy radical or $CCl_3O_2\cdot$ might have been the species trapped.

Experiments carried out using a similar model system (microsomes NADPH and CCl_4 or lipoxygenase) gave rise to the same signal (Kalyanaraman *et al.*, 1979a). The signal obtained, a poorly resolved six-line spectrum, was assigned to a MNP–lipid dienyl type radical, along with a spectrum of four sharp lines attributed to *t*-butyl hydronitroxide, related to the reduction of MNP operated by microsomes. The authors concluded that the assignment by Ingall *et al.* (1978) was inaccurate. Later experiments carried out on NADPH-supplemented microsomes has led to the assignment of the signal to a mixture of DTBN and carbon-centered radicals (Tomas *et al.*, 1980).

Lai and Piette (1979a), in an attempt to provide evidence for OH radical production in the Fenton reaction, added MNP to the iron–hydrogen peroxide mixture; an ESR spectrum consisting of a quartet with intensity 1:2:2:1 was observed and assigned to the MNP–OH adduct. In probably the first of the spin-trapping papers containing words such as *caution, reappraisal, reconsideration,* and *artifactual,* Kalyanaraman *et al.* (1979b) emphasized that an identical spectrum was obtained by chemical or enzymatic reduction of MNP. The species responsible for the quartet was reassigned to *t*-butyl hydronitroxide. Further studies (Mottley *et al.*, 1981) observed a similar artifactual reduction of nitrosobenzene to its hydronitroxide in the presence of catechols, zinc, and titanous ion.

It appears clear that hydroxyl radicals can not be detected with nitroso spin traps because the spin adducts are too short-lived due to a primary disproportionation reaction.

2.1.3. Nonradical Reactions

Nitroso compounds undergo an -ene addition, usually an unwanted reaction in spin trapping. Mason *et al.* (1980) observed this reaction in the prostaglandin synthetase oxidation of arachidonic acid in the presence of MNP. This nonradical reaction forms a hydroxylamine, which in turn may undergo oxidation giving rise to a nitroxide. The extraction in organic solvents facilitates the reaction, since the concentration of oxygen is higher in the organic solvent. Ene addition may explain the apparently anomalous results observed (Albano *et al.*, 1982), where extraction in organic solvents of the reaction products of MNP, fatty acids, and lipoxygenase resulted in trapping carbon-centered radicals, whose signal amplitudes were, however, independent of incubation time.

However, in a carefully designed experiment, the use of MNP in a biological model system is possible; e.g., in liposomes, adducts were observed when saturated and unsaturated phospholipids were mixed in various molar fractions (Feix and Kalyanaraman, 1989). Here, the radical versus nonradical mechanisms of spin-adduct formation were separated using radical scavengers.

2.1.4. Decay of Spin Adducts

A number of MNP adducts are rather unstable. Pfab (1978) demonstrated that spin trapping of peroxyl radicals in the photooxidation of C-nitrosoalkanes gave rise to alkoxyl, rather than to the expected alkylperoxy–nitroxide adducts. The absence of an observable stationary-state concentration of peroxy adduct was interpreted as a rapid subsequent decay of the adduct rather than as a failure of the trap to scavenge. In conclusion, the use of nitroso compounds as spin traps in homolytic oxidation reactions will not permit discrimination between peroxyl and alkoxyl radicals since both give rise to alkoxyl radical adducts.

MNP has also been used to trap tert-butoxy radicals, generated by either photochemical or iron-induced decomposition of *t*-butyl hydroperoxide. The adduct decomposed rapidly, and the signal due to the methyl-radical adduct was recorded (Gilbert *et al.*, 1981; 1991). In a biological model system of isolated mitochondria metabolizing *t*-butyl hydroperoxide, similar results were observed (Kennedy *et al.*, 1986; Iannone *et al.*, 1992); an indication of the impossibility of demonstrating oxygen-centered radicals using MNP in such model systems.

MNP was used to demonstrate that methmyoglobin and liver microsomal preparations initiate the radical decomposition of cumene hydroperoxide (Griffin, 1982). The only product trapped was the methyl radical formed by β-scission of the cumyloxyl radical. Peroxyl and alkoxyl radical adducts of MNP are well known. They have been obtained in chemical model systems in aprotic solvents (Mackor *et al.*, 1967; Pfab, 1978); the hyperfine splitting constants are quite distinctive, since $a_N \geq 27$ G. Such adducts have never been seen in complex biological model systems; while a tert-butyloxyl-MNP radical adduct has been reported during the decomposition of *t*-butyl hydroperoxide catalyzed by purified chloroperoxidase (Chamulitrat *et al.*, 1989).

Some carbon-centered radicals also give rise to unstable MNP adducts. The $CCl_3\cdot$-MNP adduct obtained by photolysis of CCl_4 has a half-life of ≤ 1 s; a considerable yield of acylnitroxides is observed instead (Symons *et al.*, 1982).

A report on MNP-trapping cyanyl radicals arising from cyanide and hydrogen peroxide–horseradish peroxidase (Stolze *et al.*, 1989) demonstrated that the cyanyl adduct formed was unstable at neutral pH. The parasitic di-*t*-butyl nitroxide dominated the spectrum concomitantly with a sulfur center, probably due to sulfur-containing impurities, and the formamide radical adduct, probably a cyanyl radical–derived hydrolysis product. The overcomplicated picture was such that the only conclusion the interested reader might have reached was that one should carefully avoid the use of MNP for studying cyanide oxidation.

2.1.5. Toxicological Studies

MNP affects the adhesive characteristic of cultured fetal mouse-liver cells, causing them to detach from the surface of the Petri dish (Morgan, 1985). A marked decrease in cell viability was observed after exposure to 5-mM MNP for 5 min. Recent studies (Konorev *et al.*, 1991) evaluating the effect of spin traps on cardiac function, indicated that MNP is one of the more toxic spin traps tested. It was suggested that cardiovascular toxicity was inversely related to lipid solubility.

MNP (10 mM) has been reported to significantly inhibit the formation of hydrogen peroxide through the dismutation of O_2 (Britigan *et al.*, 1991).

Thus, MNP is not a popular trap in biological model systems. Its capability of providing a "fingerprint" of the radical trapped, because the added group lies closer to the unpaired electron, is widely outweighed by its thermal and photochemical instability, its incapability of producing stable oxygen-centered adduct in aqueous solution, and its toxicity. The complex relation between monomer and dimer forms, the latter being inert toward trapping radicals, prevents any serious attempt to perform quantitative analysis of the radical yield.

MNP can be used in biological applications, taking into account the above prerequisites. Some recent applications may indicate the best use for this nitroso spin trap. Iwahashi *et al.* (1991b) have used MNP in the study of reactions of linoleic, linolenic, and arachidonic acids with lipoxygenase in the presence of MNP. The usefulness of such an approach lies in the possibility of performing HPLC mass spectrometry on the adducts. Nitric oxide, a stable free radical species, regulator of the vascular tone *in vivo*, has been trapped in an apolar solvent (Arroyo and Kohno, 1991). In a biological environment, human platelets activated by a low concentration of collagen (Pròna i *et al.*, 1991), MNP gave rise to a variety of signals, none however corresponding to the MNP–nitric oxide adduct.

2.2. 3,5-dibromo-4-nitrosobenzenesulfonate (DBNBS)

2.2.1. General Considerations

Kaur *et al.* (1981) prepared and used 3,5-dibromo-4-nitrosobenzenesulfonate (DBNBS). The trap, sometimes referred to as Perkins's trap, was devised in order to overcome problems encountered in the use of lipophilic nitroso compounds. Sensitivity to light is less of a problem with aromatic C-nitroso compounds, and the introduction of sulfonate and bromine in the 3-5 position has made the new trap hydrophilic and an efficient radical scavenger in aqueous solution. Although DBNBS gives rise to a dimeric form, the concentration of the monomer, even at a low concentration of the trap, is sufficient to give spin adducts.

DBNBS remains quite stable under illumination and temperature variation, as reported by Smith and Robertson (1987); DBNBS–oxygen-centered radical adducts are unstable (Kaur *et al.*, 1981).

2.2.2. Artifacts in Biological Applications

An as yet unassigned triplet characterized by $a_N = 12.4$ G and devoid of further hyperfine splitting has been reported by various authors (Taffe *et al.*, 1987; Athar *et al.*, 1989). It was observed, along with the DBNBS–methyl adduct, during the incubation of human keratinocytes in the presence of tert-butyl hydroperoxide. A very similar signal, however, was observed (Samuni *et al.*, 1989a) arising from DBNBS in solution stored for a few days on the shelf.

An interesting feature of this trapping agent is related to its hydrophilicity. This fact can be exploited in a positive way. The addition of DBNBS to a model system containing membranes causes an uneven distribution of the trap. Samuni *et al.* (1989a) demonstrated that within several minutes of incubation time only a small amount of DBNBS was able to penetrate the cell membrane. This fact could be exploited in order to determine the location of the free radical formation.

DBNBS has been used in a field of major interest: the detection of NO·, the vasoactive mediator identifiable with the endothelial-derived relaxing factor (Arroyo *et al.*, 1990a). Nitrone and nitroso spin traps were tested for their ability to trap gaseous NO· (Arroyo and Kohno, 1991). The results suggested that the ESR spectra obtained from nitrone traps results from the hydrolysis of the traps themselves. Nitroso spin traps and in particular DBNBS gave spectra perhaps compatible with the trapping of a nitrogen-centered radical. When ^{15}NO was substituted for ^{14}NO, the spectra did not show the expected isotope effect. Spin trapping studies carried out in human platelets (Prónai *et al.*, 1991) and in mouse neuroblastoma cells (Arroyo and Forray, 1991) gave similar spectra. Several spin adducts were recorded; however, they did not correspond to those originating from the interaction of nitric oxide gas with the spin trap. The unambiguous trapping of NO· or NO-derived radicals in a biological system remains to be demonstrated. The area is of extreme interest and cautionary notes issued at the initial steps of the development in this area might help in avoiding future reappraisal papers.

2.2.3. Nonradical Reactions

For a nitroso spin trap, the possibility of -ene addition giving rise to a nitroxide, as described for MNP, should be taken into account. No reports regarding biological applications of this trap reported the problem. In a chemical system, DBNBS gave rise to nitroxides through an -ene addition mechanism in the presence of methacrylic monomers (Smith and Robertson, 1987).

2.2.4. Decay of the Spin Adduct

Cells in culture (CHO or TB), and red blood cells were able to destroy a performed DBNBS adduct in a matter of seconds. This might hamper the utilization of DBNBS in actively metabolizing cells or tissues (Samuni *et al.*, 1989b).

2.2.5. Toxicological Studies

There is little available in the literature about the toxicity of DBNBS. It appears, though (Samuni *et al.*, 1989a), that the compound is not particularly toxic to cells in culture. DBNBS turned out to be fairly toxic for the cardiac function in a perfused heart (Konorev *et al.*, 1991), the concentration effective for 50% inhibition of the rate–pressure product being 18 mM.

2.2.6. Does DBNBS Trap the Superoxide Anion Radical?

There is an interesting example of how the erroneous assignment of a DBNBS adduct has started a wealth of counterobservations, which finally resulted in a full understanding of the apparently simple case. The seminal paper was by Ozawa and Hanaki (1986). Their paper reported the trapping by DBNBS of superoxide anion generated by xanthine–xanthine oxidase in aqueous solution or by alkaline dimethyl sulfoxide. Stolze and Mason (1987) questioned the assignment of the spectrum obtained in alkaline dimethylsulfoxide. The DBNBS adduct observed was assigned to the sulfur trioxide anion radical adduct. The same spurious sulfur-centered adduct was observed in a model system where cyanide radicals were expected to be formed (Stolze *et al.*, 1989). In fact the DBNBS-sulfur trioxide radical adduct is readily formed and is extremely stable indeed (Ozawa and Hanaki, 1987; Stolze and Mason, 1987). Caution should be used when DMSO, a common organic solvent, is used at high pH, producing a variety of radical species, including methyl, superoxide, and $SO_3^{\cdot}$.

The assignment of DBNBS spectra as originating from the xanthine–xanthine oxidase system (Ozawa and Hanaki, 1986) was also questioned in a recent paper (Mani and Crouch, 1989). The signal characterized by $a_N = 12.6$ G and $a_H = 0.7$ G, being the hydrogen hyperfine splitting constant ill resolved, could not be assigned to the superoxide adduct. The addition of superoxide dismutase or catalase failed to attenuate the signal. (The observed signal actually increased under these conditions.) The authors did not make an attempt to identify the adduct.

Independent observations by Samuni *et al.* (1989a), carried out simultaneously, convincingly showed that DBNBS in the xanthine–xanthine oxidase system did not give any observable ESR signal; the authors, however, obtained the customary superoxide adduct in the same model system using DMPO as spin trap instead. The lack of any radical adduct observed in the DBNBS experiment was related to the fact that DMSO was not present in their preparation.

The matter was finally reexamined and settled in a recent paper (Nazhat *et al.*, 1990). The authors irrefutably demonstrated, by pulse radiolysis, that DBNBS reacts rapidly with $O_2^{\cdot-}$, but the product formed is unstable and goes undetected by ESR spectroscopy. They further investigated the xanthine–xanthine oxidase system in the presence of DBNBS and demonstrated that DBNBS gave rise to the signal, having about the same hyperfine splitting constants reported above ($a_N = 12.8$ G and $a_H = 0.7$ G, being the hydrogen hyperfine splitting constants ill resolved) when incubated with H_2O_2 and horseradish peroxidase. The spectrum was reasonably assigned to the radical formed by a one-electron peroxidatic reaction of DBNBS. The final step demonstrated that some xanthine oxidase preparations possess substantial peroxidase activity.

This is a clear example of how the interference of an unexpected secondary reaction gives rise to a radical adduct, which may well draw the full attention of the researcher and prevent the correct interpretation.

3. NITRONE SPIN TRAPS

Nitrones are the most popular spin traps because of their easy application to many biological model systems. They are reasonably hydrophilic, light insensitive, and not exceedingly toxic, and they give reproducible results. In addition, many adducts, and particularly carbon-centered adducts, are stable, sometimes for years.

The drawback in using nitrones is that the radical addition in β position with respect to the nitroxide center gives rise to very similar spectra. Often minor differences in the hyperfine splitting constants are interpreted as the trapping of different radical species, which may well be not the case. The opposite is also true.

3.1. N-*t*-butyl-α-phenyl Nitrone (PBN)

3.1.1. General Observations

PBN is, along with DMPO, the most popular trap. It has a limited solubility in water or in Tris and phosphate buffer. About 0.15-M PBN can be obtained in aqueous solution after stirring for several minutes, but such a concentration cannot be reached in most cell culture media, where at a concentration higher than 50 mM a precipitate is observable.

The commercial preparations do not usually need purification before use. PBN is rather stable in solution at neutral pH; however, it is advisable to use freshly prepared solutions in order to avoid the possible accumulation of benzaldehyde, a breakdown product of PBN, favored by shifts to lower pH. The intense smell of the aldehyde should be a reliable warning (and free of cost). In connection to its lipophilicity PBN readily crosses membranes and diffuses into the cells.

PBN spectral features are poorly sensitive to the radical trapped, rendering it difficult to identify the adduct from the simple coupling-constant measurements, hence the need to use supplementary strategies for the identification of the radical adduct. Isotopic substitution and chemical synthesis of the original adduct have been widely used (Poyer *et al.*, 1980; Tomasi *et al.*, 1980; Albano *et al.*, 1988). The stability of carbon-centered PBN adducts is often such that the nitroxide can be extracted, purified, and analyzed using GC–MS (Janzen *et al.*, 1990a) or HPLC–MS (Janzen *et al.*, 1990b), a conclusive and secure way to identify the radical adduct.

3.1.2. Artifacts in Biological Applications

A relatively small number of artifactual signals have been reported in the literature for PBN. In an early paper (Saprin and Piette, 1977) it was observed that a PBN adduct resulted from the activation of lipid peroxidation in rat liver microsomes. The signal, however, was dependent not only on NADPH addition but also upon the Tris buffer concentration. This suggests the addition of a Tris-derived radical adduct; the authenticity of this assignment was further supported by the fact that no signals were observed if the experiment was run in phosphate buffer.

Incubations of PBN in the presence of cells give rise to a weak signal (Poli *et al.*, 1987), usually interpreted as the result of trapping carbon-centered radicals from peroxidizing lipids.

Spin trapping *in vivo* has been developed by administering PBN to the animal. Such experiments are sometimes referred as *ex vivo* experiments, since the adduct formed *in vivo* can actually be seen only in tissue extracts. For recent applications and review, see Reinke and Janzen (1991), Mason *et al.* (1989), Knecht *et al.* (1989), and Kennedy *et al.* (1990).

If an adduct is not detected in an *in vivo* experiment, this is not necessarily due to the lack of its formation. A good example is given by Reinke and Janzen (1991) where the trichloromethyl–PBN adduct in plasma of CCl_4–PBN-treated rats was observed only in toluene extract. They demonstrated that the trichloromethyl adduct in plasma was bound to plasma proteins, which immobilized the adduct, rendering it invisible to ESR spectroscopy.

Extraction in organic solvent is a procedure commonly employed in PBN studies. It would be desirable to demonstrate the presence of the adduct in the aqueous phase, too, before proceeding to work on the organic phase. A significant solvent effect on hyperfine splitting constants has to be expected and taken into account for the correct assignment of the adduct. Shifts in a_N and a_H have been described for a number of nitroxides at the changing of the solvents (Janzen, 1980; Janzen and Coulter, 1981).

The shift in the hyperfine splitting constants of nitroxides is due to interactions of the polar solvents with the lone pair electrons of the nitroxide oxygen atom, resulting in the redistribution of the π-electrons such that there is maximum unpaired electron density on the nitrogen atom. Theoretical calculations can predict the hyperfine splitting constants of nitroxides in solvents with different polarity (Nordio, 1976). The theoretical prediction has been confirmed experimentally and a plot of the isotropic nitrogen hyperfine coupling constants resulted that is linear with respect to Reichard's solvent polarity parameter (Janzen, 1980).

Extraction in organic solvent clearly offers some advantages as compared with measuring the ESR spectra in the aqueous solutions in which they were produced: the linewidths in the degassed organic solvent are narrow, the radical adducts are usually more stable, the sensitivity can be improved by concentrating the samples, and the concurrent processing of a high number of samples is feasible.

A controversy of some years ago provides a good example of the complications connected to the extraction of the radical adduct. Bösterling and Trudell (1981) suggested the use of ethyl acetate for extraction of hydroxyl and superoxide radical adducts of PBN formed in a purified NADPH-Cytochrome *P*-450 reductase model system. Unique coupling constants were given for PBN-hydroxyl and PBN-superoxide adduct. Kalyanaraman *et al.* (1984) challenged the assignments.

However, Trudell (1987) demonstrated that the initial assignments might have been correct. The method chosen in order to support the assertion was chemical synthesis of the original adducts and recording the hyperfine splitting constants in different solvents.

An attempt to quantify the trapped adduct has been made by some researchers. Recently Hughes *et al.* (1991) quantified the PBN adduct recorded from the bile of animals treated with halothane as compared to that observed in animals treated with carbon tetrachloride. The quantitative assessment of a radical adduct, though easy in itself, will remain prone to error; at least as long as the complex kinetics underlying the formation, the reduction, and the metabolism of the adduct and the rates of reaction of the free radical with other substrates remains mostly not proven.

3.1.3. Decay of the Spin Adduct

PBN-oxygen radical adducts are reported not to be stable in aqueous solutions at room temperature (Janzen *et al.*, 1978a). There are several reports of spin-trapping peroxyl radicals with PBN (Merrit and Johnson, 1977; Ohto *et al.*, 1977; Yamada *et al.*, 1984). All ESR spectra reported in those works were obtained in aprotic solvents such as benzene where, in general, free radical adducts are more stable.

In case the trapping of a peroxyl adduct is conjectured, the hyperfine splitting constants of the peroxyl adduct are very similar to those observed for carbon-centered radicals. For example the PBN-CCl_3 coupling constants

(a_N = 13.5 G, a_H = 1.5 G) are virtually indistinguishable from those observed for PBN-$CCl_3OO\cdot$ (a_N = 13.5 G, a_H = 1.6 G) (Symons *et al.*, 1982). The peroxyl adduct in protic solvents decomposes quickly above −20 °C to give the PBN-alkoxyl adduct derivative (Howard and Tait, 1978).

The PBN-hydroxyl radical adduct is rather stable at room temperature; it is extremely stable at acidic pH; and it decays with an increase in pH giving rise to *t*-butyl-hydroaminoxyl as a degradation product: similar findings were reported for nine PBN derivatives (Janzen *et al.*, 1991, 1992).

3.1.4. Pharmacology and Toxicology

The crux of much of the ongoing research on free radicals in biological model systems is demonstrating that free radical intermediates are the cause of the pathological alteration; as a corollary, compounds inhibiting free radical mediated reactions should have a protective, pharmacological activity.

PBN is the only spin trap extensively studied in this respect. Initial studies demonstrated that PBN protects against NADPH-induced (Saprin and Piette, 1977) and CCl_4-induced lipid peroxidation (Albano *et al.*, 1982). PBN is a cytochrome *P*-450 substrate, giving rise to a Type I spectrum (Cheeseman *et al.*, 1984; 1985), and inhibits cytochrome *P*-450 (Augusto *et al.*, 1982). The *P*-450-dependent metabolic products (Chen *et al.*, 1991), tissue distribution, and excretion have been also studied (Chen *et al.*, 1990). Recently the degradation products have been characterized using mass spectrometry (Chen *et al.*, 1991; Zhdanov *et al.*, 1991) and primarily identified as hippuric acid and, for the nitroxide adduct, the hydroxylamine derivative.

PBN affects the metabolism of glucose by red blood cells and has an inhibitory effect on the hexose monophosphate pathway suppressing pyruvate and stimulating lactate formation via the Embden–Meyerhoff pathway (Thornalley and Stern, 1985). PBN (100 but not 10 mM) inhibits the production of superoxide by stimulated neutrophils (Pou *et al.*, 1989).

Novelli *et al.* (1985) and Novelli (1992) reported a significant protection exerted by PBN observed during an experimental shock and described an ameliorating effect on endurance against muscle fatigue in rat. Novelli *et al.* (1990) and Phillis and Cloughhelfman (1990) demonstrated a protection from cerebral ischemic injury in gerbils. Very similar findings have been described by Floyd and Carney in two recent papers (Floyd and Carney 1991; Carney and Floyd, 1991).

PBN was shown to be significantly toxic on the myocardial function at a concentration usually employed in various studies. PBN (10 mM) caused 50% inhibition of the rate-pressure product (an index of cardiac function) as observed in a controlled Langedorff-type perfused heart system (Konorev *et al.*, 1991). On the other hand, PBN administered after a regional cardiac ischemia was shown to dramatically reduce the vulnerability of the myocardium to reperfusion-induced ventricular fibrillation (Hearse and Tosaki, 1987). PBN prevents doxorubicin-induced cardiotoxicity *in vitro* (Monti *et al.*, 1991); its activity has been suggested as depending on the mitochondrial accumulation of the spin-trapping agent (Cova *et al.*, 1992).

Kalyanaraman *et al.* (1991) have studied the relevance of free radical intermediates in the oxidative modification of human lipoproteins, a process held to be of primary importance in forming the atherosclerotic lesion. He found that PBN, added with the intent of demonstrating the presence of free radical intermediates arising during the autoxidative process, actually prevented the oxidative modification of low density lipoproteins.

This emphasizes again how the observation of the phenomenon often significantly modifies the model system. The spin trap, which is meant to be an innocent bystander reporting on the ongoing reactions, may prevent the reaction one intends to observe.

3.2. α(4-pyridyl-1-oxide)-N-*t*-butyl Nitrone (POBN)

3.2.1. General Considerations

POBN was developed in connection with a study of water soluble analogues of PBN (Janzen *et al.*, 1978b). It was shown to give a stable and unique ESR spectrum for the hydroxyl radical adduct. More work was performed in order to ascertain the ability of POBN to trap oxygen-centered radicals. Unfortunately, this turned out not to be the case.

POBN has been particularly exploited in cell studies. It has several advantages in that it couples to an elevated hydrophilicity and can permeate cell membranes (Albano *et al.*, 1989). POBN has been the spin trap of choice in a number of studies on ethanol and hydrazine-derivative metabolism (Albano *et al.*, 1988, 1989). Hydrazine activation to a free radical intermediate has been proposed as a cytochrome *P*-450 mediated process. However, Rumyantseva *et al.* (1991) reported that phenelzine (2-phenyl-ethyl-hydrazine) metabolism, giving rise to the 2-phenylethyl radical trapped by POBN, occurred by way of a trace transition metal catalyzed reaction. This has induced workers to reconsider the results reported in previous papers on POBN hydrazine-derivative trapping (Albano and Tomasi, 1987; Tomasi *et al.*, 1987). The matter is under active investigation, and in a recent paper (Goria-Gatti *et al.*, 1992) it was demonstrated that the alkyl radical trapped during microsomal metabolism of procarbazine (an hydrazine-derivative antitumor agent) as well as in isolated hepatocytes and *in vivo* depended on the presence of specific cytochrome *P*-450 isoenzymes (*P*-450IA and IIB).

Since POBN spectral features are insensitive to the radical trapped, by analogy with its parent compound PBN, the simple coupling constant measurements are usually not sufficient to identify the adduct. Positive identifications of the adduct has been obtained using isotopic substitution with ^{13}C-ethanol (Albano *et al.*, 1988) or by the chemical synthesis of the original adduct (Albano *et al.*, 1989).

3.2.2. Artifacts in Biological Applications

A POBN lipid hydroperoxy radical adduct was claimed to be detected when lipoxygenase and linoleic acid were incubated at pH 9.0 (Rosen and Rauckman, 1981) or in rat liver microsomes NADPH at pH 7.4 in the presence of diethylenetriaminepentaacetic acid (DETAPAC). It was observed that the addition of POBN to the lipoxygenase-linoleic acid system gave rise immediately to an adduct. In the same experiment, but in the presence of a nitroso spin trap, the ESR signal was observed at later time, when the sample was anoxic. The radical trapped by the nitroso spin trap was positively identified as a carbon-centered radical. The authors suggested that the radical trapped by POBN, arising in the presence of oxygen, was oxygen-centered. The bubbling of nitrogen prior to the addition of the lipoxygenase prevented the appearance of a POBN-dependent spectrum, corroborating the interpretation.

The matter was reexamined (Connor *et al.*, 1986) and experiments were carried out in the presence of $^{17}O_2$. The adduct obtained under these conditions did not show the typical additional ESR lines expected in the case in which an oxygen-centered radical is formed. The unambiguous assignment of the POBN radical adduct obtained under the same experimental conditions was carried out recently using HPLC–mass spectrometry. The adduct has been definitely assigned to a carbon-centered adduct (Iwahashi *et al.*, 1991a).

Definitive spectroscopic evidence that a POBN oxygen-centered adduct is indeed formed and is stable enough to be recorded has been reached using $^{17}O_2$ (Mottley *et al.*, 1986). Oxygen-centered radical adducts were observed in microsomes additioned with NADPH and paraquat, and in a xanthine–xanthine oxidase model system.

When the ESR spectrum of a spin-trapped adduct is interpreted, one must consider the chemical and biochemical properties of the reactants, for otherwise the identification and assignment may lead to misinterpretation. In other words, the spin-adduct identification needs a careful and cautious interpretation. The identification of a nitrone-derived radical adduct solely from the measurement of the coupling constants is more the exception than the rule.

It has been seen (Zini *et al.*, 1992) that POBN gives rise to a signal characterized by $a_N = 15.02$ G, $a_H = 1.68$ G when dissolved at high concentration (0.25 M), in Ringer's solution. The coupling constants are reminiscent of the described POBN–OH adduct, and the addition of DETAPAC or Desferal prevents the formation of the signal.

3.2.3. Decay of the Spin Adduct

The POBN–superoxide anion and the POBN–hydroxyl radical adducts are not extremely stable at pH 7 in aqueous solutions and decompose to diamagnetic compounds (Finkelstein *et al.*, 1979). Finkelstein *et al.* (1982) also reported the decay of POBN-oxygen-centered adducts and suggested that the superoxide adduct of 4-POBN might decompose to yield a strongly oxidizing species, probably hydroxyl radical, whose existence is proven by the reaction with a scavenger, DMSO or ethanol. The scavenger-derived radicals, in turn, give rise to a trappable and stable carbon-centered radical.

3.2.4. Pharmacology and Toxicology

POBN acts as vasodilator in a Langendorff perfused heart and is rather toxic to cardiac function, in agreement with a proposed spin-trap cardiac toxicity, which appears to be inversely related to lipid solubility (Konorev *et al.*, 1991).

3.3. 5,5-dimethyl-1-pyrroline N-oxide (DMPO)

3.3.1. General Considerations

DMPO is the most useful trap for the study of oxygen-centered free radicals. Oxygen-centered radicals are believed to play a major role in mediating tissue damage. Numerous studies have appeared in the literature, ranging from the early investigations on lipid peroxidation to recent ones on the intermediate metabolism of xenobiotics and carcinogens, on atherosclerosis, and on reperfusion damages; fields of major interest in clinical research.

The DMPO-superoxide anion adduct is unique, and the DMPO-hydroxyl adduct is almost unique. The $a_N = a_{\beta H}$ 1:2:2:1 quartet is probably the most often reported nitroxide spectrum in the literature. DMPO alkoxyl and peroxyl radical adducts are also well distinguishable (Kalyanaraman *et al.*, 1983; Taffe *et al.*, 1987; Davies, 1989) and distinct from carbon-centered radical adducts (Chamulitrat *et al.*, 1991).

DMPO (and, really, nearly any other compound) reacts rapidly with $\cdot OH$ ($K = \sim 10^9\ M^{-1}\ s^{-1}$) and gives a relatively stable adduct (Finkelstein *et al.*, 1980). In contrast, rate constants reported for the reaction of $O_2^{-\cdot}$ with DMPO is quite slow ($K = \sim 1.2$ to $15.7\ M^{-1}\ s^{-1}$ for DMPO) (Finkelstein *et al.*, 1979; Yamazaki *et al.*, 1990). This rate constant might increase substantially with the decrease of pH, where $HO_2\cdot$ predominate (Finkelstein *et al.*, 1980). Furthermore, DMPO–superoxide adduct is relatively unstable, and this has resulted in using high concentration (10–100 mM) of the trapping agent in order to offset the poor reaction kinetics. The different spin trapping efficiencies for $\cdot OH$ and $O_2^{-\cdot}$ have to be taken into account in interpreting experimental results.

DMPO is readily soluble in water, and diffuses readily in biological systems. Various papers have addressed the problem concerning the intracellular diffusion of this trapping agent. An overview (Pou *et al.*, 1989) has summarized the findings and concluded that diffusibility of DMPO into cells is prompt; however, the ability of the spin trap to diffuse to specific sites of free radical formation within the cell remains in question.

The results obtained using DMPO in biological studies have been subjected to continuous reappraisals. Part of the problem is inherent in the commercially available preparations of DMPO, which usually contains contaminants and decomposition products. These, unfortunately, vary from batch to batch. DMPO is, or should be, a colorless solid with a melting point of 25 °C. It is fairly susceptible to both light and oxygen, and storage should always be under nitrogen, away from light, and at −20 °C or less. Commercial preparations are usually yellowish and with variable melting point. DMPO has to be purified before use; some laboratories have solved the problem by synthesizing DMPO, though this approach is not recommended unless a highly experienced synthesis chemist is part of the group. A simpler approach is redistillation under vacuum. The procedure is tricky; in particular, the temperature has to be strictly checked and maintained below 30 °C. Mechanical vacuum pumps might be suitable for the process, and the use of a Kugelrorh apparatus has been suggested (Turner and Rosen, 1986). Column chromatography, using charcoal-celite (90/10 wt./wt.), having dissolved DMPO in a polar solvent, is also an efficient and quick way to improve the quality of a commercial preparation; elution can be conveniently monitored at 234 nm (Finkelstein *et al.*, 1980).

3.3.2. Artifacts in Biological Applications

A hotly debated issue concerns the spin trapping of hydroxyl and of superoxide radicals, where for superoxide is intended the equilibrium mixture of $O_2^{-\cdot}$ and $\cdot O_2H$. The first observation, stating that the DMPO-superoxide spectrum is unique and easily obtainable in the aqueous phase

(Harbour *et al.*, 1974; Harbour and Bolton, 1975) was followed by an extensive literature. The trapping of oxygen-centered radicals in a biological environment was repeatedly reviewed in a series of papers (Finkelstein *et al.*, 1979, 1980; Rosen *et al.*, 1982, 1990; Rosen and Rauckman, 1984; Turner and Rosen, 1986; Pou *et al.*, 1989).

A Six-Lines Spectrum. Problems and artifacts in the use of DMPO were present since the early papers. Ferric ions or methemoglobin can oxidize DMPO to the corresponding hydroxamic acid. Subsequent oxidation would produce the corresponding nitroxide, known as DMPOX (Floyd and Soong, 1977; Thornalley *et al.*, 1983).

Chelated iron can produce pseudoradical adducts in the presence of DMPO. An unidentified and unwanted triplet of doublets, usually called "six-lines," has been seen in many studies; for instance, adding DMPO to phosphate buffer iron–EDTA, but also in Tris buffer, with iron being present in trace amounts as a contaminant, or even in blank samples, with DMPO present at high concentration (above 50 mM). The hyperfine splitting constants of the six-lines are very close to those reported for the DMPO–methyl radical adduct (Kirino *et al.*, 1981), and it is safe to say that six-lines is a carbon-centered radical adduct.

A six-lines can probably be seen in the very early papers by Lai and Piette (1977) along with the well known DMPO–OH quartet; however, no comment was made about it.

A six-lines species was tentatively assigned to a DMPO dimer (Rosen and Rauckman, 1984). A six-lines spectrum was also observed by Britigan *et al.* (1986) in stimulated neutrophils and assigned to the methyl radical adduct. The adduct might, or should, have been derived from the reaction of hydroxyl radicals with DMSO, used as a solvent. However, the paper demonstrated that stimulated neutrophils do not produce hydroxyl radicals, hence the difficulties in explaining the formation of the six-lines spectrum.

More nonvalidated six-lines spectra have been reported in heart reperfusion studies (Blasig *et al.*, 1990; Zweier, 1988). Arroyo *et al.* (1987) tentatively assigned it to a carbon-centered lipid-derived radical.

Recently Makino *et al.* (1989) delved deep into the problem using HPLC, mass spectrometry, and NMR spectrometry. They assigned the six-lines spectrum to a nitroxide derived from the oxidation of a hydroxylamine impurity having an epoxy ring located at the 2 and 3 positions. To make things worse, this impurity competes with lipids for oxygen so as to change the yield of oxygen radical adduct formation.

It appears clearly that a six-lines observed in any model system, having a_N varying between 15.4 and 15.9 G and a_H between 22.2 and 22.8 G, should be assigned with caution. The possibility of the hydroxylamine-derived spectrum has to be taken in account. Experiments varying DMPO concentration could be the simple way to rule out such possibilities. In the case of a hydroxylamine-derived nitroxide, at the increase of DMPO concentration, the intensity of the six-lines will increase equally as a function of the spin-trap concentration.

The DMPO-OH Spin Adduct. The validation of the DMPO-hydroxyl radical adduct as derived from trapping genuine ·OH radical is under considerable debate. The demonstration of the presence of a strong oxidizing species such as the ·OH radical in a biological system is not without significance.

Many reports have tried to demonstrate the presence of genuine ·OH in various biological systems using DMPO. In the early ones, NADPH-induced lipid peroxidation in liver microsomes was suggested as generating ·OH radicals (Lai and Piette, 1977; Lai *et al.*, 1978, 1979b). In these pioneering papers, the formation of the ·OH adduct was shown to depend on the presence of NADPH and EDTA-iron. A spectrum observed upon addition of ethanol was considered the proof of the reaction between ·OH and its scavenger.

It has since been demonstrated (Makino *et al.*, 1990a, b) that spin trapping with DMPO in the presence of added iron might become deceptive. DMPO forms with iron a complex that may subsequently be attacked by oxidative species, inducing degradation of the trap, including the formation of DMPO-OH; the mechanism of formation is based on the nucleophilic attack of water on the chelating DMPO. These experiments have been carried out in a chemical model system, and the demonstration of the relevance of such a mechanism in a biological model system has yet to be demonstrated. Nevertheless, in order to validate a DMPO-OH signal as originating from trapping genuine ·OH, the possibility of a nonradical mechanism has to be kept in mind by the researcher.

Still more nonradical mechanisms could give rise to DMPO-OH. Nohl *et al.* (1981) reported the formation of DMPO-OH simply by heating DMPO in buffer for 15 min at 40 °C or decreasing the pH.

DMPO-OH Adduct in the Xanthine-Xanthine Oxidase System. Controversial data have been reported on the authenticity of the DMPO-OH radical adduct observed in the xanthine-xanthine oxidase system. This enzymatic reaction produces a high flux of superoxide anion. In an early work, Buettner and Oberley (1978a) detected both superoxide and hydroxyl radicals in a xanthine-xanthine oxidase system. The authors suggested that the trapping of ·OH was dependent on the presence of contaminating iron, which initiated a Fenton-type reaction, and indicated diethylenetriaminepentaacetic acid (DETAPAC) as the most appropriate iron chelating agent, able to avoid trace metal ion interferences.

Kuppusamy and Zweier (1989) gave evidence for genuine hydroxyl radical generation in the same model system. Methyl or α-hydroxyethyl radical were trapped when dimethylsulfoxide or ethanol were added to the mixture. This was interpreted as actual hydroxyl radical production. Superoxide dismutase did not inhibit the DMPO-OH signal formation, indicating that $O_2^{-\cdot}$ was not required for ·OH generation, giving further support to the hypothesized formation of ·OH in the system.

In less than a year, though, the same journal published two rebuttals (Britigan *et al.*, 1990a; Lloyd and Mason, 1990). These responses enumerated a number of published papers that did not find evidence of hydroxyl radical generation in the reaction (e.g., Britigan *et al.*, 1986; Thornalley and Vasak, 1985). The formation of the DMPO-OH adduct had to be attributed to the presence of adventitious iron bound to the enzyme. The contaminating metal ions, in the presence of enzymatically generated superoxide, fired a Fenton-type reaction; the preincubation of the enzyme with iron chelating agents such as DETAPAC or Desferal prevented the formation of DMPO-OH signal. The observation of a DMPO-OH signal can not be interpreted as formation of true ·OH; the observation of adducts arising from the hydroxyl radical attack on so called hydroxyl radical scavengers (ethanol or dimethylsulfoxide, e.g.) is also not sufficient for the validation of ·OH formation. In this case an approach different from spin trapping could help in elucidating the problem. Indeed, through the use of other techniques, Diguiseppi and Fridovich (1980) convincingly suggested that hydroxyl radical observed in the xanthine-xanthine oxidase model system was the result of contaminant iron.

In order to have a complete picture, it also has to be taken into account that DMPO (>10 mM) acts as a competitive inhibitor of the dismutation of superoxide to hydrogen peroxide (Britigan *et al.*, 1991). $O_2^{-\cdot}$ also reacts with DMPO-OH and DMPO-OOH (Buettner, 1990; Samuni *et al.*, 1988a) to create diamagnetic degradation products. On the other hand, it has also been reported that various nitroxides have a SOD-like activity, accelerating the rate of $O_2^{-\cdot}$ dismutation to H_2O_2 (Samuni *et al.*, 1988a).

Buettner (1990, 1991) gave a recipe for obtaining a clean superoxide adduct in the xanthine-xanthine oxidase system. The buffer has to be carefully passed through Chelex; DMPO (50–100 mM) has to be highly purified; dialysis of the enzyme or preincubation with DETAPAC is recommended, since commercially available enzyme preparations contain unknown amounts of adventitious metal ions. DETAPAC (50–150 μm) is the most useful chelating agent, since EDTA is apparently not as efficient in chelating and preventing adventitious iron to enter redox reactions; Desferal reacts at a significant rate with superoxide (Davies *et al.*, 1987; Morehouse *et al.*, 1987). A high superoxide flux, given, e.g., by 12–20 mU/ml xanthine oxidase and 0.5 mM xanthine is also desirable in order to obtain the best result. Under these conditions not the smallest trace of DMPO-OH will be recorded.

The complex and contrasting effects of the spin trap on the experimental model system render any quantitative measurement risky. In the case of a model system generating $O_2^{-\cdot}$ at a low rate, the generation of H_2O_2 and derived species may be inhibited by the same spin trap, used in high concentration in order to enhance its sensitivity.

DMPO–OH in Phagocytes. The actual formation of ·OH radical during the respiratory burst of neutrophils has been, and still is, a controversial topic, in which DMPO spin trapping has played a central role. Taking into consideration the recent literature, papers have in general shown that ·OH is not a physiological product of human neutrophils or monocytes (Britigan *et al.*, 1986, 1990b; Pou *et al.*, 1989). It must be mentioned, however, that Samuni *et al.* (1988b) have demonstrated that a high superoxide flux destroys DMPO–OH and suggest that hydroxyl radical may well be produced by stimulated human neutrophils.

DMPO–OOH and DMPO–OH adducts are both observed when phagocytes are stimulated in the presence of DMPO. The type and amount of stimulation has been demonstrated to be important. Mainly DMPO–OOH is observed when phorbol myristate acetate (PMA) is used as the stimulant. However, when PMA was used at high concentration (100 ng/ml) the DMPO–OH adduct prevailed; only DMPO–OH adduct was observed using zymosan as stimulating agent (Britigan *et al.*, 1986). After addition of N-formyl-methionyl-leucyl-phenylalanine (FMLP) or concanavalin A, no spin-trap evidence of free radical production was detected (Britigan and Hamill, 1990).

The formation of the DMPO–OH adduct may result from various mechanisms. Finkelstein *et al.* (1979) initially proposed that the short-lived superoxide adduct is apparently replaced by the hydroxyl adduct, suggesting a chemical rearrangement of one species into the other (see below for further comments on the decay of the spin adduct). Various papers have demonstrated (Janzen *et al.*, 1987; Britigan and Hamill, 1989; Bernofsky *et al.*, 1990) that the DMPO–OH adduct observed in neutrophils might also originate from a nonradical reaction between DMPO and hypochlorous acid produced during the respiratory burst.

Ueno *et al.* (1989) has suggested that the formation of the DMPO–OH adduct was associated with neutrophil injuries caused by DMPO (see Section 3.3.4 for further comments). It was proposed that enzymes released from lysed cells, such as GSH-peroxidase, may accomplish the transformation of DMPO–OOH to DMPO–OH.

Ramos *et al.* (1992) tackled the problem in a novel way. They used POBN as spin-trapping agent in conjunction with ethanol. The formation of the ethanol-derived radical adduct was inhibited by SOD, catalase, and azide. Under these conditions, a positive indication of hydroxyl radical formation was obtained in stimulated neutrophils. Since no evidence was observed in myeloperoxidase-deficient neutrophils, it was suggested that activated neutrophils generate hydroxyl radical through a myeloperoxidase-dependent mechanism.

Difficulties arise when the formation of DMPO adduct(s) is used to study the kinetics of superoxide production. The DMPO-OOH adduct has a short half-life, DMPO-OH and DMPO-OOH react with $O_2^{\cdot-}$, DMPO-OOH partly decays to DMPO-OH, and DMPO-OH itself can be metabolized by the stimulated cells; a temperature effect has been also seen (see Section 3.3.3 for further comments on this point).

The suggestion (Kleinhans and Barefoot, 1987) of using the signal intensity of the DMPO-OH adduct, corrected for the decay of the adduct, is difficult to implement. It takes into account the disappearance of the adduct, but it does not take into account the perturbation of the system (including the reactivity of the adduct with oxygen-centered radicals) caused by the trap itself. In this study the neutrophil burst was calculated to peak at 3.4–14 min for PMA concentrations of 50–500 ng, giving the impression that oxy-radical production was ephemeral. In a study carried out under similar conditions, Black *et al.* (1991a) reported that the production of superoxide could be monitored for as long as 5–6 hr; the results were corroborated by independent technologies: chemiluminescence and ferricytochrome *c* reduction.

In conclusion, the presence of the DMPO-OH spectrum is not a certain index of genuine ·OH formation in phagocytes. Attention has to be paid to checking for spin-trap or hydroxyl radical scavenger-induced toxicity in the model system and for the availability of iron, indispensable in forming ·OH by the Haber–Weiss reaction. The major open question remains that of the presence of metal catalysts. Great care has to be taken in order to avoid adventitious metal taking part to the reaction. On the other hand, metal(s) could really play a role in what could be defined as a "physiological Fenton reaction."

The Use of ·OH Scavengers. ·OH reacts in a diffusion-limited way with almost any compound of biological interest; it follows that high concentration of a scavenger has to be added in order to compete for ·OH. Ethanol or DMSO are used mostly in biological system as hydroxyl radical scavengers; the secondary radical is trapped, giving evidence of a ·OH-dependent reaction.

A spectrum characterized by $a_N = 15.8$ G, $a_H = 22.8$ G was first observed when ethanol was added to NADPH–microsomes EDTA–iron peroxidizing system (Lai and Piette, 1977). The authors suggested that the adduct could possibly be assigned to the DMPO–ethoxy radical adduct and demonstrated the presence of ·OH. It should be mentioned that Janzen and Evans (1973) had at the time already demonstrated that alkoxyl adducts of DMPO always give a β-hydrogen hyperfine splitting constant smaller than the *N*-hyperfine splitting constant. The adduct was later correctly assigned to the α-hydroxy ethyl radical adduct. McCay *et al.* (1992) reexamined the classic NADPH–microsome model system in the presence of DMPO. When animals were pretreated with aminotriazole, used to inhibit hepatic catalase activity, and the method for microsome preparation was modified in order to remove contaminants (ferritin and catalase in particular) a DMPO-OH signal was recorded. In the presence of ethanol the α-hydroxyethyl radical was trapped. The authors suggest that microsomes themselves provide the metal catalyst, most likely some form of iron. The lack of observation of DMPO-OH in previous spin-trapping experiments in microsomes has been attributed to the presence of catalase contaminant. The ethanol-radical evidence becomes the pivotal proof for ·OH production in microsomes.

However, the recording of a ·OH radical scavenger spectrum might not always be a sufficient proof of ·OH radical production. For instance, in the xanthine-xanthine oxidase model system (Kuppusamy and Zweier, 1989), the ·OH radical formation might have been related to particular experimental conditions. Makino *et al.* (1990a), checking the stability of DMPO in phosphate buffer in the presence of ferrous ion (5 mM $FeSO_4$), observed a typical DMPO-OH adduct. The addition of ethanol gave rise to the α-hydroxyethyl-DMPO adduct. More recently the same group (Makino *et al.*, 1990b) has observed that in the presence of ferric ions and methanol, a nucleophilic attack by methanol on the DMPO-ferric ion complex is occurring, giving rise to the DMPO-alkoxyl adduct of the scavenger, a possible source of confusion in interpreting the spectra.

Finkelstein *et al.* (1979) and Mason and Morehouse (1988) suggested the use of SOD and catalase in order to validate the presence of the DMPO-OH adduct. If the DMPO-OH adduct is inhibitable by SOD, it is implied that the hydroxyl radical formation depends on superoxide presence. If the addition of catalase does not result in the inhibition of DMPO-OH formation, this is an indication that the DMPO-OH is formed through nonradical pathway(s). In general, it can be said that, when superoxide is produced in the reaction(s) under study, the recording of the DMPO-OH has to be carefully validated.

1:2:2:1 Spectra, Possibly Not DMPO-OH. Janzen (1980) warned that in aqueous solutions the ESR spectra of tert-butyl hydronitroxide and DMPO-OH give rise to very similar quartets; this might become relevant if DMPO hydrolyzes under oxidizing conditions so that an open-chain hydronitroxide is produced. A careful measurement of the hyperfine splitting constants could solve the problem; the hydronitroxide gives values of about 14 G, while the DMPO-OH adduct always gives values of 14.8 G and above.

A second case of a 1:2:2:1 spectrum not due to DMPO-OH has been reported (Schaich and Borg, 1990). It has been suggested that lipid alkoxyl radicals might produce a quartet identical to that of DMPO-OH. A procedure was proposed indicating a possible way to differentiate between the spectra by extracting the adduct in ethyl acetate. However, the lack of stability of the phase transfer did not give clear evidence on the existence of two distinct 1:2:2:1 spectra. Chamulitrat *et al.* (1992) have given convincing evidence against the above assignment.

3.3.3. Decay of the Spin Adduct

Decay of DMPO-OOH. The DMPO-superoxide adduct is rather unstable, with a half-life of approximately 50 s at pH 7.4 and 25 °C (Buettner and Oberley, 1978). Oddly enough, the first paper reporting the DMPO hydroxyl and superoxide spin adducts (Harbour *et al.*, 1974) indicated the superoxide spin adduct as more stable than the hydroxyl adduct, an assertion probably due to a typographical error. A controversy is still present in the literature about the fate of the DMPO-OOH adduct. Some research groups (summarized in the papers by Finkelstein *et al.* (1982) and Pou *et al.* (1989)) maintain that an important product of the DMPO-OOH adduct decay is DMPO-OH. This conclusion is based on the observation that the short-lived superoxide adduct is apparently replaced by the hydroxyl adduct, suggesting a chemical rearrangement of one species into the other. Yamazaki *et al.* (1990) studied the decay of DMPO-OOH in various model systems and concluded that the adduct decays at pH 7.4 with a half-life of 66 s, according to first-order kinetics. The decay is accompanied by a small amount of DMPO-OH formation, with the conversion ratio being 2.8%. The authors, however, gave a warning about the reaction becoming highly intricate in the presence of chelated iron, in particular iron-EDTA. Complications depend on the reaction of iron-EDTA with superoxide, hydrogen peroxide, and the DMPO-OOH adduct itself.

Buettner (1990, 1991), in carefully controlled experiments, has demonstrated that the decay of the DMPO-OOH adduct will not give rise to other aminoxyl compounds, unless metal ions (usually adventitious and unwanted) are present in solution.

In the case of trapping superoxide in cells, the possibility of enzymatic conversion, enhanced by cell lysis, also has to be remembered (Ueno *et al.*, 1989).

In conclusion, when a relevant conversion of the DMPO-OOH adduct into DMPO-OH is presumed, it has to be seen as a warning about something that might not be under control in the model system.

Decay of DMPO-OH. The DMPO-hydroxyl adduct is rather stable (lasting several hours at room temperature in aqueous solution); however, it has been demonstrated that in the presence of cells the hydroxyl adduct is rapidly metabolized. Samuni *et al.* (1986) suggested that the species might go undetected, even if formed.

It is well known that many nitroxides in biological systems undergo a single electron reduction (Iannone *et al.*, 1989a, b; Iannone *et al.*, 1990a, b; Iannone and Tomasi, 1991). This should be taken into account particularly in the case of spin trapping in a model system characterized by high metabolic activity, such as NADPH–microsomes, isolated hepatocytes, and *in vivo*. In some cases, the ESR-silent hydroxylamine could be oxidized back to nitroxide by purging the sample with oxygen; then, having blocked the reducing activity of the system, nitrogen could be purged, in order to avoid oxygen-dependent line broadening.

A complication, observed in leukocyte experiments, has been reported by two groups (Rosen *et al.*, 1988; Black *et al.*, 1991b). At 37 °C the decay rate of DMPO–OH is accelerated substantially; the reduction of the adduct to the correspondent hydroxylamine is the main observed process. The activation of enzymatic reduction or phagocyte-derived factor(s) released into the surrounding medium could explain the observed phenomenon. Samuni *et al.* (1989c, 1990) observed that the DMPO–OH adduct decays, obeying a second-order kinetics, when exposed to a superoxide flux. The superoxide-mediated destruction did not yield reoxidizable hydroxylamine, but produced unidentified diamagnetic compounds.

3.3.4. Pharmacology and Toxicology

DMPO appears to be the least toxic of the commonly used spin traps. It does not inhibit cytochrome *P*-450 (Augusto *et al.*, 1982) or cytochrome *P*-450 reductase (Kalyanaraman *et al.*, 1980).

Pou *et al.* (1989) reported that DMPO has no effect on superoxide production by neutrophils or mononuclear phagocytes in response to PMA or zymosan stimulation. This is in conflict, however, with the report of Ueno *et al.* (1989), who found that oxygen consumption and $O_2^{-\cdot}$ production were inhibited during zymosan phagocytosis by DMPO. They observed a 50% respiratory inhibition after adding 100 mM DMPO. DMPO significantly inhibited the phagocytosis of zymosan particles at 45 mM; 90 mM of DMPO (a usual concentration in DMPO–neutrophils experiments) inhibited it completely; furthermore a significant increase in trypan blue dye–staining cells was observed. Britigan and Hamill (1990) reported that DMPO exhibited a concentration-dependent inhibition of $O_2^{-\cdot}$, however, they did not observe a loss in neutrophil viability.

In experiments carried out on endothelial-cell suspensions the trapping of oxygen-centered radicals formed during reoxygenation of endothelial cell suspensions previously subjected to anoxia has been reported (Zweier *et al.*, 1988). The DMPO–OH adduct observed was attributed to genuine ·OH production. The positive identification was based on the detection of the α-hydroxyethyl radical adduct after addition of the ·OH scavenger, ethanol. In a paper reassessing the results (Arroyo *et al.*, 1990b), it was noted that DMPO at the concentration used (50 mM) caused a significant cell viability loss (~50%), hence the possibility of enzymatic reduction of the superoxide adduct to pseudo DMPO–OH. This point is of major importance. When the experimental conditions heavily interfere with the function of the biological system under observation, the question of the significance of the results can not be eluded.

DMPO has been reported to decrease the incidence of reperfusion-induced ventricular fibrillation in rat heart (Hearse and Tosaki, 1987) and did not inhibit the rate-pressure product as observed in a controlled Langedorff-type perfused heart system (Konorev *et al.*, 1991).

4. CONCLUSION

While it is generally acknowledged that the literature of spin trapping contains many suspect interpretations of the results, the technique has been and still is widely used. The increased interest and applications are explained by the tumultuous development of a new area in biological and medical studies that could be named "free radical pathology." It appears that free radical-mediated reactions stay at the core of the inner mechanisms of life and pathology, hence the need to understand its underlying mechanisms. The spin-trapping technique might offer the much-wanted insight, without requiring, in most cases, a profound knowledge of ESR spectroscopy. The risk is that the easier a technique is to perform, the more subject to abuse the interpretation of the results are. It is certain, though, that the more people use the technique, the more advantages, also, will be discovered.

Attention has been mainly centered on the pitfalls and on the in a way unsatisfactory performances of the technique. This strategy has been based on the conviction that a sound spin-trapping experiment might be more easily planned, knowing the difficulties and complications of the technique. As a consequence, the extensive subject of the excellent examples of spin-trapping applications in biological environment has been only partially covered.

REFERENCES

Albano, E., and Tomasi, A., 1987, *Biochem. Pharmacol.* **36**:2913–2920.

Albano, E., Lott, K. A. K., Slater, T. F. A., Stier, A., Symons, M. C. R., and Tomasi, A., 1982, *Biochem. J.* **204**:593–603.

Albano, E., Tomasi, A., Goria-Gatti, L., and Dianzani, M. U., 1988, *Chemico-Biol. Interactions* **65**:223–234.

Albano, E., Tomasi, A., Goria-Gatti, L., and Iannone, A., 1989, *Free Rad. Biol. Med.* **6**:3–8.

Arroyo, C. M., and Forray, C., 1991, *Eur. J. Pharmacol.* **208**:157–161.

Arroyo, C., and Kohno, M., 1991, *Free Rad. Res. Commun.* **14**:145–155.

Arroyo, C. M., Forray, C., Elfakahany, E. E., and Rosen, G. M., 1990a, *Biochem. Biophys. Res. Commun.* **170**:1177–1183.

Arroyo, C. M., Carmichael, A. J., Bouscarel, B., Liang, J. H., and Weglicki, W. B., 1990b, *Free Rad. Res. Commun.* **9**:287–296.

Arroyo, C. M., Kramer, J. H., Dickens, B. F., and Weglicki C., 1987, *FEBS Lett.* **221**:101–104.

Athar, M., Mukhtar, H., Bickers, D. R., Khan, I. U., and Kalyanaraman, B., 1989, *Carcinogenesis* **10**:1499–1503.

Augusto, O., Beilan, H. S., and Ortiz de Montellano, P. R., 1982, *J. Biol. Chem.* **257**:11,288–11,295.

Bernofsky, C., Bandara, B. M. R., and Hinojosa, O., 1990. *Free Rad. Biol. and Med.* **8**:231–239.

Black, C. D. V., Samuni, A., Cook, J. A., Krishna, M., Kaufman, D. C., Malech, H. L., and Russo, A., 1991a, *Arch. Biochem. Biophys.* **286**:126–131.

Black, C. D. V., Cook, J. A., Russo, A., and Samuni, A., 1991b, *Free Rad. Res. Commun.* **12–13**:27–37.

Blasig, I. E., Ebert, B., Hennig, C., Pali, T., and Tosaki, A., 1990, *Cardiovascular Res.* **24**:263–270.

Boesterling, B., and Trudell, J. R., 1981, *Biochem. Biophys. Res. Commun.* **98**:569–575.

Britigan, B. E., and Hamill, D. R., 1989, *Arch. Biochem. Biophys.* **275**:72–81.

Britigan, B. E., and Hamill, D. R., 1990, *Free Rad. Biol. Med.* **8**:459–470.

Britigan, B. E., Rosen, G. M., Chai, Y., and Cohen, M. S., 1986, *J. Biol. Chem.* **261**:4426–4431.

Britigan, B. E., Pou, S., Rosen, G. M., Lilleg, D. M., and Buettner, G. R., 1990a, *J. Biol. Chem.* **265**:17,533–17,538.

Britigan, B. E., Coffman, T. J., and Buettner, G. R., 1990b, *J. Biol. Chem.* **265**:2650–2656.

Britigan, B. E., Roeder, T. L., and Buettner, G. R., 1991, *Biochim. Biophys. Acta* **1075**:213–222.

Buettner, G. R., 1990, *Free Rad. Res. Commun.* **10**:11–15.

Buettner, G. R., 1991, The basic chemistry of spin trapping: Pitfalls and subtleties, in *Proceedings of the 3rd International Symposium on Spin Trapping and Aminoxyl Radical Chemistry* (K. Makino, E. G. Janzen, and T. Yoshikawa, eds.), Kyoto, November 22–24, p. 21.

Buettner, G. R., and Oberley, L. W., 1978, *Biochem. Biophys. Res. Commun.* **83**:69–74.

Buettner, G. R., Oberley, L. W., and Chan-Leuthaser, S. W. H., 1978, *Photochem. Photobiol.* **28**:693–695.

Carney, J. M., and Floyd, R. A., 1991, *J. Mol. Neurosci.* **3**:47–57.

Chamulitrat, W., Takahashi, N., and Mason, R. P., 1989, *J. Biol. Chem.* **264**:7889–7899.

Chamulitrat, W., Hughes, M. F., Eling, T. E., and Mason, R. P., 1991, *Arch. Biochem. Biophys.* **290**:153–159.

Chamulitrat, W., Iwahashi, H., Kelman, D. J., and Mason, R. P., 1992, *Arch. Biochem. Biophys.* **296**:645–649.

Cheeseman, K. H., Albano, E., Tomasi, A., and Slater, T. F., 1984, *Chem.-Biol. Interactions* **50**:143–151.

Cheeseman, K. H., Albano, E., Tomasi, A., Dianzani, M. V., and Slater, T. F., 1985, *Life Sci. Rep.* **3**:259–264.

Chen, G. M., Bray, T. M., Janzen, E. G., and McCay, P. B., 1990, *Free Rad. Res. Commun.* **9**:317–323.

Chen, G., Janzen, E. G., Bray, T., and Lindsay, D. A., 1991, Metabolism of PBN, in *Proceedings of the 3rd International Symposium on Spin Trapping and Aminoxyl Radical Chemistry* (K. Makino, E. G. Janzen, and T. Yoshikawa, eds.), Kyoto, November 22–24, p. 66.

Cohen, M. S., Britigan, B. E., Hassett, D. J., and Rosen, G. M., 1988, *Rev. Infect. Dis.* **10**:1988–1096.

Connor, H. D., Fischer, V., and Mason, R. P., 1986, *Biochem. Biophys. Res. Commun.* **141**:614–621.

Cova, D., De Angelis, L., Monti, E., and Piccinini, F., 1992, *Free Rad. Res. Commun.* **15**:353–360.

Davies, M. J., 1989, *Biochem. J.* **257**:603–606.

Davies, M. J., Donkor, R., Dunster, C. A., Gee, C. A., Jonas, S., and Willson, R. L., 1987, *Biochem. J.* **246**:725–729.

Diguiseppi, J., and Fridovich, I., 1980, *Arch. Biochem. Biophys.* **205**:323–329.

Dodd, N. J. F., 1990, Free radical studies in biology and medicine, in *Electron Spin Resonance*, Vol. 12A (M. C. R. Symons, ed.) Royal Society of Chemistry, Cambridge, pp. 136–178.
DuBose, C. M., Janzen, E. G., Oehler, U. M., in *Proceedings of the 3rd International Symposium on Spin Trapping and Aminoxyl Radical Chemistry* (K. Makino, E. G. Janzen, and T. Yoshikawa, eds.), Kyoto, November 22–24, p. 23.
Evans, C. A., 1979, *Aldrichim. Acta* **12**:23–29.
Feix, J. B., and Kalyanaraman, B., 1989, *Biochim. Biophys. Acta* **992**:230–235.
Finkelstein, E., Rosen, G. M., Rauckman, E. J., and Paxton, J., 1979, *Mol. Pharmacol.* **16**:676–685.
Finkelstein, E., Rosen, G. M., and Rauckman, E. J., 1980, *Arch. Biochem. Biophys.* **200**:1–16.
Finkelstein, E., Rosen, G. M., and Rauckman, E. J., 1982, *Mol. Pharmacol.* **21**:262–265.
Floyd, R. A., and Carney, J. M., 1991, *Arch. Geront. Geriat.* **12**:155–177.
Floyd, R. A., and Soong, L. M., 1977, *Biochem. Biophys. Res. Commun.* **74**:79–84.
Gilbert, B. C., Marshall, P. D., Norman, R. O. C., Pineda, N., and Williams, P. S., 1981, *J. Chem. Soc. (Perkin Trans. II)*: 1392–1400.
Gilbert, B. C., Stell, J. K., Whitwood, A. C., Halliwell, C., and Sanderson, W. R., 1991, *J. Chem. Soc. (Perkin Trans. II)* **5**:629–634.
Goria-Gatti, L., Iannone, A., Tomasi, A., Poli, G., and Albano, E., 1992, *Carcinogenesis* **13**:799–805.
Griffin, B. W., 1982, *Can. J. Chem.* **60**:1463–1473.
Harbour, J. R., and Bolton, J. R., 1975, *Biochem. Biophys. Res. Commun.* **82**:680–684.
Harbour, J. R., Chow, V., and Bolton, J. R., 1974, *Can. J. Chem.* **52**:3549–3553.
Hartgerink, J. W., Engberts, J. B. F. N., and deBoer, Th. J., 1971, *Tetrahedron Lett.* **29**:2709–2712.
Hearse, D. J., and Tosaki, A., 1987, *Circulation Res.* **60**:375–383.
Howard, J. A., and Tait, J. C., 1978, *Can. J. Chem.* **56**:176–178.
Hughes, H. M., George, I. M., Evans, J. C., Rowlands, C. C., Powell, G. M., and Curtis, C. G., 1991, *Biochem. J.* **277**:795–800.
Iannone, A., and Tomasi, A., 1991, *Acta Pharmaceutica Jugoslavica* **41**:277–297.
Iannone, A., Bini, A., Swartz, H. M., Tomasi, A., and Vannini, V., 1989a, *Biochem. Pharmacol.* **38**:2581–2586.
Iannone, A., Hu, H., Tomasi, A., Vannini, V., and Swartz, H. M., 1989b, *Biochim. Biophys. Acta* **991**:90–96.
Iannone, A., Tomasi, A., Vannini, V., and Swartz, H. M., 1990a, *Biochim. Biophys. Acta* **1034**:285–289.
Iannone, A., Tomasi, A., Vannini, V., and Swartz, H. M., 1990b, *Biochim. Biophys. Acta* **1034**:290–293.
Iannone, A., Jin, Y.-G., Tomasi, A., and Vannini, V., 1993, *Free Rad. Res. Commun.*, in press.
Ingall, A., Lott, K. A. K., and Slater, T. F., 1978, *Biochem. Soc. Trans.* **6**:962–964.
Iwahashi, H., Albro, P. W., Mcgown, S. R., Tomer, K. B., and Mason, R. P., 1991a, *Arch. Biochem. Biophys.* **285**:172–180.
Iwahashi, H., Parker, C. E., Mason, R. P., and Tomer, K. B., 1991b, *Biochem. J.* **276**:447–453.
Janzen, E. G., 1971, *Acc. Chem. Res.* **4**:31–40.
Janzen, E. G., 1980, A critical review of spin trapping in biological systems, in *Free Radicals in Biology*, Vol. IV (W. A. Pryor, ed.), Academic Press, N.Y, pp. 115–153.
Janzen, E. G., 1990, *Free Rad. Res. Commun.* **9**:163–167.
Janzen, E. G., and Coulter, G. A., 1981, *Tetrahedron Lett.* **22**:615–619.
Janzen, E. G., and Evans, C. A., 1973, *J. Am. Chem. Soc.* **95**:8205–8206.
Janzen, E. G., Evans, C. A., Liu, J. I-P, 1973, *J. Magn. Reson.* **9**:513–516.
Janzen, E. G., Nutter, D. E., and Davis, E. R., 1978a, *Can. J. Chem.* **56**:2237–2242.
Janzen, E. G., Wang, Y. Y., and Shetty, R. V., 1978b, *J. Am. Chem. Soc.* **100**:2923–2925.
Janzen, E. G., Jandrisits, L. T., and Barber, D. L., 1987, *Free Rad. Res. Commun.* **4**:115–123.

Janzen, E. G., Krygsman, P. H., Lindsay, D. A., and Haire, D. L., 1990a, *J. Am. Chem. Soc.* **112**:8279–8284.

Janzen, E. G., Towner, R. A., Krygsman, P. H., Haire, D. L., and Poyer, J. L., 1990b, *Free Rad. Res. Commun.* **9**:343–351.

Janzen, E. G., Kotake, Y., and Hinton, R. D., 1991, Stabilities of hydroxyl radical spin adducts of PBN-type spin traps, in *Proceedings of the 3rd International Symposium on Spin Trapping and Aminoxyl Radical Chemistry* (K. Makino, E. G. Janzen, and T. Yoshikawa, eds.), Kyoto, November 22–24, p. 28.

Janzen, E. G., Kotake, Y., and Hinton, R. D., 1992, *Free Rad. Biol. Med.* **12**:169–173.

Kalyanaraman, B., Mason, R. P., Perez-Reyes, E., and Chignell, C. F., 1979a, *Biochem. Biophys. Res. Commun.* **89**:1065–1072.

Kalyanaraman, B., Perez-Reyes, E., and Mason, R. P., 1979b, *Tetrahedron Lett.* **50**:4809–4812.

Kalyanaraman, B., Perez-Reyes, E., and Mason, R. P., 1980, *Biochim. Biophys. Acta* **630**:119–130.

Kalyanaraman, B., Mottley, C., and Mason, R. P., 1983, *J. Biol. Chem.* **258**:3855–3858.

Kalyanaraman, B., Mottley, C., and Mason, R. P., 1984, On the use of organic extraction in the spin-trapping technique as applied to biological systems, in *Reviews in Biochemical Toxicology*, Vol. 4 (E. Hodgson, J. R. Bend, and R. M. Philpot, eds.), Elsevier Biomedical, New York, pp. 73–139.

Kalyanaraman, B., Joseph, J., and Parthasarathy, S., 1991, *FEBS Lett.* **280**:17–20.

Kaur, H., Leung, K. H. W., and Perkins, M. J., 1981, *J. Chem. Soc.* 142–143.

Kennedy, C. H., Pryor, W. A., Winston, G. W., and Church, D. F., 1986, *Biochem. Biophys. Res. Commun.* **141**:1123–1129.

Kennedy, C. H., Maples, K. R., and Mason, R. P., 1990, *Pure Appl. Chem.* **62**:295–299.

Kirino, Y., Ohkuma, T., and Kwan, T., 1981, *Chem. Pharm. Bull.* **29**:29–34.

Kleinhans, F. W., and Barefoot, S. T., 1987, *J. Biol. Chem.* **262**:12,452–12,457.

Knecht, K. T., Connor, H. D., LaCagnin, L. B., Thurman, R. G., and Mason, R. P., 1989, *In vivo* detection of free radical metabolites as applied to carbon tetrachloride and related halocarbons in *Intermediary Xenobiotic Metabolism in Animals: Methodology, Mechanisms and Significance* (D. H. Hutson, J. Caldwell, and G. D. Paulson, eds.), Taylor and Francis, London and New York, pp. 375–380.

Konorev, E. A., Baker, J. E., Joseph, J., and Kalyanaraman, B., 1991, Effects of spin traps on cardiac function, in *Proceedings of the 3rd International Symposium on Spin Trapping and Aminoxyl Radical Chemistry* (K. Makino, E. G. Janzen, and T. Yoshikawa, eds.), Kyoto, November 22–24, p. 74.

Kuppusamy, P., and Zweier, J. L., 1989, *J. Biol. Chem.* **264**:9880–9884.

Lai, C-S., and Piette, L. H., 1977, *Biochem. Biophys. Res. Commun.* **78**:51–59.

Lai, C-S., and Piette, L. H., 1978, *Arch. Biochem. Biophys.* **190**:27–38.

Lai, C-S., and Piette, L. H., 1979, *Tetrahedron Lett.* **9**:775–778.

Lai, C-S, Grover, T. A., and Piette, L. H., 1979, *Arch. Biochem. Biophys.* **193**:373–378.

Leaver, L. H., Ramsay, G. C., and Suzuki, E., 1969, *Aust. J. Chem.* **22**:1981–1987.

Li, A. S. W., de Haas, A. H., Buettner, G. R., and Chignell, C., 1989. A Database for Spin Trapping Implemented on the IBM PC/AT, ESR Dept., Laboratory for Molecular Biophysics, NIEHS.

Lloyd, R. V., and Mason, R. P., 1990, *J. Biol. Chem.* **265**:16,733–16,736.

Mackor, A., Wajer, Th. A. J. W., and de Boer, Th. J., 1966, *Tetrahedron Lett.* **19**:2115–2123.

Mackor, A., Wajer, Th. A. J. W., and de Boer, Th. J., 1967, *Tetrahedron Lett.* **5**:385–390.

Makino, K., Suzuki, N., Moriya, F., Rokushika, S., and Hatano, H., 1981, *Radiat. Res.* **86**:294–310.

Makino, K., Imaishi, H., Morinishi, S., Hagiwara, T., Takeuchi, T., and Murakami, A., 1989, *Free Rad. Res. Commun.* **6**:19–28.

Makino, K., Hagiwara, T., Imaishi, H., Nishi, M., Fuji, S., Ohya, H., and Murakami, A., 1990a, *Free Rad. Res. Commun.* **9**:233–240.
Makino, K., Hagiwara, T., Hagi, A., Nishi, M., and Murakami, A., 1990b, *Biochem. Biophys. Res. Commun.* **172**:1073–1080.
Mani, V., and Crouch, R. K., 1989, *J. Biochem. Biophys. Meth.* **18**:91–96.
Mason, R. P., 1984, Spin trapping free radical metabolites of toxic chemicals, in *Spin Labeling in Pharmacology* (J. L. Holtzman, ed.), Academic Press, New York, pp. 87–129.
Mason, R. P., and Morehouse, K. M., 1988, Spin Trapping—The Ideal Method for Measuring Oxygen Radical Formation? in *Free Radicals: Methodology and Concepts* (C. Rice-Evans and B. Halliwell, eds.), Richelieu, London, pp. 157–168.
Mason, R. P., Kalyanaraman, B., Tainer, B. E., and Eling, T. E., 1980, *J. Biol. Chem.* **255**:5019–5022.
Mason, R. P., Maples, K. R., and Knecht, K. T., 1989, In vivo detection of free radical metabolites by spin trapping in *Electron Spin Resonance*, Vol. 11B (M. C. R. Symons, ed.), Royal Society of Chemistry, London, pp. 1-9.
McCay, P. B., Noguchi, T., Fong, K-L, Lai, E. K., and Poyer, J. L., 1980, Production of radicals from enzyme systems and the use of spin traps, in *Free Radicals in Biology*, Vol. IV (W. A. Pryor, ed.), Academic Press, New York, pp. 153–186.
McCay, P. B., Reinke, L. A., and Rau, J. M., 1992, *Free Rad. Res. Commun.* **15**:335–346.
Merritt, M. V., and Johnson, R. A., 1977, *J. Am. Chem. Soc.* **99**:3713–3719.
Monti, E., Paracchini, L., Perletti, G., and Piccinini, F., 1991, *Free Rad. Res. Commun.* **14**:41–45.
Morehouse, K. M., Flitter, W. D., and Mason, R. P., 1987, *FEBS Lett.* **222**:246–250.
Morgan, D. D., Mendenhall, C. L., Bobst, A. M., and Rouster, S. D., 1985, *Photochem. Photobiol.* **42**:93–94.
Mottley, C., and Mason, R. P., 1989, Nitroxide radical adducts in biology: Chemistry, applications, and pitfalls, in *Biological Magnetic Resonance*, Vol. 8 (L. J. Berliner and J. Reuben, eds.), Plenum Press, New York and London, pp. 489–546.
Mottley, C., Kalyanaraman, B., and Mason, R. P., 1981, *FEBS Lett.* **130**:12–14.
Mottley, C., Connor, H. D., and Mason R. P., 1986, *Biochem. Biophys. Res. Commun.* **141**:621–628.
Nazhat, N. B., Yang, G., Allen, R. E., Blake, D. R., and Jones, P., 1990, *Biochem. Biophys. Res. Commun.* **166**:807–812.
Nohl, H., Jordan, W., and Hegner, D., 1981, *FEBS Lett.* **123**:241–244.
Nordio, P. L., 1976, General magnetic resonance theory, in *Spin Labeling: Theory and Practice* (L. J. Berliner, ed.), Academic Press, New York, pp. 5–52.
Novelli, G. P., 1992, *Lab. Invest.* **20**:499–507.
Novelli, G. P., Angiolini, R., Tani, G., Consales, L., and Bordi, A., 1985, *Free Rad. Res. Commun.* **1**:321–327.
Ohto, N., Niki, E., and Kamiya, Y., 1977, *J. Chem. Soc. (Perkin Trans.)* **11**:1770–1774.
Ozawa, T., and Hanaki, A., 1986, *Biochem. Biophys. Res. Commun.* **136**:657–664.
Ozawa, T., and Hanaki, A., 1987, *Biochem. Biophys. Res. Commun.* **142**:410–416.
Packher, L., Maguire, J. J., Mehlhorn, R. J., Serbinova, E., and Kagan, V. E., 1989, *Biochem. Biophys. Res. Commun.* **159**:229–235.
Perkins, M. J., 1980, *Adv. Phys. Org. Chem.* **17**:1–64.
Pfab, J., 1978, *Tetrahedron Lett.* **9**:843–846.
Phillis, J. W., and Cloughhelfman, C., 1990, *Neurosci. Lett.* **116**:315–319.
Poli, G., Albano, E., Tomasi, A., Cheeseman, K. H., Chiarpotto, E., Parola, M., Biocca, M. E., Slater, T. F., and Dianzani, M. U., 1987, *Free Rad. Res. Commun.* **3**:251–255.
Pou, S., and Rosen, G. M., 1990, *Anal. Biochem.* **190**:321–325.
Pou, S., Hassett, D. J., Britigan, B. E., Cohen, M. S., and Rosen, G. M., 1989, *Anal. Biochem.* **177**:1–6.

Poyer, J. L., McCay, P. B., Lai, E. K., Janzen, E. G., and Davis, F. R., 1980, *Biochem. Biophys. Res. Commun.* **94**:1154–1160.
Prònai, L., Ichimori, K., Nozaki, H., Nakazawa, H., Okino, H., Carmichael, A. J., and Arroyo, C. M., 1991, *Eur. J. Biochem.* **202**:923–930.
Ramos, C. L., Pou, S., Britigan, B. E., Cohen, M. S., and Rosen, G. M., 1992, *J. Biol. Chem.* **267**:8307–8312.
Reinke, L. A., and Janzen, E. G., 1991, *Chem.-Biol. Interact.* **78**:155–165.
Rosen, G. M., 1985, *Adv. Free Rad. Biol. Med.* **1**:345–375.
Rosen, G. M., and Rauckman, E. J., 1981, *Proc. Natl. Acad. Sci. USA* **78**:7346–7349.
Rosen, G. M., and Rauckman, E. J., 1984, Spin trapping of superoxide and hydroxyl radicals, in *Methods in Enzymology*, Vol. 105 (L. Packer, ed.), Academic Press, New York, pp. 198–198.
Rosen, G. M., Finkelstein, E., and Rauckman, E. J., 1982, *Arch. Biochem. Biophys.* **215**:367–378.
Rosen, G. M., Britigan, B. E., Sohen, M. S., Ellington, S. P., and Barber, M. J., 1988, *Biochim. Biophys. Acta* **969**:236–241.
Rosen, G. M., Cohen, M. S., Britigan, B. E., and Pou, S., 1990, *Free Rad. Res. Commun.* **9**:187–195.
Rumyantseva, G. V., Kennedy, C. H., and Mason, R. P., 1991, *J. Biol. Chem.* **266**:21,422–21,427.
Samuni, A., Carmaichael, A. J., Russo, A., Mitchell, J. B., and Riesz, P., 1986, *Proc. Natl. Acad. Sci. USA* **83**:7593–7597.
Samuni, A., Krishna, C. M., Riesz, P., Finkelstein, E., and Russo, A., 1988a, *J. Biol. Chem.* **263**:17,921–17,924.
Samuni, A., Black, C. D., Krishna, C. M., Malech, H. L., Bernstein, E. F., and Russo, A., 1988b, *J. Biol. Chem.* **263**:13,797–13,801.
Samuni, A., Samuni, A., and Swartz, H. M., 1989a, *Free Rad. Biol. Med.* **7**:37–43.
Samuni, A., Samuni, A., and Swartz, H. M., 1989b, *Free Rad. Biol. Med.* **6**:179–183.
Samuni, A., Murali-Krishna, C., Riesz, P., Finkelstein, E., and Russo, A., 1989c, *Free Rad. Biol. Med.* **6**:141–148.
Samuni, A., Krishna, C. M., Mitchell, J. B., Collins, C. R., and Russo, A., 1990, *Free Rad. Res. Commun.* **9**:241–249.
Saprin, A. N., and Piette, L. H., 1977, *Arch. Biochem. Biophys.* **180**:480–492.
Schaich, K. M., and Borg, D. C., 1990, *Free Rad. Res. Commun.* **9**:267–278.
Sergeev, G. B., and Leenson, I. A., 1978, *Russian J. Phys. Chem.* **52**:312–315.
Smith, P., and Robertson, J. S., 1987, *Can. J. Chem.* **66**:1153–1158.
Stolze, K., and Mason, R. P., 1987, *Biochem. Biophys. Res. Commun.* **143**:941–946.
Stolze, K., Moreno, S. N. J., and Mason, R. P., 1989, *J. Inorg. Biochem.* **37**:45–53.
Symons, M. C. R., Albano, E., Slater, T. F., and Tomasi, A., 1982, *J. Chem. Soc. (Faraday Trans.)* **78**:2205–2214.
Taffe, B. G., Takahashi, N., Kensler, T. W., and Mason, R. P., 1987, *J. Biol. Chem.* **262**:12,143–12,149.
Thornalley, P. J., and Stern, A., 1985, *Free Rad. Res. Commun.* **1**:111–117.
Thornalley, P. J., and Vasak, M., 1985, *Biochim. Biophys. Acta* **827**:36–44.
Thornalley, P. J., Trotta, R. J., and Stern, A., 1983, *Biochim. Biophys. Acta* **759**:16–22.
Tomasi, A., Albano, E., Lott, K. A. K., and Slater, T. F., 1980, *FEBS Lett.* **122**:203–206.
Tomasi, A., Albano, E., Botti, B., and Vannini, V., 1987, *Toxicol. Pathol.* **15**:178–183.
Trudell, J. R., 1987, *Free Rad. Biol. Med.* **3**:133–136.
Turner, M. J., III, and Rosen, G. M., 1986, *J. Medicinal Chem.* **29**:2439–2444.
Ueno, I., Kohno, M., Mitsuta, K., Mizuta, Y., and Kanegasaki, S., 1989, *J. Biochem.* **105**:905–910.
Wajer, Th. A. J. W., Mackor, A., deBoer, Th. J., and van Voorst, J. D. V., 1967, *Tetrahedron Lett.* **23**:4021–4026.

Yamada, T., Niki, E., Yokoi, S., Tsuchiya, J., Yamamoto, Y., and Kamiya, Y., 1984, *Chem. Phys. Lipids* **36**:189–196.

Yamazaki, I., Piette, L. H., and Grover, T. A., 1990, *J. Biol. Chem.* **265**:652–659.

Zhdanov, R., Chen, G., Janzen, E. G., and Reinke, L. A., 1991, Metabolism of PBN spin adducts, in *Proceedings of the 3rd International Symposium on Spin Trapping and Aminoxyl Radical Chemistry* (K. Makino, E. G. Janzen, and T. Yoshikawa, eds.), Kyoto, November 22–24, p. 67.

Zini, I., Tomasi, A., Grimaldi, R., Vannini, V., and Agnati, L. F., 1992, *Neurosci. Lett.* **138**:279–283.

Zweier, J. L., 1988, *J. Biol. Chem.* **263**:1353–1357.

Zweier, J. L., Kuppusamy, P., and Lutty, G. A., 1988, *Proc. Natl. Acad. Sci. USA* **85**:4046–4050.

Contents of Previous Volumes

VOLUME 1

VOLUME 3

VOLUME 4

VOLUME 6

VOLUME 7

VOLUME 8

VOLUME 12

Index